ATE DUE

THE PICTURE OF THE TAOIST GENII PRINTED ON THE COVER
of this book is part of a painted temple scroll, recent but traditional, given to
Mr Brian Harland in Szechuan province (1946). Concerning these four divinities,
of respectable rank in the Taoist bureaucracy, the following particulars have been
handed down. The title of the first of the four signifies 'Heavenly Prince', that
of the other three 'Mysterious Commander'.

At the top, on the left, is Liu *Thien Chün*, Comptroller-General of Crops and
Weather. Before his deification (so it was said) he was a rain-making magician
and weather forecaster named Liu Chün, born in the Chin dynasty about +340.
Among his attributes may be seen the sun and moon, and a measuring-rod or
carpenter's square. The two great luminaries imply the making of the calendar, so
important for a primarily agricultural society, the efforts, ever renewed, to reconcile
celestial periodicities. The carpenter's square is no ordinary tool, but the gnomon
for measuring the lengths of the sun's solstitial shadows. The Comptroller-General
also carries a bell because in ancient and medieval times there was thought to be
a close connection between calendrical calculations and the arithmetical acoustics
of bells and pitch-pipes.

At the top, on the right, is Wên *Yuan Shuai*, Intendant of the Spiritual Officials
of the Sacred Mountain, Thai Shan. He was taken to be an incarnation of one of
the Hour-Presidents (*Chia Shen*), i.e. tutelary deities of the twelve cyclical characters
(see p. 262). During his earthly pilgrimage his name was Huan Tzu-Yü and he was
a scholar and astronomer in the Later Han (b. +142). He is seen holding an
armillary ring.

Below, on the left, is Kou *Yuan Shuai*, Assistant Secretary of State in the Ministry
of Thunder. He is therefore a late emanation of a very ancient god, Lei Kung.
Before he became deified he was Hsin Hsing, a poor woodcutter, but no doubt an
incarnation of the spirit of the constellation Kou-Chhen (the Angular Arranger),
part of the group of stars which we know as Ursa Minor. He is equipped with
hammer and chisel.

Below, on the right, is Pi *Yuan Shuai*, Commander of the Lightning, with his
flashing sword, a deity with distinct alchemical and cosmological interests. According
to tradition, in his earthly life he was a countryman whose name was Thien Hua.
Together with the colleague on his right, he controlled the Spirits of the Five
Directions.

Such is the legendary folklore of common men canonised by popular acclamation.
An interesting scroll, of no great artistic merit, destined to decorate a temple wall,
to be looked upon by humble people, it symbolises something which this book has
to say. Chinese art and literature have been so profuse, Chinese mythological
imagery so fertile, that the West has often missed other aspects, perhaps more
important, of Chinese civilisation. Here the graduated scale of Liu Chün, at first
sight unexpected in this setting, reminds us of the ever-present theme of quanti-
tative measurement in Chinese culture; there were rain-gauges already in the Sung
(+12th century) and sliding calipers in the Han (+1st). The armillary ring of
Huan Tzu-Yü bears witness that Naburiannu and Hipparchus, al-Naqqās and
Tycho, had worthy counterparts in China. The tools of Hsin Hsing symbolise that
great empirical tradition which informed the work of Chinese artisans and technicians
all through the ages.

SCIENCE AND CIVILISATION
IN CHINA

中國科學技術史

李約瑟 著

莫朝鼎

JOSEPH NEEDHAM

SCIENCE AND CIVILISATION IN CHINA

VOLUME 5

CHEMISTRY AND CHEMICAL TECHNOLOGY

PART I: PAPER AND PRINTING

BY

TSIEN TSUEN-HSUIN, PH.D.

PROFESSOR EMERITUS OF CHINESE LITERATURE AND
LIBRARY SCIENCE AND CURATOR EMERITUS
OF THE FAR EASTERN LIBRARY AT THE
UNIVERSITY OF CHICAGO

CAMBRIDGE
UNIVERSITY PRESS

Published by the Press Syndicate of the University of Cambridge
The Pitt Building, Trumpington Street, Cambridge CB2 1RP
40 West 20th Street, New York, NY 10011–4211, USA
10 Stamford Road, Oakleigh, Melbourne 3166, Australia

First published 1985
Reprinted 1985
Third printing, revised 1987
Reprinted 1993

Printed in Great Britain at the
University Press, Cambridge

Library of Congress catalogue card number: 54–4723

British Library Cataloguing in Publication Data
Needham, Joseph
Science and civilisation in China.
vol. 5: Chemistry and chemical technology
Pt. 1: Paper and printing
1. Science—China—History
I. Title II. Tsien, Tseun-hsuin
509′.51   Q127.C5
ISBN 0 521 08690 6

To

three eminent scholars who have contributed to
our knowledge of the History of Printing

L. CARRINGTON GOODRICH

Dean Lung Professor Emeritus of Chinese
Columbia University

HOWARD W. WINGER

Professor Emeritus of Library Science
The University of Chicago

and

KWANG-TSING WU

Former Head, Chinese and Korean Section
The Library of Congress
Washington, D. C.

this volume is dedicated

CONTENTS

LIST OF ILLUSTRATIONS

FOREWORD

It was in 1948, some thirty-six years ago, that the writing of this series of volumes began, after I came back from China and from Unesco. My first collaborator was Wang Ling[1] (Wang Ching-Ning[2]); and then eight years later Lu Gwei-Djen[3] came over from Paris to occupy the same position of Adjutant-General of the Project which she still fills. While Wang Ling's background was rather mathematical and chemical, Lu Gwei-Djen's is more medical and biological. Then, some fifteen years ago, we found ourselves faced with a great dilemma, whether to peg away alone for the rest of our lives, or whether to associate with ourselves a group of collaborators, who would bring us nearer to seeing the conclusion in our own lifetime. We decided on the latter course. It was a veritable turning-point.

Now, with the present Volume, we see the first-fruits of this plan. We were able to persuade our dear friend Professor T. H. Tsien (Chhien Tshun-Hsün[4]), of the University of Chicago, and one of the world's most eminent authorities on the subject, to accomplish this task in our series. We greatly admire what he has done. Of course, the book of T. F. Carter (1) has long been a classic, but it is now quite old, having seen the light first in 1925 and not revised since 1955; since then many archaeological finds, and many literary perspectives, have intervened, so that a new synthesis was urgently necessary.

We expect that in due course other Volumes not written personally by us will be appearing in our series; notably next Vol. 6, pt. 2, on the 'History of Agriculture in China', which has been written by Ms. Francesca Bray. This again deals with a subject of the highest importance for the general theme of the history of science and technology in China, and we feel fortunate that her work for some ten years past on our staff has proved so successful. We now anticipate that still further Volumes will be contributed by our collaborators.

I suppose that no theme could be more important for the history of all human civilisation than the development of paper and printing. Francis Bacon, 'the bell that call'd the wits together', fully recognised this.[a] In the present Volume readers will be able to follow all the vicissitudes of paper and printing during those Chinese centuries when Europe knew nothing of such arts. I always suspected that the Chinese Buddhists may have had something to do with the technique of documentary reproduction, because of their mania for the infinite replication of images, as I had plenty of opportunity to observe on the walls of the Thang cave-temples at Chhien-fo-tung[5] (Tunhuang[6]).[b]

---

[a] Cf. Vol. 1, p. 19.   [b] Vol. 1, p. 126.

[1] 王鈴   [2] 王靜寧   [3] 魯桂珍   [4] 錢存訓   [5] 千佛洞
[6] 敦煌

It has always been extremely hard to believe that Johann Gutenberg about +1454 knew nothing (even by hearsay) of the Chinese printed books which had been circulating in large numbers in China for five previous centuries—and there are some sources fairly contemporary which aver that he did.[a] It seems perhaps less likely that he ever knew of his predecessor Pi Shêng[1], the artisan who had anticipated him as the inventor of movable type by some four centuries. We have already alluded to the celebrated passage about Pi Shêng in the *Mêng Chhi Pi Than*[2], and we have illustrated a rotating 'case' for the type sorts depicted by Wang Chen[3] later.[b] Many printers in Korea as well as China, subsequently made use of this invention, but of course it was much more inviting to use movable type for the alphabetic languages needing only twenty-six letters, than for the ideographic ones where as many as 53,500 characters, with some 400 radicals, would be involved.

Be that as it may, however, there is plenty of evidence that Chinese printing and book-production were the envy of the whole world that knew about them before Gutenberg's time.[c] Even long afterwards we have Jesuit relations which show their great admiration for the books of China, which enshrined so much of human learning and knowledge.[d] True it is that, as Francis Bacon again said: 'The wits and knowledges of men remain in books, exempted from the wrong of time, and capable of perpetual renovation.' Let us pray that no evil fire-storm will be let loose upon the world to destroy in an instant much, if not all, of the accumulated culture of the ages, and put an end to that most glorious of human achievements, printing and the paper on which it may be done. So let us wish all success to this story told by Tsien Tsuen-Hsuin.

JOSEPH NEEDHAM

Cambridge
January 1984

[a] Cf. Vol. 1, p. 244 (*b*).
[b] Vol. 4, pt. 2, pp. 33, 533 (Fig. 685). We also came across Pi Shêng in the context of chemical industry; Vol. 5, pt. 3, p. 187.
[c] For example the Arabic scholar Dāwūd al-Banākitī in +1317; Needham (88) repr. in (64), pp. 22–4.
[d] See vol. 4, pt. 2, p. 439. The words were those of Matteo Ricci himself.

[1] 畢昇   [2] 夢溪筆談   [3] 王禎

AUTHOR'S NOTE

This volume is devoted to the study of the origin and development of papermaking and printing in Chinese culture from their earliest known beginnings to the end of the 19th century, when both handicrafts had gradually been replaced by modern technology. It is intended to be a comprehensive study, covering all periods in their history and all aspects of their techniques, aesthetics, applications, and worldwide spread and influence, based on extensive investigation of literary records, archaeological discoveries, scientific reports, as well as artifacts available for examination. Previous contributions to the field are either limited in scope or outdated in many respects. Numerous Chinese documents concerning paper had not been explored and many important topics had not been covered in Western scholarship. Details of printing, on the other hand, are rarely documented in Chinese literature, but hundreds and thousands of items printed from the +8th century onward are available for investigation. The present work tries to bring up to date or fill gaps in previous studies; it also offers new interpretations based on fresh data and new evidence. An attempt in this direction is briefly explained at the end of the Introduction.

The project started in late 1968 at the request of Dr Joseph Needham to contribute a section on 'Paper and Printing' to his great series of volumes *Science and Civilisation in China*. His proposal actually suggested a sequel to my previous work on pre-paper and pre-printing records of China.[a] A travel grant from the American Council of Learned Societies for 1968–9 enabled me to visit Cambridge and many libraries and museums in Europe and America for a preliminary exploration of opportunities. During the next few years, basic sources were collected and screened, a series of lecture-discussion sessions was held on several university campuses, and three sections on paper were then drafted. To pursue further the study of printing, a seminar was set up at the University of Chicago in 1972 and again in 1974 for a systematic examination of the sources and a discussion of various problems involved. But the actual writing of subsequent sections was interrupted by the increasing load of my duties at the University. With the generous support of the U. S. National Science Foundation and National Endowment for the Humanities from 1977 to 1980, I was relieved of part of my teaching and administrative responsibilities at the University. This allowed me to devote more time to research. Additional support from the East Asian History of Science Trust in Cambridge made it possible for me to complete the entire work by the end of 1982. The original plan was to write a section of about 100 pages, but subsequent studies

[a] *Written on Bamboo and Silk: The Beginnings of Chinese Books and Inscriptions.* University of Chicago Press, Chicago, 1962.

went deeply into the sources, and this resulted in a volume much broader in scope and size and one which consumed more time than originally planned.

The present work consists of ten sections, with a bibliography of nearly 2000 entries and about 200 plates and figures for illustration. Three sections each deal separately with paper, printing (including inkmaking and bookbinding), and their worldwide spread and influence, in addition to an Introduction as a summary and orientation. The last section on the contributions of paper and printing to world civilisation serves as a conclusion. Each part on paper and printing is treated both chronologically and topically. Special attention is given to technical and artistic aspects of each subject as well as to their role in society. Where previous studies are available, a brief description of the topic or events is usually given; where gaps exist in earlier scholarship, fuller discussion is provided. For special editions of works in Chinese and Japanese cited in this volume, their publisher, date, or inclusion in a series are given in parentheses following the specific title in the footnotes and bibliographies. Abbreviations for titles of journals, large collections, symposia, and *tshung-shu* are found in a list preceding the bibliography.

In the preparation of this volume, many of my friends, colleagues, and former students have contributed significantly. I am especially indebted to three eminent scholars and specialists in the field who served as consultants of the project, Professor L. Carrington Goodrich, Professor Howard W. Winger, and Dr K. T. Wu[1], for their constant advice, reading and criticism of the entire manuscript. Professor and Mrs H. G. Creel also gave generously of their time to read over the final draft. All their comments from different perspectives have been most useful in improving this work. I wish also to express my gratitude for the valuable help in many ways from Dr Poon Ming-Sun[2], Mr James K. M. Cheng[3], Mr Ma Tai-Loi[4], Mr John Grobowski, Dr Michael Finegan, and Miss June Work, who served at one time or another as research assistants in the project.

To all my colleagues at the University of Chicago, especially in the Center for Far Eastern Studies, the Department of Far Eastern Languages and Civilisations, the Graduate Library School, and the University Library, I am grateful for advice and continuing support. In the course of my writing, I made use of several excellent papers and theses developed in my seminars by Miss Constance Miller, Mr Edward Martinique, Mrs Lily Chia-Jen Kecskes, and Dr Poon Ming-Sun, whose contributions are acknowledged at various points in the book with reference to their respective works in the bibliography.

For supply of materials for my study and photographs for illustrations, I am indebted to the librarians and curators of the British Museum and British Library, the Cambridge and Oxford University Libraries, the Bibliothèque Nationale, the Musée Guimet, the Austrian National Library, the Museum für Völkerkunde, the Royal Ontario Museum, the Dard Hunter Paper Museum, the Field Museum of Natural History, the Metropolitan Museum of Art, the Fogg Art Museum, the

---

[1] 吳光淸   [2] 潘銘燊   [3] 鄭烱文   [4] 馬泰來

Printing Department of Newberry Library, the Asian Division of the Library of Congress, the Spencer Collection of New York Public Library, the Harvard-Yenching Library, and the East Asian Library of Columbia University. However, the basic sources of information for my study were drawn largely from the Far Eastern Library of the University of Chicago. All credits to figures in this volume are acknowledged.

During my travel to the Far East in 1979, I benefited greatly from visiting libraries and museums, meeting and discussing with specialists, and collecting additional materials for my final revision of the draft. Certain old paper specimens and samples of rare printing were examined in China and Japan. Valuable experience was gained by my visits to the printing shops of Jung-Pao-Chai[1] in Peking and Tuo-Yün-Hsien[2] in Shanghai, where woodblock cutters and printers were interviewed and tools and accessories collected. I am grateful to Mr Ku Thing-Lung[3], Director of Shanghai Library, for his advice and cooperation in making a series of photographs and drawings, based on my interviews, showing step-by-step procedures for carving and printing. My thanks are also due to Mr Phan Chi-Hsing[4] of the Institute of the History of Science, Academia Sinica, for his expert advice and reading of the three sections on paper; to Mr Chang Hsiu-Min[5] formerly of the Peking Library and Mr Hu Tao-Ching[6] of Shanghai, for their counsel and information; and to Mr Chhang Pi-Te[7] of the Palace Museum in Thaipei for his generous assistance in sending me many photographs of old and rare samples of printing now kept in Taiwan.

I feel particularly fortunate to have been associated with the East Asian History of Science Library in the Needham Research Institute at Cambridge, where my hardworking friends have been patiently watching the slow progress of my study. My gratitude goes to Dr Lu Gwei-Djen for her gracious advice, Dr Michael Salt for sending me useful materials, Dr Colin Ronan for his skilled editing of the manuscript, and Mr Peter Burbidge and the staff of the Cambridge University Press for their helpful advice in the publication of this volume. My deepest debt is due to the architect of the project, Dr Joseph Needham, who provided timely help and advice on many problems in the process of my writing. Without his inspiration, guidance, and constant support, this enterprise would not have been accomplished.

Finally, I owe a debt of gratitude to my wife, Hsü Wen-Chin[8], who has not only given her encouragement and support to my study over many long years, but also contributed a piece of her calligraphy for the oldest poem on 'Paper' appearing in Figure 1230, which adds a special feature to this volume. To many others who cannot be named here, I wish to offer my thanks for help of various kinds.

The University of Chicago   T. H. TSIEN
October 1983

[1] 榮寶齋   [2] 朵雲軒   [3] 顧廷龍   [4] 潘吉星   [5] 張秀民
[6] 胡道靜   [7] 昌彼得   [8] 許文錦

32. PAPER AND PRINTING

(a) INTRODUCTION

OF all the products from the ancient world, few can compare in significance with the Chinese inventions of paper and printing. Both have played a profound role in shaping world civilisations; and both have exerted a far-reaching impact for a very long time on the intellectual as well as the daily lives of countless people everywhere. Paper has proved to be the most satisfactory material on which human thoughts are committed to writing, and when printing came to be allied to it, the ideas of one individual could be communicated to a multitude of others separated across great stretches of space and time. In short, the printed message has brought about changes in the intellectual mode of the human mind, and paper has provided the most economical and convenient means for its transmission. But of course paper has other uses than for writing and publishing; it has penetrated into every corner of ancient and contemporary society to become an indispensable article in daily life. Even though new media of communication have developed in recent times, the unique combination of paper, ink and printing are still the basic, permanent, portable, and perhaps the least expensive and most accessible communication device known to us today.

(1) ORIGIN, DEVELOPMENT, AND MIGRATION OF PAPER AND PRINTING

It is common knowledge that paper was invented in China some time before the Christian era. From early in the +2nd century its manufacture became improved, using new materials and superior techniques. By the +3rd century it had become widely used in China itself and had begun to migrate across the Chinese borders; it reached the Western world only just prior to the modern age. Printing from woodblocks was first practised by the Chinese around +700, and movable type several centuries earlier than Gutenberg. Even the indelible ink of lampblack, prized by scholars and artists throughout the centuries in the East as well as the West, and which has been manufactured in the West under the misnomer 'Indian Ink', can be traced back to antiquity in Chinese civilisation. It was the introduction of these ingenious elements that made possible mass production of written records for wide circulation. Of the materials and techniques for the modern book, printed with black ink on white paper, the Chinese have contributed most to its development.

Paper is a felted sheet of fibres formed from a water suspension process using a sieve-like screen. When the water escapes and dries, the layer of intertwined fibres becomes a thin matted sheet which is called paper. Over the span of the two millennia which have elapsed since the inception of the idea of papermaking, the

I

craft has changed and the tools have become more complex, yet the basic principles and processes remain the same.

Traditionally, the invention of paper was attributed to Tshai Lun early in the +2nd century, but recent discoveries of very ancient paper fragments in North and Northwest China have pushed back the date of this invention at least some two to three centuries before him. Indeed, as we shall see, the invention of paper in China is now believed to have originated from a process of pounding and stirring rags in water several centuries before the start of our Era. It is very likely that an accidental placing of fibres from the rags on a mat with water draining away, may have suggested the idea of making a thin sheet of paper. But paper was not invented expressly for writing, as has often been presumed. It was extensively used in China in the fine and decorative arts, at ceremonies and festivals, for business transactions and records, monetary credit and exchange, personal attire, household furnishings, sanitary and medical purposes, recreations and entertainments, and so on. What is more, all these non-literary applications were common in Chinese society before paper was introduced into Europe in the +9th century.

Paper was not used for writing until perhaps early in the +1st century, and even then did not entirely replace the more cumbersome bamboo and wood slips as the chief materials for making books until the +3rd century. But when it came, the use of paper enabled books to be cheaper and more portable, though their extensive production and wide distribution was not possible until the invention of printing. It is uncertain when and where the first book was printed in China and who was the earliest printer, but probably the process developed gradually.

There is a long history of pre-printing techniques in China, including the use of seals for stamping on clays and later on silk and paper, of stencils to duplicate designs on textiles and paper, and of the inked impressions taken from stone inscriptions. All these processes gradually led to more efficient methods of the mechanical multiplication of copies and, as archaeological and literary evidence indicates, by the +7th century or around +700, printing began in China. Movable type was introduced by the middle of the 11th century and multi-colour printing some time in or before the 12th century. The movable type was first made of earthenware, but various other materials, including wood, metal, and a variety of other ceramics, were also adopted repeatedly and intermittently in the following centuries.

Because of the great number of characters in written Chinese, woodblock printing was used far more often than movable type for book production in China until recent times. Wood blocks were simpler and more economical, and could be stored easily and were readily available when a reprint was needed; movable type was preferred only for large-scale production of voluminous books. Nevertheless, both wood blocks and movable type have gradually given way, since the mid-19th century, to modern printing methods.

After papermaking was perfected, it not only became popular in China but spread in all directions throughout the world, first eastwards in the +2nd century, then westward during the +3rd century. However, it did not reach India until the +7th century, and only became popular there in the 12th. Paper arrived in Western Asia in the middle of the 8th century, and in Africa in the 10th. The Arabs monopolised paper-making in the West for some five centuries. Only in the 12th century was it manufactured in Europe, and it did not reach America until the 16th century and Australia in the 19th. Thus it took more than fifteen hundred years for paper to spread from China to almost every part of the world.

Whether or not typography in Europe was influenced by the Chinese is controversial, but it is certain that Chinese printing and printed materials from China were known in Europe before printing began there. As might be expected, there are many theories about how printing reached Europe. Some suggest that it travelled from China to Europe along routes similar to those taken by paper; others, emphasising the differences between European and Chinese printing, suggest that European typography was independent in origin. However, there is strong evidence from cultural considerations of a close connection between them. Certainly there is no doubt that paper-making originated in China, and was already a fully developed craft before it spread over the rest of the world. It is probably the most complete of all the inventions China has given to civilisation.

(2) FACTORS CONTRIBUTING TO THE EARLY INVENTION OF PAPER AND PRINTING IN CHINA

The prerequisites for a useful invention include both the physical and the mental readiness for the event; besides a creative mind and a popular demand, proper materials and the essential basic techniques must be available. Since all the material facilities for the invention were present in Europe as well as in China, several questions arise. Why did the invention occur in one civilisation but not the other? What were the factors responsible for such development? What was it that made these two great inventions appear very early in Chinese culture but only after a long delay in the West, at least a thousand years for paper, six hundred years for wood-block printing and four centuries for movable type? In an attempt to find the answer, we shall discuss and compare the conditions that led to these developments.

The key elements for the manufacture of paper are water, fibres, and a mould. The first was present almost everywhere and fibres were available from rags or hemp or linen just as soon as textiles were woven in the ancient world. The use of the two together was common enough, but not so the process of turning rags into separate fibres through maceration, and using a screen mould to hold these fibres while allowing the water to drain away. Perhaps, as will become evident later,[a] the Chinese tradition of washing rags in water and allowing the fibres to form a felted sheet on the mat was responsible for this discovery in ancient times. The earliest

---

[a] See pp. 36 ff. below.

mould is believed to have been made of a piece of cloth stretched with frames to support the macerated fibres and to let the water escape through its meshes.

The invention of paper-making was, of course, a continuing process rather than a single event. An important step came with the introduction of new and fresh raw fibres, allowing unlimited production. Here the discovery of the suitability of the paper mulberry (*Brousonetia papyrifera*) was certainly significant. It is a plant that is native to China, though it has been cultivated extensively in many other temperate and tropical zones throughout the world. Its bark, after being beaten into a cloth, was used for clothing in China as well as in other regions along the equator, and ancient Chinese literature provides evidence that it was manufactured and traded by native tribes in the southern part of China, as we shall see.[a] The invention of paper-making with tree bark attributed to Tshai Lun in the early +2nd century was possibly influenced by the acquaintance of the people in his area with the paper mulberry.[b] Tshai Lun was a native of Lei-yang in what is now Hunan province, and it was here that the bark was made into cloth by beating and then into bark paper after maceration.[c] Since, then, the maceration process of turning rags into pulp was already known in China, it was very likely that the people in the south of the country were the first to convert paper mulberry bark into a pulp for paper-making. Neither paper mulberry nor bark cloth was, it seems, used in Europe, where its cultivation appears to have been unknown, even in the +18th century; indeed among the numerous kinds of plant tested for paper-making by European scientists at this time, paper mulberry was not included.[d] Furthermore, it was described with curiosity by the early Jesuit missionaries to China, and they suggested its transplantation to France.[e]

The popular demand for a better writing material was another important factor leading towards the invention and utilisation of paper. In China, paper was a much cheaper and more ideal writing medium than expensive silk and the clumsier bamboo or wood. But in Europe, paper did not have too many advantages over papyrus or parchment. Papyrus was plentiful, simple to prepare, inexpensive, and perhaps as light and convenient as paper. Parchment, although it cost more,[f] had a smoother surface and was more durable than paper. Indeed, in the early days, paper was not much cheaper than parchment, in contrast to silk, and not any more portable than papyrus, in contrast to bamboo and wood. Because of its fragility,

---

[a] See pp. 56 ff. below.

[b] See Ling Shun-Sheng (7), pp. 1 ff.

[c] Lu Chi of the +3rd century said that the bark of paper mulberry was used by the people south of the Yangtse River to make cloth or was pounded for making paper; see discussion of *tapa* and paper clothing, pp. 109 ff.

[d] Searching for new materials for paper-making, Dr Jacob Schäffer (1718–90) tells of how he tested over thirty kinds of raw material including moss, asbestos, potato, wood, and various other plants, for use in paper making to be mentioned in his six-volume work published between 1765 and 1771, but he did not include paper mulberry or bamboo, which had been major raw materials used in paper making in China and other nations in east and south Asia; see Hunter (9), pp. 309 ff.

[e] See Batteux & de Bréquigny (1), ed., *Mémoires...* (15 vols., Paris, 1776–91), vol. II, p. 295.

[f] In 1367, 31 quires of parchment, each containing three dozen sheets, cost 76 livres, 5 sous, 8 deniers in Tours; in 1359, two quires of paper cost 18 deniers; and in 1360, four quires of paper cost 2 shillings, 4 deniers; see Blum (1), pp. 62–3, note 2.

paper was even banned for official documents in Europe.[a] It was also not a welcome commodity when it was first introduced to Europe from the Arab world, since Europeans were distrustful of anything from a hostile land during and after the Crusades; its use was even attacked by clergymen like the abbot of Cluny.[b] Not until the spread of printing in Europe during the second half of the fifteenth century did a great demand for paper arise, although it had gradually come into use for manuscripts and household records before then. The situation was so different in China, where paper established its supremacy as a popular medium for writing even before it was officially adopted by the court in the early +2nd century.

The basic materials needed for block printing included wood, ink, and paper. The same kinds of wood, including pear, boxwood, or other deciduous trees, were used for woodblocks for printing in both China and the West. Ink of lampblack was probably discovered very early by people of all civilisations, since soot was naturally collected when fire was controlled. The use of black ink or a carbon mixture in China can be traced back to remote antiquity,[c] and a similar ink of lampblack mixed with an aqueous solution of vegetable gum was used by Egyptian scribes as early as −1300; it spread to Western Asia a little later.[d] The Greeks also made an ink of soot consisting of the same basic ingredients of lampblack and gum and in the same solid form as the Chinese ink.[e]

Of the three basic necessities for printing, paper was perhaps the most important. Without a soft and absorbent medium, it would have been impossible for printing to develop, and the prior use of paper by the Chinese certainly contributed to the early invention of printing in China. Clearly, the late introduction of paper to Europe had a significant effect on the slow development of printing in the West.

However, paper was certainly not the only essential prerequisite for this invention, for printing did not appear in China until paper had been used for writing for at least six or seven hundred years, and there were still four printless centuries after the arrival of paper in Europe. Printing developed quite naturally from techniques developed in making and using seals and stamps, engraving on stone and metals, and taking inked rubbings from stone and other inscriptions. Religious and secular demand for a great numbers of copies, however, called for some mechanical means to replace hand copying.

Seal inscriptions in a mirror image, from which a correct position was obtained by stamping on clay and later on paper, embody the technique closest to that which eventually led to the invention of printing. The use of seals began in antiquity in both Chinese and Western civilisations. In China, seals cast in bronze with designs and inscriptions in relief survive from the Shang dynasty. Other seals

[a] Paper was forbidden for official use by King Roger of Sicily in 1145 and again by Emperor Frederick II of Germany in 1221; cf. Blum (1), pp. 23, 30.
[b] Cf. Blum (1), p. 30.
[c] See further discussion on pp. 233 ff. below.
[d] Cf. J. H. Breasted (1), pp. 230–49; Wiborg (1), p. 7.
[e] Cf. Wiborg (1), pp. 71–2.

made of metals or carved on stone, jade, ivory, horn, earthenware, and wood have continued in use until this day. They are characterized in general by a flat surface, square, oblong, or occasionally in other shapes, bearing inscriptions of characters in relief or intaglio of personal names or official titles, always in reverse. They have been used always to indicate ownership, authenticate documents, and establish authority.[a]

The use of seals in Western culture began and flourished in Mesopotamia and Egypt, perhaps even before the invention of writing.[b] These seals made of stone, ivory, shell, or metals were of two principal types, cylinders and stamps. The cylindrical type was used in Mesopotamia and in areas under Babylonian influence. Their designs, primarily of deities, heroes, animals, celestial bodies, instruments, and emblems, were impressed by rolling the cylinder over a flat surface of clay, mortar, cement, or wax.[c] The stamp seals have a variety of shapes. Those used in Egypt were of scaraboid form with a beetle on the back, a sacred symbol of resurrection and immortality. Their bases are flat and engraved with designs or inscriptions of mottoes, personal names and titles of officials.[d] These had strong religious overtones as well as practical functions. Both the cylindrical and stamp forms of seals were also used in Asia Minor, Syria, and Palestine. Their use was discontinued after the fall of the Western Roman Empire but revived in the second half of the +8th century. Since then, round or oval seals engraved with designs and legends have been employed in the West until modern times.

Generally speaking, the seals developed in Chinese and Western cultures bear some similarities and differences. They were both made of the same kinds of materials, impressed originally on the same kind of surfaces, and used primarily for the same purposes. But there were some major diversities which led perhaps to development in different directions. Chinese seals were mostly made in a square or rectangular shape with a flat base, engraved with characters in reverse, and used to stamp on paper. These characteristics are very close to those of block printing. Although the surface and inscriptions of most seals were small or limited, some wooden seals were as large as printing blocks and were inscribed with texts more than one hundred characters long.[e]

The seals of the West, on the other hand, were cylindrical or scaraboid, round or oval, and inscribed primarily with pictures or designs and only occasionally with writing. The cylindrical seals used to roll over clay had no potential to develop into a printing surface.[f] While the scaraboid seals were flat-based, their primarily religious nature was predominant over their functional aspects as a tool of multi-

[a] The use of seals in China is considered to be one of the most important technical prerequisites for the invention of printing in China; Cf. Carter (1), pp. 11 ff; see also pp. 136 ff. below.

[b] Cf. Chiera (1), p. 192.

[c] For the development of cylindrical seals, see Eisen (1), Frankfort (1), and Wiseman (1).

[d] For scarab seals and their religious meaning, see Newberry (1) and Ward (1).

[e] See *Paa Pho Tzu, Nei Pien (SPTK)*, ch. 17, p. 23a; tr. Carter (1), p. 13.

[f] This could be considered as the forerunner of the rotary press, which was certainly not developed directly from this practice.

plication.[a] Furthermore, seal inscriptions always took a positive position, the impressions being made primarily on stiff material such as wax, rather than on a flexible medium such as parchment or paper. Such different usage discouraged any development in the West of the idea of printing from seals.

The use of seals as symbols of authority and authenticity was similar in standing to that accorded to coins. In ancient times, the circulation and acceptance of metal money depended upon official sponsorship, which was usually indicated by marking on the coins their value, place of minting and, sometimes, the official symbol of approval. These numismatic inscriptions were made either by casting in a mould, or by stamping or punching the face of the coins. From very early times in China coins in spade, knife, plate, and circular shapes were cast from moulds.[b] But in the West, they were first made by stamping and later by casting, a technique which was subsequently borrowed by bookbinders, who cast separate metal characters for stamping titles on a book. This craft was eventually adopted by printers to cast metal type and thus was the forerunner of typography in the West.[c]

The technique of engraving on stone tablets is close to that of carving on wood blocks, and taking inked squeezes or rubbings from stone inscriptions is very similar to the process of block printing. Inscribing on stone was developed very early in both China and the West. Chinese inscriptions on stone survive from the Chou dynasty, and subsequently stone became the most popular medium for commemorative and sacred writings, and for the preservation and standardisation of the canonical texts. The Mesopotamians also used stone in addition to clay tablets for writing, the Egyptians used it for tomb inscriptions and it was adopted for monuments by the Romans as well as other peoples in the ancient world,[d] but inscriptions in the West were neither as extensive nor as refined as they were in China, and were never in scale with those of the Chinese, where hundreds of thousands of characters of Buddhist, Taoist, and Confucian texts were carved on stone throughout many centuries.[e] Moreover, stone was used in the West more as an artistic material than, as it was in China, for writing. Such differences in the nature, scope, and content of stone inscriptions caused them to develop in divergent directions in China and in the West.

Taking inked squeezes or rubbings from stone inscriptions is similar to printing in principle and purpose, but different in process and end-product.[f] Both result in duplication on a sheet of paper of an engraved object, but their different methods

---

[a] See Miller (1), pp. 14–26.
[b] The earliest metal coins cast with inscriptions may have first been made in the Shang or early Chou; see Wang Yü-Chhüan (1), p. 114; for the development of Chinese numismatic inscriptions, see Tsien (2), pp. 50 ff.
[c] Cf. Blum (2), p. 21.
[d] Cf. Diringer (2), pp. 44–5, 82, 358.
[e] Cf. Tsien (2), pp. 64 ff.
[f] A. C. Moule, in his review of Carter (1) in *JRAS* (1926), p. 141, expressed doubts about the influence of rubbing on printing because the two processes are essentially different. However, the difference in one respect does not preclude influence in another, as can be seen in the fact that special reference to stone inscriptions was made when Confucian classics were first printed from woodblocks in the + 10th century; see discussions on pp. 143 ff. and p. 370 below.

result in different kinds of reproduction.[a] The technique of taking inked rubbings from stone, and eventually from all kinds of hard surfaces, can be traced back to the +6th century or earlier in China. Yet it does not seem to have been used in the West until perhaps the 19th century, when antiquarians and artists began to experiment with the use of a crayon-like agent in tracing designs from memorial brasses, tombstones, brick walls, carved wood, and sewer-plates.[b] The duplications they obtained were, however, far less sophisticated than those of the inked squeezes originally made by the Chinese, and it was the combination of this skill in making duplications by inking and rubbing on a sheet of paper coupled with the art of carving seals with a mirror image in relief that resulted in the methods of block printing.

Besides the necessary materials and techniques, there were also social and cultural factors which had a great effect on the application or rejection of printing. Since printing is a mechanical extension of writing, the system of writing used is one of the most important factors affecting the development of printing. Chinese writing was from the very beginning characterised by an ideographic script which is basically composed of numerous separate strokes of different shapes.[c] Since each character has a definite and distinct form, the writing of characters tends to be elevated to an art and is thus more complicated and time-consuming than alphabetical writing, especially when a special style is sought in a formal and respectable text. On the other hand, Western writing, ever since the Phoenicians developed the rudiments of an alphabetic language, has evolved into a system of symbols representing sounds. Its written components are merely substitutes for their spoken counterparts, and have tended to evolve into simple signs composed of continuous lines.[d] Copying in an alphabetic language is easier than in an ideographic script. It is likely, therefore, that the slower and more complicated process of copying Chinese resulted in a greater demand for mechanical aid in duplication in China than in the West. It is also natural that movable type was more acceptable to an alphabetic language, while block printing was more suitable to the Chinese writing system.

Religion is another cultural factor that has played a great part in the long history of the development of printing; religious zeal in spreading sacred writings to all believers has created a demand for a ready means of reduplication, and Buddhism, Islam, and Christianity all exerted an influence. Buddhism even teaches that mass production of its *sutras* is a way to receive blessing from the Buddha. Indeed, the Buddha is said to have remarked, 'Whoever wishes to gain power from the *dharani*

[a] See discussion of the methods of making inked squeezes in Tsien (2), pp. 86 ff., and on pp. 143 ff. below.
[b] Cf. K. Starr (2), p. 3. An archaeologist said in 1930 that she learned a most satisfactory method employed by Orientalists to make rubbings from inscriptions and decorations, 'even the finest of lines appear most distinctly'; see Margaret Ashley (1).
[c] Chinese characters have been composed of one to more than thirty independent strokes or dots, straight or cursive lines, and squares since their development into the clerical and regular styles from around the advent of Christian era in the Chhin and Han dynasties.
[d] For a comparative study of word-syllabic and alphabetic systems, see Diringer (1), Gelb (1); for development of written forms, see Anderson (1).

[charms] must write seventy-seven copies and place them in a pagoda.... This *dharani* is spoken by the ninety-nine thousand *koti*[a] of Buddhas and he who repeats it with all his heart shall have his sins forgiven.'[b] The enthusiasm of the Buddhist devotees for producing a great multitude of sacred texts was highly influential in the birth of printing in China, which occurred during the high tide of Buddhism in the early Thang. This religious motivation is further confirmed by the earliest printings of the *dharani* discovered in Japan and Korea.

In the West, on the other hand, the demand for multiple copies was not strong as early as it was in China. Hand-copying by slave scribes could produce more texts than were needed in the Roman Empire. In the Middle Ages the reading public was very small, and the copyist tradition was carried on in monasteries and churches. Such demand as there was for books could be met with handmade copies prepared by scribes; there was no incentive to produce them in large quantity. Not until the renaissance and the Reformation did the demand for Bibles and other reading materials significantly increase.

Another factor in the relatively late development of printing in Europe may have been the influence of the growth of various kinds of craft unions and guilds. First organised in Greece and Rome to facilitate the sharing of common interests by skilled men, by the Middle Ages they had gained political power and took on the role of protecting the professional skills and livelihood of their members.[c] These guilds naturally became very strict and exclusive as far as their membership was concerned. For instance, the block printers who engraved and printed playing cards and religious images belonged to the company of painters or artists, which represented such craftsmen as scribes, illuminators, sculptors, stonecutters, glass-makers, and wood-engravers. Typographers were not admitted as members of that society.[d] As late as 1470, guilds of scribes and illuminators in France still forbade multiplication of religious images by means other than by hand-copying.[e] And between 1485 and 1590, among all the early typographers of Antwerp, it seems that only one was probably admitted to guild membership as a wood-engraver, and this most likely on account of his illustrations printed with his text.[f] This power of the guilds in the Middle Ages to limit membership to certain crafts may well have had a negative effect on the early development of printing in Europe.[g]

To sum up, then, the early use of printing in China was chiefly due to the early invention of paper, the specialised use made of seals and rubbings for duplication, the greater need for mechanical aid in duplicating texts written in a complex ideographic script, the standardisation of Confucian texts used for civil service examinations and, finally, the demand for great quantities of copies of Buddhist

[a] A *koti* is variously put at one hundred thousand, one million, and ten million; cf. Carter (1), p. 53, note 15.
[b] See translation of the *dharani* cited in Carter (1), p. 50.
[c] Cf. Frey (1), pp. 9–17.
[d] Chatto & Jackson (1), p. 121.
[e] Bliss (1), pp. 10–11.
[f] Chatto & Jackson (1), p. 122.
[g] Cf. C. R. Miller (1), pp. 53 ff.

scriptures which could not be met by hand-copying. In the West, paper was not introduced until a rather late date, seals were not used as duplication devices, rubbing was not known until fairly recently, while printers were restricted by craft unions or guilds, and added to all this, the relative simplicity of the alphabetic script lessened the need for a mechanical duplication aid. Thus the materials and techniques necessary for the invention of printing were either not developed, or did not lead in the direction of a printing process. Furthermore, there was no such incentive or demand for huge quantities of copies as developed in connection with Buddhism; the needs that did exist could be met by hand-copying. Until all these factors were changed in the middle of the 15th century, the threshold for the invention of printing was not reached in Western society.

(3) INFORMATION ON PAPERMAKING AND INKMAKING IN CHINA

Sources of information for the study of paper and papermaking include paper specimens, scientific and field reports, early records, and secondary sources. We shall take these in order.

Paper specimens are important because they can be subjected to microscopic, chemical, and physical analyses for determining their composition, technique of manufacture, and other features. Since the turn of the century, tens of thousands of early paper specimens have been found within China and outside it, including some fragments from the −2nd century which are at present the oldest known papers in the world. A few specimens bearing characters of perhaps the +2nd century attest to the use of paper for writing before or at this time.[a]

Paper fragments and documents found in modern Sinkiang by various expeditions were primarily products of the Three Kingdoms, Chin, and Southern and Northern Dynasties from the +3rd to 6th centuries, when paper began to be used widely and to travel across Chinese borders.[b] Paper rolls dating from the 4th to the 10th centuries discovered at Tunhuang represent the best examples of paper and paper books in a roll form before and during the Thang period.[c] From this time on, specimens of different varieties of paper survive in books, documents, works of painting and calligraphy, in stationery, paper cutting, and other paper products. In addition, certain old paper documents extant outside China testify to their early diffusion worldwide.[d]

---

[a] Up to the present time (1983), at least seven discoveries of old paper fragments of the Han period have been reported, but only one or two said to have been dated to the Later Han bear some writing; see discussion on pp. 38 ff. below.

[b] Cf. Tsien (2), pp. 142–58; see also discussion on pp. 43 ff. below.

[c] A grotto library of some 30,000 rolls of paper books and documents in Tunhuang was first visited by Aurel Stein in 1907 and later by many others; see Stein (4), Giles (13), Pelliot (60), Chhen Yüan (5), and a summary of the documents by Fujieda (2).

[d] Some 12,500 paper documents discovered in Egypt, dating from +800 to 1388, are now preserved in the Erzherzog Rainer Collection at Vienna, and many Chinese papers of the Sui and Thang periods survive in Japan and Korea.

Samples of these old papers have been scientifically analysed, and reports are available concerning fibre composition, substances used for sizing and coating, and such physical qualities as thickness, strength, opacity, absorbency, and watermark, if any. In 1885–7, Joseph Karabacek (2) and Jules Wiesner (1, 2) made the first analysis of Arabic paper documents from the +9th to the 14th centuries, which had been found in Egypt. Then, in 1902–11, Wiesner (3, 4, 5) studied the papers found by Aurel Stein in the course of his first two expeditions to Chinese Turkestan and Tunhuang. It had formerly been believed in the West that cotton papers were first made by the Arabs in the 8th century and that the making of rag papers was invented by the Europeans in the 13th century, but the findings of these studies confirmed that paper was invented in China at least in the early Christian era, as recorded in Chinese histories. Moreover, using the scientific data obtained from these analyses and with the support of documentary sources, the history and routes of the migration of paper step by step from China westwards over a period of more than a thousand years have been reconstructed.

The Tunhuang papers at the British Museum were further studied by Robert Clapperton (1) in 1934, M. Harders-Steinhäuser (1) in 1969, and Jean-Pierre Drège (1) in 1981, and those in the Peking collection by Phan Chi-Hsing (2) in 1966. Samples of other discoveries have also been examined by Chinese scientists. A piece with characters of the Later Han period found at Chü-yen in 1942 was analysed by Wu Yin-Chhan,[1] a botanist, and reported by Lao Kan (1). Specimens of the +4th to +8th centuries found in Sinkiang by Chinese archaeological teams in recent years and now at the Sinkiang Museum, as well as samples of papers used for calligraphy and paintings dating from the +3rd to the +12th centuries now kept at the Palace Museum in Peking, have also been studied and reported on by Phan (4, 5, 7). The data collected on their physical appearance, the fibres used in manufacture, and the techniques for treating them, in addition to descriptions in ancient literature, have been used in reconstructing the methods of old papermaking in China (Fig. 1052).[a]

Literary sources on paper in Chinese can be divided into two major categories. One consists of general works, such as historical documents, local gazetteers, literary writings, and classified encyclopedias; the other of chapters or books specifically on paper. For instance, the beginnings of paper are recorded in such historical works as the *Tung Kuan Han Chi*,[2] a contemporary official record compiled about +120, and the *Hou Han Shu*,[3] a standard history of the Later Han dynasty based on earlier sources. The subsequent development of paper and its manufacture in different periods are described in dynastic and other histories, and also in works on administrative codes of successive dynasties, such as the *Thang Liu*

---

[a] Experiments of papermaking with hemp fibres according to analyses of old specimens and descriptions of the Han period were conducted in 1965 by the Institute of the History of Natural Sciences, Academia Sinica, and the results were reported to be successful; see Phan Chi-Hsing (6).

[1] 吳印禪   [2] 東觀漢記   [3] 後漢書

(a)

(b)

(c)

Fig. 1052. Old method of papermaking in China. A modern drawing of the ancient process according to steps described in early literature, showing (a) cutting, stamping, and washing the raw material; (b) cooking, pounding, and mixing fibres; and (c) dipping and lifting the mould; drying and sorting the sheets. From Phan Chi-Hsing (9).

Tien[1] (*c.* +738) of the early Thang period, in which official positions of those in charge of the manufacture, acquisition, processing, and use of paper in various branches of government are listed. Local materials, products, or tributes are recorded in local gazetteers or regional descriptions, including the *Yüan-Ho Chün Hsien Thu Chih*[2] (+814), a Thang geography; *Chia-Thai Kuei Chi Chih*[3] (+1201), a local history of Kuei-chi, Chekiang; *Chiang-Hsi Sheng Ta Chih*[4] (+1556), a provincial history of Chianghsi; and many of other periods. Occasional references to paper are found in literary collections by noted writers of the Thang and Sung periods, whose poems acknowledging gifts made of paper are included. Descriptions of paper and paper products or the use of paper at festivals or ceremonies are found in such memoirs as the *Tung-Ching Meng Hua Lu*[5] (+1148), about the Northern Sung capital of Khaifeng, and such collections of miscellaneous notes as *Kai Yü Tshung Khao*[6] (+1750). Finally, special chapters or sections devoted to paper are found in classified encyclopedias, including the *Thai Phing Yü Lan*[7] (+983) and the *Thu Shu Chi Chheng*[8] (+1726), in which quotations from histories, anecdotes, poems, prose, and miscellaneous items on paper are systematically arranged, even though they may not always agree with the original texts.

The other category of literary sources consists of chapters or works exclusively on paper and papermaking. The earliest one, *Wen Fang Ssu Phu*,[9] a general treatise on implements for writing in a scholar's studio compiled by Su I-chien (+953–96),[10] contains a section on paper which is divided into four parts: history, manufacture, anecdotes, and literature selected from earlier sources, chiefly of the Thang period, many of which have since been lost. Another work limited to local descriptions, the *Shu Chien Phu*[11] on Szechuan papers written by Fei Chu[12] of the 14th century, includes information on local products, papermakers, and designers in that particular province. Many similar works by noted men of letters, such as the *Phing Chih Thieh*[13] by the Sung artist Mi Fu,[14] describe the qualities and appreciation of paper. The most important and only early work on the technology of papermaking is included in the *Thien Kung Khai Wu*[15] by Sung Ying-Hsing[16] (1587–*c.* 1660), in which an illustrated chapter is devoted to methods of papermaking with bamboo and paper mulberry.[a] Later works of a similar nature include an eyewitness account of papermaking by Huang Hsing-San[17] of *c.* 1850,[b] a systematic treatise on paper by Hu Yün-Yü[18](*1*), whose description of the manufacture of the famous *Hsüan chih*[19] in Anhui is especially useful; a technical work on papermaking with bamboo by Lo Chi[20](*1*) published in 1935; and a recent illustrated work on various plant fibres used in papermaking in China by Yü Chheng-Hung[21] & Li Yün[22](*1*).

---

[a] Cf. tr. Sun & Sun (*1*), pp. 223 ff.   [b] Cited in Yang Chung-Hsi (*1*), ch. 5, pp. 39*a*–40*b*.

[1] 唐六典   [2] 元和郡縣圖志   [3] 嘉泰會稽志   [4] 江西省大志
[5] 東京夢華錄 [6] 陔餘叢考 [7] 太平御覽 [8] 圖書集成   [9] 文房四譜
[10] 蘇易簡   [11] 蜀箋譜   [12] 費著   [13] 評紙帖   [14] 米芾
[15] 天工開物 [16] 宋應星 [17] 黃興三 [18] 胡韞玉 [19] 宣紙
[20] 羅濟   [21] 喩誠鴻   [22] 李澐

The contributions of modern scholarship to Chinese papermaking include the scientific studies of old papers already mentioned, and historical research into the origin, development, and migration of paper, and field surveys of traditional papermaking in modern societies. The historical studies were initiated primarily by sinologists like Stanislas Julien (13), Friederich Hirth (29), Edouard Chavannes (24), Berthold Laufer (48), and especially Thomas Carter (1), whose chapter on paper in his book on printing is still an authority on its dissemination westwards. Since then, works by paper historians and experts, including André Blum (1), Armin Renker (1), Henri Alibaux (1), R. H. Clapperton (1), and especially Dard Hunter (9), have given the Chinese invention a proper place in the history of papermaking. Hunter was not an expert on China, but his field investigations of handmade paper mills in China, Korea, Japan, Indochina, Siam, and India, and his personal experience with handmade paper, have added a new dimension to the comparative study of traditional papermaking in Asian civilisation.[a]

The few articles on papermaking published by Chinese scholars during the first half of the 20th century largely consisted of translations of Western sources on the subject or expositions of traditional opinions. An early study by Yao Tshung-Wu[1] (1) on the introduction of paper to Europe, published in 1928, was primarily based on Western scholarship supplemented with documentation from Chinese sources.[b] Lao Kan[2] (7), writing on the origin of paper in 1942, reaffirmed and further elucidated a theory advanced by the Chhing scholar Tuan Yü-Tshai[3] (1735–1805) that paper originated from the use of silk fibres and from washing rags in water, and this has since been followed by many Chinese and Western scholars.[c]

During the second part of the 20th century, little has been added to our knowledge of Chinese papermaking in Western languages, but more contributions have been made in Chinese and Japanese. Works in three major areas may be mentioned. On the origin of paper, reports and studies of new discoveries of what are now the oldest known paper specimens, identified with the Former Han period, have not only pushed back the date of invention at least two or three hundred years before Tshai Lun, but also confirmed that early papers were made of hemp and not silk fibres. Also, a theory was advanced by Ling Shun-Sheng[4] (7) that the origin of paper might have been influenced by the *tapa* culture widespread in southern China, the Pacific, and other tropical regions in ancient times. Although this theory is not new and some of his conclusions are doubtful,[d] his thesis, supported by

---

[a] Hunter published some twenty monographs on paper, most of which were printed by himself on his own handmade paper; see partial list in Bibliography; the specimens collected during his trips are exhibited at the Dard Hunter Museum, Appleton, Wisconsin.

[b] The article was first published under the name Yao Shih-Ao[5] in 1928 and was reprinted under the pseudonym Shang-Yin[6] in 1966–7; see Bibliography B under Yao Tshung-Wu.

[c] See the discussion of the definition of paper in the *Shuo Wen Chieh Tzu* on pp. 35 ff, below.

[d] Ling's theory that paper money, paper armour, and Chin-Shu paper were all made of bark cloth and not bark paper has been proved to be wrong; literary records or recent tests of the products confirm that they were made of real paper.

[1] 姚從吾   [2] 勞榦   [3] 段玉裁   [4] 凌純聲   [5] 姚士鰲
[6] 善因

full documentation from Chinese sources, offers a new interpretation of this subject.

The development of papermaking during different periods following its beginnings in the Han, has been studied by Chang Tzu-Kao[1] (2, 7) and Yüan Han-Chhing[2] (2) on the early period, Wang Ming[3] (1) on the Sui and Thang, Shih Ku-Feng[4] (1) on the Sung, and Phan Chi-Hsing[5] (1–12) on all aspects of Chinese paper, especially its origin and technology. Recent discoveries of stone tablets inscribed with regulations for paper mills in the +18th century have provided original documents for the study of the social and economic conditions of papermaking in Chinese history.[a] Finally, several general histories of Chinese papermaking have recently been published. These include a popular work by Hung Kuang[6] & Huang Thien-Yu[7] (1), a brief account by Liu Jen-Chhing[8] (1), and a comprehensive treatise by Phan Chi-Hsing (9). Phan's work, which consists of eighteen chapters on historical development in different periods, topical studies based on scientific analyses of old paper specimens, and field surveys of papermaking by Chinese minority peoples and with different materials, is by far the most complete and detailed study of the technology of Chinese papermaking in any language.

Japanese sources for the study of handmade papers include artifacts, old documents, and modern compilations. The Shōsōin in Nara has a collection of many old paper specimens from China and Japan.[b] Early historical documents, including the *Kojiki*[9] (+712), *Nihongi*[10] (+720), and *Engishiki*[11] (+927), contain records of the early introduction of paper into Japan, the offices in charge of papermaking, the transplanting of paper mulberry, the manufacture of a great variety of papers, and the uses of paper for writing, wrapping, garments, making screens, and mounting on walls and houses. Important excerpts from these documents, together with quotations from Chinese works, are arranged in chronological order in a collection edited by Seki Yoshikuni[13] (2). He has also published a history of handmade paper, appended with sources from Korean documents,[c] and a companion volume of pictures illustrating the papermaking processes and paper shops of different periods and different countries.[d] Numerous monographs and articles on Japanese handmade paper by modern authors can be found in bibliographies[e] and the works of Jugaku Bunshō[14] (1). Japan remains noted for its continuing manufacture of fancy handmade papers, and specimens have been assembled and published in several collections, including Thomas & Harriet Tindale (1), Seki Yoshikuni (1), Mainichi Shimbun[15] and Takeo Kabushiki Kaisha[16]. The set by

[a] See report of Liu Yung-Chheng (3) and discussion on pp. 50 ff. below.

[b] See *Shōsōin no Kami*,[12] or *Various Papers Preserved in the Shōsōin*, ed. by the Shōsōin Jimusho, with text in Japanese, specimens, plates, and English introduction.

[c] See Seki Yoshikuni (4), pp. 411–16.

[d] See Seki Yoshikuni (3), which includes over 200 illustrations from Japan, China, Korea, Europe, and America.

[e] See, for example, a bibliography of Japanese paper compiled by Sorimachi Shigeo (1), which consists of 422 items in different languages collected by the late Frank Hawley now located in the Tenri Library, Nara.

[1] 張子高   [2] 袁翰青   [3] 王明   [4] 石谷風   [5] 潘吉星
[6] 洪光   [7] 黃天佑   [8] 劉仁慶   [9] 古事記   [10] 日本書紀
[11] 延喜式   [12] 正倉院の紙   [13] 關義城   [14] 壽岳文章   [15] 每日新聞
[16] 竹尾株式會社

Mainichi Newspapers, *Tesuki Washi Taikan*,[1] or *Comprehensive Collection of Handmade Japanese Paper*, published in five large boxed sections in 1973–4 with text in Japanese and English, consists of samples of raw materials and some 1000 specimens of various kinds of handmade paper.[a] The set by the Takeo Co., *Kami*,[2] or *Handmade Paper of the World*, provides articles in Japanese and English and samples of handmade papers from twenty-three countries throughout the world, including twenty-nine samples from China.[b]

Together with paper, there is also a great deal of information for the study of Chinese ink, which has been closely associated with the writing brush, the inkstone, and with paper, these comprising the 'four treasures of the scholar's studio'. The literary records include general treatises and history of ink, biographies of ink-makers, recipes and procedures for inkmaking, albums of ink designs, catalogues of ink dealers and connoisseurs, collections of works on ink, and modern studies in different languages. Besides these, artifacts, ink traces on ancient relics and on paintings and in calligraphy are also available for examination to help determine the composition of old ink.

The first general treatise on ink, as on paper, is included in the + 10th-century work, *Wen Fang Ssu Phu*, which also contains sections on other writing materials. Since then, numerous monographs have been written exclusively on ink and inkmaking, and these include at least five or six titles written by Sung and Yüan authors between 1100 and 1330, and nearly two dozen such works by Ming and Chhing authors between 1400 and 1900. Many interesting but not easily accessible writings can be found in the two comprehensive collections on ink: the *Shih Liu Chia Mo Shuo*,[3] which contains sixteen titles mostly by Ming-Chhing writers, compiled by Wu Chhang-Shou[4] in 1922; and the *She Yüan Mo Tshui*,[5] which includes twelve works by authors from the Sung to the Republican period, published by Thao Hsiang[6] between 1927 and 1929.

Interest in ink has continued into modern times, especially in its new techniques and the artistic appreciation of its use. This is exemplified in two Japanese monographs, one by Watanabe Tadaichi[7] (*1*) on colour pigment, crayon, and inkmaking which has been translated into Chinese, and another by Tomori Soshinan[8] (*1*) on the connoisseurship of Chinese, Korean, and Japanese inks. A more recent treatise in Chinese by Mu Hsiao-Thien[9] (*1*) contains an historical account of inkmaking in She-hsien, Anhui, which has been the centre for manufacturing the four scholar's 'treasures for writing' for many centuries. A catalogue of four modern ink collections in Peking was published as late as 1956.[c]

The earlier studies in Western languages of Chinese ink include those by Stanislas Julien & Champion (2) in 1833 and J. Goschkewitsch (1) and Maurice

---

[a] See description under Anon. (*243*) in Bibliography B.
[b] Cf. Takeo Eiichi (*2*).
                                            [c] See Yeh Kung-Chho (*2*).

[1] 手漉和紙大鑑   [2] 紙   [3] 十六家墨說   [4] 吳昌綬
[5] 涉園墨萃   [6] 陶湘   [7] 渡邊忠一   [8] 外守素心庵   [9] 穆孝天

Jametel (1), who translated into German in 1858 and French in 1869 respectively the Chinese work on inkmaking *Mo Fa Chi Yao*[1] by Shen Chi-Sun of the 14th century. The most comprehensive translation of Chinese sources on ink was made by Herbert Franke (28), who in 1963 published in German the full text of four monographs on Chinese ink by Sung and Yüan authors, in addition to parts on ink in seventeen other works and verses on ink by six poets from before the 16th century.[a] A few monographs on Chinese chemical arts, printing, and writing also contain some information on ink.[b] Other accounts include those by Berthold Laufer, who contributed five chapters on Chinese, Japanese, Central Asian, and Indian ink in F. B. Wiborg (1); by Wang Chi-Chen (2) on the appreciation of Chinese ink, based on the fine examples at the Metropolitan Museum of Art in New York City; and by van Gulik (9, 11) on the connoisseurship of Chinese ink. Finally, a recent study by John Winter (1) on ink traces in old Chinese paintings, which uses scanning electron micrography, represents a new approach to the study of this subject.

(4) SOURCES FOR THE STUDY OF CHINESE PRINTING

Information on Chinese printing is available primarily from artifacts, printed materials, descriptions in book catalogues, and other literary records. The artifacts include printing blocks and movable types made of various materials from different periods and various kinds of tools for carving and printing. These objects provide us with details about some of the technical aspects of the craft which are not normally described in literary sources. Only a few such specimens pre-date the Ming period. One wooden block of an Amida Buddhist *sutra* said to have been made in the Northern Sung period is kept in an American collection.[c] Two other pieces with human figures, perhaps of the same period, are now kept in the Chinese Historical Museum, Peking (see Fig. 1053).[d] Over 1200 wooden blocks from the 16th century are in the Thien I Ko[2] Library in Ningpo.[e] Many more of the Chhing and Republican periods are kept in various libraries and publishing houses in Peking, Nanking, Hangchow, Szechuan, and other places.[f] Bronze blocks for printing

[a] Also, the *Thien Kung Khai Wu*, a 17th-century work on Chinese technology, includes a chapter on vermilion and ink; see tr. Sun & Sun (1), ch. 16, pp. 279–88.

[b] For example, such works as Stanislas Julien & P. Champion (2), Carter (1), Li Chhiao-Phing (1), and T. H. Tsien (2) include a chapter on ink.

[c] The block is said to have been found in Chü-lu in Hopei. It was formerly in a Japanese private collection and is now kept in the Spencer Collection of the New York Public Library; see a picture and short description in *Toyo Kodai Hanga Shu* (Tokyo, 1913).

[d] Both blocks are said to have belonged to Northern Sung and also to have been found at Chü-lu. One is a carved picture of a well-dressed woman in the Thang attire with hands clasped. The other is a picture of three women under a curtain with inscriptions on both sides. The one on the left says: 'Good luck for raising silkworms by three women'; the one on the right says: 'Good luck for collecting a thousand pounds and a hundred ounces.' Apparently this is a picture for worshipping patrons of silk culture; see *WWTK*, 1981 (no. 3), pp. 70–1.

[e] See list of titles for the blocks in Feng Chen-Chhün (1), ch. 6, Appendix 1.

[f] Information supplied by Chang Hsiu-Min in his letter of 2 Jan. 1980.

[1] 墨法集要　　[2] 天一閣

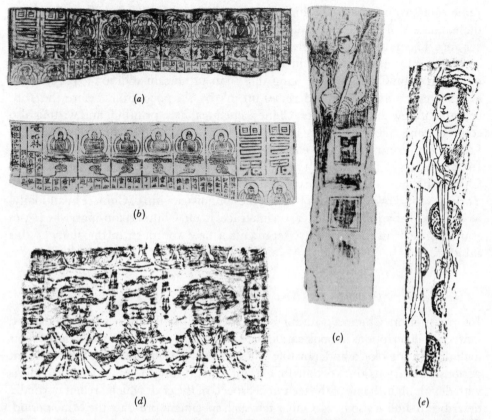

Fig. 1053. Earliest extant carved blocks said to be of the Northern Sung period found in Chü-lu, Hopei Province. (a) Block with negative Buddhist images and scriptures, 12.5 × 43.1 cm. (b) Rubbing of the above reproduced in positive. (c) Picture of a sitting monk on the back of the block. (d) Rubbing from a block with three patroness of silk culture, 13.8 × 26.4 cm. (e) Rubbing from another block with woman in Thang attire, 15.3 × 59.1 cm. Figures (a, b, c) from Spencer Collection, New York Public Library; (d, e) from Museum of Chinese History, Peking.

paper money in the Sung and Yüan periods still survive.[a] Also one composite block made of water-buffalo horn in the late 19th century, for printing credit notes, is extant.[b] Samples of carved blocks of different periods are found in public and private collections throughout the world.

As for movable type, very few Chinese specimens have survived to the present time, but many of non-Chinese origin have been preserved. The oldest is a font of several hundred wooden types in Uighur of about 1300 found in Tunhuang early in this century.[c] Over half a million bronze, iron, wooden, and clay movable types made by Koreans from the late 18th century on are reputed to be kept in Korea, and specimens of earlier dates are collected by many libraries and museums.[d] The only types made in China known to survive include a set of earthenware types and

[a] The earliest known sample extant is a copper plate of c. +1024–1108 (see Fig. 1080).
[b] Britton (3), pp. 99 ff.; the block is now kept in the collection of the American Numismatic Society, New York.
[c] Cf. Carter (1), pp. 218–19. According to Dr Goodrich, Mme Pelliot and her lawyer informed him several years after her husband's death that the Uighur type could no longer be found.
[d] Cf. Carter (1), pp. 230, 235, n. 27. A collection of several boxes of early bronze movable type from Korea is said to have been preserved in the Museum of Natural History in New York; see Natural History (Sept. 1980), p. 74.

clay moulds of the early 19th century found recently in Hui-Chou, Anhui, which are now kept in the Institute of the History of Natural Science, Academia Sinica, in Peking.[a] A small number of wooden types of later date are also on hand.[b] Aside from these, types made of clay, wood, or bronze are virtually no longer extant. Failing also to survive are the tools for carving, brushing, and printing used in former times, but some of those used by modern craftsmen are believed to be similar.[c]

While few blocks and movable types survive, printed books and single sheets are available for study. These range in date from the earliest use of block printing to the early years of this century, when traditional block printing was still in use.[d] Specimens of woodblock printing from the 8th to the 10th century are extant in and outside China.[e] Some 2000 printed works of the Sung, colour printing from the Yüan,[f] and about thirty titles printed with bronze type from the late Ming period are known to be kept in public and private collections throughout the world.[g] In North America alone, more than one-half of the four million volumes of Chinese books in various library collections are believed to be traditionally printed and bound editions, including over 100,000 volumes printed in the Sung, Yüan, and Ming periods.[h]

Besides these original editions, many such books have been reproduced in facsimile using woodcut, lithography, offset, or a photo-duplication processes. The facsimiles provide near-exact copies of the originals, except for paper and ink, and permit study of their calligraphy, format, and other details.[i] There are also several compilations of selected samples of rare editions in facsimile, such as the woodcut specimens by Yang Shou-Ching[1] (*1*), the photolithographic Sung, Chin, and Yüan samples by Chhü Chhi-Chia[2] (*1*), and Ming editions by Phan Chheng-Pi[3] and Ku Thing-Lung[4] (*1*). Also notable in this regard is the *Chung-Kuo Pan Kho Thu Lu*,[5] an illustrated catalogue compiled by the Peking Library, which includes

[a] Cf. Chang Ping-Lun (*1*), pp. 90–2; see also Fig. 1141, the photographs for which were supplied by him.

[b] A set of wooden types from Chhang-chou were sent to the Leipzig exhibit by the Peking Library in 1958.

[c] Samples are found in woodblock printing shops in Peking and Shanghai; see Figs. 1135–6.

[d] The craft survives on a limited scale in China today, since multicolour block prints have been revived and some rare books are reproduced by woodcut in Shanghai and several other cities.

[e] These include the *dharani* of *c.* +704–51 from Korea, the one million *dharani* of *c.* +770 from Japan, the *Diamond sutra* of +868, two calendars of +877 and +882, over a dozen printed books, and many single sheets of Buddhist images from Tunhuang, another *dharani* in Sanskrit of *c.* +850–900 from Chheng-tu, and three versions of an invocation *sutra*, dated to +956, +965, and +975, from the Wu-Yüeh Kingdom. All these were made prior to the Sung period.

[f] Several colour pictures attributed to the Thang period from Tunhuang are extant, but the colour appears to have been added by hand to the printed outline; samples of a Kuan-yin image coloured on a printed outline and on the back sheet are kept at the Bibliothèque Nationale, Paris.

[g] For a list of the Sung editions, see Poon Ming-Sun (*2*), appendix; for Sung editions in Japan, see Liang Tzu-Han (*1*); for Ming bronze movable type editions, see list in Chang Hsiu-min (*9*); Chhien Tshün-Hsün (*2*), appendix.

[h] A census of rare Chinese editions in American libraries as at 1957 indicates that twenty-eight titles in 887 volumes of the Sung, thirty-five titles in 2445 volumes of the Yüan, and 4518 titles in 92,899 volumes of the Ming were kept in thirteen libraries in the United States and Canada; see Tsien (*11*), p. 10.

[i] These include large sets or rare editions such as those reprinted in the *Ssu Pu Tshung Khan*[6] and those photo-duplicated in microform.

[1] 楊守敬   [2] 瞿啓甲   [3] 潘承弼   [4] 顧廷龍   [5] 中國版刻圖錄
[6] 四部叢刊

samples of all kinds of Chinese printing by woodblock, movable type, and multi-colour processes from different periods.[a]

Where original works no longer survive, much information about them can be gathered from bibliographical descriptions in catalogues of public and private collections, in which dates, names of printers, and descriptions of formats are sometimes given. Unlike the great deal of information that can be found for paper, there is very little on printing in Chinese literature. Almost nothing about such technical matters as how blocks were carved or used in printing is mentioned in any pre-modern documents, except for occasional records made by foreign observers.[b] Descriptions of printed editions or publishing records are often found in standard histories, histories of institutions, individual literary collections, miscellaneous writings, local histories and sometimes clan records,[c] but they are generally very scattered and sketchy.

Two early systematic studies of Chinese printing of great influence may be mentioned here. One is the *Shu Lin Chhing Hua*[1] by Yeh Te-Hui,[2] first published in 1911 and with a supplement in 1923. It consists of an initial chapter on the moral obligations of printing, a discussion of bibliographical terms, and topical remarks on manuscripts, printing, publishers, and dealers arranged under broad chronological groupings through the Chhing period. Another is the *Chung-Kuo Tiao Pan Yüan Liu Khao*[3] by Sun Yü-Hsiu,[4] published under the pseudonym Liu-An[5] in 1916. This is a collection of quotations from various sources grouped under the broad topics of printing, paper, and bookbinding, without critical comments. Besides these two early monographs, information on printing is generally included in works on historical bibliography, or *pan pen hsüeh*,[6] such as Chhü Wan-Li[7] and Chhang Pi-Te[8] (*1*), Chhen Kuo-Chhing[9] (*1*), and Mao Chhun-Hsiang[10] (*1*) and in general histories of books or publishing such as Liu Kuo-Chün[11] (*1, 2, 3*). Concerning the origin and development of Chinese printing, major contributions were made by Hsiang Ta[13] (*1–3*), Wang Kuo-Wei[14] (*3–7*), and Li Shu-Hua[15] (*4–11*) for the Thang and Five Dynasties, and by Chang Hsiu-Min[16] (*1–19*), Chhang Pi-Te (*2–8*), and others on various subjects from the Sung to the Chhing period. As for the artistic and technical aspects of Chinese printing, numerous studies were made on movable type by Chang Hsiu-Min (*7, 9–12*), on woodcuts and book illustrations by Cheng Chen-To[17] (*6*), Kuo Wei-Chhü[18] (*1*), and Wang Po-Min[19]

---

[a] See description under Anon. (*229*) in Bibliography B.

[b] The only article on printing procedures was written by Lu Chhien[12] (*1*) in 1947. The earliest records on carving and printing were made by the Persian historian Rashīd al-Dīn in about 1300 and the Jesuit missionary Matteo Ricci in 1600; see discussion on pp. 306 ff. below.

[c] For example, biographies of such bronze movable type printers as Hua Sui, Hua Chheng, and An Kuo of the 15th and 16th centuries are found in their respective family records, but this is due to their political or scholarly influence rather than their careers as printers.

[1] 書林清話	[2] 葉德輝	[3] 中國雕板源流考		[4] 孫毓修
[5] 留庵	[6] 版本學	[7] 屈萬里	[8] 昌彼得	[9] 陳國慶
[10] 毛春翔	[11] 劉國鈞	[12] 盧前	[13] 向達	[14] 王國維
[15] 李書華	[16] 張秀民	[17] 鄭振鐸	[18] 郭味渠	[19] 王伯敏

(*1–2*), and on the evolution of the physical book and book-binding by Ma Heng[1] (*1*), Li Wen-Chhi[2] (*1*), and Li Yao-Nan[3] (*1*). These represent the best of modern Chinese scholarship in the field.

Japanese sources for the study of Chinese printing are more numerous and important than those in any other language besides Chinese. Japan's long tradition of collecting and reprinting books in Chinese has left us with many useful descriptive catalogues of Chinese books, facsimiles of old and rare Chinese editions, and reprints of Chinese woodcuts and book illustrations. Lists of Chinese books can be traced back to the +8th or 9th centuries, when large-scale importation of Chinese Buddhist and Confucian classics commenced. During the 17th and 18th centuries, several catalogues of Chinese and Japanese movable type editions were compiled.[a] However, systematic studies of Japanese and Chinese printing were not launched until the beginning of the 20th century, when contributions were made by such scholars as Shimada Kan[4] (*1*) in 1905, Asakura Kamezo[5] (*1*) in 1909, and Nakayama Kyoshirō[6] (*1*) in 1930. Their studies laid the foundation for further investigations in both China and Japan. Shimada's critical notes on old Chinese manuscripts and printing include sections on Chinese bookbinding and printing which are considered to be pioneer studies in the field, despite some shortcomings.[b] The work by Asakura is the first systematic study of Japanese printing and that of Nakayawa is still the most comprehensive treatise in the field, with rather full coverage of China and Japan and a brief account of Korea, although it is out of date in some respects.[c] The most important contributions of Japanese scholars to our knowledge of Chinese and Japanese printing made in the last fifty years or so are those of two eminent scholar-bibliographers, Kawase Kazume[7] (*1–5*) and Nagasawa Kikuya[8] (*3–12*). Their studies are both intensive and extensive. While Kawase's contributions are primarily in the areas of Japanese books and printing, especially movable type and Gozanban printing, Nagasawa's over two dozen monographs and numerous articles cover more aspects of the Chinese book, bibliography, and printing. His works on the history of Japanese and Chinese printing, published in 1952, and an illustrated history of Japanese and Chinese printing, published in 1976, are both resourceful and critical. His studies of Sung and Yüan block cutters have suggested a new method for the identification of old printing.

Not many Korean sources on Chinese printing are to be found, but numerous artifacts and several secondary sources are available for the study of Korean printing, especially movable-type printing. One of the modern authorities on early movable type in Korea is Kim Won-Yong[9] (*1*), which deals with the historical

[a] See Nagasawa Kikuya (*4*), pp. 14 ff.
[b] For example, Shimada Kan insisted that printing was invented in the Sui dynasty on the basis of invalid evidence; see discussion on pp. 148 ff. below.
[c] The work was intended to consist of three volumes on a universal history of printing, but the last one, on Western printing, has not been published.

[1] 馬衡　　[2] 李文裿　　[3] 李耀南　　[4] 島田翰　　[5] 朝倉亀三
[6] 中山久四郎　[7] 川瀬一馬　[8] 長澤規矩也　[9] 金元龍

development of movable type and includes a table of different fonts and a summary in English. There are also collections of samples of Korean movable type printing. One by McGovern (1) includes twenty-two reproductions of original or facsimile pages, and another by Sohn Pow-Key (2) with text in Korean and English also includes samples of facsimile pages. Both of these were made by using deeply etched photo-engravings taken from original pages. Several kinds of handmade papers and water-based ink were used with metal plates in imitation of the original process. These specimens look and feel more authentic than facsimiles made by offset processes or photographic duplication.

Western scholarship on Chinese printing has focused on two major aspects of this subject, namely its beginnings and its spread westwards. Writings on these subjects by early European travellers and missionaries before the end of the 18th century will be discussed later.[a] Detailed discussions and scholarly researches on Chinese printing began only in the 19th century.[b] These include works on the history of printing by Isaiah Thomas (1) in 1810, Robert Curzon (2) in 1858, and Theodore De Vinne (1) in 1876; remarks on Chinese printing in general histories of China by John F. Davis (1) in 1836 and S. Wells Williams (1) in 1848; and particularly the monograph on the history of Chinese printing by Stanislas Julien (12) in 1847. Julien's studies of Chinese block printing and movable type, despite their incorporation of quotations from inaccurate Chinese sources,[c] laid down the foundation for all later studies by Western scholars.

With the discoveries of manuscripts and printed specimens in Tun-huang, Central Asia, and Africa around the turn of the 20th century, studies by such sinologists as Hermann Hülle (2) in 1923, Thomas F. Carter (1) in 1925, Berthold Laufer (48) in 1931, and Paul Pelliot (41), whose notes were published posthumously in 1953, have added substantially to our knowledge of the subject. Especially significant was the work by Carter, which was revised by L. Carrington Goodrich in 1955. It synthesised all previous researches and further elucidated the subject in the context of Chinese-Western contacts. This work, which has had a significant influence on Chinese and Western scholarship on printing, remains a classic in the field. More than three quarters of a century following its first appearance, no work of comparable magnitude on Chinese printing has been published. In recent years, major contributions have primarily been made by Goodrich (30–32) in his critical studies of new discoveries and his revision of Carter; Richard Rudolph (14, 15) in his translation and study of the Wu Ying Tien manual on movable type of the Chhing dynasty; and K. T. Wu (6, 7, 8, 9) in his works on the development of Chinese printing from the Sung to the modern period. In the fields of woodcuts and

---

[a] See discussion on pp. 313 ff. below.

[b] Many works on printing deal also with paper; see also the recent researches on the origin and development of Chinese paper cited on pp. 293 ff.

[c] The attribution by Julien (12) of the date of the Chinese invention of printing to +593 was based on unreliable secondary sources, and was followed by all later works on the subject, including the 11th ed. of the *Encyclopedia Britannica*, until a correction was made by Arthur Waley (29) in 1919 and Carter (1), 1925 ed., p. 202, n. 13, based on Yeh Te-Hui (2), p. 20.

colour printing, several monographs have been produced by Max Loehr (1), Josef Hejzlar (3), and especially Jan Tschichold (1–7) whose reproductions of a series of Chinese colour prints using modern techniques provide a new source for appreciation and study of this Chinese art.

Despite the contributions to the study of Chinese paper and printing of this band of international scholars whose specialties cover various disciplines, many gaps in this field are still open. Systematic investigations of their social, economic, and intellectual roles and influences in Chinese history are lacking; comparative studies of their origins in and impact on China and the West are especially needed, while certain questions have never been raised for discussion. For example, why were paper and printing invented early in China and not in other civilisations of the world? What effects did these inventions have on changes in Chinese society compared with those in the West?

In studies of Chinese paper, local histories of manufacture and distribution are scarce and the origins of the various uses to which it has been put have not been fully and systematically covered. For example, paper clothing and furnishing have never been mentioned in Western literature. Wallpaper and paper-folding are said to have originated in China, but further evidence is needed to substantiate this claim. On the other hand, watermarks are said to have been invented in the West in the 13th century and marbled paper in the 16th, but both artifacts and documentary sources indicate that they were made in China several centuries earlier. Also, hundreds of trade-names of paper derived from materials, methods, quality, locality, and makers or designers, for the most part incomprehensible to laymen, need to be collected and explained.

In previous studies of Chinese printing, emphasis has been placed on its origin and spread westwards, but its development and contributions have either been oversimplified or underestimated. Many technical and artistic aspects that have been ignored, especially the procedures involved in preparing and printing from both blocks and movable types, need detailed step-by-step descriptions accompanied with illustrations. Also needed are analytical studies of calligraphy, formats, materials, and methods used in printing that could lead to the establishment of new criteria for dating and authenticating old printings. Although an attempt has been made here to fill some of these gaps in the present study, many questions are still waiting for satisfactory answers.

(b) NATURE AND EVOLUTION OF PAPER

(1) PRE-PAPER MATERIALS FOR WRITING

BEFORE paper was used for writing at the beginning of the Christian era, the Chinese selected a great variety of hard and soft materials for documents, historical records, for personal communication. These included such animal products as bones, shells, ivory, and silk; minerals such as bronze, iron, gold, silver, pewter,

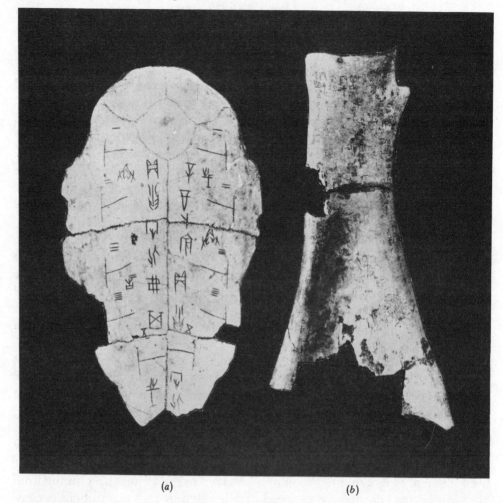

Fig. 1054. Oracle bone inscriptions of the Shang dynasty. (a) A whole tortoise plastron, c. −12th century, 20 × 12 cm. Academia Sinica. (b) A large piece of ox bone consisting of 25 separate divinations on harvest and raining, c. −12th century, 28 cm. long. Research Institute of Humanistic Studies, Kyoto University.

stone, jade, and clay; and vegetable matter like bamboo and wood. Although many of these materials were also used by the peoples of other ancient civilisations,[a] their use in China was much more common, extensive, and refined. However though animal skins and plant leaves were also used extensively by others, they were never adopted for writing by the Chinese. Generally speaking, writings on ceramics in China can be traced back to as early as Neolithic times; bone, shell, ivory, bronze, and bamboo to the Shang dynasty; stone, jade, silk, and several metals to the early Chou; and wood to the Han. While some of these hard and durable materials were used primarily for permanent records and monumental

---

[a] For writing materials used by peoples of the ancient world, see Diringer (1).

inscriptions, such perishable materials as bamboo, wood, and silk were used extensively for books, documents, and other writings of daily life. The former group of materials was intended for vertical communication across generations and the latter, primarily, for communication among contemporaries. [a]

Bone and shell are the oldest surviving materials on which the earliest known Chinese writings were inscribed. The bone primarily used was the ox scapula, which provides a wide, smooth surface for writing. The shells were the plastron and carapace of the tortoise (Fig. 1054). They were used for divination by the royal house of the Shang dynasty, and oracular messages were usually recorded on them after divination was performed. The inscribed materials so far discovered cover about 250 years from the early — 14th to the late — 12th century.[b] The Chou court continued to use bones and shells for divination,[c] but oracular inscriptions were, in general, written separately on bamboo and silk, except perhaps for the early Chou period.[c] The inscriptions recorded such natural phenomena as eclipses, rain, wind, snow, or clear skies; forecasts of happenings during the coming evening, day, ten-day week, or year; predictions for forthcoming travel, hunting, fishing, and military campaigns; human fortunes such as birth, illness, or death, and the evidence of dreams, as well as sacrifices to ancestors, deities, and other spiritual beings. The inscriptions were carved with a stylus, but a few were written with brush and ink made of lampblack or cinnabar. The carved grooves were sometimes illuminated with pigments or inlaid with turquoise for decoration. Since very few pieces of literature transmitted from the Shang dynasty survive today, these inscriptions are the most important documents for the study of the history and institutions of ancient China.[d]

Inscriptions were also made on various kinds of metal objects, ceramics, and clay materials. The most important and extensive inscriptions are found on bronzes dating from the Shang to the Han period.[e] These inscribed bronzes include sacrificial vessels, musical instruments, military weapons, standards of measurement, mirrors, coins, seals, and other articles, but most of the inscriptions of historical significance were cast on ritual vessels, especially those of the Chou dynasty. These contain from a few to as many as some 500 characters each; the latter equalling in length the chapter in an ancient book.[f] Bronze inscriptions of

[a] For detailed discussion of pre-paper materials, see Tsien (2).

[b] A total of over 100,000 pieces were found in private diggings and official excavations from 1899 to 1937; diggings have continued since 1950, and some 5000 pieces were found near Anyang in 1973.

[c] Individual pieces dated to the Chou were found in Hung-chao, Shansi, in 1954, Feng-hsi in 1956, and Peking in 1975, and a score of pieces from the early Chou period were found in Chhi-shan, Shensi, in 1977; see *WWTK*, 1979, no. 10, pp. 38–43, pls. 4–7; also *Wen Wu Khao Ku Kung Tso San Shih Nien, 1949–1979*, pp. 4–5, 126.

[d] For further discussion, see Creel (1), pp. 1 ff.; Tsien (2), pp. 19 ff.; Keightley (1), pp. 134 ff.

[e] Liu Thi-Chih, *Hsiao-Chiao-Ching-Ko Chin Wen Tho Pen* (1935) contains about 6500 bronze inscriptions reproduced from rubbings; numerous inscribed objects have been found since that publication.

[f] The most significant finds in recent years include a series of 64 bronze bells inscribed with some 2800 characters on music, found in Sui-hsien, Hupei; and two Chan-Kuo bronzes containing the longest inscriptions of that period ever discovered: a large tripod with 469 characters and a vase with 448 characters from a tomb dated to c. −308, of the state of Chung-shan in modern Phing-shan, Hopei, found in 1974; cf. *WWTK*, 1979, no. 10, p. 6; *KKHP*, 1979, no. 2, pp. 147 ff.

Fig. 1055. Bronze *kuei* vessel of the Western Chou dynasty with cast inscription on inner surface, *c.* − 11th century (diameter 27 cm.). British Museum.

the Western Chou include long narrative compositions such as records of military campaigns, covenants, treaties, appointments, rewards, cermonial events, and other political and social affairs (Fig. 1055). Inscriptions from the later Chou are in general shorter, more apt to follow a formula, occasionally composed in rhyme, and written in ornamental style, sometimes in 'bird script'.

Mirror inscriptions appear on the inner or outer part of circular designs on the backs. The earlier ones, from the Warring States and Han periods, include expressions of human desires for spiritual and material satisfactions, greetings and good wishes, political messages, and allusions to various folk beliefs. Inscriptions of the Sui and Thang periods are mostly formulae consisting of a few characters.[a] Numismatic inscriptions appear on almost all early and late metal coins, which were shaped like spades or knives or were round with a square hole in the centre. These inscriptions are primarily the names of places of manufacture and numerals indicating denominations of the coins.[b] Ancient seals were of various material, including cast metal. They were used to stamp inscriptions on small lumps of clay for authentication before they were used to make impressions on silk and paper. Pottery, bricks, and tiles also bear inscriptions, while symbols and numerals are found on Neolithic pottery, with later vessels being inscribed with names of makers or owners, official titles, place and date of manufacture, and, sometimes, imperial edicts (Fig. 1056). Inscriptions on bricks consist mostly of dates, names, and

[a] Cf. Karlgren (18); Liang Shang-Chhun, (1), (2); Tsien, (2), pp. 47 ff.
[b] Cf. Wang Yü-Chhüan (1); Tsien (2), pp. 50 ff.

(a)

(b)

Fig. 1056. Inscriptions on pottery vessels. (a) A measuring vessel with imperial decree of the First Emperor of Chhin, dated −221, ordering standardisation of weights and measures. (b) Rubbing from a vessel with imperial edict of Wang Mang, +9.

miscellaneous records. Decorative inscriptions or pictures on roof tiles include lucky formulae, and the names of palaces, temples, mausoleums, granaries, or other public or private buildings, to commemorate their construction.[a]

Inscriptions on bronze and stone are the two major categories of material for the study of Chinese epigraphy and archaeology. While bronze inscriptions may be more ancient, those on stone are more numerous, longer, and more readily accessible. This is because stone is more abundant, more permanent, and makes possible a wider surface for inscriptions, with the result that, from the +2nd or +3rd century onwards, stone was extensively used not only for monumental and commemorative inscriptions but also as a permanent material for preserving all the canonical literature of China.

Among the earliest inscribed stones of historical significance are the ten drum-shaped boulders known as the Stone Drums (*shih ku*[1]), dated variously from the −8th to −4th century.[b] The inscriptions were written in verse and concerned hunting and fishing expeditions on certain memorable occasions; they originally comprised some 700 characters.[c] In addition, during the reign of the First Emperor of Chhin, between −219 and −211, seven monumental stone tablets were inscribed to praise the achievements of his administration.[d] All these earlier stones were in the form of crudely truncated rocks, but a more refined style of flat stele, known as a *pei*,[2] was used since the Han dynasty. From that time, numerous tablets have been inscribed to commemorate historical occasions, to preserve the memory of individuals, and to standardise sacred texts in correct and permanent form.

One of the most gigantic projects for the preservation of standard texts was the engraving on stone of the entire collection of Confucian classics. No fewer than seven different editions were inscribed from the late +2nd until the end of the 18th century. The first edition, of seven classics in over 200,000 characters, was carved on both sides of forty-six steles, from +175 to +180 (Fig. 1057). The last, including all the Thirteen Classics, was made from 1791 to 1794 under the Chhing dynasty.[e] Engravings of Buddhist canons on stone came later than the second century, but achieved much greater scope and size. The Buddhists selected stone because it was the best material for preservation of their sacred texts; as one devotee said: 'Silk will decay, bamboo is not permanent, metal seems hardly eternal, and skin and paper are easily destroyed.'[f] A grotto library, started in +605 and continued until +1091, consisted of 105 Buddhist *sutras* of over four million words carved on more than 7000 steles. This collection is still preserved in the Mountain of Stone *Sutras*

---

[a] For inscriptions on seals, sealing clay, and pottery vessels, see Tsien (2), pp. 54 ff.; also below, pp. 136 ff.
[b] Dating of these stones is controversial; see Kuo Mo-Jo (*14*); Ma Heng (*2*); Thang Lan (*1*); Akatsuka Kiyoshi (1); and a summary in Tsien (1), pp. 73 ff.
[c] The original stones are kept in Peking, but only about 300 characters are extant.
[d] Cf. *Shih Chi* (*ESSS/TW*) ch. 6, pp. 14b–27a; ch. 6, p. 30b.
[e] See Chang Kuo-Kan (*1*); Ma Heng (*3*); Wang Kuo-Wei (*10*); and a summary in Tsien (2), pp. 73 ff.
[f] Cited in *Shan Tso Chin Shih Chih* (1797 ed.), ch. 10, p. 21b.

[1] 石鼓   [2] 碑

Fig. 1057. Largest surviving specimen of the stone classics of the + 2nd century. Text of the *Kung-Yang Commentary* fragment, front 49 × 48 cm. and back 48 × 47 cm., found in Loyang, Honan province in 1934.

(Shih Ching Shan[1]), near Fang-shan, Hopei, after well over a thousand years.[a] The engraving of Taoist literature on stone was much later still. The earliest Taoist stele was erected at I-chou, Hopei, in + 708, and several were engraved in the following centuries.

Not all the inscriptions engraved on hard materials are considered books. The direct ancestry of the Chinese book is in the tablets of bamboo or wood which were connected by thongs and used like the paged books of modern times. Bamboo and wood were the most popular materials for writing before paper, and their use has had a most significant and far-reaching effect on the tradition of the Chinese book and culture. Not only are the vertical arrangement of Chinese writing and its movement from right to left believed to have been derived from this system, but also the format and terms for a physical book, used until today, are supposed to have originated from those for the tablets.[b]

[a] See *Chhin Ting Jih Hsia Chiu Wen Khao* (1774 ed.) ch. 131, pp. 4 a ff.; *Shun Tien Fu Chih* (1885 ed.), ch. 128, pp. 9 a–10 b; 42 a–64 a; Vaudescal (1), pp. 375 ff.; and a summary in Tsien (2), pp. 79–83.
[b] For further discussion, see Tsien (2), pp. 90 ff., 183–4.

[1] 石經山

The succession of various book materials in China may be divided into three periods: bamboo and wood from the earliest times to the +3rd or +4th century; silk from the −7th or −6th to the +5th or +6th century; and paper from the +1st century to the present.[a] Thus the uses of bamboo and of silk overlapped by about 1000 years, those of silk and of paper by 500 years, and those of bamboo and paper by 300 years. The old-fashioned materials were replaced by new ones only gradually, and not until after the +3rd century were bamboo and wood entirely superseded by paper.

Since the end of the 19th century, no fewer than 40,000 tablets of bamboo and wood have been unearthed from various locations in China.[b] They cover a span of almost 1000 years of Chinese history. The important discovery sites include those in Hunan, Hupei, Honan, and Shantung in the central plain of China; Tunhuang, Chü-yen, Chiu-chhüan, and Wu-wei in the northwest; and the ruins of Loulan, Khotan, and Turfan in modern Sinkiang. Among these, Chhang-sha, Hsin-yang, and Yün-meng yielded the oldest bamboo tablets, dating back to the Warring States and Chhin periods; Chü-yen supplied the largest quantity and the most important core of wooden tablets of the Han dynasty; and all documents from Loulan belong to the Chin dynasty. These tablets include official documents, private letters, calendars, lexicons for beginners, laws and statutes, medical prescriptions, literary texts, and miscellaneous records.[c] The most important finds in recent years include a group of 490 wooden tablets, including seven chapters of the I Li[1] or Book of Rituals (Fig. 1058), recovered from a Later Han tomb at Wu-wei, Kansu, in 1959; over 4490 bamboo tablets on military classics from a Western Han tomb at Lin-i, Shantung, in 1972; 400 bamboo tablets on taxation and economic matters from a Western Han tomb at Chiang-ling, Hupei, in 1973; 600 bamboo tablets of funerary inventories from Western Han tombs at Ma-Wang-Tui, Chhang-sha, Hunan, in 1973; 1100 bamboo tablets of legal documents of the Chhin state from Yün-meng, Hupei; and some 20,000 wooden tablets dated from c. −119 to +26, from Chü-yen, Kansu, in 1972–6. Among the most interesting items found in Chü-yen are some seventy-five complete or nearly complete documents on connected tablets strung together on two or three lines of hemp threads, and in the original format.[d]

The preparation of tablets involved several steps. The bamboo stem was first cut into cylinders of a certain length and then split into tablets of a certain width. After the external green skin was scraped off, the tablets were dried over a fire to prevent any quick decay. Writing was carried out on the outside surface and sometimes on the inner surface of the stem as well. Wood, on the other hand, was cut into large

---

[a] See Ma Heng (1), pp. 201–2; Tsien (2), p. 91.
[b] These include over 10,000 pieces found from 1899 to 1930, and nearly 30,000 from 1951 to 1975; see list of finds from ancient times to 1975 in WWTK, 1978, no. 1, p. 44.
[c] See a checklist in WWTK, 1978, no. 1, p. 44; Loewe (14), pp. 101 ff.
[d] See report in WWTK, 1978, no. 1, p. 7, pl. 8, illus. 35.

[1] 儀禮

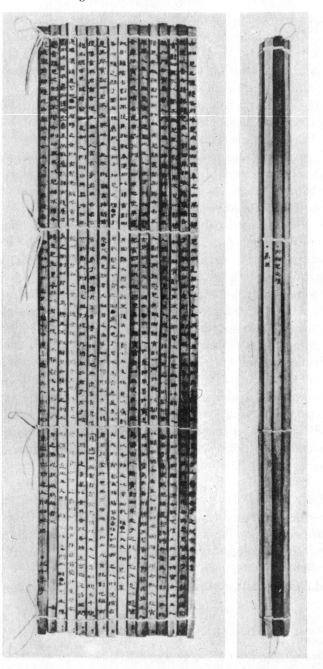

Fig. 1058. Partial text of *I-Li*, or *Book of Rituals*, on long wooden tablets of the Later Han period found at Wu-wei, Kansu province in 1959. Spread out on the left and rolled up at right, each tablet ranges from 54 cm. to 58 cm. long and 1 cm. wide and consists of 60 to 80 characters. From *Wu-wei Han Chien*, 1964.

slabs which were split into pieces of various sizes, and the surfaces smoothed for use as documents. Most of the wooden tablets are of pine, willow, poplar, and Chinese tamarisk, all noted for whiteness, light weight, fine texture, and absorbency. Old tablets could be re-used after the writing on them had been removed with a book knife, which was also used to erase errors.[a]

Bamboo tablets were narrow strips from eight inches to two feet four inches long,[b] used for classics, literary compositions, and ceremonial documents. The length of wooden tablets was fixed at from five inches to two feet; they were used primarily for official documents, personal correspondence, and short messages. The tablets are almost invariably narrow, in most cases not more than an inch wide. A single bamboo tablet was called *chien*[1] and a wooden one *tu*.[2] Several tablets bound together with cords to form a physical unit were called *tshe*.[3] Writing of a certain length which formed a literary unit comparable to a chapter was called *phien*.[4] Square or rectangular wooden pieces, called *fang*[5] or *pan*,[6] were sometimes made with a wider surface to accommodate more than one line of characters, or for maps and illustrations.

Characters were usually written with brush and lampblack ink, on one side, but in some cases on both sides. Each column of a tablet contains from a few to as many as eighty characters, but in general the average number is about thirty characters. The vertical arrangement of Chinese writing since very ancient times is believed to have been influenced by the vertical position in which the tablet was held, the grain of the bamboo or wood, and the soft brush which wrote more easily in a downward direction. Interestingly, modern studies indicate that vertical lines can be read faster than horizontal ones.[c] The habit of a right-handed person, holding a narrow tablet in his left hand and writing on it with a brush in his right hand, would be to lay the finished tablet to his left, near at first and successively farther to the left as later tablets were finished. This might have resulted in the right-to-left direction of later Chinese writing and reading.[d]

As we have seen, silk was first used for writing, along with bamboo, no later than the −6th or −7th century, and continued in use after the +3rd or +4th century, after paper had become popular as a writing material. Yet it could have been used much earlier, since silk cloth, brush, and ink were available as early as Shang. Silk continued in use for documents as late as the Thang, being valued for its softness, light weight, durability, and absorbency. The term *chu po*,[7] 'bamboo and silk,' was

[a] For the use of the book knife in the Han dynasty, see Chhien Tshun-Hsün (*1*), tr. John Winkelman (1), pp. 87 ff.
[b] All measurements mentioned here refer to original standards of the Han period; for details of the system, see Tsien (2), pp. 109–11.
[c] See a UNESCO study by William S. Gray (1), p. 50; also Tsien (2), pp. 183–4; Chhien Tshun-Hsün (5), pp. 176–7, 181–2.
[d] For further discussion, see Tsien (2), pp. 90ff., 183–4.

[1] 簡   [2] 牘   [3] 冊   [4] 篇   [5] 方
[6] 版   [7] 竹帛

used extensively in ancient literature, to refer to written documents.[a] Since silk was much more expensive than bamboo, but had a wider surface, it was used only where bamboo or wood did not suit the special purpose. We know that silk was employed for the final edition of a book, while bamboo tablets served for drafts.[b] Silk was used in particular for books on divination and occultism, for illustrations appended to books of tablets, for maps,[c] for inscriptions for sacrifice to spirits and ancestors, for recording the sayings of kings for transmission to posterity, and for commemoration of meritorious achievements of great statesmen and military heroes.[d]

Although writing on silk is frequently mentioned in ancient literature, few specimens survive today. Until recently, only remnants of silk materials bearing long or short messages had been discovered in several sites in China and in central Asia.[e] Two of the earliest specimens on silk came from Chhang-sha, one bearing an illustrated text and the other a drawing. The silk document consists of two paragraphs of text in about 600 characters. Each is upside down in relation to the other, and they are surrounded by strange animal and human figures in colour on all four sides of the piece.[f] The other piece is a painting depicting a woman with wasp waist at the centre with a strange animal over her head. Both of these seem to indicate the mysterious nature of the Chhu culture. The most significant find in recent archaeological excavations is a collection of ancient books copied on silk fabric with ink, in the small seal and *li* styles of calligraphy. These include more than a dozen pre-Chhin works such as the *Lao Tzu* (Canon of the Virtue of the Tao) (Fig. 1059), the *Chan Kuo Tshe* (Intrigues of the Warring States), and the *I Ching* (Book of Changes), several other titles, and some old maps. The total number of characters comes to over 120,000, including many parts of ancient books which have been long lost and are not included in the extant editions.[g] This is the first major cache of silk documents to be discovered.

Silk used for writing is generally called *su*,[1] a plain white fabric without design or dye. Its varieties included the *chüan*,[2] a thin and gauze-like material used especially for painting and calligraphy; the *chien*,[3] a fine and closely woven textile made of double threads and yellowish in colour; and the *tseng*,[4] a thicker and darker fabric

[a] Pre-Chhin philosophers frequently mention that ancient sage-kings recorded their beliefs on bamboo and silk; see *Mo Tse*, tr. Mei Yi-Pao (1), p. 167; *Huai Nan Tze*, ch. 13, pp. 20a–b.

[b] It is said that texts by Liu Hsiang (c. −80 to −8) were first written on bamboo and then copied on silk when completed; cited in *TPYL*, ch. 606, p. 2a.

[c] Three −2nd-century maps on silk were recently found in the Western Han tombs at Ma-Wang-Tui, Chhang-sha.

[d] See citations on the special usefulness of silk for documents in Tsien (2), pp. 127 ff.

[e] For earlier finds by Stein, see Chavannes (12a), nos. 398, 398A, 503.

[f] For progress in the study of the silk documents from Chhangsha, see Chhien Tshun-Hsün (5), pp. 112 ff., n. 24.

[g] Numerous articles on the discovery, decipherment and interpretation of these silk documents have been published in Chinese journals; see especially those on the content, text, and historical significance by a group of experts in *WWTK*, 1974, no. 9, pp. 40 ff.; 1975, no. 2, pp. 26 ff.; also a survey by Loewe (14), pp. 116–25.

[1] 素   [2] 絹   [3] 縑   [4] 繒

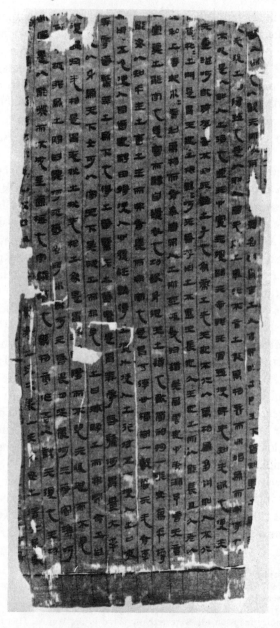

Fig. 1059. Silk book of the Former Han dynasty from Ma-Wang-Tui, Changsha. One of two earliest extant versions of the *Lao Tzu*, *c*. +2nd century, written on plain silk fabric 24 cm. in height.

probably made from wild silk. When they were used for books and writing, they could be cut as needed and rolled up as a physical unit called *chüan*[1] (roll). This system was inherited almost entirely by paper when it was substituted for silk for writing. As the historical record says, 'silk was expensive and bamboo heavy,'[a] and a light but less expensive material called *chih* (paper) was finally introduced as an ideal medium for writing.

(2) DEFINITION AND NATURE OF PAPER

Paper, as we have seen, is a matted or felted sheet of fibres formed on a fine screen from a water suspension. When the water drains away, the remaining mat of fibres must be removed from the screen and dried.[b] This definition applies to what 'paper' is today and also to what it was yesterday. Sheets of paper were made from disintegrated fibre upon a flat mould before the time of Christ and paper is still formed in this fashion; the only difference lying in the construction of the moulds and the treatment of the fibre.[c] Thus the fundamental principle of papermaking involves two basic factors, the fibres and the mould. These two principal elements were clearly given in a definition of paper contained in an old Chinese lexicon compiled during the time when paper was beginning to be popularly used in the early years of the Christian era.

In the *Shuo Wen Chieh Tzu*[2] (Analytical Dictionary of Characters), compiled by Hsü Shen[3] around +100, the word *chih*[4] for paper is defined as 'a mat of refuse fibres' (*hsü i chan yeh*[5]).[d] Here the key words are *hsü* (refuse fibres) and *chan* (mat).[e] According to the definitions given in the same ancient lexicon and its commentaries by later scholars, *hsü* means fibrous remnants obtained from rags or from boiling cocoons, and *chan* meant a mat made of interwoven rushes which was used for covering.[f] Thus the fibres and the water-draining mat have been the two basic factors in papermaking since ancient times. The definition given in the old Chinese dictionary, which mentions these two basic elements, corresponds very well with what is described today.

---

[a] *Hou Han Shu* (*ESSS/TW*), ch. 108, p. 5a.

[b] See definitions given in American Paper and Pulp Association, *The Dictionary of Paper* (1940), p. 246; 3rd ed. (1965), p. 323; Browning (1), p. 1; (2), p. 18. Since materials for more recent papermaking include all kinds of fibres, the new edition of *The Dictionary of Paper* has changed its definition of paper from 'a sheet of vegetable fiber' to 'all kinds of matted or felted sheets of fiber (usually vegetable but sometimes mineral, animal or synthetic).' We shall follow this new definition in our discussions below.

[c] See Dard Hunter (7), p. 10.

[d] See collected commentaries on the definition in Ting Fu-Pao (2), *Shuo Wen Chieh Tzu Ku Lin*, p. 5902.

[e] The character *chan*[6] is written with the grass radical in the earlier editions of the *Shuo Wen Chieh Tzu* (see reprint of a Sung edition in the *SPTK*), but this was changed by later commentators to the bamboo or the bamboo plus the water radical in later editions, to suit their own interpretations; see Chhien Tshun-Hsün (5), p. 127, n. 8.

[f] This translation of the definition of *chan*[6] is based upon the form with the grass radical in the earlier editions; see forms reproduced in Ting Fu-Pao (2), p. 5902; supplement, p. 896.

[1] 卷   [2] 說文解字   [3] 許慎   [4] 紙   [5] 絮一苫也
[6] 苫,笘,箈

It has been suggested that the manufacture of paper in China originated from the process of pounding and stirring rags in water, after which the wadded fibres were collected on a mat.[a] The treatment of rags in water was probably an old practice in China many centuries before the Christian era, for ancient literature frequently testifies to the washing, pounding, or stirring of rags in water by women. Wu Yün[1] (−6th to −5th century), a political refugee who fled from the Chhu to the Wu state, is said to have stopped by the Lai River where a woman who fed him with food was pounding rags (chi hsü[2]).[b] The Chuang Tzu (−3rd century) says that a family of the Sung state had a recipe for salving chapped hands and from generation to generation his family made their living by pounding and stirring rags in water (phing phi huang[3]).[c] Ssu-ma Chhien also says that Han Hsin[4] (d. −196), Lord of Huai-yin, was fishing outside of the city where he witnessed many women washing rags in the Huai River; one of them worked continuously for several tens of days.[d] Apparently, the treatment of refuse silk, the re-use of old fibres in quilted clothes, and the washing of rags of hemp and linen required such constant activities with fabrics in water. It is very likely that an accidental placing and drying of refuse fibres on a mat suggested the idea of making a thin sheet of paper.

Before papermaking, there were other ways of turning fibres into a sheet. It could be done by matting and pressing into a felt, by spinning and weaving into a textile, and by soaking and beating into a bark cloth or tapa. The art of felting is one of the oldest methods of making fabrics and earlier than that of weaving and spinning; felt being used for clothing and covering by the inhabitants of northern and middle Asia from very ancient times.[e] How early felt was used in China is uncertain, but the Chou Li[5] (−3rd century or earlier) records that felt (chan[6]) was made of animal hair by the officials of the Chou court in charge of leather (Chang Phi[7]).[f] No evidence is known to suggest that the invention of papermaking in China was influenced by the method of felt making, but the techniques of both are quite similar.

There is a close relationship between textiles and paper. Not only were they made of the same kinds of raw materials at the beginning of their manufacture, but they also had a similarity of physical forms and properties. Even their uses were often interchangeable. Textiles were sometimes employed for writing and painting, while paper was substituted for textiles for clothing and furnishing. Indeed, it is

---

[a] See commentaries on the definition of chih by Tuan Yü-Tshai (+1735−1805) and others in Ting Fu-Pao (2), p. 5902; also discussions by Lao Kan (1), pp. 489−92; Chhen Phan (12), pp. 257−65; and Chhien Tshun-Hsün (5), pp. 126−8.

[b] Yüeh Chüeh Shu (SPTK), ch. 1, p. 3b.   [c] Chuang Tzu (SPTK), ch. 1, p. 15b.

[d] Shih Chi (ESSS/TW), ch. 92, p. 1b.

[e] See Appendix C, 'Manufacture and Use of Felting by the Ancients,' in C. G. Gilroy (1), pp. 414ff.; Laufer (24), pp. 1 ff.

[f] Cf. Chou Li (SPTK), ch. 2, p. 24a. The Chinese never utilised wool for fabrics in early times, and felt-making could have been learned by the Chinese from their northern neighbours.

[1] 伍員   [2] 繫絮   [3] 洴澼絖   [4] 韓信   [5] 周禮
[6] 氈   [7] 掌皮

generally known that silk cloth had long been used for writing before it was
replaced by a thin sheet made of refuse fibres, which were obtained either from the
remnants from boiling silk cocoons or by pounding rags in water. When the sources
of raw materials expanded to include such new fibres as those of raw hemp and tree
bark, it opened a new page of papermaking with fresh vegetable fibres for unlimited production.

While the use of raw hemp was likely to have evolved from that of rags of hemp or
linen, the adoption of tree bark for papermaking could have been inspired by the
prior use of bark cloth made from the paper mulberry. This tree has been cultivated
extensively in China and the bark cloth is known to have been made and used for
clothing in south China from very ancient times.[a] The bark of paper mulberry,
after it was beaten into a thin sheet, had also been used for clothing, covering, and
hanging by primitive peoples in the temperate and tropical zones throughout the
world.[b] It has been suggested that the cradle of bark cloth was in China, and that
manufacture spread perhaps from the southern part of China by way of the islands
in the South China Sea eastward to the farthest regions of Pacific and Central
America, and westward through the Indian Ocean to reach Central Africa,
covering almost all areas along the equator.[c]

The moistened tissue of the bark can be expanded to as much as ten times its
original size, and several pieces can be joined together by gluing the overlapped
edges into a very large sheet. Although it is as white, soft, and flexible as paper, its
manufacture is much more laborious and time-consuming. Only one to three sheets
could be made by one worker per day, while 2000 sheets of paper could be made by
the same labour. It is very likely that the native people in south China, who were
familiar with the making of bark cloth, made use of the same fibres for papermaking. When the beating process was replaced by maceration and felting, it was
natural that the same material could be turned into a thin sheet of paper.

Besides sheets manufactured from fibres, there were other natural materials such
as animal skins, leaves, and papyrus which were all used as a medium on which to
write, though none were ever used in China. In Europe and the Middle East from
the −2nd century onwards until after the arrival of paper, skins of sheep, goats,
kids, calf and other animals were made into parchment. The most superior quality
parchment, known as vellum, was especially prized for its fineness, whiteness and
smoothness as a writing material.[d] It was stronger than paper, but more expensive
as the skins of some two hundred animals were needed to provide enough sheets for
a single book.

Of all the materials that were used for writing, probably the earliest were leaves.[e]

---

[a] Paper mulberry was grown in the Yellow River valley from Shang times and the use of bark cloth for personal
attire dates to the Chhun Chhiu period in the −6th century; see discussion of papermaking with paper mulberry
(pp. 56 ff. below) and use of bark paper for clothing and furnishings (pp. 109 ff. below).
[b] See Dard Hunter (9), pp. 29–47.   [c] Cf. E. G. Loeber (2), pp. 87 ff.   [d] Cf. Diringer (2), pp. 170 ff.
[e] Pliny says the first Cretan writings were on palm leaves which were followed by tree barks; see Diringer (2), p.
42.

palm leaves, which are thick, narrow, and sometimes as long as three feet, were used in India and other nations in South and Southeast Asia. They were incised by a stylus and rubbed with black ink or other pigments. The strips of leaves were then bound by stringing on to cords. Papyrus was used in Egypt as early as the third millennium before Christ, and was made from the inner bark of the papyrus plant (*Cyperus papyrus*). The bark was split into pieces which were placed crosswise in several layers with an adhesive between them, and then pressed and dried into a thin sheet which was polished for writing.[a] Scholars of both East and West have sometimes taken it for granted that paper and papyrus were of the same nature; they have confused them as identical, and so have questioned the Chinese origin of papermaking.[b] This confusion resulted partly from the derivation of the word *paper*, *papier*, or *papel* from papyrus and partly from ignorance about the nature of paper itself. Papyrus is made by lamination of natural plants, while paper is manufactured from fibres whose properties have been changed by maceration or disintegration.

(3) The Beginnings of Paper in the Han

A number of specimens of paper from the −2nd century onwards have recently been discovered in various parts of China, a witness to the origin and development of paper in the Han dynasty. The oldest paper extant today is probably the specimens discovered in 1957 in Pa-chhiao,[1] near Sian in Shensi province, in a tomb dated no later than the period of Wu Ti (r. −140 to −87) of the Former Han dynasty. The specimens, including one large piece about 10 cm. square (Fig. 1060a) and many fragments found under three bronze mirrors, are said to be light yellow, thick and uneven, coarse and crude, with some textile impressions on the surface.[c] On one of the specimens some loops of fibres are visible, and on another a small remnant of thin, two-ply hemp cord. These seem to indicate that this paper was made of rags or other previously used materials made of hemp, probably dried on a mat woven like a piece of fabric.

The existence of paper of vegetable fibres in the Former Han may be supported by several other old specimens from archaeological discoveries.[d] One fragment, 10 by 4 cm., dated on circumstantial evidence to around −49, was found in the

---

[a] Cf. Diringer (2), pp. 126 ff.

[b] Joseph Edkins (18), pp. 67–8, questioned the validity of the Chinese invention; the Chinese historian Chien Po-Tsan (1), p. 511, says that paper existed in Athens and Alexandria 400 years earlier than in China; and the noted Egyptologist Jaroslav (1), p. 31, n. 2, considered that the Chinese invention was influenced by acquaintance with Egyptian papyrus; see discussion on the Western origin of paper in Tsien (2), pp. 140–2.

[c] This discovery was first reported in the *WWTK*, 1957, no. 7, pp. 78–9, 81; illus.; the specimens were said to have consisted of fibres similar to silk; but later microscopic study reveals it was made of hemp; see Phan Chi-Hsing (3), pp. 45–7.

[d] Questions concerning dating to the Former Han period of recently discovered paper specimens have been raised and discussed by paper specialists in China; see Wang Chü-Hua & Li Yü-Hua (1), Cheng Chih-Chhao & Jung Yüan-Khai (1), and Phan Chi-Hsing (12).

[1] 灞橋

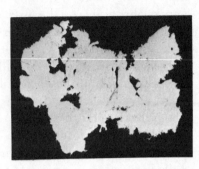

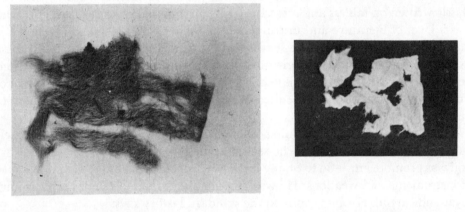

(a)

(c)

(b)

(d)

Fig. 1060. Oldest paper specimens of the Former Han period. (a) A large piece of hemp paper, 10 × 10 cm., dated −2nd century from Pa-Chhiao, Shensi province. (b) Fibres of the above enlarged 4 times. (c) Paper fragment from Chin-Kuan, Chü-Yen, 21 × 19 cm., c. −2nd century. (d) Paper specimen from Fu-feng, Shensi, 6.8 × 7.2 cm., c. −1st century. Courtesy of Institute of the History of Science, Academia Sinica, Peking.

ruins of a watchtower in Lopnor by a member of the Mission of the Northwestern Expedition of China in 1934.[a] Two larger pieces dated to be of the second half of the −1st century, were found in 1974 at a watchtower in Chin-kuan,[1] near Chü-yen[b] (Fig. 1060c). A few other pieces, mostly fragments attached to a lacquer utensil with coins of the Hsüan Ti period (−73 to −49), were found in 1978 in an

[a] See Huang Wen-Pi (1), p. 168, pl. 23, fig. 25. It is reported that this piece was destroyed during the war in the 1930's.
[b] See a report in *WWTK*, 1978, no. 1, pp. 1–14, pl. 1–9, fig. 1–42; on the same and other nearby sites some 20,000 wooden tablets of the same period were discovered.

[1] 金關

underground vault in Chung-yen[1] village, Fu-feng, Shensi province[a] (Figs. 1060d; 1064). All these pieces are of later date but are similar in quality to the Pa-chhiao paper, being made of hempen fibres, yellowish, thick, coarse, and uneven with loops of fibre visible on the surface. They provide some evidence of the beginning of the art, especially those discovered in the more precisely dated sites located in the central plain of China.

Besides the archaeological evidence, paper before Tshai Lun's time is mentioned in several places in the ancient literature. In one story laid in −93, an imperial guard advises a prince to cover his nose with a piece of *chih* (paper).[b] Another relates that in a murder case in −12 the poisonous 'medicine' was wrapped in *ho-thi*,[2] which means a thin piece of *chih* of red colour, according to the commentator Ying Shao[3] (*c.* +140–206).[c] The official history records that, in the reorganisation of the imperial secretariat by Emperor Kuang-Wu (r. +25–56), 'the Assistant of the Right (*Yu Chheng*[4]) was responsible for the seals and cords of the office, and for paper, brush, and ink.'[d] It also says that a scholar was summoned to the court in +76 to give instruction to twenty students, who were each given a copy of the classic written on tablets and *chih*.[e] In +102, an imperial consort, née Teng, who was a lover of literature and instrumental in Tshai Lun's presumed invention, is said to have asked that *chih* be sent as tribute from various countries.[f] All these stories recorded in official histories and other documents indicate that *chih* existed before +105, the traditional date of the invention of paper by Tshai Lun who, since his own time, has been credited as the inventor or sponsor of the methods of papermaking.

Tshai Lun[5] (d. +121), *tzu* Ching-Chung,[6] a native of Kuei-yang[7] (modern Lei-yang, Hunan), was a eunuch who served at the imperial court in or before +75, and was promoted in +89 to Shang Fang Ssu,[8] an office in charge of manufacture of instruments and weapons. He was described as a man of talent and learning, loyal and careful. His biography in the standard history says:

In ancient times writings and inscriptions were generally made on tablets of bamboo or on pieces of silk called *chih*.[9] But silk being costly and bamboo heavy, they were not convenient to use. Tshai Lun then initiated the idea of making paper from the bark of trees, remnants of hemp, rags of cloth, and fishing nets. He submitted the process to the emperor in the first year of Yüan-Hsing [+105] and received praise for his ability. From this time, paper has been in use everywhere and is universally called 'the paper of Marquis Tshai'.[g]

Similar records are included in the *Tung Kuan Han Chi*,[10] an official history written by a group of contemporary historians from +25 to +189,[h] and in other sources,

---

[a] See two reports in *WWTK*, 1979, no. 9, pp. 17 ff.
[b] *San Fu Ku Shih* (*EYT*), p. 9a.   [c] *Chhien Han Shu* (*ESSS/TW*), ch. 97b, p. 13a.
[d] *Hou Han Shu* (*ESSS/TW*), ch. 36, p. 7b; cf. Hans Bielenstein (3), p. 55.
[e] *Hou Han Shu*, ch. 66, p. 17a.   [f] *Hou Han Shu*, ch. 10a, p. 19b.
[g] *Ibid.* ch. 108, pp. 5a–b; cf. tr. Blanchet (1), pp. 13–14; Carter (1), p. 5; Hunter (9), pp. 50–3.
[h] A briefer version appears in the reconstructed text of the *Tung Kuan Han Chi* (*SPPY*), ch. 20, p. 2b; see tr. in Tsien (2), p. 136. The reconstructed text is slightly different from that cited in *Pei Thang Shu Chhao* and *Chhu Hsüeh Chi*.

[1] 中顏	[2] 赫蹏	[3] 應劭	[4] 右丞	[5] 蔡倫
[6] 敬仲	[7] 桂陽	[8] 上方寺	[9] 紙	[10] 東觀漢記

concerning Tshai Lun's paper.[a] All provide information about the kinds of raw materials used, the date of the official presentation, and the life of Tshai Lun himself.

The existence of paper before Tshai Lun does not necessarily contradict the story of his contribution as recorded in the official history. It is possible that he was an innovator who used new raw materials in papermaking. Indeed, the term *tsao i*[1] used in his biography can be read as 'to initiate the idea' of introducing new materials, especially tree bark (*shu fu*[2]) and hemp ends (*ma thou*[3]), which was not a second-hand material previously used for some other purpose. As to the rags of cloth (*pho pu*[4]) and fish nets (*yü wang*[5]), they may still have been mentioned as common or officially approved materials for papermaking, even if they had been so used before. In any case, rags and other second-hand materials were probably used first, but their supply was limited compared with that of fresh fibres from trees or other plants, which made possible the large-scale production of paper. It is also suggested that the use of tree bark for papermaking by Tshai Lun may have been influenced by the bark cloth culture which existed in China before the manufacture of paper by a process of felting.[b]

Several recent archaeological discoveries support the literary evidence of the official records. A specimen of paper with writing contemporary with Tshai Lun was found under the ruins of an ancient watchtower in Tsakhortei near Chü-yen (Khara-khoto) by a party from the Academia Sinica in 1942.[c] This remnant of paper, which is said to be made of vegetable fibre, is coarse and thick, with no clear screen marks, but it bears about two dozen readable characters in a *li* style (Fig. 1061), said to be similar to that appearing on a piece of pottery dated +156.[d] From historical evidence, this piece of paper can be dated between +109 and 110, when the watchtower was abandoned by Chinese defenders because of the rebellion of the Hsi-chhiang tribe.[e] This was the first find and the earliest sample of paper with writing on it. Other specimens of the Later Han period found in recent years include a plain piece found near a mummy in a tomb in Min-feng,[8] Sinkiang, in 1959,[f] and several pieces with some writing, not all of which is legible, from a tomb in Han-than-pho,[9] Wu-wei, Kansu province, in 1974.[g] Some of the latter specimens are said to have been nailed, in three layers, with wooden strips on both sides of an ox cart when they were discovered. Made of hemp, they have proved to be more advanced than other finds, since some of them are white, much thinner than

---

[a] See Tung Pa[6] (+3rd century), *Yü Fu Chih*,[7] quoted in *TPYL*, ch. 605, p. 72.
[b] See Ling Shun-Sheng (7), pp. 11–16, 40–3; also discussion of *tapa* on pp. 110 ff. below.
[c] Lao Kan (1), pp. 496–8. A bundle of wooden tablets dated +93 to 98 was discovered at this site, which was explored earlier by the Sino-Swedish Expedition in 1930; see Bergman (4), pp. 146–8.
[d] See Phan Chi-Hsing (1), p. 48.
[e] Lao Kan, who found this paper, dated it around +98 in his article (1) in 1948, but later changed his dating to +109 or 110 on historical grounds; see Chhien Tshun-Hsün (5), pp. 183–4.
[f] See a report in *WWTK*, 1960, no. 6; the finding of a paper specimen at this site is mentioned in Phan (9), p. 63.
[g] See Phan Chi-Hsing (10), pp. 62–3.

[1] 造意　　　[2] 樹膚　　　[3] 麻頭　　　[4] 破布　　　[5] 魚網
[6] 董巴　　　[7] 輿服志　　　[8] 民豐　　　[9] 旱灘坡

Fig. 1061. Paper specimen with writing of the Later Han period. A fragment from Chü-Yen, dated *c.* +110.

the Pa-chhiao paper, and can be written on with brush and ink. It is the second find of paper specimens of the Han period with writing.

A few artisans of the Han period are known to have contributed to the improvement in the quality of paper. Tso Po[1] (+200), *tzu* Tzu-i,[2] a man of Tung-lai in Shantung who lived at the end of the Han dynasty, is said to have made paper of 'beautiful and lustrous appearance,' which matched the excellence of ink and brushes manufactured by other great masters.[a] Khung Tan,[3] a legendary figure, is reputed to have lived in the Hsüan-chheng area in the late Han period, and to have discovered the value of the bark of *than* wood in making the high grade paper (*hsüan chih*[4]) for painting and calligraphy.[b] No other names of papermakers are recorded in early literature or on artifacts, but the methods of Chinese papermaking progressed with new materials and techniques at the advent of each new dynasty.

(4) THE PROGRESS OF PAPERMAKING FROM THE CHIN TO THE THANG PERIOD

While the Han dynasty witnessed the beginning of papermaking, the period from Chin to Thang (+3rd to +10th century) was probably the most important time for discovery of new raw materials, further improvement in techniques, wider

[a] *Wen Fang Ssu Phu* (*TSHCC*), p. 53.   [b] See Mu Hsiao-Thien (*1*), p. 4.

[1] 左伯   [2] 子邑   [3] 孔丹   [4] 宣紙

application, and the more widespread use of paper. Rattan, grown primarily in southeast China was introduced as a raw material, and marked a great step forward in papermaking in this period.

The use of local raw materials for papermaking was perhaps conditioned by the political and intellectual factors of the time, especially the isolation of Eastern Chin following the move of its capital from Loyang to Nanking, and by the increasing demand for paper for writing and other uses. Although it had been used for writing as early as the +1st century, it was only from this period that Chinese books on bamboo and wooden tablets were entirely replaced by paper. We know that one of the largest discoveries of bamboo books made in the −3rd century was transcribed on paper to be kept in the imperial library. The Chin bibliography recorded all books as rolls (*chüan*[1]) instead of tablets (*phien*[2]) as was done in earlier bibliographies.[a] We have also found that from this time paper began to be made on a fine bamboo screen-mould, well sized, and treated with an insecticidal dye for permanence. It was made in many colours for stationery, cut into designs for embroideries and decorations, used to make rubbings from stone inscriptions, for documents and books, for painting and calligraphy, for visiting cards, and for such household articles as fans, umbrellas, lanterns, and kites; there were even sanitary and toilet papers.[b]

A great many paper fragments of Chin dynasty date, discovered in Central Asia since the turn of this century, illustrate the spread of paper outside the domain of the Chinese empire. Some of them, bearing dates from +252 to 310, were found by Sven Hedin in the Loulan region in 1900.[c] In the same region, Aurel Stein in 1914 found hundreds of paper fragments dated +263–80,[d] and a collection of paper documents and letters written in Sogdian in about +312–13.[e] Some paper manuscripts of this period were discovered in the region of Turfan and Kao-chhang by the Prussian Expedition of 1902–14,[f] and also by the Japanese Expedition of the Nishi-Honganji in 1909–10.[g] A great many paper books, documents, and artifacts from the +4th to the +9th century have also been found in this area by the Chinese archaeological excavations in more recent years (Fig. 1062).[h]

On the other hand, paper made in foreign lands may have been imported to China in the early Chin period. One source records that in +284 Ta Chhin[5]

[a] About three quarters of the books were recorded as *phien* in the section on literature in the *Chhien Han Shu*, ch. 30; cf. Tsien (2), pp. 92–3.
[b] See discussions on the uses of paper and paper products on pp. 84 ff. below.
[c] Conrady (1), pp. 93, 99, 101; pls. 16:1–2, 20:1, 22:8.
[d] Stein (4), vol. II, p. 674; Chavannes (12a), nos. 706–8.
[e] These documents were first dated to the 3rd year of Yüan-Chia[3] (+153) through an erroneous reading of Yung-Chia;[4] a revised dating of Yung-Chia (+307–13) is generally accepted; see Henning (2), pp. 601–15.
[f] Cf. Yao Tshung-Wu (1), pp. 27–9.
[g] Ōtani Kōzui (1), preface.
[h] Some twenty-six dated documents of +346 to 907 have been scientifically analysed and studied for the raw materials and methods of their manufacture; see Phan Chi-Hsing (4), pp. 52 ff.

[1] 卷　　　[2] 篇　　　[3] 元嘉　　　[4] 永嘉　　　[5] 大秦

Fig. 1062. Earliest extant version of the *Confucian Analects* with commentary of Cheng Hsüan, dated +716. Paper manuscript copied by Pu Thien-Shou, 27 × 43.5 cm., on the back of an account book of the Thang period, found in Sinkiang.

presented 30,000 rolls of honey fragrance paper (*mi hsiang chih*[1]) to the Chinese Emperor, who bestowed 10,000 rolls on Tu Yü[2] (+222–84) for writing his commentary on the Confucian classics.[a] Scholars question that this paper originated in Ta Chhin (the Roman Empire), but suppose that it was made in Indo-China of garco wood (*Aquilaria agallocha*), and brought to China by Alexandrian merchants in lieu of articles from their home.[b] Another story relates that Emperor Wu (r. +265–90), bestowed upon Chang Hua[3] (+232–300), for writing his *Po Wu Chih*[4] (Record of the Investigation of Things), 10,000 pieces of intricate filament paper (*chhih li chih*[5]) sent as tribute from Nan-Yüeh[7] (modern Vietnam).[c] Geographic proximity makes it likely that the craft of papermaking reached southern neighbours rather early, and paper made of local materials was brought to China as tribute to the Chinese emperor.

Political stability, economic prosperity, and official fostering of scholarship in the Thang dynasty, all encouraged the increasing production and further improve-

---

[a] See *Nan Fang Tshao Mu Chuang* (*HWTS*), ch. 2, p. 6a, attributed to Hsi Han[6] (+263–306). A recent study expresses doubt of the authenticity of its author and dating and says that stories similar to that of the honey fragrance paper occur in two Thang works; see Ma Thai-Lai (1), pp. 239–40; (1), pp. 199–202. Nevertheless, the importation of exotic papers from foreign lands in the +3rd century must not be dismissed.

[b] Hirth (1), pp. 274–5.

[c] This story appears in the +4th-century work *Shih I Chi* (*HWTS*), ch. 9, p. 7b; its authenticity also has been questioned; see Chang Te-Chün (1), p. 88.

[1] 蜜香紙   [2] 杜預   [3] 張華   [4] 博物志   [5] 側理紙
[6] 嵇含   [7] 南越

ment of paper. The government not only selected the best paper for documents and other official uses, but also ordered certain districts in the country to manufacture paper of special quality known as *kung chih*[1] (tribute paper). According to historical records, no fewer than eleven districts of the empire sent such tribute paper to the government from time to time.[a] The Imperial Library at Chhang-an and later at Loyang had all its books written on the best paper, manufactured in Szechuan,[b] while special official positions were established in various academic institutions at the court to dye, mount, and treat paper for permanent preservation. The government also established many paper factories in the south along the Yangtze. More than ninety such factories were operated in the districts of modern Chiangsu, Chekiang, Anhui, Chiangsi, Hunan, and Szechuan. A standard kind of paper, in uniform sheets, known as *yin chih*[2] (printing paper), was made for keeping accounts by shops, monasteries, and families of officials and the gentry. Stimulated by the economic growth of the empire, paper was increasingly used in foreign trade, on ceremonial occasions, in making wearing apparel, armour, household furnishings and appliances, and for other decorative and recreational uses.

Many paper documents and books from the +4th century onwards, survive in good condition. The earliest and largest collection of such books is that of more than 30,000 paper rolls dating from the +4th to the 10th century, found in a stone cave at Tunhuang at the beginning of this century.[c] The construction of this cave, decorated with Buddhist sculptures and fresco paintings, was begun in about +366 and continued for several centuries. The paper rolls found in the cave included mostly Buddhist sutras with some Taoist and Confucian texts, government documents, business contracts, calendars, and such miscellaneous materials as anthologies, excerpts, dictionaries, glossaries, and models of letter-writing and composition, which must have been used by children of the monastic schools. The greater part of these materials are in Chinese, but some are in Sanskrit, Sogdian, Iranian, Uighur, and especially Tibetan. Except for a few specimens of early printing and rubbings, most are manuscripts made during the time from the Chin to the Five Dynasties, a period of some six hundred years, with most belonging to the Sui and Thang periods (Fig. 1063).[d] All of them were concealed or sealed away behind a wall of the stone cave, perhaps in the early +11th century, thus escaping

[a] See *Hsin Thang Shu* (*ESSS/TW*), ch. 41; pp. 1a–22b; *Yüan-Ho Chün Hsien Thu Chih* (*TSHCC*), ch. 26, pp. 681–94, ch. 28, pp. 743–63.

[b] See *Chiu Thang Shu* (*ESSS/TW*), ch. 47, p. 46b.

[c] About 7000 rolls and 3000 fragments acquired by Aurel Stein in 1907 are in the British Museum, London; more than 3500 rolls collected in 1907 by Paul Pelliot are in the Bibliothèque Nationale, Paris; a few hundred rolls obtained by the Japanese expeditions in 1908–14 were kept in Ōtani Kōzui's house near Kobe and are said to have been transferred to the Ryojun Museum in Dairen, Liao-ning; about 10,000 rolls were removed in 1909 to the Imperial Library in Peking; over 10,000 rolls, mostly fragments, acquired by Sergie F. Oldenburg in 1914–15, are in the Institute of the Peoples of Asia, Leningrad; and certain minor collections are scattered throughout the world; see Fujieda Akira (1) and (2); Su Ying-Hui (4), pp. 73–83.

[d] For a summary of the content of the Tunhuang rolls, see Fujieda (2); Su Ying-Hui (4), pp. 25–68; for description of these documents see catalogues of various collections listed in Tsien (10), pp. 436–41.

[1] 貢紙   [2] 印紙

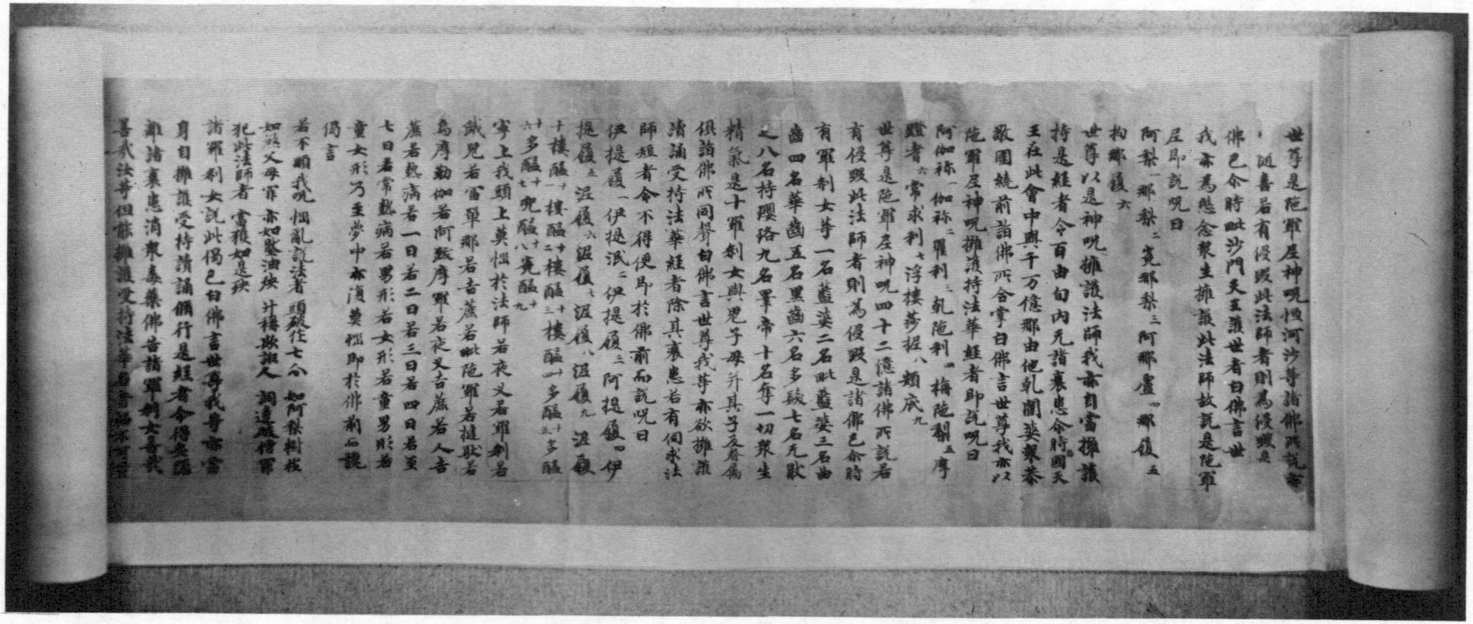

46

Fig. 1063. Fragment of the *Lotus sutra* from Tunhuang, c. +9th century, representing paper book in the roll form of the Thang dynasty. Far Eastern Library, University of Chicago.

injury from natural or artificial causes, and have remained in perfect condition.

The commonest papers in the Tunhuang collection were made of hemp and paper mulberry, with a few of ramie and mulberry.[a] Although literary sources indicate that bamboo and rattan also were used at this time, these materials were not found in Tunhuang, probably because both bamboo and rattan were grown in south China and not available in this border region. The papers of earlier periods, especially those made in the +7th and 8th centuries, are reported to be generally thin, of unvarying thickness, highly finished, well sized, and stained yellow or brownish. Those of later date, especially of the +10th century, show deteriorated quality. With rare exceptions, they are coarse, drab-coloured, and thick.[b] The rolls are made of from ten to as many as twenty-eight sheets of paper, pasted together to form a long scroll, the beginning of which was covered with a piece of thick sheet that was attached to a roller at the end. The individual sheets average about one foot wide by two feet long, while some rolls are as much as twenty-three feet long.[c] The earliest sheets of paper were narrower, but the size gradually increased in the Sui and Thang dynasties. Sizes were variously uneven in the Five Dynasties period.

(5) DEVELOPMENT OF PAPERMAKING FROM THE SUNG DYNASTY

As the supply of rattan was gradually exhausted, the Sung dynasty made extensive use of bamboo for papermaking. During this dynasty, the major manufacturing centres included those in Kuei-chi and Shan-chhi in modern Chekiang; Hsi-hsien, Hui-chou, and Chhi-chou in modern Anhui, Fu-chou in modern Chiangsi; and Chhengtu and Kuang-chhing in modern Szechuan, which had been a major centre for papermaking since the Thang. It was said that, in the +tenth century, some of its skilful papermakers were recruited by Li Yü (+937-78), ruler of the Southern Thang, to go to Nanking to make the time-honoured paper bearing the name of his studio.[d] After the fall of the Southern Thang, however, these craftsmen migrated to other cities in the lower Yangtze valley, where new centres for papermaking were developed to compete with those in Szechuan. Fei Chu (fl. +1265) said that papers manufactured in Anhui, Chiangsu, and Chekiang were sold for as much as three times the price of those made in Szechuan, but people liked the fine, thin sheets of the imported paper better than the heavy local product.[e]

For the various needs of the government, the levy of tribute paper continued. It is recorded that the prefecture of Hsin-an (modern Hsi-hsien, Anhui) sent some 1,500,000 sheets of paper of seven varieties to the capital as tribute each year before

[a] Some fifteen documents dated +406 to +991 in the British Museum, analysed by Clapperton (1), p. 18, are reported to be dominantly paper mulberry; thirty-two specimens dated +259 to +960 in the Peking collection are mostly of hemp; see Phan Chi-Hsing (2), pp. 40–1.

[b] The average thickness of papers of early Thang is measured at ·002–·005 inch, and of later Thang at ·008–·012 inch; see Clapperton (1), p. 18.

[c] See an analytical study with tabulation in Phan Chi-Hsing (2), pp. 39–42.

[d] For the making of the Chhen Hsin Thang paper for painting and calligraphy, see p. 90 below.

[e] See Chien Chih Phu (TSHCC), p. 3.

+1101. In that year, the Emperor reduced the quota because sending extra large sheets of paper caused a heavy burden on the people. To supply its needs for paper money, exchange certificates, and other uses, the government also established many large paper factories of its own. For the printing of paper money alone, several factories were operated in Hui-chou, Chheng-tu, Hangchow, and An-chhi. In +1175, the Hangchow factory, for example, employed more than a thousand daily workers.[a] Fei Chu said that in Chheng-tu there was a temple to Tshai Lun where he was worshipped by several hundred families engaged in papermaking, all of whom resided in a village some five miles south of the city.[b] Farmers attracted by the profits of papermaking were leaving their fields to take employment in paper mills because of the increasing demand for paper to be burned for spirit sacrifices, and for other uses.[c]

After printing became popular in the Sung, the need for large quantities of paper for making books further stimulated the development of the paper industry. Not only did the National Academy (Kuo Tzu Chien[1]) in Khaifeng and later in Hangchow engage in large-scale publishing, but many private families and trade agents in Chhengtu, Hangchow, and Chien-yang also engaged in printing and papermaking. Also popular in the Sung was the making of inked squeezes from stone and bronze inscriptions, since interest in traditional archaeology was developing at this time. Indeed, scholars might possess as many as several thousand rolls of such squeezes as treasures in their collections.[d] To meet the special needs of painting and calligraphy, a kind of paper in extra large sheets, known as *phi-chih*,[2] was specially made. Su I-chien (+957–95) described the making of a sheet of paper fifty feet long in Hui-chou. The hold of a ship was used as a vat, and some fifty workers joined in lifting the screen-mould in time to the beating of a drum. This paper was dried over a big brazier instead of on a wall as was usual, in order to make the sheet even.[e] Many other varieties of paper were perfected in the Sung, including the famous Golden-Grain paper (*Chin-su chien*[3]) for copying Buddhist *sutras*.[f] Comments by the famous calligrapher Mi Fu (+1051–1107) indicate that papers made at this time were generally of excellent quality, white, smooth, and absorbent, most suitable for artistic purposes.[g] It was during this time that paper was first named as one of the four treasures of a scholar's studio (*wen fang ssu pao*[4]), on which a famous treatise was written by Su I-chien in +986.

The Mongols are credited with the further spread of paper and paper products

[a] See *Hsin An Hsien Chih* (1888 ed.), ch. 2, pp. 30–31 a.
[b] *Chien Chih Phu*, p. 1.
[c] See a memorial by Liao Kang (+1071 to +1143) in *Kao Feng Wen Chi* (*SKCS*), ch. 1, pp. 15 a–17 b.
[d] Ouyang Hsiu[5] (+1007–72) listed a thousand rolls of rubbings in his *Chi Ku Lu*,[6] and Chao Ming-chheng[7] (+1081–1129) had some 2000 rolls in his collection, as noted in Li Chhing-chao's preface to the *Chin Shih Lu*.[8]
[e] *Wen Fang Ssu Phu* (*TSHCC*), p. 53.
[f] See *Chin Su Chien Shuo* (*TSHCC*), pp. 127–30.
[g] . See *Phing Chih Thieh*, also known as *Shih Chih Shuo*, in *MSTS*, vol. 6, pp. 305–7.

[1] 國子監   [2] 疋紙   [3] 金粟牋   [4] 文房四寶   [5] 歐陽修
[6] 集古錄   [7] 趙明誠   [8] 金石錄

westwards and elsewhere. Marco Polo was among the early European visitors to China who witnessed the wide circulation of paper money and the extravagant burning of paper effigies and replicas as offerings to the dead in the empire of Genghis Khan. An Arabian writer, Ahmed Sibab Eddin (+1245–1338), twice mentioned Chinese paper money in his book based on the eyewitness reports of others.[a] Through intermediate steps, papermaking was introduced to Europe in the +12th century. Subsequent to this, Mongol conquests resulted in the first issues of paper money in Persia in the 13th century, and in Korea and Vietnam in the 14th century, while Japan also used paper money during this period. About this time, too, playing cards and other paper products were introduced to Europe, possibly through the Arab world.[b]

Throughout the Ming dynasty, papermaking continued to develop to meet the demands of the government and for general use in writing, publishing, art, and daily life. Bamboo became the predominant material in Ming papermaking, especially in the wide region bordering Chekiang, Chiangsi, and Fukien, where enormous bamboo groves were grown on mountains and along streams. A Ming gazette records that no fewer than thirty paper mills were operated in the town of Shih-thang (in Chhien-shan, Chiangsi), each with one or two thousand workers; a total of some fifty or sixty thousand people were engaged in the making of paper there in +1597.[c] It was said that papermaking was the only profitable handicraft in Chhien-shan, whence paper was traded to all parts of the country.

The various departments of the government requisitioned all kinds of paper for different purposes. The administrative code of the Ming dynasty records that 314,950 sheets of all varieties of paper were requisitioned by the Board of Works. Paper for legal use was acquired each season and reported at the end of each year. For the civil service examinations, 16,800 sheets of paper known as *pang chih*[2] were requisitioned each year, with 1400 additional sheets required if an intercalary moon occurred. A total of 1,200,000 sheets of such paper, specified to be 4 feet 4 inches long and 4 feet wide, were requisitioned from all provinces every ten years to be kept in storage for official use. If this were not sufficient, other kinds of paper, of irregular sizes, might be substituted. In 1537,[d] the price of *pang chih* was 0·1 tael of silver per 100 sheets, and of *lung li*[3] paper was 0·4 tael per 100.

For printing exchange certificates for tea, salt, and other commodities, a total of 1,500,000 sheets was required to be sent in by various producing provinces. The quota set in 1393 included 380,000 sheets from Chihli, 250,000 from Chekiang, 200,000 from Chiangsi, 170,000 from Hunan and Hupei, 150,000 from Shensi, 100,000 each from Shansi and Pei-phing, 55,000 each from Shantung and Honan,

[a] See translation by Ch. Schefer (2), pp. 17, 20.
[b] See Carter (1), pp. 183–8; also pp. 99 ff. below.
[c] See *Chhien Shu*[1] (Wan-Li ed.), ch. 1; Pheng Tse-I (*1*), I, p. 11.
[d] See *Ta Ming Hui Tien* (1589 ed.), ch. 195, pp. 4*b*–5*b*.

[1] 鉛書   [2] 榜紙   [3] 龍瀝

and 40,000 from Fukien. When the papers were delivered they were carefully checked by the specifications and then deposited in storage until needed. In 1422, it was decreed that if the quality of paper did not meet the standards the same amount would be required for replacement. In 1424, the paper from Fukien did not meet the specifications, and the official responsible for requisitions was punished by the provincial judge.[a]

The first detailed description of the manufacturing process was made at the end of the Ming dynasty, when a chapter on papermaking was included in the *Thien Kung Khai Wu* (The Exploitation of the Works of Nature), written around 1637 by Sung Ying-Hsing.[b] This is probably one of the most important records of the technology of papermaking ever made. At about the same time as this book, several albums of ornamental stationery with multi-coloured designs were produced. Poems, notes, private correspondence, and certain contracts were all written on such elegantly and beautifully designed note papers. All these testify to the progress toward artistic and technical excellence of papermaking in the Ming dynasty.

Numerous books and documents written or printed on a great variety of paper in the Ming and Chhing dynasties survive today, as well as works of painting and calligraphy, and examples of articles made of various kinds of paper. For documentary use in the Manchu court, highly decorated fancy papers were specified in detail in the administrative code of the Chhing government. In 1644, when the first Manchu emperor came to the throne, it was ordered that stiff yellow paper in two layers should be used for the announcement of successful candidates in the palace examination for the *chin-shih* degree. Three grades of ornamental paper were used for imperial orders: golden dragon fragrant paper in four layers mixed with sandalwood sawdust; painted dragon paper with fragrant ink in three layers, and paper of dragon edge design in two layers printed with fragrant ink. For scrolls for bestowing honours to be inherited, 1000 rolls of paper were ordered in 1738. Three hundred of these were 30–40 feet long, 100 were 100 feet, and some were as long as 500 feet.[c] The compilation and production under imperial auspices of the famous grand encyclopaedia *Thu Shu Chi Chheng*[1] in 5050 volumes, presented to the emperor in 1725, and of the huge collection of four branches of literature, *Ssu Khu Chhüan Shu*[2] in some 36,275 volumes of over two million double pages, with its subsequent duplication in seven hand-written copies in the 1780s, required extensive supplies of high-grade paper. A special kind of fine and sturdy white paper made in Khai-hua, Chekiang, was especially selected for printing the books at the Wu Ying Tien,[3] which were known as palace editions (*tien pan*[4]).

The recent discovery of some stone tablets containing regulations for the management of paper factories has revealed working conditions and wages of paper

---

[a] *Ta Ming Hui Tien*, ch. 195, pp. 4*b*–5*b*.
[b] See tr. Sun & Sun (1), pp. 223–32, and discussion on pp. 69 ff. below.
[c] See *Ta Chhing Hui Tien Shih Li* (1899 ed.), ch. 940, p. 4*a,b*.

[1] 圖書集成   [2] 四庫全書   [3] 武英殿   [4] 殿板

workers in the early Chhing period. One tablet from Soochow, of 1794, says that some thirty-six paper factories employing more than 800 workers were operating in the three districts of Shang-yüan, Chhang-chou, and Wu-hsien in the prefecture of Soochow. Most of the workers came from nearby Chiang-ning (Nanking) and Chen-chiang. Every three to six factories in and outside the city were organised into a ward for the purpose of enforcing the regulations and inspecting the working conditions within each ward. The workers included permanent technicians, temporary employees, and apprentices, who were managed by a superintendent and foremen and watched over by inspectors. Workers received a monthly allowance of 7.2 *chhien*[1] in 1757 and 1.2 tael of silver in 1794, in addition to room and board.[a] Wages were counted by the piece at the rate of 600 sheets per working day, not by the number of days in a month. A monthly bonus of ·45 *chhien* was awarded to workers who put in extra hours or exceeded their quota of production. Apprentices received a stipend and were qualified after an indenture of three years. The names of all workers were registered in a factory roster, and they were not allowed to transfer to another factory if dismissed for violation of working regulations. They were required for moral reasons to stay in the factory dormitory at night.

Another tablet, dated 1757, concerning an agreement by thirty-four paper factories of the same three districts in Soochow reveals that each paper dyer was paid ·24 *chhien* per day for dyeing 700 sheets of green paper, and a bonus of ·5 *chhien* for every additional 700 sheets. For 1000 sheets of red paper the pay was ·21 *chhien*, with ·5 *chhien* for every additional 1000 red sheets. The tablet records some twenty-three different rates for dyeing different kinds of paper.[b]

The time-honoured handicraft of papermaking in China suffered a rapid decline after the market was invaded by foreign papers in the middle of the 19th century. As the historian Liu Chin-Tsao[4] remarked, production of handmade paper was slow, and its price was high; it was not suitable for machine printing and could not compete with machine-made papers from foreign countries. The greatest consumption of paper at that time was for printing newspapers and for wrapping, but handmade paper, which could not be printed on both sides, was not suitable for newspapers. Moreover, imported wrapping paper at fifteen cents a pound was cheaper than *mao-pien*[5] at twenty-six cents. In Chhien-shan (in Chiangsi), for example, paper production was worth half a million taels of silver each year before the end of the 19th century. After the introduction of foreign papers, the business dropped to less than 100,000 taels. In Shih-chheng (Chiangsi), eighty to ninety per

---

[a] The complete text of this tablet is given in Liu Yung-Chheng (*3*), pp. 85–7; according to a local record of about the same period, the factory employed a team of four workers for each vat—one each to lift, pound, check, and dry. Another source records that monthly wages were 5 *chhien* in Fukien in 1769; 900 *wen* in Chiangsi in 1783 and in Chekiang in 1807; and 1200 *wen* in Shensi in 1815; see Pheng Tse-I (*1*), pp. 396–7. In the Chhing dynasty, the unit *chhien*[1] equalled one-tenth of one tael (*liang*,[2] or Chinese ounce), and 1000 pieces of cash (*wen*,[3] or copper) equalled one tael of silver.

[b] See partial text reported in *WWTK*, 1957, no. 9, pp. 38–9; the cost of rice was mentioned in one tablet as 1.5 tael of silver per picul in +1716.

[1] 錢   [2] 兩   [3] 文   [4] 劉錦藻   [5] 毛邊

cent of the handmade paper mills had closed by the beginning of the twentieth century.[a]

(c) TECHNOLOGY AND PROCESSES OF PAPERMAKING

THE TECHNICAL aspects of papermaking include the materials, tools, and methods of manufacture. From ancient times Chinese papermakers had wisely selected almost all kinds of plants known to the modern paper industry as producing the best of fibres and yet being most economical in cost. The application of chemical agents for sizing, loading, coating, colouring and dyeing was also known to them not long after paper was invented. The use of water as an inexpensive agent contributed not only to the swelling and bonding of the fibres but also to the increase of the mechanical strength of the paper. Such utensils as the vat, mould, and press used many centuries ago in China are still basic to modern papermaking, as well as the processes of maceration, washing, lifting, pressing, and drying. Historians of paper have agreed that the ancient principles and practices of papermaking are the basis upon which modern paper machines are designed and operated. The following pages will examine in detail some of these traditional materials, tools, and methods used before the coming of the machine age in Chinese papermaking.

(1) RAW MATERIALS FOR PAPERMAKING

According to studies of existing specimens and documentary evidence, a broad variety of vegetable fibres was used as raw material for papermaking in China. Almost all plants produce fibres, but only those rich in cellulose, abundant in supply, easy to treat, and cheap in cost are most suitable. Especially satisfactory are those plants containing higher yields of long cellulose but lower in binding substances, which must be eliminated in the process of maceration. These materials include the bast plants, such as hemp, jute, flax, ramie, and rattan; tree bark of mulberry and paper mulberry; grasses, such as bamboo, reeds, and stalks of rice and wheat; and such fibres as cotton. Hemp and cotton are probably the best, producing the highest yields of pure fibres, but as they were needed primarily for the textile industry, paper mulberry and bamboo became the chief raw materials for papermaking in China.

Chronologically speaking, hemp was probably the earliest material used for papermaking from the Former Han ($-206-+8$), followed by paper mulberry from the Later Han ($+25-220$), rattan from the Chin ($+265-420$), bamboo from the middle of the Thang ($+618-906$), and straw probably from before the Sung dynasty ($+960-1280$). Except for hemp, which was no longer used in large quantity after the Thang, and for rattan, the supply of which was exhausted since

[a] See Liu Chin-Tsao (1), pp. 11306, 11314-15, 11419-20.

Fig. 1064. Hemp and its fibres (× 67, from the Former Han specimen found at Fu-Feng, Shensi in 1978).

the early Sung, these materials have been continued in use until today. The use of raw fibres for papermaking varied a great deal according to the local production of the materials. Su I-Chien[1] (+957–95), author of the first treatise on paper, said that hemp was used in Szechuan, bamboo in Chiangsu and Chekiang, mulberry bark in the north, rattan in Shan-chhi, and seaweed by the people in the south. Paper made of wheat stalks and rice straw by the people of Chekiang was brittle and thin; and that of wheat stalks mixed with rattan from Yu-chhüan was the best.[a] This seems to be true both of earlier as well as later times.

(i) Hemp, jute, flax, and ramie

The plants which yield the richest and strongest bast fibres were the earliest materials used in Chinese papermaking known to us. The major varieties of the bast-yielding group are hemp (*Cannabis sativa*), known in Chinese as *ta ma*[2], jute (*Corchorus capsilaris*) or *huang ma*[3], flax (*Linum perenne*), *ya ma*[4], and ramie or China grass (*Boehmeria nivea*), *chu ma*[5]. They were grown in ail parts of China, especially in the northern and western regions. Ancient Chinese documents refer to all of these varieties as *ma*, which has generally been rendered as hemp (Fig. 1064). This was probably the earliest fibre plant used for clothing in China before the extensive use of cotton fibres for textiles ever since the Ming dynasty. Ramie or China grass is a perennial plant. Sung Ying-Hsing (+1587–c. 1660) described two varieties of ramie, green and yellow. Their stems could be cut two or three times each year; from these fibres were obtained materials for making summer clothing, curtains,

---

[a] *Wen Fang Ssu Phu* (*TSHCC*), p. 53.

[1] 蘇易簡   [2] 大麻   [3] 黃麻   [4] 亞麻   [5] 苧麻

and mosquito nets.[a] Most of the ancient papers were made either of used materials or raw fibres from these bast plants.

The oldest paper specimens of the Han dynasty discovered in Lop-nor, Pa-chhiao, Chü-yen, and other places were all made of hemp[b]. Those papers from the +3rd to the +8th century found in Sinkiang consist of, besides mulberry bark, chiefly raw and fabricated fibres of hemp, flax, and China grass.[c] The manuscripts found in Tunhuang, dating from the +4th to the +10th century, were also made chiefly of hemp, jute, and China grass.[d]

Hemp paper, which is described as pliable but tough, fine and waterproof, was especially popular for use in calligraphy, bookmaking, and official documents. It was used for writing by noted artists and for manuscripts in the Chin dynasty. Those produced in Szechuan in different sizes and colours were especially chosen by the Thang court for writing decrees, daily instructions and orders, and other official documents.[e] It is said that the scholars in the Academy of Assembled Worthies (Chi Hsien Shu Yüan[1]) were provided every month by the court with 5000 sheets of hemp paper made in Szechuan.[f] In the Khai-yüan period (+713–42), all the books in the imperial collections in the two capitals were written on hemp paper made in I-chou[2] (modern Szechuan).[g] No specific mention of hemp paper is found after the Thang dynasty. It is assumed that since then hemp has not been the chief material for papermaking. The reason why it was the first material used in papermaking was the discovery that hemp and similar fibres drained on a mat gave paper. Hemp was also used for papermaking in Europe before the early 19th century, when wood pulp was commercialised, though even today, many high-quality papers are still made of hemp. However, being more in demand in China as a material for textiles, ropes, and other uses, it was gradually replaced by rattan and especially bamboo since the time of the Thang dynasty.

(ii) *Rattan*

The climbing rattan (*Calmus rotang*) (Fig. 1065) is known to have been used for making paper in certain regions in China, especially in the southeastern part corresponding to modern Chekiang and Chiangsi, where paper made of this plant, known as *thêng chih*[3], was popular for almost a thousand years. The origin of the use of rattan for papermaking may be traced back to the +3rd century at Shan-chhi[4] (modern Chhêng-hsien, Chekiang), where rattan plants were said to have spread

[a] *Thien Kung Khai Wu* (*KHCP*), p. 39; cf. tr. Sun & Sun (1), p. 63.
[b] Phan Chi-Hsing (3), pp. 47–9.
[c] Cf. Hoernle (1), pp. 665 ff.; Carter (1), pp. 6–7; Phan Chi-Hsing (4), pp. 54–5, reports over 90 per cent of paper specimens found in this area are made of hemp materials.
[d] See the analytical studies of Tunhuang papers by Clapperton (1), p. 18; Giles (17); Phan Chi-Hsing (2), pp. 40–1.
[e] *Thang Liu Tien* (1836 ed.), ch. 9, p. 66a.   [f] *Hsin Thang Shu* (*ESSS/TW*), ch. 57, p. 2b.
[g] *Chiu Thang Shu* (*ESSS/TW*), ch. 47, p. 46b.

[1] 集賢書院   [2] 益州   [3] 藤紙   [4] 剡溪

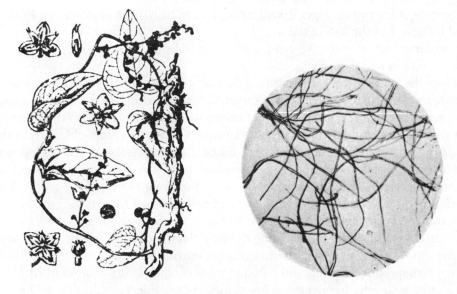

Fig. 1065. Rattan and its fibres (×50).

over hundreds of miles on the mountains along the Shan-chhi river. The old paper made of rattan from Shan-chhi has been called *Shan theng*[1], or rattan paper of Shan-chhi. Fan Ning[2] (+339–401), a native of Honan and an official who served in the capital, said that locally made paper was not suitable for official documents and it was ordered that rattan and bark paper be used instead.[a]

Rattan paper became most popular in the Thang dynasty and the area of its production was greatly extended beyond Shan-chhi to many neighbouring districts in Chekiang and Chiangsi. During the first part of the +8th century, it was recorded in official gazetteers and other documents that paper was an item of local tribute from some eleven districts, including Hangchow, Chhü-chou, Wu-chou (in modern Chekiang), and Hsin-chou (in modern Chiangsi),[b] from which rattan paper was exclusively sent, some of the districts being said to have sent as many as 6000 sheets at one time. A special variety of rattan paper made in Yu-chhüan[3] village of Hangchow, known as 'Yu-chhüan paper' was especially popular.

Rattan paper, described as smooth, durable, with fine texture, and in different colours, was selected for bookmaking, documents, calligraphy, and other uses. The administrative codes of the Thang dynasty specified that the white rattan paper be used for decrees on bestowing, requisition, and punishment; blue for sacrificial messages at the Taoist temple Thai Chhing Kung[4]; and yellow for imperial instructions and orders.[c] The famous calligrapher Mi Fu[5] (+1051–1107) said:

[a] Quoted in *Pei Thang Shu Chhao* (1888 ed.), ch. 104, p. 5*b*.
[b] See *Yüan-Ho Chün Hsien Thu Chih* (*TSHCC*), ch. 26, pp. 681–94; ch. 28, pp. 743–63.
[c] *Han Lin Chih* (*BCSH*), p. 3*a*.

[1] 剡藤   [2] 范寧   [3] 由拳   [4] 太清宮   [5] 米芾

'The back of the rattan paper from Thai-chou can be written on, since it is smooth and hairless. It is the best in the world and can never be matched.ᵃ It was also used for making bags to preserve tea-leaves after roasting, because its firm texture prevented loss of flavour.ᵇ

Since the rattan plant grew naturally in a limited area and its growth was slow as compared with that of hemp, which can be harvested in one year, or of paper mulberry, in three years, the supply of rattan was gradually exhausted. However, the process was gradual, and when rattan was exhausted in Shan-chhi during the Sung dynasty, the centre of production shifted from the western to the eastern part of Chekiang. Rattan paper made in Thien-thai, known as *Thai theng*¹, then became popular, as well as that of Yu-chhüan. After the Sung there followed a gradual decline of the use of rattan; this was due to several reasons. One was the growing use of bamboo which replaced rattan and hemp as the chief raw material for paper-making after the middle of the Thang dynasty. Another was simply the exhaustion of the supply of the material due to excessive cutting without proper cultivation. Many writers lamented this, and a Thang scholar-official, Shu Yüan-Yü² (d. 835), satirised people who frequently wrote millions of worthless words, thus killing the growth of rattan.ᶜ

(iii) *Paper mulberry and mulberry*

Paper mulberry (*ku*,³ *chhu*,⁴ or *kou*,⁵ᵈ *Broussonetia papyrifera*) is a shrub which grows naturally in many parts of China (see Fig. 1066). Chinese records reveal its cultivation, manufacture, trade, and use for making cloth from very early times; indeed, Ssuma Chhien says in the Annals of Yin that 'mulberry and paper mulberry were grown together'.ᵉ Poems mentioning paper mulberry written in the −9th or −8th century are included in the *Book of Poetry*.ᶠ In commenting on this classic, Lu Chi⁶ (+3rd century) says:

Paper mulberry was called *ku sang*⁷ or *chhu sang*⁸ by people in Yu-chou (in North China); *ku* in Ching, Yang, Chiao, and Kuang provinces (in South China); and *chhu* in Chung-chou (in Central China). Both mulberry and paper mulberry were grown together during the Chung-Tsung period [−1637 to −1563] of the Shang dynasty. Nowadays, the people in the south of the Yangtze River use its bark to make cloth and also pound it into paper,

ᵃ *Shu Shih* (*TSHCC*), p. 20.
ᵇ *Chha Ching* (*HCTY*), ch.2, p. 2b.
ᶜ See his essay, 'The Lament of the Old Rattan of Shan-chhi' in *Chhüan Thang Wen* (1818), ch. 727, pp. 20a–21b.
ᵈ *Ku, chhu*, and *kou* (also pronounced *ku*) are said to be the same tree but with different names; others say they are different species from the same mulberry family but with slightly different shapes of leaves; see Laufer (1), p. 558.
ᵉ *Shih Chi* (*ESSS/TW*), ch. 3, p. 7a.
ᶠ See *Mao Shih* (*SPTK*), ch. 11, pp. 4a, 5b; cf. tr. Legge (8), pp. 297, 301.

¹ 台藤   ² 舒元輿   ³ 榖   ⁴ 楮   ⁵ 構
⁶ 陸璣   ⁷ 榖桑   ⁸ 楮桑

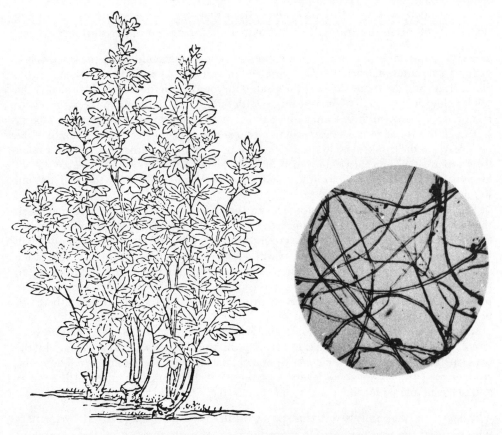

Fig. 1066. Paper mulberry and its fibres (× 50). The stumps near the roots of the tree were made by previous cutting.

called paper mulberry paper (*ku phi chih*[1]), several tens of feet in length, pure white and shining.[a]

The earliest reference to the use of tree bark for papermaking is found in the biography of Tshai Lun in the *History of the Later Han Dynasty*. It does not specify what kind of tree bark, but Tung Pa[2] of the early +3rd century said that 'The Eastern Capital (Loyang) has the paper of Marquis Tshai, which was the paper made by Tshai Lun; that made of used hemp is called hemp paper; that of tree bark *ku chih*[3] (paper mulberry paper); and that of used nets net paper.'[b]

The earliest literature which described the methods of planting and harvesting the shrub and the treatment of its bark is included in an ancient work on agricul-

---

[a] Cf. *Mao Shih Chhao Mu Niao Shou Chhung Yü Su* (*TSHCC*), pp. 29–30; Chung-Tsung[4] is the temple name of the Shang ruler Thai-Wu.[5]

[b] A quotation from *Yü Fu Chih* in *TPYL*, ch. 605, p. 72.

[1] 穀皮紙   [2] 董巴   [3] 穀紙   [4] 中宗   [5] 太戊

ture and farming by Chia Ssu-Hsieh[1] (fl. +553–9), a magistrate of Kao-yang
district in modern Shantung. The chapter on planting the paper mulberry says:

The *chhu* (paper mulberry) should be planted in very good ground along the streams in a
valley. In the autumn, when the fruits of paper mulberry are ripe, collect them abundantly,
wash, clean, and dry in the sun. Till the ground thoroughly. In the second month, after the
soil is ploughed, sow seeds of the tree mixed with those of flax and smooth the ground with
labour. In the autumn and winter, the flax is not to be cut to keep the paper mulberry
warm. [If this is not done, the tree will in most cases die of cold.] In the first month of the
next year, cut the trunks close to the ground and burn them. Thus the trees will grow taller
than a man after one year.[If not burnt, it grows leaner and slower.] The tree can be cut
for use after three years. [If cut in less than three years, the skin will be too thin and not
suitable for use.]

   The method of cutting: The best time for cutting is in the twelfth month, or the next best,
the fourth month. [If cut at other times, the paper mulberry usually withers and dies.] In
the first month of every year, burn the ground with fire. [The dry leaves on the ground will
be enough for burning. If not burnt, it will not grow luxuriantly.] In the middle of the
second month, select and weed out the weak ones. [Weeding is to enrich the trees and to
preserve enough strength and moisture in the ground.] Those transplanted should also be
planted in the second month and cut every three years. [If not cut by three years, there will
be loss of money and no profit.]

   If the trees are sold on the ground, labour is saved but profit will be less: selling the bark
after boiling and peeling is more laborious but profitable. [The wood is useful as fuel.] If the
bark can be used for making the paper by oneself, the profit will be even higher. Those who
plant 30 *mou*[2] can harvest 10 *mou* every year, thus the field can be rotated every three years
and make an annual income equal to one hundred *phi*[3a] of silk.[b]

This short passage points out that paper mulberry was domesticated by farmers
who planted the tree primarily for producing raw materials for papermaking and
that the process of boiling and peeling the tree bark was the first step in making
paper. It was a highly profitable farm subsidiary to combine planting the tree and
then for the farmer to manufacture paper as a handicraft. The use of paper
mulberry bark for papermaking may have been influenced by acquaintance with
its earlier use in China for making bark cloth.[c] Mulberry paper was very popular in
the Chin dynasty and continued in the Thang and later times, since many of the
manuscripts discovered in Tunhuang and Turfan are reported to be made of *chhu*
bark.[d] It was the chief paper used for paper money, known as *chhu chhao*,[4] for
clothing and furnishings, for mounting windows, for book covers, and for other uses
in all times.

   Mulberry (*sang*,[5] *Morus alba*) is a native of China chiefly cultivated for the culture
of silkworms. Marco Polo said that Chinese paper money was made of 'the bark of

---

   [a] One *phi* equals 40 feet, according to *Shuo Wen* and *Wei Shu* (*ESSS/TW*), ch. 110, p. 4*b*.
   [b] *Chhi Min Yao Shu* (*TSHCC*), ch. 5, pp. 92–3; cf. new commentaries and punctuation by Shih Sheng-Han (*1*).
The passages in brackets are in smaller characters in the text, apparently explanation by the author or an early
commentator.
   [c] See Ling Shun-Sheng (*7*), pp. 2–5, 29–31; also discussion on pp. 109 ff. below.
   [d] Cf. Wang Ming (*1*), p. 120.

   [1] 賈思勰   [2] 畝   [3] 疋   [4] 楮鈔   [5] 桑

certain trees, in fact of the mulberry tree'.[a] Bretschneider argued that 'He seems to be mistaken. Paper in China is not made from mulberry-trees, but from the *Broussonetia papyrifera.*'[b] To prove that bark of the mulberry tree, as well as that of paper mulberry, was used as a material for papermaking, Laufer cited a number of authorities to prove that 'Marco Polo is perfectly correct: not only did the Chinese actually manufacture paper from the bark of the mulberry-tree (*Morus alba*), but also it was this paper which was preferred for the making of paper money'.[c]

Chinese sources testify that the mulberry tree was and is still used for paper-making. Su I-Chien (+957–95) said that paper was made from the bark of the mulberry tree (*sang-phi*[1]) by the people in the north.[d] The *History of the Ming Dynasty* also specifies that paper money was 'made of mulberry fibre (*sang jang*[2]) in rectangular sheets, one foot long and six inches wide, the material being of a greenish colour,'[e] and a levy of some two million catties of mulberry bark for manufacture of paper money in 1644, apparently because of inflation, almost provoked the peasants into rebellion.[f] Sung Ying-Hsing (+1587–c. 1660) mentions that 'mulberry fibre paper (*sang jang chih*), made from the bark of mulberry trees, is extremely thick and smooth; that produced in east Chekiang is necessary to the silk producers in the lower Yangtze region for repositories for silkworm eggs'.[g] Even today, the mulberry is described as 'produced in all provinces in China and its bark is a very good material for papermaking'.[h]

(iv) *Bamboo*

This plant was extensively cultivated in China, except for the extreme northern part of the country (Fig. 1067). In ancient times it was grown probably as far north as the provinces along the Yellow River, but was later driven much farther south by change of climate or by deforestation. It is now abundantly grown in the Yangtze valley and the provinces to the south, especially Chiangsu, Chekiang, Fukien, and Kuangtung. Because of its long fibres, rapid growth, and low cost, it has been a major source of raw fibres for papermaking ever since the middle of the Thang dynasty.[i]

---

[a] Yule (1), p. 423.   [b] Bretschneider (10), vol. 1, p. 4.

[c] Laufer (1), pp. 560–3, cites among others S. Julien, Ahmed Sibab Eddin, Aurel Stein, J. Wiesner, and some Chinese works to prove that 'good Marco Polo is cleared, and his veracity and exactness have been established again'.

[d] *Wen Fang Ssu Phu* (*TSHCC*), p. 53.   [e] *Ming Shih* (*ESSS/TW*), ch. 81, p. 1.

[f] *Ni Wen-Cheng Kung Nien Phu* (*TSHCC*), p. 60; *Jih Chih Lu* (*TSHCC*), ch. 4, p. 103.

[g] *Thien Kung Khai Wu* (*KHCP*), p. 219; cf. tr. Sun & Sun (1), p. 230; Laufer (1), p. 561, n. 1–2, quotes S. Julien that 'according to the notions of the Chinese, everything made from hemp, like cord and weavings, is banished from the establishments where silkworms are reared'; and adds 'There seems to be a sympathetic relation between the silkworm feeding on the leaves of mulberry and the mulberry paper on which the cocoons of the females are placed'.   [h] Yü Chheng-hung & Li Yün (1), p. 37; pl. XXXIV.

[i] A theory that bamboo paper existed in Chin dynasty (+265–420) is generally considered invalid. It is based on a statement, in *Tung Thien Chhing Lu* by Chao Hsi-Ku (fl. +1225–64), that 'genuine' specimens of the work of the famous calligraphers Wang Hsi-Chih and his son Wang Yin-Chih were written on bamboo paper with horizontal screen-marks, made in Kuei-chi of modern Chekiang. But the same author mentions later in the same work that these 'genuine' specimens from the two Wangs were no longer extant in his time. Thus the specimens must actually have been imitations made later.

[1] 桑皮   [2] 桑穰

Fig. 1067. Bamboo and its fibres (× 50). Painting by courtesy of Mr N. H. Chang, Thaipei.

The earliest reference to the use of bamboo for making paper in China is found in a book by the Thang historian Li Chao[1] (fl. +806–20), who said that 'bamboo paper (*chu chien*[2]) was made in Shao-chou[3] (in modern Kuangtung)'.[a] His contemporary Tuan Kung-lu[4] (fl. +850) also mentioned the use of 'bamboo-membrane paper' produced in Mu-chou[5] of modern Chekiang.[b] Since the first use of the material must have been earlier than the recorded date, it is assumed that bamboo was first used in papermaking not later than the middle of the Thang, or the second half of the +8th century. Apparently it was developed as a substitute for hemp, which was a chief material for textiles, and for rattan which, as we have just seen, was almost exhausted at the end of the Thang dynasty.

The use of bamboo probably originated in Kuangtung, where the plant grew abundantly in the warm and humid climate. The method had spread to Chekiang and Chiangsu by the Sung dynasty, but the technique of making bamboo paper seemed to be still in the initial stage of experiment and the product was not by then perfected. As Su I-Chien said, paper was made of young shoots of bamboo in the Chiangsu and Chekiang area in his time, but no one could fold it if written with small characters, as it would break when touched, and could not be refolded.[c] The

---

[a] *Thang Kuo Shih Pu* (*CTPS*), ch. 3, p. 18*b*.
[b] *Pei Hu Lu* (*HPIS*), ch. 3, p. 7*b*.   [c] *Wen Fang Ssu Phu* (*TSHCC*), p. 56.

[1] 李肇   [2] 竹牋   [3] 韶州   [4] 段公路   [5] 婺州

great poet Su Shih[1] (+1032–1101) said, 'Modern people use bamboo for making paper, which was not done in ancient times.'[a] Another Sung author, Chou Mi[2] (+1232–98), said that the use of bamboo paper began in the Shun-hsi period (+1174–89).[b] This suggests that bamboo paper was unknown in his locality until the second part of the +12th century. A local gazetteer of Kuei-chi[3] (in modern Chekiang), compiled in +1201, records: 'The name of rattan paper of Shan-chhi was the earliest to come, seaweed paper next, and bamboo paper has become popular throughout the world only recently.'[c]

From the literary records, we may conclude that bamboo paper was invented in the latter part of the +8th century and not yet perfected in the 10th century, but the product from Shan-chhi, with many different varieties and colours, became popular, especially with artists, toward the end of the 12th or beginning of the 13th century. It is still not clear, however, how bamboo paper was developed during the long period between its initiation in the 8th and its perfection in the 12th century. But the description in the gazetteer indicates that the method of preparing the bamboo fibres was apparently borrowed through the long experience in the use of rattan.

(v) Other materials

Besides the major materials discussed above, many other plants were used. The most common fibres were from the stalks of rice and wheat. The process of making straw paper (tshao chih[4]) was much simpler than that for other materials. Since these fibres are tender, less time was required for beating in preparation. A Sung author said straw was used in Chekiang, and produced the best paper if mixed with rattan.[d] Sung Ying-Hsing mentions the mixing of rice stalks with bamboo fibres to make wrapping paper.[e] It is said that the straw first receives a preliminary pounding and then, after saturation in a lime solution, is buried in a trench. When properly disintegrated, the straw is removed and placed in porous cloth bags, which in turn are suspended in a running stream so that the fibres may be cleansed of all particles of lime.[f] Straw is still one of the raw materials most used for making paper for wrapping, burning, and sanitary purposes.

The blue sandalwood (Pteroceltis tartarinowii, Maxim.), known in Chinese as chhing than[5] (Fig. 1068), was the major material for making the famous Hsüan chih[6] for painting and calligraphy. It is made from the bark of the than tree grown primarily in the Hsüan-chheng area and manufactured in Chin-hsien; both were under Hsüan-chou in the Thang dynasty. The quality of the paper depends upon

[a] Tung-Pho Chih Lin (TSHCC), p. 43.
[b] Kuei Hsin Tsa Shih (HCTY), ch. 1, p. 32b.
[c] Chia-Thai Kuei-chi Chih (1926), ch. 17, p. 42a.
[d] Wen Fang Ssu Phu, p. 53.
[e] Thien Kung Khai Wu, p. 219; cf. tr. Sun & Sun (1), p. 230.
[f] Cf. Hunter(2), p. 16.

[1] 蘇軾　　[2] 周密　　[3] 會稽　　[4] 草紙　　[5] 青檀
[6] 宣紙

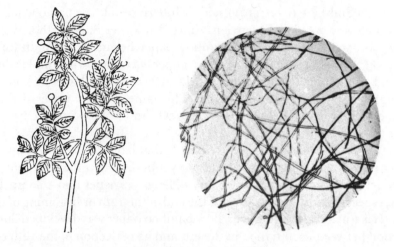

Fig. 1068. Blue sandalwood and its fibres (× 50).

the proportion of the raw material, which may either be pure bark or contain one-half or seven-tenths bark, with rice stalks making up the remainder. The more bark the better the paper.[a] A legend of the present-day Chin-hsien area, where *Hsüan chih* was made, relates that a certain Khung Tan[1] of the Later Han dynasty found by accident the bark of a *than* tree which had turned rotten and white after being soaked in a stream.[b] Since no older specimen is known of *than* bark and no earlier record mentions its use, it seems unlikely that the *than* bark was used in that early period.

Another fibre often mentioned as raw material for papermaking is the bark of the hibiscus (*Hibiscus mutabilis*), known in Chinese as *fu jung phi*[2]. It is generally believed that the famous stationery designed by the courtesan Hsüeh Thao[3] (+768–831) of Szechuan was made of hibiscus bark.[c] Sung Ying-Hsing reports that 'The bark of the hibiscus is cooked to a pulp, and aqueous extract of powdered hibiscus flower petals is added. This process was probably first devised by Hsüeh Thao and has been known by that name down to the present day. This paper is famous for its beauty, not for the quality of its material.'[d]

The use of seaweed (*Algae marina*) for making intricate filament paper, known as *tshe li chih*[4], appears frequently in early Chinese literature. Wang Chia[5] of the +4th century said: 'When Chang Hua (+232–300) presented his work *Po Wu Chih*[6] to the emperor, he was granted ten thousand pieces of *tshe li* paper, which was sent as tribute from Nan-Yüeh (modern Vietnam). The Chinese pronounced *chih li*[7] as *tshe*

---

[a] The method of making *Hsüan chih* had never been recorded until the processes were described in 1923 by Hu Yün-Yü (*1*); see also Chhen Pheng-Nien (*1*).

[b] Mu Hsiao-Thien (*1*), p. 4.   [c] Cf. *Chien Chi Phu* (*TSHCC*), pp. 1–2.

[d] *Thien Kung Khai Wu*, p. 219, cf. tr. Sun & Sun (1), p. 231.

[1] 孔丹   [2] 芙蓉皮   [3] 薛濤   [4] 側理紙   [5] 王嘉
[6] 博物志   [7] 陟貍

li by error. Since the people in the south used seaweed (*hai thai*[1]) in the making of paper with intricate and crooked lines, it was so called.'[a] Many other authors of later times continued to make reference to the use of seaweed in papermaking.[b] Su I-Chien (+957–95) said that paper was made from *thai* in the south,[c] and since seaweed contains long, strong, viscous filaments, it is possible that this material was used for making paper.[d] It is also possible that it was used for sizing, and thus the intricate hairy filaments appeared on the surface of the paper as a decorative pattern.

Cotton perhaps produces the best fibres of all but it has not been used as a major material for paper, and even modern paper manufacturers abstain from using raw cotton, probably because of its importance in the textile industry. A certain variety called cotton paper (*mien chih*[2]) was not actually made of cotton but of paper mulberry. Sung Ying-Hsing said: 'Torn lengthwise, the strong, hard-sized bark paper will show ragged edges resembling cotton fibres, hence it is called "cotton paper".'[e] Cotton stems have been used more recently for papermaking, but 'cotton paper' is not made from raw cotton.

Whether silk has ever been used as a raw material for papermaking is uncertain. Mention of the use of silk fibres is based primarily upon philological speculation without sufficient evidence. It was thought that, since the character *chih*[3] for paper bears the silk radical at its left, *chih* before Tshai Lun's time must have been made of silk fibre.[f] It is true that silk cloth was written on before the invention of paper, and the word *chih* is thought to be derived from one for silk, but the material of *chih* was not necessarily made of silk fibres. Technically, as many experts have said, silk fibres do not possess the colloidal properties which contribute so essentially to the entanglement and binding of the vegetable fibres.[g] At present, no early paper made of pure silk fibres is known to exist, nor is their use documented in literature.

It is possible, however, that silk fibres have been used in a mixture with other fibres, or that floss silk from silk cocoons has been used. Several references have been made to the use of silk cocoon paper (*tshan chien chih*[4]): one mention, in the early +8th century, says that the famous calligrapher Wang Hsi-Chih (+321–79) used silk cocoon paper in writing.[h] The Chin-su paper made in Soochow from

[a] See *Shih I Chi* (*HWTS*), ch. 9, p. 7*b*; also quoted in *TPYL*, ch. 605, p. 7*b*.

[b] See statements by Thao Hung-Ching (+451–536), Su Ching (+7th century), and others in *Wen Fang Ssu Phu*, p. 54; also Bretschneider (1), pt 3, pp. 369–70.

[c] *Wen Fang Ssu Phu*, p. 53.

[d] Seaweed was among the raw materials used by European papermakers in the 18th century; see Hunter (9), p. 316.

[e] *Thien Kung Khai Wu*, p. 219, cf. tr. Sun & Sun (1), p. 230.

[f] Chavannes (24), p. 12; Carter (1), p. 4; Lao Kan (1), pp. 489–91; Tsien (2), pp. 133–5.

[g] Armin Renker (1), doubted the feasibility of using silk fibre for papermaking, and Henri Alibaux (1), President of the Chambre Sindicale du Papier de Lyon, agreed with Renker.

[h] See *Lan Thing Chi*[5] by Ho Yen-Chih[6] of the Thang Dynasty, cited in *Lan Thing Khao* (*CPTC*), ch. 3, p. 9*a*. A Sung scholar commented: 'The so-called "silk cocoon paper" is actually silk cloth'; see *Fu Hsüan Yeh Lu* (*CPTC*), ch. 1, p. 4*a*.

[1] 海苔   [2] 棉紙   [3] 紙   [4] 蠶繭紙   [5] 蘭亭記
[6] 何延之

+1068 to 1094 for copying the *Tripitaka* was said also to have been made of silk cocoons.[a] Sung Ying-Hsing said that entangled or broken cocoons cannot be reeled into ordinary silk, but are made into wadding known as 'pot-bottom silk' (*kuo ti mien²*), used for quilting garments and bedding.[b] It is likely that silk waste or floss silk was used to make silk paper, since cocoons contain a gum which would serve to bind the fibres, but which is removed when pure silk is reeled.

The use of fresh plant fibres as raw material was apparently unknown to European papermakers when the craft of papermaking was introduced to Europe in the middle of the +12th century. For over five hundred years after the introduction, all paper of occidental origin was made from linen and cotton rags or a mixture of these second-hand fibres. After the beginning of the 18th century, as rags gradually became less plentiful and no longer economic material, European scientists looked for substitutes in order to meet the increasing need of the paper industry.[c] A great variety of vegetation, including hemp, bark, wood, straw, vines, seaweed, and grain husks, was tested and examined, even though such materials had already been in use in China for many centuries. Finally, wood pulp was widely adopted and has become, since the beginning of the 19th century, the chief raw material of the modern paper industry.[d] Because of limitations of forest resources in China, where most wood had to be used for construction, it was little utilised for paper, and even today, use of other materials than wood is encouraged.[e]

(2) THE INVENTION OF THE SCREEN MOULD

(i) *Functions of the mould*

When the disintegrated fibres are floating in water, they can be lifted up from it in a thin layer or sheet. The formation of such a layer of fibres supported on a piece of cloth or a mat was the very inception of the whole idea of papermaking, and the invention of an implement capable of picking up these matted fibres and yet permitting the water to escape was the key to the whole process. Subsequent improvements with specially designed screens resulted in an advancement of the technique, for the mould has remained throughout the centuries the essential tool in making paper by hand, and the very principle upon which the modern paper-machine is founded. Indeed, the entire development of papermaking is so closely

---

[a] Cf. *Chin-Su Chien Shuo* (*TSHCC*), pp. 1–17. Phan Chi-Hsing (*9*), p. 97, says his test shows it was not made of silk but of plant fibres.

[b] Cf. *Thien Kung Khai Wu*, pp. 33–4; cf. tr. Sun & Sun (*1*), p. 48.

[c] Cf. Hunter (*9*), pp. 312–40.

[d] The use of wood for papermaking was suggested by the French academician Réaumur in 1719 and printing on paper from wood by a German, Bruckmann, in 1727–30; wood pulp was commercialised and patented in England in 1800; see Anon (*191*), pp. 66–9.

[e] Cf. Yü Chheng-Hung & Li Yün (*1*), p. 1.

¹ 鍋底綿

Fig. 1069. Floating or woven type of mould for papermaking. A primitive type of mould found in Fu-Shan,
Kuangtung province in the 1930's. Dard Hunter Paper Museum.

connected with mould construction that it is only through a careful examination of
moulds that the origin and evolution of paper can be understood.

The mould could have been used in two distinct ways. In one, it was dipped
perpendicularly into the water which contained the macerated fibres, and was
brought up horizontally under them, lifting the matted fibres as in a sieve, allowing
the water to drain through the cloth. The other way was to hold the mould flat and
then pour the floating fibres on it. The woven material retained the fibres in a moist
sheet and at the same time allowed the water to drain through its interstices. In this
case the thin deposit of matted and felted fibres adhering to the mould was placed
in the sun to dry.

(ii) *Floating type of screen*

It is generally believed that the floating or woven type of mould was the earliest
form used by the ancient Chinese, and that the technique used was the one of
pouring the disintegrated fibres on to the mould.[a] During his travels in China in the
1930's, Dard Hunter found that this kind of woven mould was still in use in
Kuangtung province (Fig. 1069).[b] The woven screen is said to have been composed

[a] Cf. Hunter (9), pp. 78–9. An experiment in making paper with this kind of primitive mould was successfully
conducted by the Institute of the History of Science, Academia Sinica, Peking; see Phan Chi-Hsing (6), pp. 55–7.
[b] A mould of woven material from Fo-shan, Kuangtung, is kept in the Paper Museum, Appleton, Wisconsin.
F. A. McClure (2), pp. 115–29, found two villages near Canton where paper was still being made by the old
process with woven screen.

of ramie or China grass fastened to a square bamboo frame and stitched with slender bamboo strings which ran through the cloth and around the frame bars. After the moisture had evaporated from the sheet, the paper was easily stripped from the mould. The warp and woof of the mould and the stitches could have left impressions in the paper, in the same way as watermarks are formed in handmade paper today. There is some evidence to justify this assumption, derived from observation of such primitive moulds in modern south China.

Chinese sources are silent about the construction of the mould in ancient times, but the old definition of *chih* (paper) throws some light on the mould's form and the material of which it was made. As we have mentioned earlier, the character *chan* (mat), in the *Shuo Wên Chieh Tzu*, includes the radical for grass and, according to the early commentators, means a kind of cloth for covering (*kai*[1]) made of woven rushes (*pien mao*[2]).[a] It is possible that this early mat used in Han times was made of some kind of grass woven into a cloth which would support the macerated fibres and yet let the water escape through its meshes; such an appliance could have retained the primitive form without much change in its basic construction. Hunter noted with special interest that the locality where these woven moulds were found is not more than 200 miles from Lei-yang, where Tshai Lun was born.[b]

When Hunter made this statement he said that no +2nd century paper of the woven type, showing the impressions of the woven fabric upon which it had been formed, had ever been discovered in Asia.[c] It is true that no specimens found before that time bear clear screen marks, but the Pa-chhiao paper discovered in 1957 and other old specimens found in recent years, are reported to have fabric impressions on the surface. If this be true, it may very well support this theory of the earliest form of mould used in Han times.

(iii) *Dipping type of screen*

On the other hand, all those specimens of later periods examined show that another type of mould was used. This so-called dipping or laid type of mould, dipped into the vat of suspended fibres (Fig. 1070), must be a later invention. The idea of a mould from which the sheet of paper could be removed while still moist was a most important advance in papermaking, but the transfer of the wet substance from the mould to a board without damage required the construction of a very smooth and firm screen from which the moist sheet could be easily freed. For this purpose the screen was made of thin strips of rounded bamboo side by side, horizontally or vertically, and fastened together at regular intervals with strings of silk, flax, or hair from animal tail.

[a] See discussion on pp. 35 ff. above.
[b] Cf. Hunter (9), p. 83. It is said that fibres from the stems of the day lily are still used for making the screen mould in the Hopei area; see Phan Chi-Hsing (2), p. 45, n. 5.
[c] Cf. Hunter (7), p. 82.

[1] 蓋   [2] 編茅

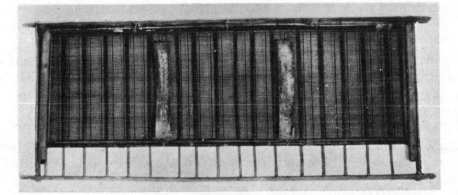

Fig. 1070. Dipping or laid type of mould. The screen is made of bamboo slivers fixed on the frame.
Dard Hunter Paper Museum.

No clear description of how the mould was made nor what it looked like, is found in early literature. The first illustration and description of a mould is found in the work by Sung Ying-Hsing of the late Ming dynasty. He said that the mould to lift fibres in papermaking was made of extremely fine bamboo strips, and that when the screen was unrolled and spread out, it was supported underneath by a frame with vertical and horizontal bars. Apparently, though, the screen was not fixed to the frame. On immersion into the fibre suspension in the vat, it picked up the matted fibres, and when the water had drained off, the mould was inverted to drop the felted fibres onto a wooden board. The sheets were then piled together, pressed, and dried with heat.[a]

The quality of the paper depended largely on the construction of the screen, the techniques of which seem to have been a secret. A late Chhing writer noted that a Thang family in the southern Chekiang area did not wish to pass on the techniques of screen-making to anyone outside the family.[b]

(iv) Impressions of screen marks

The bamboo strips in the screen left impressions in the paper, in the same way that watermarks are formed today, and these marks are useful for the determination of how the mould was made in early papermaking. It is reported of the paper specimens of the Han dynasty recently discovered in Pa-chhiao, Lopnor, and Chü-yen that the screen marks are not clear. The same is true of such papers of the +2nd and 3rd centuries discovered in Chinese Turkestan and elsewhere, but thousands of papers of the Thang and later times clearly indicate the construction of the mould at the time of their manufacture. On papers of the late +3rd and 4th centuries onwards, screen marks are clearly visible, and many paper specimens from

---

[a] *Thien Kung Khai Wu* (*KHCP*), p. 218; cf. tr. Sun & Sun (*1*), p. 227.
[b] See Yang Chung-Hsi (*1*), ch. 5, p. 39*a*.

Tunhuang are reported to have two distinct types of screen marks. Those of the Chin and Six Dynasties ($+265-581$), as well as those of the Five Dynasties ($+907-60$), have broad horizontal marks, while papers of the Sui and Thang periods ($+581-907$) show fine, close screen lines.[a] Hunter reports of some of the Thang papers he examined that there were twenty-three impressions of bamboo strips to every inch, and the hair stitchings of the mould were spaced at intervals of approximately one and one-sixteenth inches.[b]

The construction of the bamboo screen may have varied in different localities. One source from the Sung dynasty indicates that the screen was transversely laid in north China, so that the northern paper showed a horizontal grain, while in south China the matting was vertically laid, so that southern paper showed a longitudinal grain.[c] This theory has been used as a guideline for examination of old papers by artists and collectors since then, but recent analysis of surviving specimens of this period shows it is not necessarily true.[d] There is no such record for later times, though actual specimens exist for examination, showing that almost all papers from the $+4$th century are horizontally marked.

One may conclude that the woven type of mould made of cloth was used before the $+3$rd century, while the laid type of screen mould made of bamboo strips was introduced in the $+4$th century. In the former case, paper was dried directly on the woven mould and no couching was needed. For the laid mould, sheets were formed and dropped on the board without interleaving cloths between them, as has been the practice of Western papermakers. Since the damp fibres did not stick to the smooth bamboo screen-mould as they did to the coarse cloth, the new appliance made possible the continuous use of the same mould to make unlimited numbers of sheets without waiting for each one to dry on it. This was certainly the most significant step in the technique of papermaking.

(3) PROCESSES OF PAPERMAKING

Paper was made mostly by hand with the aid of natural resources, tools, utensils, and chemical agents. The workshop was usually operated in a site near a mountain, where the supply of raw materials and fuels was most convenient, and by a stream so as to make use of the water necessary for soaking, pounding, and washing the materials. The methods of papermaking varied slightly according to materials, periods, and locations, but the basic processes were more or less the same throughout the centuries.

Much of the earlier literature describes the quality and format of the paper for various uses, but none of it discloses details of how paper was manufactured, until

[a] Phan Chi-Hsing (2), p. 45.   [b] Hunter (9), p. 86.
[c] *Tung Thien Chhing Lu* (*MSTS*), p. 256.   [d] See Phan Chi-Hsing (9), p. 63.

the early +17th century when Sung Ying-Hsing[1] (c. 1600–60) wrote the *Thien Kung Khai Wu*[2] (The Exploitation of the Works of Nature). He devoted one entire chapter to the technical description and illustration of making paper with bamboo and paper mulberry. In Chapter 13, on 'Killing the Green', he gives step-by-step processes of the manufacture of paper, including preparation of raw materials by soaking, pounding, boiling, washing and bleaching the fibres; lifting the pulp with the screen; pressing the sheets to squeeze out the water; and finally drying on a heated wall (see Fig. 1071). His detailed account is given as follows:[a]

(i) *Preparation of raw materials*

The making of bamboo paper is a craft of the south, especially popular in Fukien province. After the bamboo shoots have started to grow, the topography of the mountain area should be surveyed. The best material for papermaking is the shoots that are about to put forth branches and leaves. During the season of *mang-chung*[3],[b] the bamboos on the mountains are cut into pieces from five to seven feet long. A pool is dug right there in the mountain and filled with water in which the bamboo stems are soaked [Fig. 1071a]. Water is constantly led into it by means of bamboo pipes to prevent the drying up of the pool.

After soaking for more than one hundred days, the bamboos are carefully pounded and washed to remove the coarse husk and green bark. [This is called 'killing the green' (*sha chhing*[4]).][c] The inner fibres of the bamboo, with a hemp-like appearance, are mixed with high-grade lime in a thick fluid and put into a pot to be boiled over a fire for eight days and nights. The pot for boiling bamboo, four feet in diameter, is enclosed in a wooden cask, measuring fifteen feet in circumference and more than four feet in diameter [Fig. 1071b]. The pot is attached to the cask with the aid of mud and lime and has a capacity of some ten catties of water. The cask is covered for boiling for eight days.

After the fire has been put out for one day, the bamboo fibres are taken from the cask and thoroughly washed in a pool with clean water. The bottom and four sides of the pool are lined securely with wooden boards to keep out dirt. [This is not necessary in making coarse paper.] When the fibres have been washed clean, they are soaked in a solution of wood ashes and put again into a pot, pressed to flatten the top, and covered with about an inch of rice straw ashes. When the water in the pot is heated to boiling, it is poured into another cask and strained with the solution of wood ash. If the water cools off, it is boiled again to repeat the straining. After some ten days of such treatment, the bamboo pulp naturally becomes odorous and decayed. It is then taken out to be pounded in a mortar [water-powered pestles are available in mountain regions] until it has the appearance of clay or dough, and the pulp is then poured into a vat for use.

[a] See *Thien Kung Khai Wu* (*KHCP*), pp. 217–19; cf. tr. Sun & Sun (1), pp. 224–7. Original notes to the text in small characters are translated in brackets.

[b] A solar term, 'Grain in Ear', about the sixth day of the sixth month.

[c] This is an old term for removing the green skin of bamboo in preparing tablets for writing. Apparently this term was borrowed for this purpose since in both cases the process was part of the preparation of material for writing.

[1] 宋應星   [2] 天工開物   [3] 芒種   [4] 殺青

Fig. 1071. Chinese papermaking as illustrated in a +17th-century book on technology. The panels above show various steps of the process. (*a*) Cutting and soaking bamboo twigs. (*b*) Cooking inner mass of bamboo in a pot. (*c*) Dipping the mould and lifting pulp from a vat. (*d*) Pressing moist paper sheets to release water. (*e–f*) Drying paper sheets on a heated wall. From *Thien Kung Khai Wu*, +1637 ed.

(ii) *Operations in forming sheets*

The pulp vat is a box-shaped tank, the size of which is determined by that of the screen-mould, which in turn is determined by the size of the paper to be made. When the fibres are ready and floating in the water of the vat, a solution for treating the paper is added as the fibrous mass gradually settles about three inches below the surface of the water. [The solution is made from a material similar to the leaves of the peach-bamboo and has no definite name as it varies locally.] When the paper is dried, it will turn brilliant white.

The screen is made of a mat woven of finely split and polished bamboo strips. When it is open it is supported by a rectangular frame. The screen is held with both hands and submerged in the vat to stir up the suspended fibres [Fig. 1071c]. When it is lifted up, the fibres are caught on top of the screen. The thickness of the paper depends on the way in which the screen is manipulated. Shallow submerging results in a thin sheet, while a deeper dip produces a thick one. Water from the pulp drains off around the screen's edges and back into the vat. The screen is then inverted and the paper is dropped onto a wooden board until many such sheets have been piled together.

When the number is sufficient, the sheets are covered with another board and the boards are tied with rope with the aid of a pole placed over the top board, as in a wine press, and all the water is thoroughly squeezed from the sheets [Fig. 1071d]. Then each sheet is lifted by means of a small pair of tweezers.

To dry the paper, a double wall of earthen bricks is erected, with the ground between the two rows covered with bricks. Holes in the lower part of the wall are left by the spaced omission of bricks. A fire is lighted at the first hole and the heat travels through the apertures and spreads to the wall surfaces where the bricks become hot. The wet sheets of paper are spread on to the wall one by one, baked dry, and then taken off as finished sheets [Fig. 1071e].

(iii) *Steps in making bamboo paper*

About two centuries after Sung Ying-Hsing, an eyewitness description of the processes of making bamboo paper was again given by a scholar, Yang Chung-Hsi[1] (1850–1900), in a supplement to his collected notes. His description is somewhat similar to that of Sung Ying-Hsing, except for a few remarks which may be useful to supplement Sung's account. Yang said that from the cutting of twigs to the drying by heat, the raw paper material changes hands seventy-two times before it becomes paper. A proverb in the paper trade says: 'A sheet of paper does not come easily; it takes seventy-two steps to make.'[a]

Yang also said that a man named Huang Hsing-San[2] of Chhien-thang visited Chhang-shan[3] (in modern Chekiang), where people on the mountain told him that the craft of papermaking required twelve major steps. The following is Yang's record of the processes according to the twelve steps given by Huang:

1) *Cutting the shoots.* The young and tender bamboo that has not yet sprouted twigs is selected. Its shoots are snapped off and, over a month, chopped into short pieces.

2) *Refining the fibres.* They are thoroughly soaked in lime until the bark and husk are sloughed off completely and only the fibres remain, tangled up like hemp. This is the stuff for papermaking.[b]

3) *Steaming.* The fibres are broken in two, arranged in bundles, and again soaked. Then they are put into a pot and subjected to very hot steaming.

[a] See Yang Chung-Hsi (*1*), ch. 5, pp. 39*a*–40*b*.
[b] This process is to extract the fibres and to remove impurities from the material.

[1] 楊鍾義   [2] 黃興三   [3] 常山

4) *Cleansing*. Afterward, they are washed and cleansed with water. When the cleansing is done, they are sunned.

5) *Sunning*. The sunning must be done on a flat area of several acres, paved with pebbles and sprinkled with green vitriol to inhibit the growth of weeds. For this reason, the area for sunning paper may not be cultivated.

6) *Soaking with ashes*. After sunning they are soaked and steamed three more times, until the yellow turns white. For these soakings the seeds of the *tung* tree or wood ashes of the yellow thorn [*Vitex negundo*], known in Chinese as the *huang-ching*[1], must be used for whitening.

7) *Pounding*. The fibres are then put under water and pounded. In a day three catties can be made and the fibres turned into a pulp.

8) *Purifying*. If the pulp is not pure, it is poured into a fine-meshed cloth bag and lowered into a deep stream. A board is projected into the bag to stir it occasionally. In this way the ashes are got rid of completely and the pulp is rendered as white as snow.

9) *Making the tank*. The tank is made of chiselled stone. It should be slightly larger than the sheet of paper.

10) *Weaving the screen*. The bamboo is woven into a screen, the size of which is measured by the size of the tank. The craft is extremely exact, and is practised only by a Thang family in the mountain, who do not allow it to be taught to others. When the screen is ready, the paper pulp is poured into the tank, and water is added, mixed with gum and the sap of the *mu-chin*[2] [*Hibiscus syriacus*] for binding. Two men lift the screen[a] from the pulp and stir it up as it forms into a sheet of paper.

11) *Repelling the water*. The sheet is then placed on top of a stone. After a hundred sheets are piled up, they are pressed together to squeeze out the moisture.

12) *Drying by heat*. The sheets are then lifted and dried by heat on a wall. The centre of the wall is hollow and a fire is placed in it. The men holding the paper place the sheets closely together on the wall. By the time the next one is ready, the previous one is dried. The processes of straining and drying require varying degrees of timing, but once the technique is mastered it can be carried out easily.

(iv) *Methods of making bark paper*

For making paper of paper mulberry bark, Sung Ying-Hsing describes the cultivation of the plant and the mixing of the bark with bamboo and rice stalks. Except for describing the cutting of paper mulberry, he gives no details of separate steps of the process, which apparently were similar to those for making bamboo paper. He only added:

In making bark paper, sixty catties of paper mulberry bark are added to forty catties of very tender bamboo. They are soaked together in the pool, mixed with fluid lime, and boiled in a pot to be macerated. Recently, a more economical method uses seventy per cent of bark and bamboo mixed with thirty per cent of rice stalks. With a proper formula added to the pulp, the paper can still be brilliantly white. [b]

[a] Only one person is required if the screen is small.
[b] *Thien Kung Khai Wu* (*TSHCC*), p. 219; cf. tr. Sun & Sun (1), p. 230.

[1] 黃荊   [2] 木槿

A similar method of making bark paper is given by Dard Hunter from his personal observation of the method of papermaking in China. He says that before the bark pulp can be moulded into paper, a mucilaginous gum made from the leaves of deciduous trees is added.[a] In general, the treatment of tree bark is a more arduous and painstaking process than that of preparing bamboo for papermaking.

(4) TREATMENT OF PAPER

Before the formation of a sheet of paper, an adhesive solution and certain insoluble materials were usually added to the pulp, in order to improve the physical and chemical qualities of the finished product. After the paper was manufactured, it was sometimes treated with special ingredients to protect it against injury from insects or for artistic purposes. The processes included sizing, loading, dyeing, colouring, and coating. Different kinds of vegetable, animal, and mineral substances were used as ingredients for such treatment and sometimes prepared and applied by elaborate procedures. All these were essential steps in the technical and artistic advancement in papermaking.

(i) Sizing and loading

Sizing is essential to make the paper suitable for writing with ink, to prevent undue absorption and running of the ink. It is necessary not only for artistic purposes in painting or calligraphy, but also for technical reasons. It helps to keep the fibres floating in the tank, thus making the sheet even and uniform, and also adds to the bonding strength of the fibres. Especially important, when the sheets are transferred from the screen to the board to be pressed and dried, is the fact that the sizing prevents the sheets from sticking together. Loading with certain finely powdered materials improves the opacity, texture, and weight of the finished paper.

The earliest papers were sheets of fibres without sizing or loading, but these additions are believed to have been introduced as early, probably, as before the +3rd century. The specimens of the Chin dynasty (+265–420) found in Sinkiang are well sized and loaded. These papers were first coated with gypsum and later sized with gum or glue made of lichen. Subsequently starch flour was used to make paper much stronger and harder.[b] Recent studies of ancient papers by Chinese scientists have found that specimens of the late +4th and early 5th centuries are coated with starch on the front side and smoothed with a stone. Those of early 5th century from Tunhuang and Sinkiang are sized with starch in the pulp.[c] Paper sizing in modern Kuangtung is made by boiling the leaves and twigs of an evergreen shrub called hsi yeh tung chhing[1] (Ilex pubescens), or from shavings of the

---

[a] See Hunter (2), pp. 15–16.   [b] Clapperton (1), p. 8.
[c] Phan Chi-Hsing (9), pp. 61–2.

[1] 細葉冬青

cedar (*Machilus thunbergii*) known as *pao hua*[1], which has a mucilaginous substance that was used by Chinese women as a pomade for the hair.[a]

Another formula is found in a modern work on methods of making bamboo paper,[b] where it is said that both plant and animal materials are used in sizing and loading. The animal glue is a gelatin extracted from cowhide, which is dissolved in hot water and added together with the fine talc powder to the prepared fibres. About two or three ounces of cowhide and one or two ounces of powder are needed for every twenty catties of pulp. The plant materials, from a kind of hibiscus known in Chinese as *huang chü khuei*[2] or *chhiu khuei*[3] (*Hibiscus abelmoschus*), is preferred because it is cheaper than animal glue. Its root, after being washed clean and sliced, is soaked in cold water for one night. The sticky juice is rubbed out with the hands, and strained through a fine cloth; it is then mixed with the pulp to soften the fibres. Other plants used for sizing are the leaves of the peach-bamboo (*thao chu*[4]) and *mu-chin*[5] (*Hibiscus syriacus*), as mentioned by other authorities.

For loading in Chinese papermaking, the soya bean (*huang tou*[6]) was used. The beans were soaked in water for five to six hours and then ground into a liquid starch which was separated by draining the water through a fine cloth. After several washings, the starch was poured over the raw fibres packed tightly into a tank about one foot deep. Additional piles of fibres mixed with starch could be added if necessary. This resulted in softening the fibres and facilitated their sticking together. After washing with clean water, the fibres were stamped with bare feet or pounded using water power.[c]

(ii) *Dyeing*

The process of dyeing paper to a yellowish colour, known as *jan huang*,[7] was apparently used very early and was common when paper began to be extensively used for books in the +2nd or 3rd century. In the old dictionary *Shih Ming*[8] (Explanation of Names), compiled by Liu Hsi[9] about +200, the word *huang* was already defined as 'dyeing paper', while Meng Khang[10] of the +3rd century mentioned that paper was dyed yellow in his time.[d] In a letter to his brother Lu Chi[11] (+261–303), the noted writer Lu Yün[12] (+262–303) said: 'Of the first series of your work in twenty rolls, ten have just been completed [copied] and will be set for dyeing.'[e] Apparently it was a general practice at that time to have ordinary paper dyed to prevent damage from insects and to obtain a glossy surface, before or after it was used for writing. The ingredient was a liquid obtained from the Amur cork tree, known as *huang po*[13] (*Phellodendrum amurense*), which has a fragrance and a

[a] Hunter (17), p. 24.   [b] Lo Chi (1), pp. 89–94.   [c] Lo Chi (1), pp. 77–81.
[d] *Chhien Han Shu* (*ESSS/TW*), ch. 97b, p. 13a.
[e] *Lu Shih-Lung Wen Chi* (*SPTK*), ch. 8, p. 51.

[1] 刨花   [2] 黃菊葵   [3] 秋葵   [4] 桃竹   [5] 木槿
[6] 黃豆   [7] 染潢   [8] 釋名   [9] 劉熙   [10] 孟康
[11] 陸機   [12] 陸雲   [13] 黃蘗

toxic effect in keeping insects away.[a] It was prepared by soaking its inner bark, which is yellow and bitter, to produce a liquid used for dyeing.

The methods of preparation and application of this liquid are described by Chia Ssu-Hsieh[1] (+5th century):

The paper to be treated should be unsized, for it is tough and thick, and especially suitable for dyeing. When the whiteness is diminished through the treatment, it should not be dyed too deeply, or its colour will turn dark in the course of time.

When the *huang po* is thoroughly soaked, if one throws away the dregs and uses the pure liquid only, it is wasteful. After soaking the *huang po*, the dregs should be pounded and boiled, pressed in a cloth sack, and again pounded and boiled, three times. The liquid is then added to and mixed with the pure juice. Thus four times as much liquid is saved, and the paper so dyed will be bright and clear.

Writing on a book should be treated after the lapse of one summer, then the seams will not be loosened. Those newly written should have the seams pressed with a flat iron; only thereafter may they be dyed. Otherwise, the seams will become loose.[b]

Many papers dyed with this liquid are found among the manuscripts from Tunhuang surviving today. The earliest example of known date is a *sutra* written in +500, which is about twenty-six feet long and dyed yellow, except at the very end, where the original whitish colour remains.[c] Other examples of such dyed paper are found among the manuscripts especially of the +7th and 8th centuries some of which, treated by this process, are said to have been preserved in better condition than others which were not. In some cases, the name of the dyer is given in the colophon, indicating the importance of such artisans in producing books.

Some twenty *sutras* written in +671–7 name the dyers as Hsieh Shan-Chi[2], Wang Kung[3], Hsü Chih[4], and Fu Wên-Khai[5]. A few mention the dyer, but omit his name.[d] The dyers, known as *chuang huang chiang*[6], also served in various departments of the court, along with such other artisans as scribes, makers of inked rubbings, and brush-makers. The Thang administrative codes of +723–38 and the *History of the Thang Dynasty* both record that official positions for paper mounter-dyer and paper-sizer existed in various academic agencies, including nine in the Chancellery (Men Hsia Sheng[7]), six in the Palace of Assembled Worthies (Chi Hsien Tien[8]), three in the Academy of Respecting Literature (Chhung Wen Yüan[9]) and ten in the Imperial Library (Pi Shu Sheng[10]), for duties connected with the treatment of paper for documents.[e] A decree of +675 said: 'Since the issuing of decrees and orders is a permanent institution and since white paper has generally been dam-

---

[a] See chemical formula in Phan Chi-Hsing (2), p. 46.
[b] *Chhi Min Yao Shu* (*TSHCC*), ch. 3, p. 57; cf. Tsien (2), p. 152; van Gulik (9), pp. 136–7.
[c] Giles (17), p. 813; no. S. 2106.
[d] See the list of colophons in Giles (13), p. xv; also discussion in Phan (2), p. 46.
[e] *Thang Liu Tien*, ch. 8, p. 43 b; ch. 9, p. 3 b; ch. 10, p. 2 a; ch. 26, p. 4 a; *Chiu Thang Shu*, ch. 43, p. 40 a; *Hsin Thang Shu*, ch. 47, pp. 8 b–9 a.

[1] 賈思勰	[2] 解善集	[3] 王恭	[4] 許芝	[5] 輔文開
[6] 裝潢匠	[7] 門下省	[8] 集賢殿	[9] 崇文院	[10] 秘書省

aged by insects, hereafter let the Grand Secretariat be instructed to order that yellow paper be used by the various government offices and all the districts and prefectures.'[a] This practice of dyeing paper was continued until some time in the Sung dynasty, when book format was changed.

Another method of treating the paper with insecticide was the use of litharge or red lead (*hung tan*[1] or *chhien tan*[2]), which is a mixture of lead, sulphur, and saltpetre. Paper treated with a solution of these chemicals turns a bright orange colour called *wan nien hung*[3] (ten-thousand-years red) and is toxic to bookworms.[b] Many books printed in the Kuangtung area in the Ming and Chhing period and bound with such papers have been preserved in perfect condition without being damaged by insects. The manufacture of the red lead is described by Sung Ying-Hsing of the 17th century, as follows:

The ingredients for making lead litharge are: ten ounces of native sulphur, one ounce of nitre and a catty of lead. Melt the lead first. While it is in the molten state, add some drops of vinegar, then add a piece of sulphur while the molten mass is steaming. Shortly afterward, a small bit of nitre is added. When the steaming subsides, more vinegar is added and the process is repeated with the nitre and sulphur being added little by little. Litharge is obtained when the mass turns into powder.[c]

This powder was mixed with water and vegetable glue and heated into a solution, which was then applied over the white paper. After drying, the treated sheets were used as endpapers inside the covers for the protection of the untreated paper from damage by bookworms.[d] After the book was changed from the roll to the flat format, it was impossible to dye the entire book in the former way. The use of paper treated with red lead solved this problem and proved to be much simpler, easier, and more effective than the old method.

(iii) *Colouring*

While dyeing of paper was primarily for insecticidal reasons and permanency, colour was also added for artistic purposes. The earliest known coloured paper was probably the *ho-thi* of the Han dynasty, which was described by Meng Khang of the +3rd century as a kind of 'silk-paper dyed red for writing, as it is dyed yellow today'.[e] If this is correct, red paper must already have been used as early as the

---

[a] *Wen Fang Ssu Phu* (*TSHCC*), p. 51.
[b] For a study of the litharge-treated paper, its chemical analysis, and bookworms, see a report by Chou Pao-Chung and others (1), pp. 194–206.
[c] See *Thien Kung Khai Wu* (*KHCP*), p. 235; cf. tr. Sun & Sun (1), p. 256.
[d] See the test report by the Study Group for Moth-Proof Paper of the Museum of Chinese History, Peking, in *WWTK*, 1977, no. 1, pp. 47 ff.
[e] *Chhien Han Shu*, ch. 97b, p. 13a.

¹ 紅丹   ² 鉛丹   ³ 萬年紅

— 1st century, and yellow came into vogue in the +3rd. At the court of the Later Han (+25–220), the princes were given 100 sheets each of maroon and bright red hemp papers when they were invested. The use of yellow paper continued in the following centuries and probably reached a peak in the Thang, as official documents were ordered to be written on such papers. Other writings intended for permanence, such as Buddhist *sutras*, were written on them.

Paper in various other colours also was used early and became plentiful and popular in the Thang dynasty. In the +4th or 5th century, 'peach-blossom paper', in bright green, blue, and red, was used in Szechuan,[a] and in the Thang, the stationery papers of Szechuan were dyed in ten different colours: maroon, pink, apricot pink, bright yellow, dark blue, light blue, dark green, light green, bluish green (verdigris), and 'light clouds'.[b] Besides these, there were fancy varieties of artistic papers in different colours and patterns, specially made for writing and decorations on various occasions.[c] Also, a small notepaper known as 'Hsüeh Thao paper', dyed red, was designed in the Thang and continued to be imitated for many centuries.

Some papers apparently were coloured by adding pigments in the pulp, but many were dyed after the paper was manufactured. One source indicates that ten sheets were piled in a stack, with a bamboo clip attached to one end of each stack. Then the liquids of various colours were applied for dyeing the sheets. When they dried, the papers looked lustrous and beautiful.[d]

To imitate aged paper, especially for making forgeries of old paintings or calligraphy, paper might be dyed or smoked. Sometimes incense ashes were scattered over the surface of paper, then brushed off with a stiff-haired brush, or water mixed with dust was used. Thus the paper obtained a yellowish or greyish tinge, and looked as if it had aged in the course of centuries. However, the colour of paper artificially aged by dyeing also appeared on the reverse side; thus it might be distinguished from the true aged paper which was discoloured only on the surface. The Sung connoisseur Chao Hsi-Ku (c. +1200) says: 'Those who sell calligraphy often produce fakes by dyeing old paper so as to give it a dark colour. But they are entirely unaware of the fact that water mixed with dust will penetrate both the face and the back of the paper. Genuine darkened paper is discoloured only on the face, while the reverse is as new.'[e] Mi Fu (+1051–1107) also commented that scrolls 'aged by exposing them to smoke will retain an odour. The dark tinge of all genuine scrolls will be deep on the obverse, but shallow on the reverse. Moreover, genuine old paper or silk has a peculiar fragrance of its own'.[f]

[a] *Wen Fang Ssu Phu (TSHCC)*, p. 49.
[b] *Chien Chih Phu (TSHCC)*, p. 2.
[c] See discussion of ornamental stationery on pp. 91 ff. below.
[d] *Wen Fang Ssu Phu (TSHCC)*, p. 53.
[e] *Tung Thien Chhing Lu (MSTS)*, ch. 8, p. 257; cf. van Gulik (9), pp. 100–1.
[f] *Hua Shih (TSHCC)*, p. 63; cf. van Gulik (9), p. 184.

(iv) *Coating*

To make the paper glossy, stiff, and translucent, a kind of yellow wax was applied to its surface. The wax was called *ying huang*[1] (stiff and yellow), or by some scholars *huang ying*.[a] After the application of the wax to the surface with a hot iron, the paper became smooth, stiff, and translucent, and could be used for tracing paintings or calligraphy. The process was also used to brighten paper material if darkened by age. It was commonly used from the Thang through several dynasties for these purposes; thus Chang Yen-Yüan[2] (*c.* +840) mentioned that 'art lovers should keep a hundred sheets of Hsüan paper to be prepared and coated with wax for making traced copies'.[b] Later, Chang Shih-Nan[3] (+13th century) said that paper could be evenly coated with beeswax by means of a hot iron. Although the paper became somewhat stiff, it was glossy, smooth, and translucent like *ming-chüeh*[4] (translucent horn) so that the slightest particles could be seen through it.[c] A Ming scholar, Li Jih-Hua[5] (+1565 to 1635) wrote:

Ying huang paper is made because people dislike the opacity and rough surface of ordinary paper. Therefore they heat paper over a hot flat iron, and then wax it evenly with yellow wax. Although the paper then becomes slightly stiff, it is glossy and translucent, resembling flakes of fish bone or sheets of transparent horn. If one lays a sheet of this paper over something, even the smallest details of such an object will be perfectly discernible. Generally antique autographs from the Wei and Chin periods, left by calligraphers like Chung Yu[6] [+151-230], So Chhing[7] [+239-303], or Wang Hsi-Chih[8] [+321-379], are treated in this way, because in the course of the centuries they have grown dark.[d]

Apparently there were two kinds of stiff yellow paper. One was dyed with a yellowish insecticidal substance and used for writing, the other was coated with yellow wax and it was this that was used for tracing painting or calligraphy. Tung Yu[9] (fl. +1127) wrote: '*Ying huang* was used by the Thang people for tracing writing. There was another kind of paper for copying *sutras*. These two kinds of paper were similar, but different in quality; the stiff and heavy kind was not [suitable] for *sutra* paper.'[e] The paper called *ying huang* is generally characterised as thick, stiff, glossy or translucent and heavy; it was to be found in smaller sheets. A Chhing scholar said that the *ying huang* paper of the Thang dynasty was two feet and 1.7 inch long and 7.6 inches wide, weighing 6.5 ounces.[f]

Besides *ying huang*, the term *tzhu huang*[10] should be mentioned. As used in early writings it meant a mineral (As_2S_2), similar to *hsiung huang*[11] or orpiment (As_2S_3),

---

[a] Cf. Rudolph (13), pp. 15-17.
[b] See *Li Tai Ming Hua Chi* (*TSHCC*), pp. 75-6; also *Wen Fang Ssu Phu* (*TSHCC*), p. 56.
[c] *Yu Huan Chi Wen* (*TSHCC*), ch. 5, p. 28.
[d] *Tzu Thao Hsüan Tsa Chui* (1768 ed.), ch. 3, p. 6b; cf. van Gulik (9), p. 137.
[e] *Kuang Chhuan Shu pa* (*TSHCC*), ch. 6, p. 73.   [f] *Chin Hsiang Shuo* (*KHHK*), p. 22a.

[1] 硬黃   [2] 張彥遠   [3] 張世南   [4] 明角   [5] 李日華
[6] 鍾繇   [7] 索靖   [8] 王羲之   [9] 董逌   [10] 雌黃
[11] 雄黃

poisonous and insecticidal. It is insoluble in water, and was prepared by grinding into powder and mixing with gum to form a solid stick. For use, *tzhu huang* was ground with water into a liquid, as solid ink is ground, and applied to paper to protect it from bookworms. Chia Ssu-hsieh (+5th century) gives a recipe for treating book rolls with *tzhu huang*,[a] which was also used to correct errors in writing. Since most documents were written on yellow paper one could easily obliterate erroneous characters by painting them over with orpiment. Shên Kua (+1030–94) said: 'If mistakes occurred in the clean copies in the official palaces, they could be deleted by painting them over with orpiment.'[b] Thus *tzhu huang* was generally used for the purpose of erasing.

(5) PRESERVATION OF PAPER

Chinese paper is usually very thin, and only one side is used for writing or printing. Various methods have been used to increase its durability. Paper bearing artwork is usually mounted by attaching an additional piece or pieces of paper to its back to increase its weight, eliminate wrinkles, and especially to improve its artistic quality. When the paper becomes old, it can be remounted to freshen its appearance. If a sheet of paper is worn or torn, the damage can be patched, and another sheet of paper can be inserted between the folded leaves of a book for reinforcement. To protect paper from damage by insects and to increase its permanency, it may be exposed to light during certain seasons of the year to adjust its dryness and temperature. Special warnings were given by many early book collectors concerning careful handling of books and scrolls in order to prolong their life. Many examples of painting or calligraphy on paper, or of books made a thousand years ago are still in very good condition, and their survival is largely due to the work of preservation through mounting, restoration, and special care given to them.

(i) *Mounting*

The earliest known reference to the mounting of paper dates back to the +4th or 5th century when paper was extensively used for writing and bookmaking. The Thang connoisseur Chang Huai-Huan[1] wrote in +760 that up to the Chin dynasty mounting on the back of the paper was not satisfactory because the paper used for backing developed creases. However, Fan Yeh[2] (+398–445), author of the *Hou Han Shu*, made some progress in mounting, and the Emperor Hsiao-Wu (+451–64) of the Sung was able to charge Hsü Yüan[3] with re-mounting scrolls with ten sheets of paper in one roll, limiting each roll to twenty feet.[c] Gradually the

---

[a] *Chhi Min Yao Shu* (*TSHCC*), ch. 3. pp. 57–8; cf. van Gulik (9), p. 137, n. 2.
[b] *Meng Chhi Pi Than* (*TSHCC*), ch. 1, p. 4.
[c] Quoted in *Fa Shu Yao Lu* (*TSHCC*), p. 64; see also van Gulik (9), pp. 139–40.

[1] 張懷瓘   [2] 范曄   [3] 徐爰

technique was improved, and mounting became one of the most important processes in the decorating and preservation of works of art on paper. In the preface to a work on mounting, the Ming scholar Chang Chhao[1] says: 'Mounting is to a scroll what make-up is to a woman.' Without it, 'although this will not detract from her charm, yet it will make her insignificant'.[a]

Mounting is a very specialised technique that requires skill and a knowledge of art to handle different varieties of material—mounting of new paper stock, remounting of old works on paper, or restoring an antique work of art. In any case, it involves many steps of careful study of the work and days or weeks of working time to complete a piece of mounting. It is generally agreed that the quality of workmanship determines the length of life of a picture on paper, and the mounter is said to be 'the arbiter of the destiny of scrolls'.[b]

For mounting, a new piece of paper is usually treated with a thin alum solution first, to prevent the ink or pigments from running. The crystals of alum are first dissolved in cold water, then the solution is brushed over the coloured portion of the paper, on the right side. When this is thoroughly dried, the treatment is repeated on the reverse. Next, a thin but tough piece of backing paper is spread out and moistened with a large, soft-haired brush dipped in clean water, and placed over the original paper. A very thin layer of paste is applied with a broad pasting brush to attach the two sheets of paper, and they are made to adhere by softly tapping with a stiff-haired brush (Fig. 1072). If one sheet of backing is insufficient, a second or more may be added in the same way. When the paste has had time to penetrate well into the paper but has not yet dried thoroughly, the scroll must be taken from the table, transferred to the boards on the wall, and left to dry for a week or longer. When this is done, a last sheet of backing may be added before the stave, roller, and suspension loop are attached to complete the scroll.

When an antique piece of art on paper is remounted, the scroll is spread face down on the table and thoroughly soaked by going over it repeatedly with a large, soft-haired brush dipped in clear water. After a while, the old backings are peeled off one by one with a bamboo spatula and tweezers while the scroll is still moist. Holes and tears in the paper are patched by pasting thin strips of paper of matching colour on the reverse. If the surface of the paper has accumulated dust in the course of years, this is washed off with a pure extract of acacia pods or of loquat seeds (*Gleditschia sinensis*), both of which contain detergent.[c] When the dust is thus removed, a fresh and bright surface will be obtained without fading of its colours. After the paper has dried thoroughly, new backing is added as is done with new paper, and additional work such as re-touching is done.

For mounting sheets of inked squeezes, there are different methods: depending

---

[a] Introduction to Chou Chia-Chou (*c.* +1590–1660), *Chuang Huang Chih* (*TSHCC*); see text and translation in van Gulik (9), p. 289.

[b] Cf. van Gulik (9), p. 8.

[c] Washing with acacia pods is as early as the Thang dynasty, when Chang Yen-Yüan mentioned it in his *Li Tai Ming Hua Chi* (*TSHCC*), ch. 3, p. 107. The use of the loquat seeds is mentioned in the *Chuang Huang Chih* (*TSHCC*), pp. 2–3; see van Gulik (9), pp. 114, 293–4.

[1] 張潮

Fig. 1072. Picture mounting in operation at Jung Pao Chai, Peking.

on whether the original is cut or uncut, and is to be rolled in a sheet or album, or folded. Mounting of an uncut piece is primarily to strengthen the sheet by backing with a thin layer of bark paper. A large sheet can be folded to be kept in a box, or made into a scroll with rollers, like a painting for hanging. Mounting of a cut piece requires a special technique. A sheet of inked squeezes from a large tablet may be cut into long vertical strips like the columns of a traditional book and then mounted in book form as an album. The technique depends on the skill with which the strips are cut. Vertical cutting should be done in such a way that all pages neatly fit to each other in an unbroken row, while the horizontal cutting should result in a perfectly straight top and bottom; if this is expertly done, the work is as good as finished.[a]

(ii) *Preparation of the paste*

The success of mounting depends largely upon the proper preparation of paste, especially its viscosity and its insecticidal effect. The basic ingredient is flour or rice

---

[a] For mounting of inked squeezes, see *Chuang Huang Chih* (*TSHCC*), p. 7; cf. van Gulik (9), p. 305.

starch, but the addition of a plant root called *po-chi*[1] (*Bletilla striata*) is said to be best for increasing viscosity. Thao Tsung-I[2] (+1320–99) cites in his *Cho Keng Lu*[3] a conversation between a scholar, Wang Ku-Hsin,[4] and an 84-year-old monk, Yung-Kuang,[5] the custodian of old *sutras* in a Buddhist temple. When he visited Wang's studio, the monk was asked: 'How is it possible that the *sutra* rolls of former dynasties, although their seams are as narrow as a thread, have after so many years not become loose?' He answered: 'The old method was to use sap of paper mulberry, fine flour, and a powder of *po-chi*. When the three are mixed into a paste and used to fasten the sheets of paper together, they will never become loose, for this paste is as sticky as glue or varnish.'[a]

The *po-chi* is an orchidaceous plant, which tastes bitter and contains a large proportion of mucilaginous substance. The sap of its root has been used in pharmacy as well as in preparing paste, as is mentioned in the *PaoPhu Tzu* by Ko Hung (+4th century), who said in a chapter on medicine for immortality that 'the *po-chi* is for making paste'.[b] Sometimes other ingredients such as pepper, incense, and alum were added for fragrance, preservation, and an insecticidal effect. The recipe by Chou Chia-Chou says that the paste is made of *po-chi* and white alum, to which is added a little frankincense, yellow wax, and also *hua-chiao*[6] (*Xanthoxylum pipertum*) and *po-pu*[7] (*Stemona tuberosa*), which were put into boiled water. Both *hua-chiao* and *po-pu* have the effect of keeping insects away, because the paste is never eaten by insects, while the paste itself will never allow the sheets to become loose or be subject to other deficiencies.[c]

(iii) *Restoration*

The methods of repairing and restoring the torn pieces of paper of a traditional book are similar in part to those for remounting antique art pieces. The paper of Chinese traditional books has usually been damaged by insects attracted by the rice starch paste, by fungi caused by heat and moisture, and other hazards such as water, dust, and smoke. Insects perforate the paper, while fungi leave discoloured spots, and also weaken the paper. To repair insect perforations or tears and other worn places, a piece of thin but strong bast paper may be pasted on the reverse. If the paper to be restored is yellowed, the discoloured patches may be dyed with an infusion of tea to which a disinfectant is added, while if the perforations are minor, they may be repaired individually. The page is placed face down on a waxed board and paste is applied to the edges of the perforation, and the patch is applied. After drying, the page is carefully peeled off the board. If the perforations are com-

[a] *Cho Keng Lu* (*TSHCC*), ch. 29, p. 8a.
[b] *Pao Phu Tzu, Nei Pien* (*SPTK*), ch. 11, p. 2a, b.
[c] *Chuang Huang Chih* (*TSHCC*), pp. 8–9. The methods of preparing paste are translated by van Gulik (9), pp. 307–8; also discussed in Nordstrand (1), pp. 129–30.

[1] 白芨   [2] 陶宗儀   [3] 輟耕錄   [4] 王古心   [5] 永光
[6] 花椒   [7] 百部

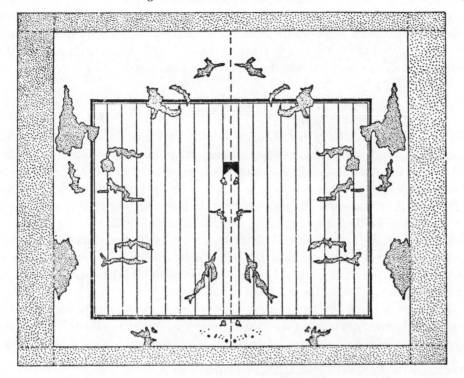

Fig. 1073. Restoration of book leaf damaged by bookworms by mounting a new sheet of paper on the back and cutting the superfluous edges after drying. From Nordstrand (1).

plicated, the patch may be placed on the board and the damaged page placed over it, face up. Then the edges of the perforations are placed in their correct positions on the pasted patch with a pair of tweezers. The page is worked into perfect contact with the patch, and is left to dry on the board (Fig. 1073), from which it is carefully peeled away when completely dry.[a]

If the paper is weak and brittle, or extensively damaged, it is necessary to mount the pages on sheets of thin, strong bast paper, as described earlier for mounting pieces of artwork. An extra, usually longer, sheet of paper is inserted inside the fold of the double leaf. If a white paper is inserted into a yellow one, it is known as *chin hsiang yü*[1], 'gold inlaid with jade'.

Apparently restoration of paper was practised very soon after paper became used extensively for writing. Chia Ssu-Hsieh (+5th century) said that when book rolls were damaged, if thick paper were used to patch them, these patches would become hard like a scab or scar, which would in turn damage the book itself, but if a

---

[a] The description here is taken primarily from Nordstrand (1), pp. 112–28, on the restoration and conservation of Chinese double-leaved books. The account is generally based on traditional techniques practised by Chinese artisans for centuries.

[1] 金鑲玉

piece of paper as thin as a scallion leaf were used, the patch could hardly be seen unless it were looked at against the light. If the damaged spot was curved, the patch should be cut to fit, but if too large a piece were used, and not cut to the shape of the damaged area, the torn parts of the paper would twist and shrink.[a]

(iv) *Conservation*

For preservation of the paper, Chia suggested that musk and quince (*Cydonia sinensis*) be placed in bookcases to prevent insects from breeding there. During the humid and hot season of the fifth month, bookworms are hatched. If the books are not unrolled during summer, there are sure to be insects in them. Between the fifteenth of the fifth month and the twentieth of the seventh month, book rolls must be unrolled and rolled three times. This should be done on a clear day in a spacious house which is aired and cool, and books should not be exposed directly to the sun, for it will turn the paper brownish. Rolls heated by the sun quickly attract insects, and rainy and humid days should especially be avoided. If books were cared for in this way, they would last for several hundred years.[b]

Chia also warned readers against carelessness in handling paper rolls. He suggested opening and closing the books slowly and protecting the roll with extra wrappings. He said:

When a book is unrolled for reading, the extending paper at the beginning of the roll should not be opened in haste; if it is it will be creased, which in turn will cause tears. If a ribbon is wrapped around the extending paper of the roll, it is certain to become damaged. If a few sheets of paper are added before the ribbon is wrapped around the upper and lower parts of the roll, the roll will be kept tight and will suffer no damage.[c]

These are some of the external factors which influenced the permanence of paper. While the best fibres were used to ensure its durability, all the physical, biological, and atmospheric factors to enhance the longevity of paper were taken into consideration rather early in China, especially for artistic work and for graphic records.[d]

(d) USES OF PAPER AND PAPER PRODUCTS

PAPER has always been a cheap and convenient substitute for more expensive materials or more clumsy objects which have other uses, and it is sometimes suitable for uses in which other materials will not serve. Paper was apparently not invented for writing, but as time went on writing on paper developed into a special branch of art, and for both calligraphy and painting, paper turned out to be the best medium for artistic expression. The use of paper made further progress where

---

[a]   *Chhi Min Yao Shu* (*TSHCC*), p. 57; cf. van Gulik (9), p. 142.
[b]   *Chhi Min Yao Shu* (*TSHCC*), p. 58.
[c]   *Chhi Min Yao Shu* (*TSHCC*), p. 57; cf. van Gulik (9), p. 141.
[d]   See the discussion on the factors influencing the durability or permanence of paper in Browning (1), pp. 31–3.

it became available in fancy colours and delicate designs as stationery and for decorative purposes. Being cheap and light, it has been used as a medium of exchange in place of heavy valuable metals, or as a substitute material for personal furnishings, household articles, and recreational objects. Paper has also been chosen for the craft of making replicas or models of treasured objects for ceremonial and festive occasions. Today, paper and paper products have hundreds of uses in communication, business, industry, and household operations; they are found everywhere in daily life. Yet many of these uses can be traced back to centuries ago when paper was used as extensively and variously in China as it is elsewhere in the world today.

Generally speaking, paper was probably used for wrapping from the moment of its invention in the Western Han; for writing from the Later Han; for cutting into designs, making stationery, fans, and umbrellas from the third or fourth century onwards; for clothing, furnishings, visiting-cards, kites, lanterns, napkins, and toilet purposes no later than the fifth or sixth century; for family ceremonies in the seventh; for state sacrifices and making replicas of real objects from the eighth; and for playing cards and in lieu of metal as a medium of exchange from the ninth century. In other words, all these uses for the graphic and decorative arts, for commercial and ceremonial occasions, and for household and recreational purposes existed in China before paper was known to the West.

The progress of papermaking is reflected in the increasing varieties of and names for paper, which have many different origins. Some of the names denote the raw materials from which paper was made; others refer to places where it was manufactured; and still others are the names of designers or of the studios which the product has made famous. Papers are also named for methods of treatment, as sizing, coating, dyeing, or treating with spices; for their appearance or size; and for the use for which a variety is particularly made. The following pages will trace the origin and development of some special kinds of paper and paper products used for different purposes as recorded in literature or found in existing specimens or artifacts.

(1) PAPER FOR GRAPHIC ARTS AND STATIONERY

Paper was used very early as a substitute for bamboo and silk as writing material. No written characters, however, are found on the earliest paper specimens so far discovered, and no reference to the use of paper for writing was made in the Former Han period, though paper was certainly used for books and writing from the Later Han. It is recorded that paper was used together with brush and ink at the court during the reign of Hsüan Ti (+25–56),[a] and that a copy of the *Chhun Chhiu Tso Chuan*[1] on paper was given to students who studied the classic at the Han court in

[a] *Hou Han Shu* (*ESSS/TW*), ch. 36, p. 7*b*.

[1] 春秋左傳

+ 76.[a] Some two dozen characters are found on a remnant of paper from Chü-yen dated around + 110 (see Fig. 1061), and samples of the late + 2nd century recently discovered in Han-Than-Pho, Kansu, also have a few characters on them.[b] Numerous paper documents discovered in Chinese Turkestan bear dates from the + 3rd century onwards, while Hsün Hsü (+231−89), custodian of the imperial library of the Chin dynasty, wrote that the bamboo books discovered in the Wei tomb in +280 were copied on paper and kept in three separate collections.[c] The increasing use of paper for books is also reflected in the records in earlier historical bibliographies. From such evidence, it may be concluded that paper was adopted for writing from the + 1st century, but not extensively used for books and documents until the late + 2nd or + 3rd century.

The earliest extant example of a complete book on paper is probably the *Phi Yü Ching*[1] (Parable *sutra*) written in +256 on *liu ho chih*[2] (Fig. 1074), which is said to be a paper made of six different materials or in Liu-ho in northern Chiangsu.[d] This kind of paper may have continued in use in the Sung dynasty, for the noted artist Mi Fu (+ 1051−1107) remarked that the *liu ho* paper had been used since the Chin dynasty.[e] Other early book rolls extant today are generally written on papers of hemp, paper mulberry, ramie, or a mixture of these materials.[f] Some of them survive in excellent condition (Fig. 1075).

The most common papers used in the Thang dynasty were made of hemp, paper mulberry, and rattan, as is testified by both analytical studies of the paper specimens and literary records. Some sixty pieces of paper from Tunhuang and made between the + 5th and the + 10th century, were photomicrographically examined by Clapperton,[g] who says that the earliest papers are all thin, transparent, and almost without exception made from carefully prepared and well beaten materials, while the sheets themselves are even and free from 'pinholes' or thick or thin patches. They generally exhibit a high degree of skill on the part of the vatman or papermaker. In addition, most are well sized and can be written on with ease with modern ink and a steel pen. But after the middle of the + 8th century the quality of the paper rapidly deteriorated, becoming thick, flabby, uneven in texture, with poor resistance to ink. This change is generally attributed to the political and economic chaos during the later period of the Thang dynasty.

The papers used during the Thang for copying *sutras* and other books were of two kinds, both made chiefly from hemp. The white kind, called *pai ching chien*[3] (white *sutra* paper), is in small sheets but made heavy by loading or coating. The yellow,

---

[a] *Hou Han Shu*, ch. 108, p. 5a.   [b] See report of Phan Chi-Hsing (9), pp. 62−3.
[c] See Hsün Hsü's preface to the *Mu Thien Tzu Chuan* (*SPTK*), p. 3a.
[d] This is one of the earliest dated paper manuscripts now in the Calligraphy Museum in Tokyo; see Nakamura Fusetsu (1).
[e] See *Phing Chih Thieh* (*MSTS*).
[f] For composition of papers used for the Tunhuang rolls, see Giles (6), (7), Harders-Steinhäuser (1), and Phan Chi-Hsing (2), pp. 40−1.
[g] Clapperton (1), p. 18.

[1] 譬喻經   [2] 六合紙   [3] 白經牋

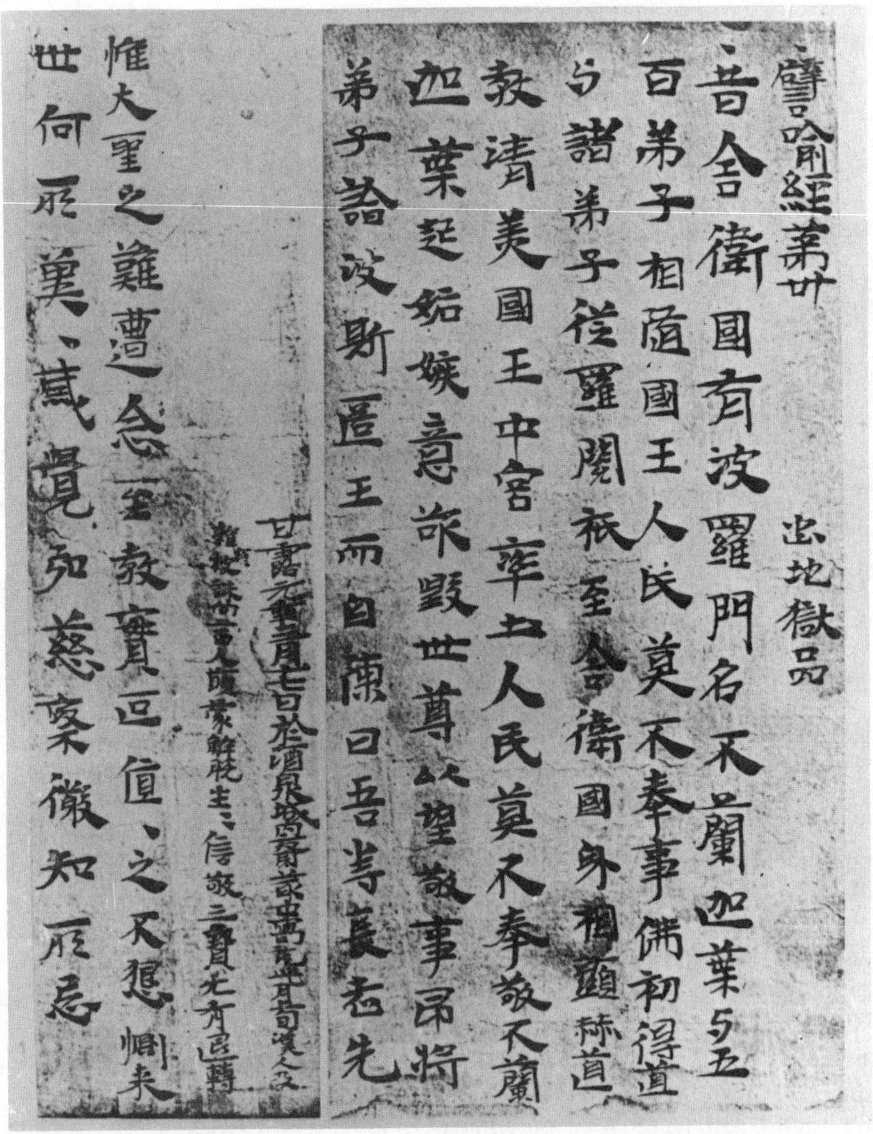

Fig. 1074. Oldest surviving book on paper, +256. Fragment of *Parable sutra* written in a transition form from the *li* to the *khai* style. Calligraphy Museum, Tokyo.

called *ying huang chih*[1] (stiff yellow paper), was treated with an insecticidal liquid which gave a glossy surface and a strong, close texture.[a] The thicker kind of paper was made in Szechuan and the thinner in Chhang-an, Loyang, and Anhui. The same sort of paper continued to be manufactured in the Sung dynasty, when *Chin-su*

[a] See *Kuang Chhuan Shu Pa* (*TSHCC*), ch. 6, p. 73; also discussion of the *ying huang* paper on pp. 78 ff. above.

[1] 硬黃紙

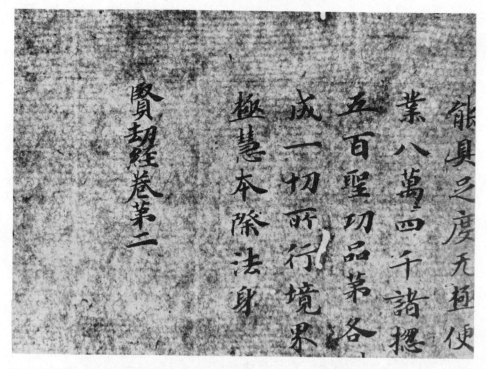

Fig. 1075. Old manuscript of the *Bhadrakalpa sutra*, transcribed in the Sui dynasty and preserved at Shōsō-in, Nara, showing waterlines of the paper using light from behind.

chien[1] (paper from the Chin-su Mountain) was specially made for copying the *Tripitaka* in the Kuang-hui Temple[2] at the foot of Chin-su Mountain, Hai-yen, on the coast of Chekiang province. This paper, made in Soochow in the latter part of the +11th century, is strong and sturdy, with no watermarks but a surface made very smooth and glossy by waxing both sides.[a] Each sheet bears a red seal with the inscription *Chin-su-shan tsang ching chih*[3] and some fifteen sheets of paper were joined into a roll,[b] the *Tripitaka* itself containing a total of over ten thousand rolls. This paper has become a collector's item and has been imitated throughout the centuries. The imitation, known as *tsang ching chien*,[4] is still being used most commonly by calligraphers or for title labels of books and scrolls.[c]

A similar well-treated, golden yellow paper, known as *chiao chih*[5] (pepper paper), was made in Chien-yang, Fukien, in the Southern Sung dynasty. It was dyed in a liquid prepared from the seeds of the pepper tree (*Zanthoxylan piperitum*), which

[a] The *Chin-su* paper is said to have been made of silk fibres, but a recent analysis indicates that it contains mulberry and hemp but no silk; see Phan Chi-Hsing (9), p. 97.

[b] See *Chin-Su Chien Shuo* by Chang Yen-Chhang (1738–1814), a native of nearby Hai-yen, who was among many collectors of this paper. The book gives well-documented sources and reproductions of the seals impressed on the front and back of the paper.

[c] A sample of a modern imitation is given in van Gulik (9), Appendix v, no. 19.

[1] 金粟牋   [2] 廣惠寺   [3] 金粟山藏經紙   [4] 藏經牋   [5] 椒紙

contains an insecticidal substance with a spicy flavour. This paper is strong, with a scent which is said to last several hundred years, and books printed on it survive today. Other papers popular for book printing in the Sung dynasty included *pei chhao chih*,[1] a lustrous white kind made probably from the creeping plant *pei hsieh*[2] (*Dioscorea quinqueloba*) in Fu-chou, Chiangsi; *Phu-chhi chih*,[3] a medium heavy paper made in Phu-chhi, Hupeh; *Kuang-tu chih*,[4] a product of the paper mulberry from Kuang-tu, Szechuan; *Yu-chhuan chih*,[5] a rattan paper made in Yu-chhuan village of Hangchow, Chekiang; and *Chi-lin chih*,[6] a very smooth and heavy paper made in Korea.[a]

From the Sung dynasty on, printing became popular and papermaking made further progress. Publications by the National Academy and various local government agencies, monasteries, private families, and trade publishers greatly increased in such places as Khaifeng, Chhengtu, Hsüan-chheng, Hangchow, and Chien-yang, which were also known as centres for papermaking. A close examination of the extant Sung, Yüan, Ming, and Chhing editions printed on various kinds of paper shows the high quality of the paper used for printing, which was, generally speaking, thin, soft, light, and fine. The raw materials were mostly bamboo and paper mulberry, with sometimes a mixture of rice stalks and other substances. In the Yüan and Ming dynasties, a kind of extra wide paper was made of bamboo for writing, and was known as *ta ssu lien*[7] (large fourfold), whilst in some districts, fine bamboo was made into an especially heavy and sturdy paper called *kung tu chih*[8] (official document paper); it was kept primarily for official documents. Of the paper used for books during the Ming, the best is said to have been the white, sturdy *mien chih*[9] (cotton paper), which was made of paper mulberry from Yung-feng, Chiangsi; next was the soft, heavy *chien chih*[10] (stationery paper) from Chhang-shan, Chekiang; then came the less expensive *shu chih*[11] (book paper) from Shun-chhang, Fukien, and last was the *chu chih*[12] (bamboo paper) from Fukien, which was short, narrow, dark, and brittle, and lowest in quality and price.[b] The *Khai-hua chih*,[13] a paper of extraordinary quality made in Khai-hua, Chekiang, was especially selected by the Chhing court for the printing at the Wu Ying Tien of the *tien pen*,[14] the Palace editions.

Most popular among the many kinds of paper for artistic uses, especially painting and calligraphy, has been and still is the *Hsüan chih*,[15] a fine, white soft paper made in Hsüan-chou (modern Hsüan-chheng, Anhui). This paper was first mentioned in the Thang documents as an article of tribute from Hsüan-chou; since then it has been continuously used and praised by artists. Not all papers made in the Hsüan-chou area, however, were of high grade and suitable for artistic use.

---

[a] For various titles of books printed on the different kinds of paper discussed here, see Yeh Te-Hui (2), pp. 163–6.
[b] See *Shao Shih Shan Fang Pi Tshung* (Peking, 1964), p. 57.

[1] 草鈔紙	[2] 草薢	[3] 蒲圻紙	[4] 廣都紙	[5] 盂拳紙
[6] 鷄林紙	[7] 大四連	[8] 公牘紙	[9] 棉紙	[10] 柬紙
[11] 書紙	[12] 竹紙	[13] 開化紙	[14] 殿本	[15] 宣紙

Many other kinds were made of bamboo or straw, and used for wrapping, burning, or such handicrafts as making umbrellas; only those of pure bark or bark and straw are suitable for calligraphy and painting. These high-grade papers include *yü pan*[1] (jade tablet), a very large sheet of white, heavy paper; *hua hsin*[2] (picture heart), one of the artists' treasures; and *lo wen*[3] (silk stripes) which apparently was made with textile patterns. Unlike writing or printing, works of art sometimes require very large sheets of paper, and the *Hsüan chih* is especially noted for its extraordinary size, normally twelve by eight feet, with one to three layers in one sheet. Some of the sheets, known as *phi chih*,[4] were as long as fifty feet.[a] A sheet of one layer is made by lifting the screen once from the vat of pulp, of two layers by lifting twice, and of three layers by lifting three times, by two or more vatmen operating the large screen. These sheets are soft, absorbent, smooth, strong, and elastic, suitable for books, documents, stationery, rubbings, and especially for calligraphy and painting.

Another time-honoured paper for artists is the *Chheng-Hsin-Thang chih*[5] (paper from the Pure Heart Hall), especially made in Anhui for the royal poet Li Yü[6] (+937–78), the last ruler of the Southern Thang dynasty. The name of this paper is apparently derived from that of his royal studio. Its raw material was paper mulberry, but its fibres were made extremely pure and fine through additional processing. Its surface was polished and waxed after manufacture to increase its quality and beauty, and was described as fine, thin, glossy, smooth, and absorbent; it was considered the best kind of paper at that time. The Ming connoisseur Chang Ying-Wen[7] (16th century) said that 'the paper is thin like the membrane of an egg, tough and clean as jade, and covered with a fine and brilliant coat'.[b] Another Ming collector, Thu Lung[8] (1542–1605) claimed that the paper was so extremely good that many famous calligraphers and painters of the Sung dynasty used it,[c] and Ouyang Hsiu[9] (+1007–72) is supposed to have used it for his draft of the *History of the Five Dynasties*. Apparently the formula for making this paper was still used in the Sung and later times, even though the royal studio was no longer extant by then.

Paper was certainly the most popular medium for artistic use by the Chinese as well as peoples in other nations in East Asia from very early times. They adopted it not only for painting and calligraphy,[d] but also for making rubbings from inscriptions and engraved designs and for many different kinds of decorative arts. One example was its use for reproduction of painting and calligraphy by weaving very fine paper strips in colour or in black and white into a sheet as if it were made in fabric with other fibres. A set of forty-six pictures of the *Keng Chih Thu*,[10] showing the processes of tilling and weaving, accompanied by handwritten poems of the

---

[a] The making of the extra-large paper sheets is described in *Hsin-an Hsien Chih* (1888 ed.), ch. 10, p. 17a; *Hsüan chih* is produced in many other places today.
[b] *Chhing Pi Tshang* (*MSTS*), p. 216.   [c] *Chih Mo Pi Yen Chien* (*MSTS*), p. 136.
[d] See discussion on pp. 361 ff. below.

[1] 玉版   [2] 畫心   [3] 羅紋   [4] 匹紙   [5] 澄心堂紙
[6] 李煜   [7] 張應文   [8] 屠隆   [9] 歐陽修   [10] 耕織圖

Fig. 1076. Art work made by interwining black and white paper strips. Above is one of 46 pictures on tilling and weaving, *Keng Chih Thu*, with poems by Emperor Khanghsi. From Strehlneek (1).

Emperor Khang-Hsi, was manufactured in this fashion. All the pictures were made by the extremely skilful and delicate intertwining of long and thin strips of paper with no help from brush or ink[a] (Fig. 1076).

For many centuries in China, paper has been especially designed with a variety of patterns, plain or coloured, for writing letters, poems, and commercial docu-

[a] This set of pictures, measuring 9 3/4 in. × 11 3/8 in., was acquired by a collector in Stockholm; see E. A. Strehlneek (1), pp. 238–57.

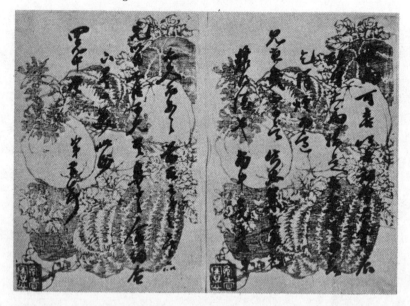

Fig. 1077. Ornamental stationery in multi-colour paper for letter writing. Handwriting of the Chhing scholar Phan Tsu-Ying (+1830–90) addressed to Li Hung-Tsao (+1820–91), imperial tutor of Emperor Thung-Chih. Far Eastern Library, University of Chicago.

ments. These papers were dyed a single colour, printed with multicoloured paintings (Fig. 1077), embossed in patterns, or sprinkled with gold or silver dust, in order to make them as elegant and pleasing as possible. The making of such papers was an art in itself, and many artists and poets contributed to the advancement of techniques for such ornamentation. For example, violet notepaper was used by the famous calligrapher Wang Hsi-Chih[1] (+321–79), and notepapers decorated with a peach blossom pattern in bright green, blue, and red were designed by Huan Hsüan[2] (d. +404) in Szechuan,[a] a centre especially distinguished over many centuries for its decorative notepapers.

According to early records, stationery papers in ten different colours were made by a Thang official, Hsieh Shih-Hou[3] of Szechuan.[b] There were also fancy varieties of paper called by such names as pine flower, golden sand, bright sand, rosy clouds, golden powder, dragon and phoenix (red with gold), peach-pink with golden spots. However, the most famous notepaper throughout the centuries was probably the small reddish note sheets designed by Hsüeh Thao[4] (+768–831), a courtesan well versed in poetry who exchanged poems on notepaper of her own design (Fig. 1078) with such well-known poets as Yüan Chen[5] (+779–813) and Po Chü-I[6]

[a] *Wen Fang Ssu Phu* (*TSHCC*), ch. 4, p. 49.
[b] *Chien Chih Phu* (*TSHCC*), p. 2.

[1] 王羲之   [2] 桓玄   [3] 謝師厚   [4] 薛濤   [5] 元稹
[6] 白居易

吳友如百美畫譜　製箋載詩

三十九

下集

Fig. 1078. Hsüeh Thao, a Thang courtesan, designed her own note paper for writing poetry. Drawing by Wu Yu-Ju in *Pai Mei Hua Phu*, printed in lithography in Shanghai, 1926.

(+772–846). This paper was made of hibiscus skin mixed with powdered hibiscus flower petals to increase its lustre.[a] It continued to be made, bearing the famous name 'Hsüeh Thao notepaper,' for many centuries, not only in Szechuan but all over the country.

[a] *Thien Kung Khai Wu* (*KHCP*), p. 219; cf. tr. Sun & Sun (1), p. 231.

The earliest notepaper with multi-coloured pictures was probably developed before the Sung dynasty. The family of Yao Chhi[1] (fl. +940) made lustrous letter papers in beautiful coloured designs of mountains, rivers, forests, trees, flowers, fruits, lions, phoenixes, insects, fishes, 'old father', 'the Eight Immortals', and ancient seal characters.[a] A poem written about this time describes the designs of a landscape painting of wild geese, reeds, and a setting sun on letter paper.[b] Apparently, paper with embossed designs, with watermarks, and even a marbled paper were also developed at this time or earlier. Su I-Chien (+957–95) said that the people of Szechuan made stationery papers with decorations by pressing them on wooden blocks to make such designs as flowers, trees, unicorns, and phoenixes in numerous styles. He also described a 'fish-eggs notepaper' (*yü-tzu chien*[2]), made with starch on a piece of closely-woven cloth from which a hidden design like fish eggs resulted,[c] and in fact many specimens of paper with hidden or translucent designs survive from the +10th century, including one example used by the noted calligrapher Li Chien-Chung[3] (+945–1108). These translucent designs are considered forerunners of the watermark.[d] Another kind was 'drifting sand notepaper' (*liu sha chien*[4]), the design of which was printed from a flour paste sprinkled with various colours, over which the paper was placed to become stained; thus the design was free and irregular. Sometimes, paste was prepared from honey locust pods (*Gleditschia sinensis*) mixed with croton oil and water, with black and coloured inks on its surface. Colours were scattered when ginger was added and gathered if dandruff was applied with a hair brush. The various designs which looked like human figures, clouds, or flying birds were transferred from the surface of the liquid to the paper, and in this way a marbled paper was made.[e] Western authorities have set the origin of watermarks in +1282 in Europe, and of marbled paper in 1550 as 'a Persian invention',[f] but the literary record as well as existing specimens show that the Chinese made such papers at least three to five hundred years earlier.

A close relationship between the manufacture of writing paper and the art of colour prints developed in the late Ming dynasty. The most famous example is the manual of ornamental letter papers designed at the Ten Bamboo Studio, known as the *Shih-Chu-Chai Chien Phu*[5], published by Hu Cheng-Yen[6] about 1645.[g] The manual included various designs printed in multicolour or by embossing without colour from wood blocks. The earliest extant designs can be seen in the *Yin Shih*

---

[a] *Chhing I Lu* (*HYHTS*), ch. 2, p. 34*b*.
[b] Li Shao-Yen (1), pp. 135–7.
[c] *Wen Fang Ssu Phu* (*TSHCC*), p. 53.
· [d] See examples at the Palace Museum, Peking examined by Phan Chi-Hsing (7), pp. 38–9; also Shih Tao-Kang (1), pp. 51 ff.
[e] *Wen Fang Ssu Phu*, p. 53.
[f] Cf. Labarre (1), p. 260; Hunter (9), pp. 474, 479.
[g] The preface of the manual is dated +1644, but one orchid design, in *chuan* 2, bears the date +1645. The work was reprinted in facsimile by the Peking Society of Woodcuts in 1935; see also below, pp. 283 ff.

[1] 姚顗   [2] 魚子牋   [3] 李建中   [4] 流沙牋   [5] 十竹齋箋譜
[6] 胡正言

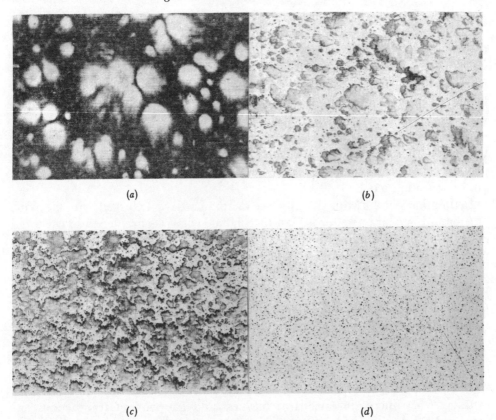

(a)   (b)

(c)   (d)

Fig. 1079. Chinese marbled papers in various colours and designs, +19th century. (a) 'Tiger skin' in yellow.
(b) 'Betel-nut' in green. (c) 'Betel-nut' in pink. (d) sparkling gold.

Chien Phu[1] and the *Lo Hsüan Pien Ku Chien Phu*,[2] published in the early 17th century.[a] They also include similar but simpler ones which make it apparent that both the techniques and the artistic schemes of Hu Cheng-yen were influenced by similar works which had appeared earlier.

Besides letters, ornamental stationery was also used for commercial papers, in which case it is commonly known as *chien thieh*.[3] These specially designed papers were created by block-printers of Hui-chou (in Anhui) during the late Ming and early Chhing dynasties. Contracts were probably first written on ordinary note paper, but later a special kind of stationery was manufactured with designs more in folk style instead of the landscapes of archaic paintings preferred by the literati.[b]

[a] See Shen Chih-Yü (2), pp. 7–10; the second volume of this manual was reproduced in facsimile in Tokyo in 1923. This art of making ornamental stationery has been handed down to the present day; a great deal of such stationery is still in use and several excellent manuals of this type have been printed by Jung Pao Chai, Peking.

[b] See Lai Shao-Chhi (1), which reproduces forty colour designs of contract papers, including one written in 1640.

[1] 殷氏箋譜   [2] 蘿軒變古箋譜   [3] 簡帖

A piece of paper used to present the visitor's name when calling was known as *ming tzhu*[1] (name card) or *pai thieh*[2] (visiting-card). This custom was derived from the use in the Han dynasty of a strip of wood on which the personal name was written. The wood was replaced by paper probably in the +5th or 6th century, when such information as the visitor's name, native town, and official title was included. The card, about two to three inches wide, was white, but was replaced by a red one in the Thang, when the price of red paper is said to have increased over ten times. Also during this time, a custom was introduced of writing a note about the visitor's business on the card when calling on high officials in their offices or private residences. An endorsement was made on the back of the card before the visitor was admitted.[a]

In the middle of the Ming dynasty, the visiting-card was enlarged, with the name written in big characters for a Han-lin scholar, who was privileged to use a red card, while the common people used white. Sung Ying-Hsing says: 'The highest grade of stationery paper made in Chhien-shan (Chiangsi), known as *kuan chien*[3] (official stationery), was used for calling cards by high officials and members of the wealthy class. It is thick and smooth without any fibrous ribs on the surface. When used on ceremonial occasions, it was dyed red with the red flower [*Hibiscus rosa-sinensis*] after being treated with alum.'[b]

(2) PAPER AS A MEDIUM OF EXCHANGE

The use of paper to represent money originated probably in the early +9th century, when increasing needs of business and government transactions encouraged the institution of 'flying money' (*fei chhien*[4]) as a convenient way to obviate carrying heavy metal coins from one place to another.[c] Provincial merchants who sold their commodities in the capital could deposit their proceeds at an office in Chhang-an and receive a certificate for cash in the designated provinces. This institution was originally a private arrangement by the merchants but was taken over by the government in +812 as a method of forwarding local taxes and revenues to the capital. Since the 'flying money' was primarily a draft, it is generally considered a credit medium rather than a true money.[d]

The system continued in the following dynasties and gradually evolved into a true paper currency. The inconvenience of clumsy iron coins, which weighed twenty-five catties per thousand led, during the period of the Five Dynasties and early Sung, to people depositing cash in deposit houses and using their receipts for financial transactions in the Szechuan area. In the early +11th century, sixteen private houses were authorised by the government to issue notes called 'exchange

---

[a] See *Kai Yü Tshung Khao* (1750 ed.), ch. 30, pp. 24*a–b*.

[b] See *Thien Kung Khai Wu* (*KHCP*), pp. 218–19; cf. tr. Sun & Sun (1), p. 230.

[c] Sources are silent about the material for the 'flying money', but it is believed that it must have been made of paper since imitation money of paper for offering to spirits was already used in the +8th century, and the word *fei*, 'flying', implies a light material such as paper.

[d] Yang Lien-Sheng (3), pp. 51–2; Sogabe Shizuo (1), pp. 6–7; Pheng Hsin-Wei (1), p. 280.

[1] 名刺   [2] 拜帖   [3] 官柬   [4] 飛錢

Fig. 1080. Earliest extant block for printing paper money. Print from block for 'Exchange Media' of the Northern Sung, *c.* +1024–1108. From Pheng Hsin-Wei (*1*).

media' (*chiao tzu*[1]) (Fig. 1080). This also began as a private arrangement, but in +1023 the government established an official agency in I-chou (modern Chhengtu) to issue such notes in various denominations. A cash reserve was established and a period of three years, as well as a ceiling, was set for circulation. In +1107, a new note called a 'money voucher' (*chhien yin*[2]) was issued and printed with six blocks of elaborate designs in blue colours.[a] By the end of the Northern Sung period, notes worth a total of about seventy million strings[b] had been issued.

The paper money of the Southern Sung period had various names and circulated in a wider area of the country. Although the *chiao tzu* and *chhien yin* were used for a while, the most popular of the notes was the 'check media' (*hui tzu*[3]). It

---

[a] The earliest known copper plate for printing the paper note of *c.* +1024 to +1108 survives; cf. Pheng Hsin-Wei (*1*), p. 280.

[b] One string equalled 1000 pieces of cash.

[1] 交子   [2] 錢引   [3] 會子

also originated as a private enterprise in the capital, Lin-an (modern Hangchow), but was taken over by the Board of Revenue in +1160, with a similar period of circulation and a fixed ceiling for the amount to be issued for each period. The area of circulation was extended from Szechuan to provinces along the sea coast and the lower Yangtze, as well as in the Huai River valley. As had happened before, increased government expenditure towards the end of the Southern Sung caused frightening inflation because of the unlimited issue of paper notes beyond the original quota and period of circulation.[a]

Besides the circulation of paper money during the two Sung periods, other credit media made of paper were also used. There was a kind of 'exchange certificate' called *chiao yin*[1] issued by a special government bureau for salt, tea, and certain other commodities; it was handled by the appropriate merchants, could be exchanged for cash, and was transferable and redeemable in the provinces where the commodities were produced. In the Southern Sung, certificates issued by the Chiao Yin Khu[2] were made of special paper and were printed in the Treasury, which was located in the capital, Lin-an, and supervised by an assistant of the bureau.[b]

The paper notes issued from +1167 to 1179 were described as printed in colour on specially made paper with very elaborate patterns. There were characters to indicate the instalment number of the issue, the year issued, the time limit for circulation, and the ceiling for the amount to be issued during the period. Patterned seals were stamped in blue, red, and black on both sides of the paper.[c] Paper for printing the notes was at first acquired from private paper mills, but as the need grew and counterfeiting increased, the government established its own factories in Hui-chou and Chhengtu for manufacturing special paper for the notes. The material used was paper-mulberry bark, and the paper money was originally called *chhu pi*[3] or *chhu chhao*[4] (paper-mulberry money). Silk or other fibres and other ingredients were probably mixed in to make imitation difficult. The Chhengtu factory, established from +1068, was reported to have employed sixty-one paper-makers and thirty-one other workers in 1194,[d] but because the shipment of Szechuan paper to Hangchow was inconvenient, a government factory was established in +1168 incorporating one already existing at An-Chhi,[5] near Hangchow, where some 1200 workers were employed in 1175. The printing was done at the Treasury, called the Hui Tzu Khu,[6] where 204 daily workers were employed,[e] and besides wood-blocks, copper plates are known to have been used.[f]

The complexity of designs with additional signatures and seals printed or

[a] For the circulation of *hui tzu*, see Yang Lien-Sheng (3), pp. 55–7; Sogabe Shizuo (1), pp. 37–55; Pheng Hsin-Wei (1), pl. 43.

[b] See *Hsien-Shun Lin-An Chih* (1830 ed.), ch. 9, p. 8a; *Meng Liang Lu* (*TSHCC*), p. 77.

[c] Ten samples given in a Yüan work on money are illustrated in *Shu Chung Kuang Chi* (*SKCS*), ch. 67, pp. 18a–23b.

[d] *Shu Chung Kuang Chi*, ch. 67, p. 14b.

[e] *Hsien-Shun Lin-An Chih* (1830 ed.), ch. 9, pp. 7b–8a; *Meng Liang Lu* (*TSHCC*), p. 77.

[f] See *Wen Hsien Thung Khao* (*ST*), ch. 100, p. 3.

[1] 交引   [2] 交引庫   [3] 楮幣   [4] 楮鈔   [5] 安溪
[6] 會子庫

stamped in colours on specially made paper, plus heavy penalties for counterfeit-ing, must all have tended to discourage such a crime, yet cases often occurred. One in 1183, memorialised by Chu Hsi[1] (+1130–1200), involved a professional wood-block cutter, Chiang Hui,[2] who had repeatedly counterfeited paper money. He was quoted as saying that he cut a block of pear wood from a traced master copy of the *hui tzu* note for one string of cash. The imitation note, including a picture of a legendary figure, was printed with serial character and number in blue and seals in red on special paper made in the countryside of Wu-chou (in Chekiang). It took him ten days to complete the cutting. In a six-month period in 1183, some 2600 sheets were printed on about twenty occasions, 100 to 200 sheets at a time.[a]

In the north, the Chin Tartars also used paper money called 'exchange notes' (*chiao chhao*[3]), first issued in 1153. The idea must have been borrowed from the Sung. They had large and small bills in various denominations, and spoiled notes could be exchanged for new ones with a charge for the printing cost. At first, the circulation rules were carefully observed, but towards the end of the 12th century and early in the 13th, excessive military expenditure caused inevitable inflation, and the value of the depreciated notes dropped to as little as one per cent of their original value.

After the Mongol conquest of China, the Yüan dynasty issued several kinds of paper money. It was the 'silk note' (*ssu chhao*[4]) issued from 1260 which was backed by silk yarns as reserve, and later the notes of the Chung-thung era (*Chung-thung chhao*[5]), which unified the currency system of China. Two specimens of this paper money were discovered in Shanyang, Shensi in 1965[b]. Old notes issued earlier were exchanged for this new note, which not only circulated universally within the empire but also spread to other parts of the world. It reached the Uighur regions in 1280, Persia in 1294, and was introduced to many other nations in the following centuries. Paper currency arrived at Korea in 1296 and was used for circulation there in 1332. The Japanese first issued the *dō chō*[6] (paper of copper coins) in 1334, the Vietnamese printed paper money in 1396, but the use of bank notes was not begun in Western countries until the later part of the 17th century.[c] It is probable that certain European systems of banking and accounting, as well as vouchers for deposited money, were also influenced by Chinese examples obtained by merchants and travellers to China.[d]

[a] See Chu Hsi's memorials in *Chu Wen Kung Wen Chi* (*SPTK*), ch. 18, pp. 17a–32a; ch. 19, pp. 1a–27a; tr. Yang Lien-Sheng (9), pp. 216–24.

[b] Both are printed with Chinese characters of the Chih-Yüan reign (+1264–94), denomination, and issuing agency with seal impressions in Mongolian in red colour; see a report of the Shanyang Museum in *KKWW*, 1980 (no. 3), pp. 70 ff.

[c] Paper money was first issued in Sweden in 1661; America, 1690; France, 1720; Russia, 1768; England, 1797; and Germany, 1806.

[d] Max Weber said that the accounting system (Verrechungswesen) of the old Hamburg Bank was based on a Chinese model, and Robert Eisler said that the old Swedish system of banking and money deposit vouchers followed the Chinese system; see Yang Lien-Sheng (3), p. 65.

[1] 朱熹   [2] 蔣輝   [3] 交鈔   [4] 絲鈔   [5] 中統鈔
[6] 銅楮

Paper currency was a subject of great interest described by many early European writers, who were impressed by its ingenuity as a substitute for heavy and valuable media of exchange. The most detailed observation was made by Marco Polo: 'The Khan causes every year to be made such a vast quantity of this money, which costs him nothing, that it must equal in amount all the treasure in the world.' He further remarked:

All these pieces of paper are issued with as much solemnity and authority as if they were of pure gold or silver; and on every piece a variety of officials, whose duty it is, have to write their names, and to put their seals. And when all is duly prepared, the chief officer deputed by the Khan smears the Seal entrusted to him with vermilion, and impresses it on the paper, so that the form of the Seal remains printed upon it in red; the Money is then authentic. Anyone forging it would be punished with death.[a]

The Mongols certainly used paper money most effectively and circulated it on a vast scale in a broad area, but its name and issuance changed frequently within a short period. The Ming government, on the other hand, had a less effective operation but issued only one kind of note during the entire dynasty. In 1375, a new note called 'Precious note of Great Ming' (*Ta Ming pao chhao*[1]) was issued, with the Hung-wu reign title printed on the note without further change. From the very beginning, the Ming note was inconvertible; but copper coins circulated along with the paper money. Throughout some 200 years this note was the only paper money in circulation, but as it gradually lost value, silver became the major medium of exchange, and the circulation of paper money was almost suspended after the end of the 15th century. Along with paper notes, exchange certificates for tea, salt, and other commodities were also issued during the Ming. Paper needed for such certificates was requisitioned from various producing provinces.[b]

The revival of paper currency was attempted at the end of Ming, but it failed, apparently because of inflation. However, a memorial by a Ming official in +1643 enumerated its many advantages. It could be manufactured at low cost, circulated widely, carried with ease, and kept in concealment; it was not liable to suffer impurity like silver, did not need weighing whenever it was used in transactions, it could not be clipped, was not exposed to thieves' rapacity and, finally, saved metals for other uses.[c] It appears, however, that the levy of some two million catties of mulberry bark for manufacture of paper money in +1644 almost provoked the peasants into rebellion.[d]

The Manchu rulers preferred to use hard money and did not issue any paper currency on a large scale, except as an emergency measure. However, printed paper documents for commercial transactions were frequent (Fig. 1081), and in

[a] Tr. Yule (1), vol. 1, p. 424.
[b] See *Ta Ming Hui Tien* (1589 ed.), ch. 195, pp. 4b–5b.
[c] See *Ming Chi Pei Lüeh* (*KHCP*), ch. 19, pp. 15–16.
[d] See *Ni Wen-Cheng Kung Nien Phu* (*TSHCC*), p. 60; *Jih Chih Lu* (*TSHCC*), ch. 4, p. 103.

[1] 大明寶鈔

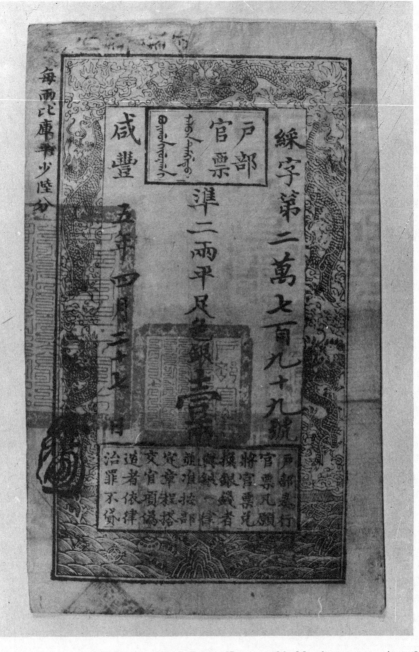

Fig. 1081. Paper note for one tael of silver issued by the Board of Revenue of the Manchu government in +1855.
Far Eastern Library, University of Chicago.

+1853, the military cost of suppressing the Thai-phing rebellion resulted in the issue by the Chhing government of paper notes called 'Official Note' (*kuan chhao*[1]) and 'Precious Note' (*pao chhao*[2]).[a] Since these were not convertible, their value dropped rapidly and the notes ceased to be used after a short period. It was not until the later part of the 19th century that a Chinese bank issued a new bank note, which was inspired primarily by Western influence.

(3) CEREMONIAL USES OF PAPER

Paper has played a significant part in many Chinese ceremonies and festivities in connection with ancestor worship, folk religion and, to some extent, the cult of scholarship. Ordinary or specially made papers were cut, folded, or decorated to represent various objects to be used or to be burned on such occasions as family ceremonies and state sacrifices. This symbolic use of paper served as an economical substitute for real but expensive objects. The objects most commonly substituted for were money, garments, utensils, vehicles, servants, livestock and buildings; they were used at funerals, festivals, and in ancestor worship. Effigies of paper were made and burned as a symbol of offerings to the spirits in the other world.[b]

The original ceremonial use of paper was probably in substitution for metallic coins at a burial. In ancient times, rich deposits of treasures as well as human and animal sacrifices were buried with the dead, though by the time of the Han dynasty, metal coins were placed in tombs as a substitute for the valuable treasures and living beings. Later, for economic or other reasons, among them the discouragement of grave robberies, paper imitations for money and real objects were used.

The paper money for the spirits consisted of imitations either of metal coins or of real paper money, but the latter had different sets of inscriptions and patterns to distinguish it from counterfeit money. Coins were usually imitated by a sheet of plain paper with designs of coins cut into it, or a small sheet of paper coated with tinfoil, folded in the form of silver or gold ingots. This was sometimes dyed yellow with a liquid from seaweed, or from the flower of the pagoda tree (*sophora japonica*).[c] The plain tin symbolised silver and the yellow represented gold. This is similar to what is described in a +7th-century Buddhist work, which says: 'When sacrifice is offered by people, the ghosts will get silver coins if we cut the coins from white paper, and gold coins if we use yellow paper.'[d] The custom of burning paper money seems to have begun with imitations of metal money; only at a later date, when real paper money was in circulation, was mock paper money used with imitation coins in making offerings (Fig. 1082). Nevertheless, offerings of paper money for the

[a] Cf. Pheng Hsin-Wei (*1*), pp. 557–8, pl. 80.
[b] For ceremonial use of paper, see Hunter (2), pp. 1–79; (9), pp. 203–17.
[c] Hunter (2), p. 24. This work includes illustrations of various kinds of joss paper and bags to contain such folded tinfoil, with messages to the spirits on the bags.
[d] See *Fa Yüan Chu Lin* (*SPTK*), ch. 48, pp. 8*b*–19*a*.

[1] 官鈔   [2] 寶鈔

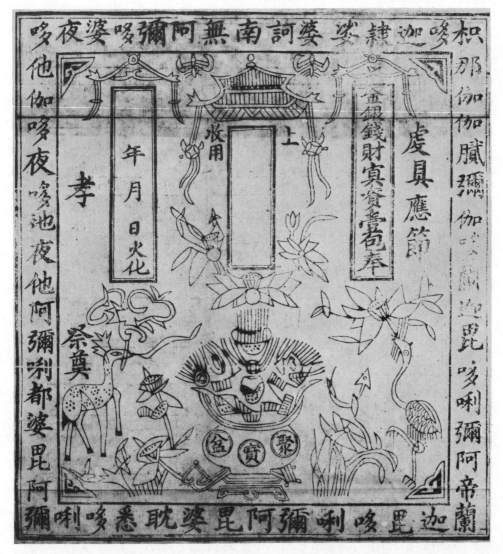

Fig. 1082. Printed messages on paper bag containing ghost money for one's ancestors with prayers on four sides, address above, and a treasure pot at the lower centre. Dard Hunter Paper Museum.

spirits existed, apparently, before real paper money was adopted in the early +9th century. An artifact of paper cut into a continuous string of cash has been found in an early Thang tomb dated +667 in Sinkiang (see Fig. 1086f),[a] while the Thang scholar-official Feng Yen[1] (+726–90?) said: 'In the past silk was buried and now the paper money is burnt. This shows people do not understand what the spirits really need.'[b]

[a] See Stein (2), IV, pl. XCIII.   [b] Cf. *Feng Shih Wen Chien Chi* (Peking, 1958), p. 55.

[1] 封演

The burning of paper money was formally introduced to the imperial sacrifice in +738 by Wang Yü, who served as Commissioner of Imperial Sacrifices and Associate Censor, and was in charge of sacrifice at the ancestral temple.[a] The adoption of this practice for state sacrifices was a subject of controversy among many officials and scholars at the time and thereafter. Some of them condemned it as absurd and others were in favour of the use of paper money as a substitute for actual silver and copper coins. This not only made the tombs less attractive to grave robbers, but also kept the actual money in circulation.

In the Sung dynasty, a minister named Liao Yung-Chung[2] (c. 1101–25) memorialised the emperor to abolish the burning of paper money. He considered such vulgarised tradition an absurd delusion and an insult to the spirits.[b] And in commenting on the family sacrifice, the noted philosopher Chu Hsi (1130–1200) said that the *Rituals of the T'ang* recorded that certain officials did not originally burn paper money as offerings to their ancestors, and this practice was followed by the gentry (*i kuan*[3], literally, clothes and caps). At the beginning of our dynasty, it was said, those who studied the rituals misunderstood the passage and used paper clothes and caps instead of paper money in the sacrifice.[c] Whether the burning of paper objects other than money was the result of misreading the ritual is not certain, but Chu Hsi's statement here seems to tell us that the burning of paper money was practised in Thang, and of paper replicas of other objects was introduced early in the Sung dynasty. Since paper clothing and paper caps are known to have been worn by men at this time,[d] it would have been natural to offer them to the spirits in lieu of silk or other textile materials (Fig. 1083). Even though the intention of the offerings was questioned by many scholars, the use of paper replicas for funeral objects was still customary in sacrifice for many centuries after that.

Meng Yüan-Lao[5] (fl. 1126–47) recalled that paper money and paper objects were offered to spirits during various festivals in the Northern Sung capital, Kaifeng.[e] He said that in the spring festival, shops selling paper replicas used paper to make pavilions and buildings on the street. He also claimed that on the fifteenth day of the seventh month, during the All Souls Festival (*chung yüan*[6]), numerous paper offerings such as boots and shoes, head-dresses, hats, belts with decorations, and colourful garments, as well as Buddhist Maudgalyàyana *sutras* (*mu lien ching*[7]), were sold in the market. A bamboo tripod was made, about three to five feet high, with a basin on top. This was called *yü lan phen*[8];[f] in it paper clothes and paper money were burned as offerings to needy spirits. This was probably one of the

---

[a] Cf. *Chiu Thang Shu* (*ESSS/TW*), ch. 130, p. 1a; a similar story is found in *Hsin Thang Shu* (*ESSS/TW*), ch. 109, p. 13a, and *Tzu Chih Thung Chien* (1956 ed.), p. 6831.
[b] See *Chiu Jih Lu*[4] (*SF*), ch. 14, pp. 4a–b.   [c] *Chu Tzu Chhuan Shu* (*CHSC* ed.), ch. 39, pp. 16b–17b.
[d] For the use of paper clothes and paper caps, see discussion on pp. 109 ff. below.
[e] *Tung Ching Meng Hua Lu* (*TSHCC*), pp. 126, 161–2.
[f] *Yü lan phen* is apparently a transliteration from the Sanskrit *ullanbana* and refers to a Buddhist ceremony concerning Buddha and the hungry mother of his disciple Maudgalyāyana; cf. Bodde (12), pp. 61–2.

| [1] 王璵 | [2] 廖用中 | [3] 衣冠 | [4] 就日錄 | [5] 孟元老 |
| [6] 中元 | [7] 目連經 | [8] 盂蘭盆 | | |

Fig. 1083. Paper robe for offering to spirits at a funeral, +20th century. From Ecke (2).

reasons why many Confucian scholars objected to the burning of paper substitutes to their ancestors, because such offerings had some association with Buddhist ceremonies.

The burning of paper effigies in connection with cremation of the dead was witnessed by Marco Polo:[a]

They take representations of things cut out of cotton-paper, such as caparisoned horses, male and female servants, camels, armour, suits of cloth of gold (and money), in great quantities, and these things they put on the fire along with the corpse, so that they are all burnt with it. And they tell you that the dead man shall have all these slaves and animals of which the effigies are burnt, alive in flesh and blood, and the money in gold, at his disposal in the next world.

The paper used for spirits, generally known as *huo chih*[1] (burnt-offering paper), was made of bamboo, the fibres of which were cooked and strained with a solution of ashes and washed with water. The process was generally the same as in making other paper, except that they were not baked dry but dried in the sun. The Ming writer Sung Ying-Hsing says:[b]

[a] Yule (1), II, p. 191; the word 'cotton-paper' is used in Moule & Pelliot (1), I, p. 337 and elsewhere. Similar customs of burning paper money and paper replicas in Tangut as described by Marco Polo no doubt show Chinese influence; see Yule (1), I, pp. 204, 207, n. 4.
[b] *Thien Kung Khai Wu (KHCP)*, p. 218; cf. tr. Sun & Sun (1), p. 229.

[1] 火紙

During the high Thang period, sacrifices to ghosts and spirits were frequent and paper money was used to substitute for burning silk fabrics, thus the particular kind of paper called *huo-chih* was made. According to recent custom in Hupei and Hunan, as much as 1000 catties of this paper was burnt for spirits on one single occasion. Actually, about seventy per cent of this kind of paper that is produced is used for burnt offerings, and thirty per cent for daily use.

During the early part of this century, the manufacture of such ceremonial paper constituted a large portion of the handmade paper industry in China,[a] while the tradition of burning paper to communicate with the spirits is still practised in certain parts of China and the Chinese communities overseas.[b] Indeed, the manufacture of the paper replicas has become a special handicraft by which almost every kind of object can be exactly and finely imitated.

Paper printed or painted with colourful images of folk gods or national heroes has played a prominent part in many Chinese households and shops. The pictures of these gods which might be hung or pasted on walls or doors of a house were used primarily for worship or for protection from evil spirits. Included among them were images of the gods of the kitchen, doors, and gates, which were among the five household spirits to be worshipped.[c] The picture of the kitchen god was hung on a kitchen wall and was sacrificed to with confectionery and paper money on the twenty-third day of the twelfth month each year. After the sacrifice, the picture was burnt to send the god to heaven. Then, on the New Year's eve, he was invited back and a new picture was put up.[d]

The most common household pictures represented the gate gods. These were pasted on both sides of the double gate at the entrance of the house. The figures were supposed to represent two military generals of the Thang dynasty, Chhin Chhiung[1] and Yü-Chhih Ching-Te,[2] who wear armour and helmet and hold weapons (Fig. 1084).[e] Other figures chosen for human satisfaction, such as the god of longevity, the god of wealth, and sometimes three gods standing together for happiness, prosperity, and longevity, were also painted or printed on paper to be pasted on walls or hung in the house.

Many other gods or national heroes were worshipped in shops or handicraft factories in honour of their contribution to the profession. Thus, the drinking poet

---

[a] *Che-chiang Chih Chi Yeh*, p. 358, says that the handmade paper produced by some 25,000 paper mills in Chekiang was put to four major classes of use: superstition, writing, wrapping, and others; and of the total production of over twenty million Chinese dollars in 1930, about thirty per cent was used for burnt offerings or other religious purposes, the largest of the four categories.

[b] Hunter (9), pp. 207–11, describes what he saw in the 1930's: along certain streets in many cities in China as well as in Chinese communities elsewhere in Asia, open shops in which paper replicas were fashioned—highly ornate cardboard chests with shiny gold and silver paper locks, flowing robes of paper painted with golden dragons and complicated patterns, shoes, hats, all manner of wearing apparel made of paper, and also full-sized carts, houses, and even automobiles.

[c] The five tutelary gods of the house are those of the kitchen, gates, doors, centre of the room or impluvium, and well; see Bodde (12), p. 4.

[d] Bodde (12), p. 98.

[e] See Werner (1), pp. 172–4; Bodde (12), p. 100.

[1] 秦瓊   [2] 尉遲敬德

Fig. 1084. Door guards on paper printed in colour, depicting two Thang military generals, Chhin Chhiung and Yü-Chhih Ching-Te, with armour and weapons, +19th century. Field Museum of Natural History.

Li Po[1] became the saint of wine shops; the legendary butcher Chang Fei[2] was worshipped in meat shops; the hero of the Three Kingdoms, Kuan Yü,[3] the god of war who warded off calamities, was the most popular tutelary god in many houses.

The most interesting god for our discussion here is Tshai Lun,[4] the supposed inventor of paper, who has become the patron saint of the profession of papermaking and has been worshipped by papermakers and others since his own day. Legends related that a stone mortar used by Tshai Lun for papermaking still exists beside a pool near his home in Lei-Yang[5] (in modern Hunan), where numerous people made papermaking their profession. Temples to him were built in his home town as well as in Lung-Thing[6] (in modern Yang-hsien, Shensi), where he was buried and received the honorific title Marquis of Lung-Thing, as well as other papermaking centres as in Chhengtu.[a] His image painted or printed on paper was hung on the walls of many paper mills and paper shops in both China and Japan. A typical example is a block-print of Tshai Lun's image, dated to the 18th century[b] (Fig. 1085). The picture, printed in six colours (green, red, yellow, pink, mauve, black), shows the hero sitting in the centre. In his hand he holds a *ju-i*,[7] a sword-like

[a] *Chien Chih Phu* (*TSHCC*), p. 1.   [b] Reproduced in Tschichold (2).

[1] 李白   [2] 張飛   [3] 關羽   [4] 蔡倫   [5] 耒陽
[6] 龍亭   [7] 如意

Fig. 1085. Paper image of Tshai Lun as patron saint of papermaking, *c.* +18th century. Picture printed in five colours with inscription at top, reading 'Patron Saint Tshai Lun'. Tshai with a black beard holding a *ju-i* is surrounded by four attendants with brushes and writing tools before him as well as a pig and a chicken for sacrifice. Legend says these two animals were the first to separate the wet paper sheets with their snout or beak. From Tschichold (2).

emblem of good wishes. He is waited on by four attendants, two of them holding paper rolls and books, and two animals are sacrificed before him. The legend at the top says, 'Patron saint Tshai Lun, Marquis of Yü-Thing[1].'

While the worship of gods drawn on paper was primarily under Buddhist influence, the Taoists for their part multiplied their potent charms with messages of good luck, written or painted on paper smeared with cinnabar to invoke protection. Sometimes large charm seals were used to impress the message on clay and later on paper with red ink to indicate authority.[a] It seems that paper was also used by Taoists as a symbol of their magic power.

Confucians also paid respect to paper on which characters had been written. As Confucian scholars enjoyed high prestige in society, what they wrote represented the sacred words of sages, worthy of respect and preservation; thus every scrap of paper bearing written or printed characters was to be revered. The phrase *ching hsi tzu chih*[2] (revered spare paper bearing characters) became a pious motto in Chinese society, where written paper was supposed not to be trampled upon or put to any indecent use. We do not know how early this tradition developed, though it was undoubtedly suggested by Confucian scholars themselves to enhance their prestige, but one early reference was made in the +6th century. Then the noted scholar Yen Chih-Thui[3] (+530–91) wrote in his family instructions that 'paper on which there are quotations or commentaries from the *Five Classics* or the names of sages should not be used for toilet purposes.'[b] The same tradition was also held in Chinese society by Buddhists, who taught that rewards will be given to those who care for and respect paper with sacred messages on it.[c]

In order to dispose of the written characters reverently, brick furnaces were built at street corners or in courtyards of temples, where scraps of written paper could be collected and placed for burning. The ashes were kept in jars and finally deposited in a river.[d] Similar instructions were given on roads leading to sacred mountains. This practice may have been a way to avoid having litter lying about, but there is no doubt that it also had had a definite connection with Confucianism and the cult of scholarship.

(4) PAPER CLOTHING AND FURNISHING

Paper is mentioned in Chinese literature as having been used for various kinds of garments, bed furnishings, and other household articles in place of woven fabric, but whether these items were all made of true paper or bark cloth is uncertain.[e]

[a] See the discussion of Taoist charm seals in Carter (1), p. 13, nn. 13–14.
[b] *Yen Shih Chia Hsün (SPTK)*, ch. 5, p. 13*b*; tr. S. Y. Teng (3), p. 21.
[c] See *Hsiu The Yü Phien (TSHCC)*, p. 1*b*; *Kung Meng Pu Fei Chhien Kung Te Lu (TSHCC)*, p. 2.
[d] Cf. Hunter (9) pp. 78–9; for illustration of a +15th-century furnace for burning papers with characters on them, see Hunter (2), p. 213.
[e] The craft of making *tapa*, a beaten bark paper, was almost universal throughout the Pacific, but it was used only for clothing and not normally for writing; see Hunter (9), pp. 27–47; cf. above, pp. 37, 56.

[1] 禹亭侯   [2] 敬惜字紙   [3] 顏之推

Early Chinese records reveal the existence of a material made of bark, known as *tha pu*[1] or *ku pu*,[2] which may have some affinity with *tapa*. Since *tapa* was made from a variety of paper-mulberry bark by a process of beating and was used for clothing, it has been called bark cloth instead of paper. The Chinese terms *tha pu* which may mean 'beaten cloth', and *ku pu*, 'paper-mulberry cloth', very probably referred to a sort of bark cloth or *tapa*.

The earliest reference to this material is found in the *Shih Chi*, in which Ssuma Chhien (*c*. −145 to −86) mentioned that a merchant in the town managed in a year to sell 'a thousand piculs of *tha pu*'.[a] The same material, called *ku pu*, appears in several other sources as early as the +3rd century. Lu Chi[3] (+3rd cent.) said that people south of the Yangtze River used the bark of *ku* (paper-mulberry) to make cloth and also pounded it to make paper, called *ku phi chih*[4] (paper-mulberry bark paper).[b] Apparently, the inner bark of paper-mulberry can be prepared in different ways and used for different purposes. Since all these items for wearing and bedding are described in Chinese records as made of *chih* (paper), we may assume that they were made of bark paper.

There are in Han literature several references to the use of paper-mulberry for hats and headdress. Han Ying[5] of the −3rd century mentioned that a disciple of Confucius named Yüan Hsien[6] (−6th century) of the Lu State wore a paper-mulberry hat (*chhu kuan*[7]),[c] which may not, of course, have been of true paper, and in the Later Han, it was fashionable for men to wear headbands called *hsiao-thou*[8] or *chhiao-thou*,[9] made of paper-mulberry bark in red or other colours.[d] During the Thang and Sung dynasties, paper-mulberry hats were worn by Taoist priests and were fashionable among scholars and poets; indeed a poem titled 'Taoist Fashion' by Wang Yü-Chheng[10] (+954−1001) says: 'Paper-mulberry hat, serge coat, and black gauze kerchief'[e], and Lu Yu[11] (+1125−1210) mentions in his poems that he has newly had made two paper-mulberry hats, 'emphatically imitating the Taoist fashion'.[f] Many other poems testify to this Taoist habit of wearing of paper-mulberry hats at this time, while several hats of stiff paper covered with plain black silk were found in a Thang tomb in modern Sinkiang.[g] Another hat, a paper belt, and a paper shoe dated +418, made of hemp fibre, yellowish and thick with a textile pattern, were among the objects recently discovered at Turfan (Fig. 1086).[h] Paper was also widely used as lining in cloth shoes.

---

[a] *Shih Chi* (*ESSS/TW*), ch. 129, p. 15 *b*. Earlier commentators called *tha pu* a coarse or foreign cloth; B. Watson (1), II, p. 494, translates the term as 'fabric made of vegetable fiber'; Ling Shun-Sheng (7), p. 30, calls it 'bark cloth'; see also detailed discussion in Pelliot (47), I, pp. 445–7.
[b] *Mao Shih Tshao Mu Niao Shou Chhung Yü Su* (*TSHCC*), pp. 29–30.
[c] See *Han Shih Wai Chuan* (*SPTK*), p. 4.
[d] See *Hou Han Shu* (*ESSS/TW*), ch. 71, p. 32 *a*; ch. 73, p. 7 *a*.
[e] See *Hsiao Chhu Chi* (*SPTK*), ch. 8, p. 22.   [f] See *Chien Nan Chi* (*SPTK*), ch. 37, p. 2.
[g] See Stein (2), IV, pl. XCIII.   [h] See Phan Chi-Hsing (4), p. 54; (9), p. 135.

[1] 搨布	[2] 穀布	[3] 陸璣	[4] 穀皮紙	[5] 韓嬰
[6] 原憲	[7] 楮冠	[8] 絹頭	[9] 幧頭	[10] 王禹偁
[11] 陸游				

(a)

(b)

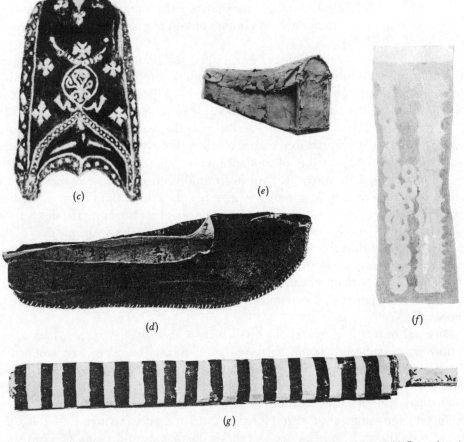

(c)

(e)

(d)

(f)

(g)

Fig. 1086. Paper articles of the Thang dynasty found in Sinkiang. (a, b, c) Paper hat or crown; (d) Paper shoe; (e) Paper coffin; (f) Paper money for spirits; and (g) Rolled paper flag with black stripes. (a, d, f, g) from Stein (2), (b, c, e) from the Institute of History of Science, Academia Sinica, Peking.

The most common paper apparel included the paper clothing (*chih i*[1] or *chih ao*[2]), used as early as the Han dynasty. The *History of the Later Han Dynasty* reports that the native tribes at Wu-ling (in modern Hunan) made bark into cloth and dyed it with grass seeds,[a] such bark cloth being a local product presented to the court as a tribute by the non-Chinese tribes who lived in the south and southwest regions of the empire. Phei Yüan[3] (+3rd century), author of the *Kuang Chou Chi*,[4] and Thao Hung-Ching[5] (+456–540), a noted physician, also mentioned that the bark of paper-mulberry was used by the people in Wu-ling for clothing which was very durable and fine.[b] In the Ta-Li period (+766–79) of the Thang dynasty, a Zen Buddhist monk wore no silk or fabric other than paper and was called Zen Master of Paper Cloth,[c] while testimony by many poets of the Sung dynasty indicates that paper clothing was worn not only in summer but during all seasons by poor people as well as by Buddhists.[d]

One Sung writer, Su I-Chien[6] (+953–96), said that those who lived in the mountains (i.e. Buddhist and Taoist priests) often wore paper clothing, probably because of the Buddhist tradition against wearing silk. The clothing was very warm but, he claimed, was bad for the health because it did not allow circulation of air. His description of its manufacture describes how the material was boiled with one ounce of walnut and frankincense (*gum olibanum*) for every 100 sheets, or steamed with an occasional sprinkling of frankincense or other liquid. When cooked and ready to be dried, it was rolled up horizontally on a stick, and then pressed vertically into wrinkles, apparently to give it some elasticity to prevent it being easily broken. Su also said that in his time some people of the I and Hsi districts (in modern Anhwei) made sheets of clothing paper as large as the size of one door of the main gates, and that this had been worn by many scholar-officials during their travels as a protection against the cold.[e] Marco Polo remarked that 'they manufacture stuffs of the bark of certain trees which form very fine summer clothing.'[f] No specimen of old paper clothes is known to have survived in China, but many are in Japan (Fig. 1087).

Personal outfits and household articles were also made of paper, and paper furnishings, including screens, curtains, bed-nets, and blankets, were frequently mentioned in Thang and Sung poems as well as in other writings. Su I-Chien mentioned that Yang Hsü[7] (+2nd century), who set an example of thrift to his subordinates, used paper curtains (*chih wei*[8]) and cloth blankets when he was prefect of Nan-yang (in modern Honan).[g] And in a poem written in +1085, the noted poet Su Shih[9] (+1036 to +1101) said that an old monk of Chin-shan enjoyed the warmth of a paper curtain when he was on board a ship travelling on

[a] *Hou Han Shu*, ch. 86, p. 1 *b*.
[b] See quotations in *Pen Tshao Kang Mu* (Peking, 1975), ch. 36, pp. 78–9.
[c] *Thai Phing Kuang Chi* (Peking, 1959), p. 2297.
[d] See poems quoted in Hsiung Cheng-Wen (*1*), pp. 34–5.
[e] *Wen Fang Ssu Phu* (*TSHCC*), p. 55.   [f] Yule (*1*), II, p. 191.   [g] *Wen Fang Ssu Phu*, p. 55.

[1] 紙衣	[2] 紙襖	[3] 裴淵	[4] 廣州記	[5] 陶弘景
[6] 蘇易簡	[7] 羊續	[8] 紙幃	[9] 蘇軾	

Fig. 1087. Paper cloth of Japan (*kamiko*) made of specially treated sheets of paper.

the Yangtze River between the Chin and Chiao mountains.[a] A Ming author, Thu Lung[1] (+1542–1605), related that 'paper bed-curtains (*chih chang*[2]) are made by fastening rattan skin and cocoon paper on wooden sticks and tightening them with a string. The paper is wrinkled and then sewn together with thread without the use of paste. The curtain top is made of loosely woven cloth instead of paper for ventilation. The curtains may be painted with plum blossoms or butterflies, appearing extremely elegant and delightful.'[b] Generally, paper curtains were used for warmth in winter and to keep out mosquitoes in summer.

---

[a] *Chi Chu Fen Lei Tung Pho Shih* (*SPTK*), ch. 23, p. 424; similar poems on paper bed-curtains are found in Su Chhe,[3] *Luan Chheng Chi* (*SPTK*), ch. 4, p. 4*b*, and in works of Ming authors.
[b] *Khao Phan Yü Shih* (*TSHCC*), p. 73.

[1] 屠隆   [2] 紙帳   [3] 蘇轍

Paper blankets (*chih pei*[1]) and paper mattresses (*chih ju*[2]) were used primarily by Buddhist and Taoist monks, as well as by some scholars, to keep them warm in winter. Apparently they were cheap but not common, for acknowledgment of gifts of paper blankets is occasionally found in literature. Thus, the monk Hui-Hung[3] (+1071–1128) of the Northern Sung dynasty wrote a poem to acknowledge the gift of a paper blanket from the Zen master Yü-Chhih,[4] describing it as white as snow, soft as cotton, and better than a blanket made of exquisite white cloth (*po tieh*[5]) or one of purple fox felt (*tzu jung chan*[6]).[a] In thanking the philosopher Chu Hsi (+1130–1200) for a gift of a paper blanket, the poet Lu Yu (+1125–1210) wrote: 'I passed the day of snow by covering me with a paper blanket. It is whiter than fox fur and softer than cotton.'[b] He enjoyed its warmth and softness, and on several occasions said he had slept comfortably and soundly under a paper blanket. Both blankets and mattresses are said to have been produced in Fukien, Yunnan, as well as many other locations.[c]

A defensive covering made of paper, known as *chih chia*[7] or *chih khai*,[8] was used to protect the body and arms in battle. It was light, convenient, and especially suitable for foot soldiers in the south, where the terrain prevented the use of such heavy armour as was normally worn by soldiers on horseback or on ships. Since paper armour is not mentioned among the thirteen kinds of armour listed in the Thang administrative codes, *Thang Liu Tien*,[9] compiled in +722–38, it is believed that its use began after this time. It was probably adopted from the late Thang dynasty, for when Hsü Shang[10] (fl. +847–94) was appointed governor of Ho-tung (in modern Shansi), he organised and kept an expeditionary army of one thousand troops in a state of readiness; they were clothed with pleated paper armour which could not be pierced by strong arrows.[d] Later, when Li Thao[11] (d. +968), a captain in the imperial army of the Sung dynasty, attacked the city of Ho-tung, he found the defenders were dressed in yellow paper armour, which appeared white in the light of bright flames.[e] While paper armour was primarily for foot soldiers, it was also used by the navy, and Hung Kua[12] (+1117–84), a commissioner of military affairs, mentioned in a memorial on the amnesty of pirates, that no less than 110 sets of paper armour were found alongside the weapons in two ships surrendered by the enemy.[f] In another memorial, Chen Te-Hsiu[13] (+1178–1235), a magistrate of Chhüan-chou, said that weapons at his fort were sufficient for the defence of the coast, except that fifty sets of paper armour were needed for his navy, for which he would exchange one half of the 100 sets of iron armour in his possession.[g]

---

[a] *Shih Men Wen Tzu Chhan* (*TSHCC*), ch. 13, p. 128; cf. discussion of *po tieh* in Pelliot (47), I, pp. 442–56.
[b] *Chien Nan Chi* (*SPTK*), ch. 36, p. 8; see also ch. 21, p. 14; ch. 74, p. 8.
[c] *Fu-chien Thung Chih* (1737 ed.), ch. 11, p. 4a.
[d] *Hsin Thang Shu* (*ESSS/TW*), ch. 113, p. 10a.   [e] *Sung Shih* (*ESSS/TW*), ch. 271, p. 10b.
[f] *Phan Chou Wen Chi* (*SPTK*), ch. 42, p. 4a.   [g] *Chen Hsi-Shan Wen Chi* (1665 ed.), ch. 8, p. 16a.

[1] 紙被	[2] 紙褥	[3] 惠洪	[4] 玉池	[5] 白氎
[6] 紫茸氈	[7] 紙甲	[8] 紙鎧	[9] 唐六典	[10] 徐商
[11] 李韜	[12] 洪适	[13] 眞德秀		

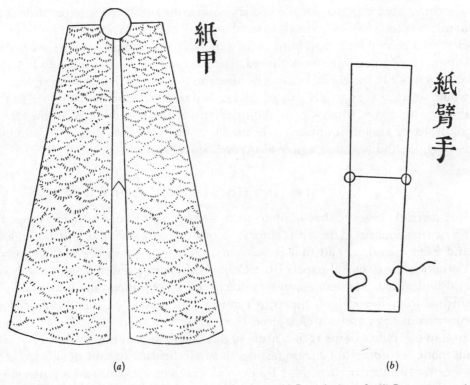

紙甲

紙臂手

(a)   (b)

Fig. 1088. Paper armour of the Ming dynasty, c. +17th century. (a) Paper body outfit. (b) Paper arm guard. From *Wu Pei Chih*. 1621 ed.

However, the best description and illustration of paper armour (Fig. 1088) are found in the Ming work recording armaments, written in 1621 by Mao Yüan-I[1] (d. 1629), in which he says:[a]

Armour is the basic equipment of soldiers, with which they are able to endure without suffering defeat before sharp weapons. The terrain in the south is dangerous and low, and where foot soldiers are generally employed they cannot take heavy loads on their backs when travelling swiftly. If the ground is wet or there is rain, iron armour easily rusts and becomes useless. Japanese pirates and local bandits frequently employ guns and firearms, and even though armour made of rattan or of horn may be used, the bullets can nevertheless pierce it. Moreover, it is heavy and cannot be worn for too long. The best choice for foot soldiers is paper armour, mixed with a variety of silk and cloth. If both paper and cloth are thin, even arrows can pierce them, not to say bullets; the armour should, therefore, be lined with cotton, one inch thick, fully pleated, at knee length. It would be inconvenient to use in muddy fields if too long and cannot cover the body if too short. Heavy armour can only be used on ships, since there soldiers do not walk on muddy fields. But since the enemy can reach the object with bullets, it could not be defended without the use of heavy armour.

[a] *Wu Pei Chih* (1621 ed.), ch. 105, pp. 17b–18a.

[1] 茅元儀

The same work says that for the protection of arms and hands a paper arm-and-hand cover (*chih pei shou*[1]) was also used. Each pair of these covers used four layers of cloth of a certain length on both outer and inner sides, plus a certain amount of cotton, cocoon paper, and silk thread. The paper armour was similar to the iron armour made in the north, but was flexible and convenient, light and ingenious. A whole sleeve was generally made, thicker in the upper part and thinner in the lower, with a very thin place in the middle to facilitate movement of the elbow joint.[a] Paper armour continued to be used by some of the native tribes in Yunnan, Kweichow, and Kwangsi as late as in the Chhing dynasty.[b]

(5) WALLPAPER AND HOUSEHOLD USE OF PAPER

It is generally believed that wallpaper was first brought from China to Europe by French missionaries in the 16th century, then later from Canton by Dutch, English, and French traders, and that it was imitated in Europe in the 17th century.[c] Certainly, the colourful papers from China with hand-painted designs of flowers and birds, landscapes, and scenes of domestic life were especially fashionable in Europe from the 17th to the 19th century (Fig. 1089). It was introduced to America in 1735 and manufactured there some fifty years later. Before the use around the middle of the 19th century of machinery for printing wallpaper, it was all made according to Chinese fashion in small sheets with unit designs printed successively either by stencils or by woodblocks to give a continuous pattern. As Laufer says: 'We owe to China in particular also our paper-hangings or wallpaper.'[d]

The earliest mention of Chinese wallpaper was a reference in 1693 in England to Queen Mary's Chinese and Indian cabinets, screens, and hangings, the last of which is believed to refer to Chinese painted papers.[e] Then, in about 1772, John Macky described the Palace of Wanstead as 'finely adorned with China paper [showing] the figures of men, women, birds and flowers the liveliest [the author] ever saw come from that country'. Some of these papers were so accurately drawn that 'a man need go no further to study the Chinese than the Chinese paper. Some of the plants which are common in China and Java as bamboo, are better figured there than in the best botanical authors that I have seen.'[f] Even in this century, the Chinese hand-painted wallpapers are still considered the most excellent and beautiful of all, and a leading British architect has said: 'No experience could be more delightful than to waken in a bedroom hung with "painted paper of Pekin".'[g]

***

[a] *Wu Pei Chih*, ch. 105, p. 19 *a–b*.
[b] See *Kuang-Hsi Thung Chih* (1801 ed.), ch. 278, p. 22 *b*.
[c] For Chinese wallpapers in Europe, see Ackerman (1), pp. 11–20; Entwisle (1), pp. 43–8; Sanborn (1), pp. 14–29; for a chronological development of the art, see Entwisle (2), pp. 11 ff.
[d] Cf. Laufer (48), p. 19.   [e] Cf. Entwisle (1), pp. 21, 43–4.
[f] See Entwisle (2), pp. 13, 23, 49.   [g] See Sitwell (1), p. 196.

[1] 紙臂手

Fig. 1089. Old European wall-paper in five colours designed by Jean Papillon (+1661–1723) showing Chinese influence of the century. Printed in yellow, black and red with blue and green done with a brush (42 in. × 21 in.). From McClelland (1).

Fig. 1090. Carved block with unit design for printing wallpaper. Dard Hunter Paper Museum.

The history of wallpaper in China is not as clear as that in Europe or America. All writings about wallpaper by Western authors indicate that it originated in China, but no clear trace can be found in Chinese sources earlier than the +17th century. Both wallpaper historians and sinologists agree that the so-called 'flock paper' made by a French printer in Rouen in 1630 and by the English at about the same time, was inspired by coloured papers imported from China.[a] Some accounts from later European visitors mention the use of wallpaper in North China from the beginning of the 17th century, when the Chhing emperors, especially Khang-Hsi (r. +1662–1722), showed a great deal of interest in developing the decorative arts, including wallpaper.[b] Certainly, many Chinese wallpaper designs were similar to the patterns of the Chinese porcelain that came to Europe, perhaps made by the same group of artist craftsmen who specialised in this style primarily for the foreign trade, and later on, unit patterns (Fig. 1090) were used to print decorative designs continuously on one large sheet of paper (Fig. 1091).

[a] Chinese influence is mentioned in Laufer (48), pp. 19–21, which cites Kate Sanborn, *Old Time Wallpaper* (New York, 1905), pp. 14–16; this in turn is based on the *Histoire générale de la tapisserie en Italie, en Allemagne, en Angleterre, en Espagne* by Muntz (3 vols. Paris, 1878–84).
[b] Ackerman (1), pp. 10–20.

Fig. 1091. Printed wallpaper in continuous pattern of design made with unit block. Dard Hunter Paper Museum.

No information has been found on how early wallpaper was used in Chinese houses. Most room partitions in Chinese buildings were wooden panels or plastered walls, and coloured designs were sometimes painted directly on such walls or ceilings.[a] Wallpaper must have been used in China in the 16th or 17th century, as indicated by Chinese works which say that paper covering on walls was vulgar and

[a] See quotations on wall paintings in *Thu Shu Chi Chheng* (Thaipei, 1964), ch. 98, p. 56; also, painting on ceilings and walls in a Ming house of a Wu family was recently recovered in Hui-chou (modern Hsiu-ning, Anhui); cf. Chang Chung-I (*1*), p. 32, figs. 75–80.

not liked by people of good taste. Wen Chen-Heng[1] (1585–1645), a not
landscape artist and calligrapher, said: 'Small rooms should not be partitioned
the middle', and 'Walls should not be pasted with paper': his opinion was th
neither painting nor writing directly on them could be compared with plain wall
Li Yü[2] (1611–80), another noted author and the owner of the Mustard Se
Garden, opposed the use of white paper for covering walls. He suggested th
brown paper might be used as a basic wall covering, and then green writing pap
torn into pieces and pasted on it.[b] This seems to be an early example of collage

Li further suggested that walls should neither be too bare nor too fanciful. F
was in favour of displaying scrolls by great artists, which should be pasted direct
on to the walls rather than hung, so that gusts of wind might not make them swa
and so be liable to damage, but also said that the scrolls might be cut and pasted o
a wooden board having frames on all sides.[c] The origin of wallpaper may possibl
be traced to such Chinese decoration of walls with hanging scrolls of variou
subjects, such as landscapes or flowers and birds. Indeed, what the early mis
sionaries brought to Europe may have been such pictures, which were at first i
frames but were later pasted on walls instead of being hung. One early Europea
reference to applying wallpaper says that the old method was to fasten a woode
framework over the surface of the bare walls; this was fixed to wooden wedge
driven into the brick or stone, thus leaving an air space between. On these frame
canvas was stretched, and on the canvas the wallpapers were fixed, and it is for thi
reason that, in many cases, it has been possible to remove them.[d] These wallpaper
may have been hanging scrolls of paintings such as were often used to decorate
Chinese houses.

The paper screen used as a movable partition of the room has been a very
important item for interior decoration in Chinese houses since Thang times. There
were two major forms of such screens, folding and stiff, both of which were
originally made of wooden board and painted sometimes on lacquered surfaces.
When paper became popular, the wooden panels were replaced with paper and
decorated with calligraphy, the screens thus made known as *shu phing*[3] (calligraphy
screen) or, with painting, *hua phing*[4] (painting screen). Panels with works by
celebrated artists are said to have been extremely expensive; a Thang work
mentions that one single panel of a folding screen cost as much as 20,000 pieces of
gold and another of medium quality was sold for 15,000.[e] One folding screen with
six panels of brocade stretched over a frame of lacquered wood survives in Japan,[f]

[a] See *Chhang Wu Chih* (*MSTS*), p. 5. This information was supplied by Dr Hsia Nai, Director of the Institute of
Archaeology, Academia Sinica, Peking, in a letter of 17 July 1972.

[b] *Li Yü Chhüan Chi* (Taipei, 1970), vol. 6, pp. 3403–40.

[c] *Ibid.* p. 2398; cf. also tr. by van Gulik (9), pp. 258–9.

[d] See discussion in Chhien Tshun-Hsün (6), p. 94.

[e] These folding screens are mentioned in the *Li Tai Ming Hua Chi* (*THCC*), ch. 2, p. 81.

[f] See Ishida Mosaku & Wada Gunishi (1) (ed.), *Shōsōin: an Eighth-century Treasure House* (Tokyo, 1954), vol. 1,
Northern Section, pp. xvii, 36. Room partitions with paper screens have been popularly used in Japanese and
Korean houses.

[1] 文震亨   [2] 李漁   [3] 書屏   [4] 畫屏

Fig. 1092. Designs of wooden window frames to be covered with translucent paper for houses in Western China, *c.* +19th century. From D. S. Dye (1).

while a stiff screen with one panel, known as *chang tzu*[1] (shields), or *hua chang*[2] (painting shield), was also used at this time. The pictures on mounted scrolls were often transferred to the screen, or from the screen back to scrolls as needed.[a]

It has been common to use white paper in lieu of glass for windows and doors in Chinese houses ranging from imperial palaces to peasants' homes. Windows were designed with lattices[b] on which paper was pasted to admit a softened sunlight (Fig. 1092). Living room doors were similarly designed with lattice on the upper part and solid panels below. Gauze was used in ancient time, though later thin but strong paper in large sheets took its place. This was generally made of paper-mulberry bark mixed with bamboo and sometimes rice stalks, the strong and hard-sized bark paper being difficult to tear cross-wise. The highest grade of this paper, used for windows in imperial palaces, was called 'window-gauze paper' (*ling chhuang chih*[3]). That used in the Ming dynasty was produced in Kuang-hsing (in modern Chiangsi); each sheet was over seven feet long and more than four feet wide, and some were dyed in various colours.[c]

Paper used to cover windows in the imperial palaces was detailed in the administrative codes of the Chhing government from the beginning of the dynasty.[d] It was specified that tributary paper sent from Korea be used for the windows of four palaces (Thai-ho, Pao-ho, Chung-ho, and Wen-hua) every year, and that yellow silk fabric be used to mend the seams every two years. Requisitions of paper for the windows and lanterns of the imperial altars and temples were made

[a] Cf. van Gulik (9), p. 159.
[b] The use of lattices in windows in Chinese architecture has a long history, and some of these attached to old wooden buildings have survived from the Ming dynasty. For a brief history and designs of Chinese lattice, see D. S. Dye (1).
[c] See *Thien Kung Khai Wu* (*KHCP*), p. 219; cf. tr. Sun & Sun, p. 230.
[d] See *Ta Chhing Hui Tien Shih Li*, ch. 940, pp. 5*a–b*.

[1] 障子   [2] 畫障   [3] 櫺窗紙

by the Court of Sacrificial Worship, and for those of hostels for foreign visitors by the Board of Colonial Affairs. And in the Chhing dynasty it was noted that as many as 108,000 sheets of Korean paper were traded, among other things, in a market-place called Chung-chiang, in the spring and autumn of every year.[a] Even today, paper is still commonly used for windows and doors in Chinese houses, as well as in those in Japan and Korea.[b]

A pair of sheets of red paper inscribed with good-luck characters or a poetic couplet, known as *men thieh*[1] (door scroll), was pasted on the double doors of Chinese houses on the eve of a new year. The origin of the custom is unknown, but it probably derived from the use of peach-wood tablets called *thao fu*[2] which were placed on doors as charms against evils in Han times or earlier. Since Thang times they seem to have been replaced by red paper, for one early reference is from the +10th century, mentioning that door scrolls with four characters from the *One Thousand Character Classic* were written for the palace doors on the eve of the new year. This practice was also observed during the Sung and Ming dynasties, and when Emperor Thai-Tsu of the Ming (r. +1368–98) was in Nanking, he ordered that such scrolls be added to all doors of officials and commoners.[c] Their function was somewhat similar to that of the door gods: primarily for protection from evil, but partly for decoration.[d]

Paper was probably used for wrapping before it was used for writing. The oldest paper of the Former Han dynasty found in Pa-chhiao, Shensi, in 1957 seems to have been used for wrapping or padding bronze mirrors, for it was discovered with these in a Han tomb dated to the −2nd century. An early kind of paper called *ho thi*,[3] believed to have been made of the bark of paper mulberry, is mentioned in the official history as having been used to wrap some poisonous 'medicine' in the −2nd century.[e] Since paper was soft and cheap, it was natural to use it for wrapping many other things, and in the Thang, paper made of rattan from Shan-chhi (in modern Chheng-hsien, Chekiang) was folded and sewn into square bags (*chih nang*[4]) to preserve the flavour of tea leaves. At this time, tea was served from baskets made of rushes which held tea cups with paper napkins (*chih pha*[5]) folded into squares,[f] and a set of several tens of paper cups (*chih pei*[6]) in different sizes and colours with delicate designs is said to have been seen in the possession of a Yu family of Hangchow.[g] In the Southern Sung dynasty, gift money for bestowing upon officials by the imperial court was wrapped in paper envelopes (*chih pao*[7]).[h] Sung Ying-Hsing says that the thickest and coarsest wrapping paper was made

---

[a] See Hsü Kho (*1*), *Chhing Pai Lei Chhao* (1917), ch. 14, pp. 13–14.
[b] For pictures of paper windows and doors, see Hunter (9), p. 221, figs. 193–4.
[c] *Kai Yü Tshung Khao* (1750 ed.), ch. 30, pp. 22*a*–23*a*.
[d] See also New Year pictures on pp. 287 ff. below.
[e] *Chhien Han Shu* (*ESSS/TW*), ch. 97*b*, p. 13*a*.
[f] *Chha Ching* (*HCTY*), ch. 2, p. 1*b*.   [g] See Shih Hung-Pao (*1*), p. 125.
[h] *Thung Su Pien* (reprint, 1977), p. 513.

[1] 門帖   [2] 桃符   [3] 赫蹏   [4] 紙囊   [5] 紙钯
[6] 紙杯   [7] 紙包

of bamboo fibres mixed with rice straw.[a] During modern times the same raw materials have been used and wrapping paper makes up more than twenty per cent of the total production.[b]

The use of paper for toilet purposes must have been practised no later than the +6th century. Although Chinese sources are generally silent about the use of paper for cleaning the body after elimination, one reference dated as early as the +6th century refers to the prohibition of paper with characters being used for such purposes. Indeed, the noted scholar-official Yen Chih-Thui (+531-91) said in his family instructions, written about +589, 'Paper on which there are quotations or commentaries from *Five Classics* or the names of sages, I dare not use for toilet purposes',[c] and an early Arab traveller to China, who was obliged by his religion to perform purifying ablutions, commented curiously upon this use of paper. In his report of +851, he says: 'They (the Chinese) are not careful about cleanliness, and they do not wash themselves with water when they have done their necessities; but they only wipe themselves with paper.'[d]

Toilet paper (*tshao chih*[1]) was made from rice straw, the fibres of which were tender and required less time and labour to process; it thus cost less than any other kind of paper. Great quantities of such paper were needed for daily use, and for the imperial court alone, it was specified in +1393 that the Bureau of Imperial Supplies (Pao Chhao Ssu[2]) manufactured 720,000 sheets, two by three feet in size, for the general use of the court and 15,000 sheets, three inches square, light yellow, thick but soft, and perfumed, for special supply to the imperial family. The quantity manufactured every year was so great that the refuse of straw and lime which accumulated in the imperial factory formed a mound that was called Elephant Mountain (Hsiang Shan[3]).[e] Even early in this century, the annual production of paper for toilet use in Chekiang alone amounted to ten million packages of 1000 to 10,000 sheets each.[f]

(6) PAPERCRAFT AND RECREATIONAL USE OF PAPER

Paper has great potential as a creative material for recreational purposes. It may be cut into designs to be pasted on windows, doors, lamps, and other surfaces, and on clothing and shoes in place of embroidery. It may be folded into flat or three-dimensional forms for art or entertainment; and making paper flowers by folding, cutting, and pasting is a popular amusement. Because of its lightness, paper is also especially suitable for making kites. Numerous articles for popular use were origi-

---

[a] *Thien Kung Khai Wu* (*KHCP*), p. 218; tr. Sun & Sun (1), p. 229.
[b] *Che-Chiang Chih Chih Yeh*, pp. 358-9.
[c] *Yen Shih Chia Hsün* (*SPTK*), ch. 5, p. 13b; cf. tr. Teng (3), p. 21.
[d] See tr. by Renaudat (1), 1718, p. 17; also Reinaud (1), p. 23; Sauvaget (2), p. 11.
[e] *Cho Chung Chih* (*TSHCC*), p. 113.
[f] Estimate based on the *Che-Chiang Chih Chih Yeh*, pp. 236, 245-6, 276-83.

[1] 草紙   [2] 寶鈔司   [3] 象山

nally made of far more expensive materials such as silk, leather, horn, or ivory, but later these were replaced with paper. Sturdy paper or papier-maché was also substituted for many more expensive materials for games, toys, and other objects for enjoyment. Paper was used in China for some of these purposes as early as the +3rd or 4th century, and for all of them by the +6th or 7th.

Cutting paper into various designs with scissors and knives, for making decorations at festivals or on other occasions, is a folk art of China with a history of many centuries. It probably derives from the custom of cutting out human figures, flowers, or landscapes in silk at the spring festival. Tsung Lin[1] (+6th century) says: 'On the seventh day of the first moon, which is called the man's day, seven kinds of vegetables are used for soup, and sheets of silk or gold foil are cut into human figures to paste on screens or hang in women's hair at the sides.'[a] It has been said that this custom derived from Madame Li, wife of Chia Chhung[2] (+217–82) of the Chin dynasty (+265–420). A similar custom of cutting paper into small banners, butterflies, or pieces of money, to place on women's heads or on flowers at the beginning of spring, was popular in the Thang,[b] and a beautiful geometric design cut out of a round sheet of paper, from the +5th or +6th century, was recently found in Sinkiang (Fig. 1093a). Again, silhouettes of shrines cut out of buff paper and pasted on a blackened sheet, and several artificial flowers made of paper cut into various shapes to form the petals, etc., were found in Tunhuang,[c] while the noted poet Tu Fu mentioned in one of his poems: 'Cutting paper to summon my souls.'[d] These, then, are some of the earliest examples of, or references to, paper-cutting known to us today. Many stories concern the skill of artists in paper-cutting and the excellence of their work. The Sung scholar Chou Mi[3] (+1232–98) mentioned several paper-cutters who cut paper with scissors into a great variety of designs and characters in different styles, and a young man who could even cut characters and flowers inside his sleeve.[e] But though most of the stories tell of male artists, many of the cutters were women who did this in their leisure time on farms.

The subjects of paper-cutting included scenes from farm life: tilling land, weaving, fishing (Fig. 1093d), or tending cattle; symbols of good luck or blessings; legendary stories and theatrical figures; and flowers and plants, birds and animals. The design might be one independent picture, a symmetrical pair, or multiple sets of from four to as many as twenty-four. If to be used in corners, a set of four triangular designs was usually made; and for a ceiling a multiple design round in shape was used. Unlike paintings, the composition of paper cuts was generally symmetrical, well balanced, with intricate designs covering an entire space; they had a strong local flavour.

---

[a] *Ching Chhu Sui Shih Chi* (*SPPY*), p. 3a.
[b] *Yu Yang Tsa Tsu*, quoted in *WWTK*, 1957, no. 8, pp. 13–15.
[c] See Phan Chi-Hsing (9), p. 56; Stein (4), II, p. 967; IV, pl. XCVII.
[d] See Hung Yeh (1), 58/19/32.
[e] *Chih Ya Thang Tsa Chhao* (*YYT*), ch. 1, p. 38a.

[1] 宗懍   [2] 賈充   [3] 周密

(a)

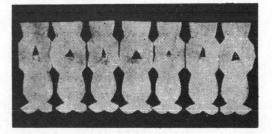

(b)

(c)

(d)

Fig. 1093. Old and modern paper cuttings. (a) The oldest extant paper-cut in geometric design of the Six Dynasties, c. +5th or +6th century, from Sinkiang. (b) Continuous human figures by paper-cut, c. +7th century. (c) Picture of shrines cut out and pasted on a blackened sheet from Tunhuang, from Stein (4), pl. XCVII. (d) A modern paper-cut of fishermen.

The process of cutting paper involved several steps. A master design was first cut and fastened over a piece of white paper upon a wooden board. The paper was then moistened with water and blackened with smoke, and when the master design was taken off, a white design appeared on the paper against a black background. A pile of sheets of white paper was then laid under the design, and fastened with paper thread at the corners and the centre before cutting. For symmetrical designs, the paper was folded and cut with scissors to duplicate the design, but only a few sheets

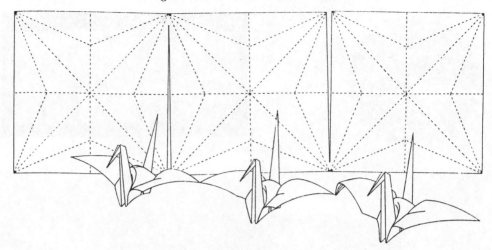

Fig. 1094. Paper folding of triple cranes. From Honda Iso (1).

can be cut in this way at a time. Independent designs were usually cut with a knife through as many as sixty or seventy sheets of paper at once. Lines could be cut out to make a positive impression, or left between cuts to make a negative impression, simple lines being cut with an ordinary blade, but delicate ones with a specially sharp, small, round knife. The inside lines were cut first and then the outside ones. After cutting was completed, the paper could be dyed with colours mixed with white wine or arsenic, with as many as forty sheets being coloured at a time. Multiple colours were applied separately, a fresh colour being added when the previous one had dried. For decoration of houses, red and multiple colours were usually used for auspicious occasions and blue for mourning.[a]

A flat piece of paper can be folded into various shapes and forms such as figurines, animals, flowers, garments, furniture, buildings, and numerous other objects (Fig. 1094). Paper-folding (che-chih,[1] origami) is probably one of the most interesting folk arts. It helps train nimble fingers, cultivates a sense of balance and symmetry, and can be used to provide visual illustrations to explain modern physics and geometry.[b] Indeed, many mathematicians have demonstrated their scientific interest in paper-folding, especially in dealing with three-dimensional problems and to show the geometric construction of regular polygons.[c]

Although paper-folding probably flourished in China for many centuries before it spread worldwide, there is no clear indication of how early it began. From all available evidence, its origin probably was not later than early in the Thang

[a] Cf. A Ying (3), pp. 1–9.
[b] Shen (1), pp. 7–8. Dr Shen, an expert in paper-folding, has provided much information on its worldwide popularity.
[c] Row (1) is completely devoted to the use of folding in geometry; see also Cooper Union Museum, *Plane Geometry and Fancy Figures: an Exhibition of Paper Folding* (Philadelphia, 1959), introduction by Edward Kallop.

[1] 摺紙

Fig. 1095. Decorated paper flowers from Tunhuang, c. 10 cm. in diameter. British Museum.

dynasty,[a] for several artificial flowers of folded and cut paper have been found in Tunhuang (Fig. 1095), and they show highly sophisticated techniques in paper-craft.[b] Today, paper-folding is one of the most popular crafts and pastimes for teaching children in classrooms, and among adults throughout the world; it is especially popular in Japan, Europe, and America, with extensive literature in different languages.[c]

Flying paper kites (*chih yüan*[1]) is a somewhat athletic pastime enjoyed by children in spring and autumn. It was said that when their kites flew in the sky, children lifted their heads, opened their mouths, and breathed deeply, which was good for their health, and the ninth day of the ninth month of each year, or the 'double ninth' festival, was especially devoted to this amusement. The paper kite, consisting of a light bamboo frame covered with sturdy paper, and with a string attached, was made in the forms of butterflies, men, birds, or other animals, often in colour. Kites were, perhaps, originally made of light wood or silk before paper became common, and how early paper was used for making them is unknown. However, a story about sending a message to a rescue mission by flying a paper kite, in c. +549, indicates that it must have been earlier than this date.[d]

Other Chinese literary sources frequently tell of the use of kites for measuring distances, testing the wind, lifting men, signalling, and communicating for military purposes. The earliest known reference to their use for amusement or pleasure tells of someone in a palace in the +10th century fastening a bamboo whistle to a kite,

[a] Vacca (1), p. 43, says that *che chih* is mentioned in a poem by Tu Fu, but the original reference has been found to be a mistake. One poem by Tu Fu does mention 'paper-cutting' (*chien chih*[2]) but not 'paper-folding' (*che chih*), see p. 124 ff. above.
[b] Stein (4), II, p. 967; IV, pl. XCII.
[c] Some 200 entries are included in Legman (1), pp. 3–8.
[d] *Kai Yü Tshung Khao* (1750 ed.), ch. 40, p. 25a.

[1] 紙鳶   [2] 剪紙

so that it made a musical sound in the wind. From this the term for the aeolian harp (*feng cheng*[1]) was derived.[a] Kite-flying diffused very early to all other nations of East and Southeast Asia, especially Korea, Japan, Indo-China, and Malaysia, and was sometimes associated with religious practices. It was introduced to Europe as a Chinese contrivance at the end of the +16th century.[b]

Lanterns in China generally consisted of wooden or bamboo frames covered with a variety of such translucent materials as horn, silk, or skin, but those of paper are said to have been especially elegant and skilfully made. They were lighted with candles, and were hung indoors or outdoors as decorations, or carried as aids for walking at night. Especially interesting was the massive display of lanterns at the annual lantern festival around the fifteenth of the first moon each year (Fig. 1096*b*), a festival which was not instituted until the Thang dynasty, although poems about lanterns were written in China as early as the +6th century.[c]

How early paper was employed for lanterns is not clear, but it was certainly used in the Thang, for an account book of the time from a monastery and found at Khotan records that 'two sheets of white paper were bought, each sheet fifty cash, for mounting lanterns'.[d] Chou Mi (+1232–98) said that in the Hsiao-Tsung period (+1163–1189) there was great variety of lanterns at the lantern festival in Hangchow, and that the best came from Soochow and Fuchow, while the latest fashions from Hsin-an were extremely extravagant. There were such varieties as 'boneless' lanterns without skeletons, lanterns of fish-egg and pearl designs, lanterns of deerskin and of silk fabrics, and lanterns of coloured waxed paper with revolving figures on horseback, spun swiftly by the heat of a candle; some of the figures cut from paper by skilful ladies were especially graceful.[e] In his memoirs, Meng Yüan-Lao[2] (fl. +1126–47) recalls annual customs in the Northern Sung capital, Khaifeng, saying that several tens of thousands of lanterns in a great variety of forms were displayed along the main street. Long poles, installed in an area enclosed with a thorny fence, were wrapped with colourful silk. Numerous paper figures of dramatic personages were hung on the poles, moving in the wind like flying fairies.[f]

A similar custom existed in the Ming dynasty, when the festival extended from two days before to five days after the fifteenth of the first moon. Starting on the thirteenth night, bamboo awnings erected from house to house were decorated with numerous lanterns hanging over the streets; those made of paper are said to have been especially attractive and well made.[g]

Fans were frequently used in daily life as a shield against dust and the sun; they were first made of feathers but, later, silk, bamboo, ivory, bone, sandalwood, and

[a]　*Hsün Chhu Lu* (*TSHCC*), p. 3.　　　　　　　[b]　Cf. Laufer (4), p. 36.
[c]　*Kai Yü Tshung Khao* (1750 ed.), ch. 31, pp. 19*a–b*.
[d]　See Chavannes (12a), nos. 969, 971.　　　　[e]　*Chhien Shun Sui Shih Chi* (*SF*), ch. 69, pp. 9*a*–10*b*.
[f]　*Tung-Ching Meng Hua Lu* (*TSHCC*), pp. 110–11.
[g]　*Shao Hsing Fu Chih*, quoted in *TSCC*, vol. 98, p. 978.

[1]　風箏　　　　　[2]　孟元老

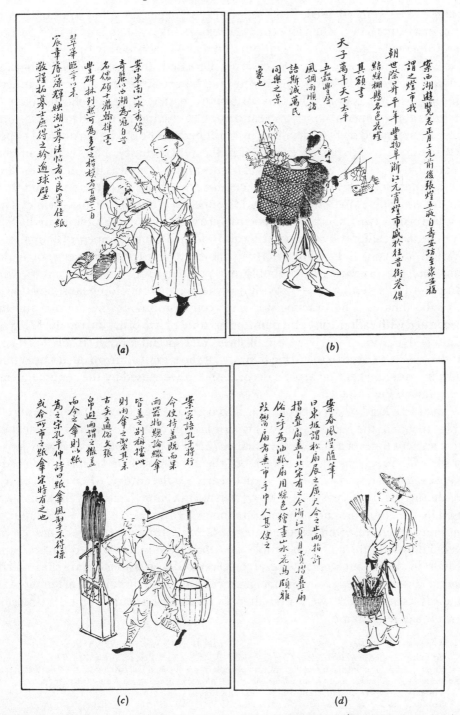

Fig. 1096. Paper articles were sold by pedlars on the street. (*a*) Rubbing of calligraphic models for a young scholar. (*b*) Paper lanterns at the lantern festival on the 15th day of the first moon. (*c*) Paper umbrellas carried by a pedlar. (*d*) Folded paper fans in the basket. Drawings by Chin Te-Yü, *Thai Phing Huan Lo Thu* (Pictures of Happy Occasions in a Peaceful World), printed by lithography in Shanghai, 1888.

palm leaves were all used. It is believed that paper fans first appeared in the Western Chin. Later, when for economic reasons silk was banned for such purposes by Emperor Hsiao-Wu (r. +373–97) and again by Emperor An (+397–418) of the Eastern Chin period, this use of paper increased, and two scrolls of calligraphy on fans by Wang Hsi-Chih (+321–79) and his son were among their writings in the imperial collections.[a] Fans were usually bestowed upon officials following lectures at the imperial court, and Emperor Che-Tsung (r. +1086–1100) is said to have been praised by his officials for his thrift because he used a fan made of paper.[b]

Circular fans of paper were popular in the Sung dynasty. The folding fan made of durable paper with various kinds of frames and designs was introduced to China from Japan via Korea in the +11th century. Su Shih (+1036–1101) said that a kind of white pine fan from Korea could be opened to over a foot and yet folded into a space of only two fingers. Many poems were written on these folding fans by Sung authors, and calligraphy and painting also adorned them.[c] Emperor Chang-Tsung (r. +1190–1200) of the Chin dynasty is known to have composed verses on folding fans, but the practice was probably not popular until the 15th century when Emperor Hsien-Tsung (r. +1465–88) wrote maxims on folding fans to bestow on his subordinates. The practice was most common among the literati, and fans decorated with calligraphy and painting became a form of art during the Ming and Chhing dynasties (Fig. 1097). Small sheets of a special kind of strong, hard-sized bark paper were used in the Ming dynasty for the manufacture of oiled-paper fans;[d] such fans usually bear no artistic decoration but were used by the common people in summer (Fig. 1096d).

The same kind of oiled paper was also used for the manufacture of umbrellas (Fig. 1096c) in the Ming dynasty. The origin of umbrellas derived from the use in very ancient times of a chariot cover called *kai*[1]. For protection against rain, a piece of silk called *san*[2] was spread above, but the use of paper umbrellas is believed to have been introduced in the late +4th or early +5th century, when the Toba tribe established its Wei dynasty (+386–532) in north China. Red and yellow ones were used by the emperor, and blue by commoners.[e] It was decreed in +1368 that silk umbrellas were reserved for the imperial family, while oiled-paper ones for rain were allowed to the common people, and they were not only used for protection from rain or sun, but were also taken on ceremonial occasions. Umbrellas called *lo san*[3] or *che yang*[4] were carried in official processions;[f] and 'umbrellas of ten thousand names' (*wan min san*[5]) were presented to specially honoured officials, inscribed with the donors' names.

[a] *Fa Shu Yao Lu* (*TSHCC*), p. 64.   [b] *Chhü Wei Chiu Wen* (*TSHCC*), p. 14.
[c] See poems quoted in *Kai Yü Tshung Khao* (1750 ed.), ch. 33, pp. 13a–14a; pictures in *San Tshai Thu Hui* (1609 ed.) ch. 21, p. 42a; for painting and calligraphy on paper fans, see such exhibit catalogues as *Chinese Fan Paintings* (Rochester, 1972), *Fan Paintings by Late Chhing Shanghai Masters* (Hong Kong Museum of Art, 1977).
[d] *Thien Kung Khai Wu*, p. 219; cf. tr. Sun & Sun (1), p. 230.
[e] See *San Tshai Thu Hui* (1609 ed.), ch. 12, pp. 22a–b.
[f] See *Ku Chin Shih Wu Khao* (*TSHCC*), p. 140.

[1] 蓋   [2] 傘   [3] 羅傘   [4] 遮陽   [5] 萬名傘

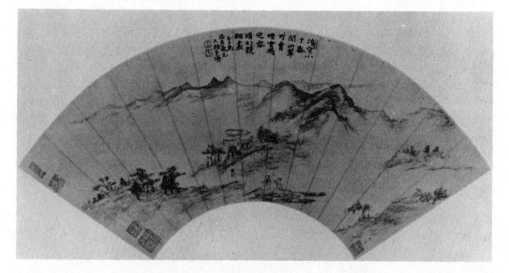

Fig. 1097. Calligraphy and painting on folded paper fans by Ming and Chhing artists Wen Cheng-Ming (+1470–1559), above, and Monk Tao-Chi (Shih-Thao, d. c. +1719).

Paper flags must have been used early. Several were found in Thang tombs in Sinkiang, one of several thicknesses of paper manuscripts pasted together, and painted with horizontal stripes of black and white; it was pasted on one side to a stick (Fig. 1086g).[a]

Playing cards made of paper, written or printed with designs, probably existed no later than the +9th century, when the relatives of a princess are said to have

---

[a] See Stein (4), IV, pl. XCIII.

played the 'leaf-game' (*yeh tzu hsi*[1]).[a] A similar date of origin is suggested by several other sources, including the noted scholar Ouyang Hsiu (+1007–72), who said that the card-game had been popular since the middle of the Thang dynasty, and related its origin to the change of book format from paper rolls to sheets or pages.[b] A book on the game, titled *Yeh Tzu Ko Hsi*,[2] was supposed to have been written by a woman toward the end of the Thang dynasty,[c] while it was mentioned by authors of successive dynasties, and numerous works on it have been written in later times.

The form of playing cards was described in a Ming work, titled *Yün Chang Ching*,[3] as about one inch wide, two inches high, and several tenths of a finger thick (Figs. 1098, 1204). It enumerated the numerous advantages of playing cards: they were convenient to carry, could stimulate thinking, could be played by a group of four without annoying conversation, and without the difficulties which accompanied playing chess or meditation. The game could be played in almost any circumstances without restriction of time, place, weather, or qualification of partners.[d] The fictional persons in the Chinese novel *Water Margin* were represented on the cards and were extremely popular toward the end of the Ming dynasty, and such characters, painted by the famous artist Chhen Hung-Shou[4] (1598–1652) are still available on playing cards today. Apparently because of the popularity of gambling, punishment was specified in penal codes of the Chhing dynasty for the manufacture or sale of more than 1000 paper cards, and for engaging in gambling by officials.[e]

Many other items made of paper for household, recreation and enjoyment are occasionally recorded in art and literature (Fig. 1096).[f] Such articles as paper chessmen (*chih chhi*[5]), were substituted for those of Yunnan stone; paper flutes, played transversely (*chih ti*[6]) and vertically (*chih hsiao*[7]), the sound of which was said to be better than that of those made of bamboo; shadow puppets; fireworks and firecrackers; and numerous kinds of toys like the paper tiger (*chih lao hu*[8]) (Fig. 1099) were some of the notable objects made of paper.

(e) ORIGIN AND DEVELOPMENT OF PRINTING IN CHINA

(1) PRE-HISTORY OF PRINTING

PRINTING is a process of reproduction with ink on paper or other surfaces from a reverse or negative image. It contains at least three essential elements: a flat surface, originally cut in relief, containing a mirror image of whatever is to be printed; the preparation of the mirror image; and the transfer of the impression of

---

[a] *Yeh Hsi Yüan Chhi* (*TMWS*), p. 1a.   [b] *Kuei Thien Lu* (*SF*), ch. 2, p. 13b.
[c] *Wen Hsien Thung Khao* (*ST*), p. 1834.   [d] Quoted in *Yeh Hsi Yüan Chhi*, p. 19a–b.
[e] *Ta Chhing Lü Li Shih Li* (1870 ed.), ch. 826, p. 13a; ch. 827, pp. 3b, 4b, 6b.
[f] *Shu Yüan Tsa Chi* (*TSHCC*), vol. 2, p. 140; *Min Hsiao Chi* (*TSHCC*), pp. 26–7; Shih Hung-Pao (*1*), p. 123; Teng Chih-Chheng (*1*), ch. 1, p. 132a–b.

[1] 葉子戲   [2] 葉子格戲   [3] 運掌經   [4] 陳洪綬   [5] 紙棋
[6] 紙笛   [7] 紙簫   [8] 紙老虎

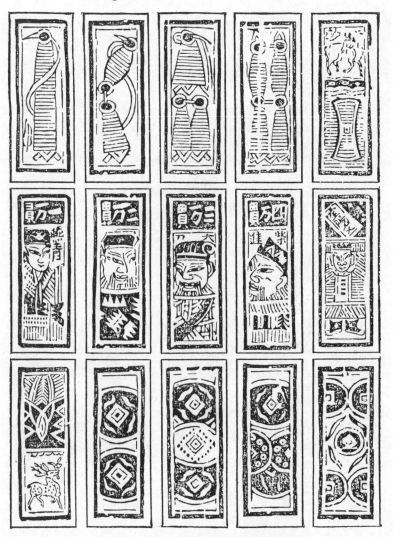

Fig. 1098. Chinese playing cards. From DeVinne (1).

this image on to the surface to be printed. In brief, the invention of printing required the development of necessary techniques for creating a proper vehicle to transfer an image on to an acceptable medium, in addition to meeting the large-scale demand for multiple copies. Before printing was used in China, many techniques for making reproductions existed. At first, of course, texts were copied by hand, but later mechanical devices were applied. These included seals for stamping on clay and, in due course, on silk and paper; the casting and engraving of inscriptions on metal and stone, the taking of inked impressions from stone inscriptions and, finally, using stencils to duplicate designs on textiles and paper. All these processes paved the way for the use of woodblock printing and later printing from movable type.

Fig. 1099. A paper tiger. Paper Museum, Tokyo.

(i) *Fingerprints and hand copying*

The desire to make a duplication of a certain image or text was probably a natural
and common practice from very early times. The characters *erh*[1] for two or double,
and *fu*[2] for second or duplicate which appear in many ancient documents, testify to
the nature of duplication and the existence of multiple copies; indeed the two-

[1] 二   [2] 副

stroke character *erh* was used as a ditto sign in the inscriptions on the stone drums dated variously from the −8th to −4th century.[a] Treaties among feudal states in the Chou dynasty were usually made in triplicate, with one copy for each of the parties and another to be filed with the spirits.[b] Then, during the Former Han dynasty, the original copy of meritorious records of marquises appointed by Empress Kao in −186 was held in the imperial ancestral hall, with a duplicate (*fu*) kept in the government office.[c]

Even before copies were made by hand, duplicate images were formed by impression from finger tips or from the palm in very early times in China. These prints have been found on pottery and clay articles as well as on documents, and are recorded in early literature.[d] The use of fingerprints for identification and authentication by the illiterate was probably a substitute for the seals of the literati, since both were intended to be unique and private; duplicates could be made only with specific authorisation.

Before printing was adopted popularly, and even after its extensive use for book production, manuscripts made by hand-copying were still common. This was because it was not only cheaper but also more convenient than printing for making single copies or a limited number of reproductions. Classics are said to have been copied by the father of Thao Hung-ching[1] (+451−536), the famous alchemist and physician, for sale at forty cash per leaf in the +5th century,[e] and the famous woman Wu Tshai-luan[2] of the early +9th century is said in many tales to have made hand-copies of rhymed books in fine calligraphy for civil service candidates, for 5000 cash per copy.[f] Many Buddhist manuscripts from Tunhuang were also copied by professional scribes with a good standard style of calligraphy, for sale to those who wished to fulfil their vows to spread the *sutras* (Fig. 1100).[g]

Official scribes were to be found at court to copy books into the imperial collections. As early as the +3rd century, the Chin Imperial Library had official scribes able to write the standard *khai* style of calligraphy for copying books on silk and paper.[h] In the Sui dynasty, it is recorded that during the reign of Emperor Yang (r. +605−17), a choice collection of 37,000 *chüan* was selected for the Imperial Library and fifty manuscript copies were made of each book to be kept in

---

[a] See transcription of the text in Kuo Mo-Jo (*14*), ch. 1, pp. 12−25.

[b] Creel (1), pp. 37−8, citing an agreement between Chin and Chhu in −578 as recorded in the *Tso Chuan*.

[c] *Chhien Han Shu* (*ESSS/TW*), ch. 16, p. 2*a*.

[d] The Chinese origin of the finger print system and its introduction to Europe and America by way of India is discussed in Laufer (51), pp. 631 ff.; for illustrations of fingerprints in documents, see Niida Noboru (1), pp. 79−131.

[e] Mao Chhun-Hsiang (*1*), p. 73.

[f] Yeh Te-Hui (*2*), pp. 285−8.

[g] The style of writing in colophons by the worshippers is quite different from that of the text, which was supposedly written by professional copyists. Shops for copying Buddhist *sutras* existed in the Thang dynasty, and an imperial order was issued in +714 to ban such shops which did not conform to Buddhist disciplines; see *Thang Ta Chao Ling Chi* (Shanghai, 1959), ch. 103; p. 588.

[h] *Wei Lüeh*[3] (*TSHCC*), p. 125.

[1] 陶宏景   [2] 吳彩鸞   [3] 緯略

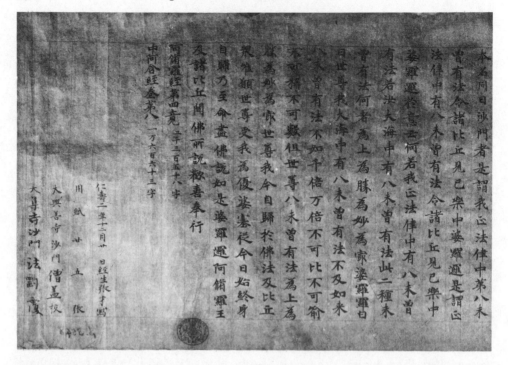

Fig. 1100. A Buddhist *sutra* from Tunhuang, *The Middle Āgama*, *chüan* 8, in 10,663 characters. It was written with 25 sheets of paper by the scribe Chang Tshai in +602 and proof-read twice by two monks. British Library.

two new buildings at the court.[a] Because so much could be done by hand, printing was not necessary unless a very large number of copies was in demand.

(ii) *Impression of seal inscriptions*

The carving and impressing of seals is considered one of the technical precursors of the invention of printing in China. The technique of carving a mirror image of characters in relief on stone, wood, or other materials in making a seal is almost the same as that of engraving characters on wood blocks or making individual types for printing. The only difference is probably the size and purpose of the carvings, and the method of casting metal seals is little different from that of casting metal types from a matrix. The stamping of seal inscriptions on clay and later on silk and paper

[a] See *Wen Hsien Thung Khao* (*ST*), ch. 174, pp. 1506–7. The source also indicates that the Imperial Library was furnished with mechanical devices to open and close the doors and curtains in the reading room. The text says: 'Fourteen rooms in front of Kuan-Wen Court served as imperial reading-rooms. The doors, windows, beds, mattresses, bookcases and curtains in these rooms were all lavishly ornamented. For every three rooms, there was a square opening with brocade curtains hanging down and two figures of flying fairies atop. A mechanical device was placed on the ground outside of the rooms. When the Emperor visited these reading-rooms, a servant holding a censer would precede. As he stepped on the device, the flying fairies would come down and pull open the curtains up to the top of the doors. The doors of the bookcases would also open automatically. When the Emperor left, the doors would close and the curtains come down again.'

was probably the earliest attempt to reduplicate writings by a mechanical process.[a]

Seals were made of almost any kind of hard-surfaced material. They were cast of bronze, gold, silver, and iron; or carved of stone, jade, clay, ivory, horn, and wood. They were usually square, though some were rectangular or round, about one or two inches in diameter. Inscriptions were carved or cast on one or more sides of the block, which was sometimes decorated with a knob on top, and with a string attached. They usually bore an inscription of a few characters to give a personal name, official title, and name of studio, or other indication of ownership, authentication, or authority.[b]

The use of seals in China can be traced back to the Shang dynasty. Three old square seals, cast in relief on a flat surface of bronze, are said to have been found at Anyang,[c] while later seals made of bronze, gold, jade, turquoise, and soapstone in various shapes and sizes, dating from the Chou, Chhin, and Han dynasties, have been discovered at various sites in China. Most of the Chou seals made of bronze were cast in moulds with inscriptions mostly in relief, and only a small number of surviving specimens have intaglio inscriptions. The seals of Chhin are similar to those of Chou except for the creation in −213 of a large imperial seal with eight characters carved on jade, to display the authority of the First Emperor. This piece was used by the successive Han emperors as the seal of inheritance of the empire,[d] though Han seals were mostly made of metal with inscriptions cast in intaglio. After the Han, however, all official seals were made with inscriptions in relief, and their size was gradually increased more or less in proportion to the rank of the officers using them (Fig. 1101).[e]

Seal inscriptions were at first stamped on clay and later impressed on silk and paper. For secrecy and authentication, documents of bamboo and wood, when ready for transmission, were covered with a board and bound with string on which a small piece of clay was affixed and impressed with a seal. After bamboo and wood ceased to be used as writing materials, seal impressions were made on soft materials, and the earliest known example with black ink is found on a piece of silk from Tunhuang dating from the +1st century (Fig. 1102).[f] Since the +5th century, seals were generally applied to paper with vermilion ink, one early reference to such impressions dating from +517, when they were stamped in red on lists of officials addressed to the court.[g] Since paper written with black ink had been commonly used for documents and books since the +2nd or 3rd century, the use of red ink for stamps in order to distinguish the seal impression from the black text may, perhaps, be earlier than the +6th century.

[a] The character *yin*[1], which was used for 'seal' in the Han, was adopted to mean printing when that began in China; cf. pp. 5 ff.; 148 ff.

[b] Cf. Tsien (2), pp. 54–8; Li Shu-Hua (4), pp. 61–73.

[c] See Yü Hsing-Wu (*1*), II, pp. 11–13.   [d] *Shih Chi* (*ESSS/TW*), ch. 6, p. 4 *a, b*.

[e] Cf. Li Shu-Hua (4), p. 65.

[f] See discussion and illustration in Tsien (2), pp. 55–6; pl. VIII, fig. E.

[g] Carter (1), p. 13, n. 10; Li Shu-Hua (4), p. 63.

[1] 印

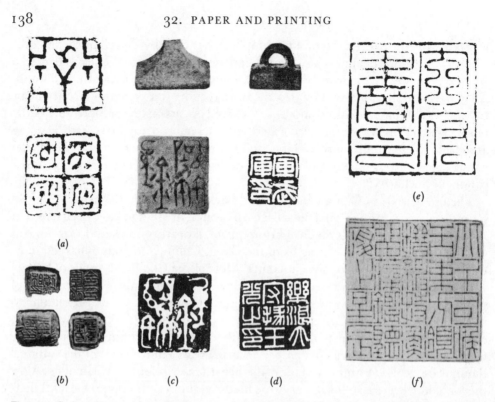

Fig. 1101. Seals, seal impressions, and sealing clays of various periods. (a) Impressions of bronze seals of the Shang dynasty. (b) Sealing clays of the Han period. (c) A carved seal of the pre-Chhin period with its back and impression. (d) Impressions from two Han seals, top one with a ring handle. (e) Seal of the Northern Sung Imperial Library carved in relief. (f) A poem in 20 characters carved on a stone seal, c. +1850.

Fig. 1102. Earliest known seal impression in positive image on silk fabric from Jen-Chheng, Shantung, c. +100, enlarged twice. From Tsien (1), pl. VIIIE.

Seals were also carved on wood, and these had a much larger surface than those of metal and other materials. As Ko Hung (+284–363) mentioned, in ancient times a seal of the Yellow God, four inches in breadth and bearing 120 characters, was used to make impressions on clay along the routes taken by travellers to keep away fierce animals and evil spirits.[a] Another source says that Taoist priests cut seals of jujube heartwood, four inches square,[b] apparently to duplicate charms on clay and then, later, on paper in vermilion ink. It is also recorded that at the court of Northern Chhi (+550–77), a large wooden seal one foot two inches long and two and a half inches wide, bearing four characters, was used to stamp the joining sheets of documents.[c] All this evidence indicates that seal-cutting on wood with mirror-image characters in relief and bearing a text of as many as over a hundred characters, can truly be considered as a forerunner of woodblock printing.

(iii) *Bronze casting and stone carving*

Other techniques which contributed directly or indirectly to the invention of printing include bronze casting and engraving on stone. Two methods are known to have been used in casting bronze vessels and their inscriptions in ancient times, namely: clay moulds and the lost-wax process. The latter, which involved the construction of a mould with characters originally traced in wax which was later to be melted and replaced with metal, may have suggested the carving of writing in reverse to obtain a positive position on the object to be cast, as was later to be done in printing.[d]

Some of the techniques of early casting may even have suggested the use of movable type to compose a long text, for it was not uncommon to use separate moulds, each with a single character or a group of characters, to make one vessel or one inscription. One of the most interesting examples still in existence is the inscription on a *kuei* vessel of Chhin (*Chhin kung kuei*[1]), probably of the −7th century, each character of which can be seen to have been cast from a separate unit, because the edges of the individual units are visible between the characters (Fig. 1103).[e] Another example is the inscription on a bronze bell of the late Chou period, with ornamental archaic characters (*Chhi tzu chung*[2]), each separately cast from an individual mould.[f] There are also examples of individual moulds bearing a group of characters instead of a single one, and a pottery container of the Chhin dynasty (*Chhin wa liang*[3]) had an inscription of forty characters made from ten separate moulds, each with four characters.[g] It would seem that this technique of using

---

[a] *Pao Phu Tzu, Nei Pien (SPTK)*, ch. 17, p. 23a.
[b] *Chhu Hsüeh Chi* (Peking, 1962), p. 624.
[c] *Thhung Tien (ST)*, p. 3586.   [d] Cf. Yetts (1), pp. 34–9.
[e] See descriptions in Lo Chen-Yü (4), p. 32b; illustrations in Jung Keng (4), I, pp. 88 (illus. 35), 158; Su Ying-Hui (8), p. 19.
[f] Jung Keng (4), I, p. 89 (fig. 89), 158.   [g] Lo Chen-yü (5), *Ping Pien*, ch. 2, pp. 2a–b.

[1] 秦公簋   [2] 奇字鐘   [3] 秦瓦量

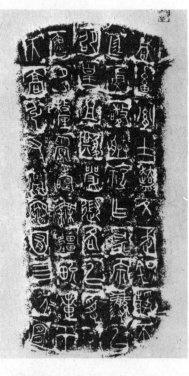

Fig. 1103. Inscription on a *kuei* vessel of Chhin cast from separate units, *c.* −7th century. From Jung Keng (4), fig. 35.

separate units in the composition of one text may be considered as the forerunner of typography.[a]

Carving inscriptions on stone is another technique which is considered as a prerequisite for engraving on wood for printing, for not only the technique of cutting, but also the change of material from stone to wood is significant for the development of printing. Cutting inscriptions on stone for commemorative and monumental purposes developed early in China. The oldest inscriptions on stone still surviving are those on ten drum-shaped boulders known as the Stone Drums of the Chhin State. They were carved with ten verses, originally with some 700 characters, although fewer than half of them are still extant.[b] After the unification of the empire, the First Emperor of Chhin erected seven stone inscriptions throughout the country between −219 and −211; these were in praise of the achievements of his administration, especially his role in the standardisation of measures and of the style of writing.[c]

From the +2nd or 3rd century, stone was extensively used not only for commemorative purposes but also as a permanent material for preserving canonical

[a] Lo Chen-yü (4), p. 32 b.
[b] For description and dating of the Stone Drums, see Tsien (2), pp. 64–7; also S. W. Bushell (5), pp. 133 ff.
[c] Tsien (2), pp. 68–9.

Fig. 1104. One of the 7000 stone tablets of Buddhist scriptures preserved in the grotto library of Fang-Shan, Hopei province, engraved since +550, each 30 metres in height. From Tsien (1), pl. XIV.

literature by Confucians, Buddhists, and Taoists.[a] A complete collection of seven Confucian classics, amounting to over 200,000 characters, was engraved on some forty-six stone tablets between +175 and 183. Since then there have been no fewer than six additional major engravings of standard texts of Confucian classics on stone; the last one was made at the end of the 18th century. And a complete set of the Thang engravings, made between +833 and 837, still survives in the Forest of Stone Tablets (Pei Lin[1]) in Sian.

The Buddhists also selected stone as a permanent material for preservation of their sutras, with the aim of avoiding destruction during periods of suppression of that religion. The most gigantic of all their stone inscriptions is probably that of 7000 stone tablets preserved in a grotto library in a mountain near Fang-shan, Hopei; all the tablets having been cut successively over many generations from the +6th to the 11th century (Fig. 1104). Though the engraving of Taoist literature on stone came later, at least eight engravings of the *Tao Te Ching*[2] are known to have been made during the Thang dynasty; in both scope and quantity, however, they are much inferior to both Buddhist and Confucian inscriptions.

Stone inscriptions were occasionally cut in relief or in a mirror image, like wood blocks for printing. They were also sometimes cut in mirror image in intaglio,

[a] For detailed discussion of the Confucian, Buddhist, and Taoist canons on stone, see Tsien (2), pp. 73–83.

[1] 碑林　　[2] 道德經

Fig. 1105. Inscription on the back of a sculpture at Lungmen, Honan with positive image cut in relief with squares, dated +5th century.

contrary to the general practice of cutting stele inscriptions in the normally direct or positive way. A few examples of this are known or are still extant, one being on the back of a sculpture at Lung-men, dating between +477 and 499, and another dating between +570 and 575; both are positive images cut in relief (Fig. 1105).[a] A stone stele with negative or mirror image inscriptions cut in relief is located near Nanking. There is also a pair of stone pillars, one cut in positive characters, reading from right to left, and the other in reverse, reading from left to right, apparently to balance the pair (Fig. 1106).[b] It is especially significant that stone inscriptions were sometimes transferred to woodblocks, as Tu Fu (+710–770) said in his poem on calligraphy, "The stele of I-shan was burned down; its inscriptions rubbed from jujubee wood are fat and distorted."[c]

[a] See Su Ying-Hui (8), p. 22.
[b] See Liu Chhao Ling Mu Tiao Chha Pao Kao (Nanking, 1935), plate 11, fig. 20 a,b; also Liu Chhao I Shu (Peking, 1981), plates 284–5.   [c] See Tu Shih Yin Te, vol. 2, p. 216.

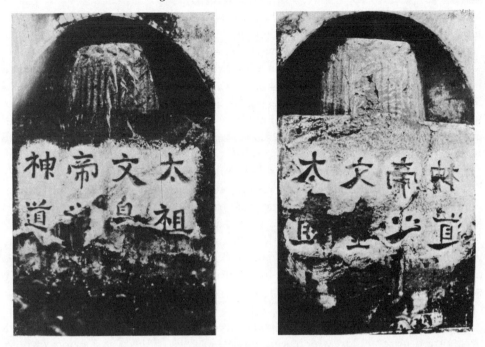

Fig. 1106. Stone pillars for the tomb of Wen-Ti of the Liang dynasty, c. +556, with inscriptions in positive image on the left and mirror image on the right. From *Liu Chhao Ling Mu Tiao Chha Pao Kao*, 1935.

(iv) *Inked squeezes and stencil duplication*

Rubbing is a process of making inked squeezes on paper from inscriptions on stone, metal, bone, or other hard-surfaced materials. The process of stone rubbing is very similar to that of block printing; the difference lies only in the methods of engraving and of duplicating. Except for very few cases, inscriptions on stone are always cut into the surface in intaglio with characters in the normal positive form. When a rubbing is made, the paper is laid on the stone and squeezed against the surface. Ink is applied to the surface of the paper, thus producing a white text on black background. The wood block, on the other hand, is always cut in relief with characters as a mirror image. When a print is made, ink is applied to the block, the paper is placed on it, and the back of the paper is brushed to obtain a black text on white background. Although the basic materials for engraving and the end products are different, the purpose of making duplications and the use of ink and paper as media are the same.

The technique of rubbing involves the processes of laying the paper on stone, tamping the paper into the intaglio, applying ink to the paper, and removing the paper from the surface after completion.[a] The whole process is much more complicated and slower than that of printing. Usually the soft paper was first folded

---

[a] For a detailed discussion of the technique, see Ma Tzu-Yün (*1*).

and then moistened, plain water or rice water sometimes being used, though the most common liquid was made by steeping slices of the dried root of *pai chi*[1] (*Bletilla byacinthina*), a tropical orchid, in water. A solution of glue and alum was also adopted on occasions, but since alum injures the stone and makes the paper fragile, it was not recommended by those experienced in making rubbings.

After the paper was properly placed, it was pressed lightly into every depression by a brush of natural fibre, usually that of the coir palm. When the paper was tight against the surface and about to dry, ink was applied with a pad, the inked pad being first struck lightly over the paper before the final application of dark ink. If the surface was plain and smooth, a light inking was sufficient and produced an inked copy considered as light as the cicada's wings; it was called *chhan i tha*[2] (cicada wing rubbing). If a dark ink was added and brushed on after the application of the light ink, the brushing action gave a dark and shining rubbing called *wu chin tha*[3] (black-golden rubbing). When the desired intensity of ink had been achieved, the rubbing was peeled off the hard surface and pressed flat, but this had to be done carefully, because distortions of the inscription resulted if the paper was stretched during peeling. The quality and thickness of the paper determined the ease or difficulty of the peeling procedure.

Rubbings were sometimes made from three-dimensional objects such as round or square bronze vessels, which were copied in perspective with a photographic effect known as *chhüan hsing tha*[4] or whole-shape rubbing (Fig. 1107). Before making such a rubbing, careful observation and study of the object was normally required, and the shape of the vessel, the curve of its surface, the distance between its front and rear, and other details were sketched. This sketch was then transferred to the paper to be moistened with the orchid liquid and placed on the vessel's surface. When the paper was almost dry, dark ink was applied to the relief portions and light ink to the intaglio parts of the design. Sometimes separate pieces of paper were used on different parts of a vessel and then pieced together to make a composite rubbing, though a single piece covering the entire vessel was sometimes used. The key technique in making whole-shape rubbings was primarily the application of the ink in the correct gradations of light and dark in accordance with the perspective sketch.[a]

The technique of rubbing is believed to have been first used in China before the +6th century, and it became well established in the following centuries.[b] Technicians in charge of rubbing, known as *tha shu shou*[5], were employed in the Thang dynasty, in academic institutions and imperial libraries, along with such skilled workers as scribes, paper-dyers, and brush-makers. A few specimens of inked rubbings from the Thang period were found in Tunhuang, including the earliest known piece extant of the inscription of the pagoda of the Hua-tu

[a] See *WWTK*, 1962, no. 11, pp. 59–60.
[b] See Tsien (2), pp. 86–9.

[1] 白芨   [2] 蟬翼拓   [3] 烏金拓   [4] 全形拓   [5] 搨書手

Fig. 1107. A composite rubbing in three dimensions of a round bronze vessel. Field Museum of Natural History.

Temple,[1] dated +632. Others include the *Wen Chhüan Ming*[2] (Fig. 1108), an inscription on the hot spring written by Thai-tsung of Thang, dated before +654, and a rubbing of the *Diamond Sutra*, made in +824. Since the time of the Sung dynasty, the technique of rubbing was broadened to include inscriptions on bronze vessels and later on pottery, bone, and other inscribed objects. It became a refined and sophisticated art for making reproductions, even excelling photography in skilfully and faithfully reproducing the exactness of inscriptions and designs on a great variety of materials.

The use of stencils was another pre-printing method for duplication. The stencil was usually made of a sheet of thick paper perforated with needles to form the designs to be reproduced. The stencil was laid on the surface and the design was transferred by applying ink to the perforations. The date of the earliest use of the stencil is unknown, but the recent discovery of silk fabrics printed in coloured patterns from the Han tomb at Ma-Wang-Tui, Chhangsha, indicates that the

---

[1] 化度寺   [2] 溫泉銘

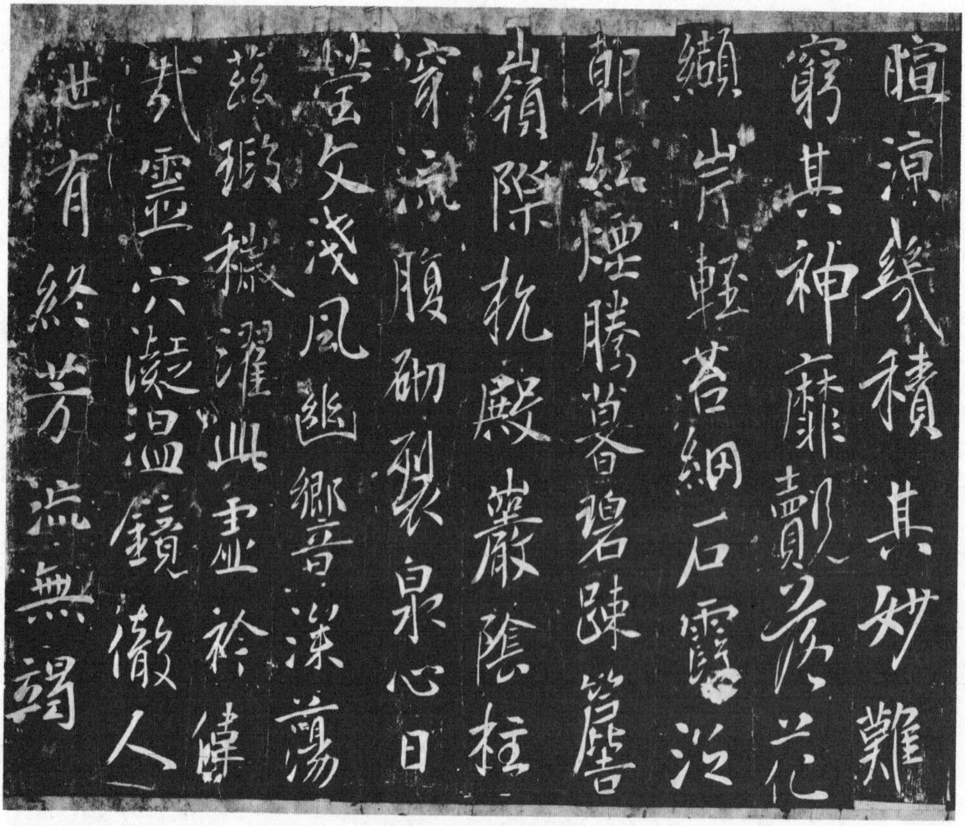

Fig. 1108. One of the earliest extant inked rubbings of a stone inscription on a hot spring, dated *c.* +654, from Tunhuang. Cut and mounted on buff paper and now preserved at the Bibliothèque Nationale, Paris.

technique can be traced back to the −2nd century. Animal skin or thin silk fabric treated with varnish or some other tree sap may have been used at this time,[a] and certainly such a use of skin and paper was common in the Thang and Sung. Several paper stencils with perforated designs of Buddhist figures have been found in Tunhuang, together with finished stencilled pictures on paper, silk and on plastered walls (Fig. 1109);[b] other paper stencils of later dates already in museums were used for the reproduction of designs on textiles.[c]

(2) BEGINNINGS OF WOODBLOCK PRINTING

Numerous dates have been suggested for the earliest use of woodblock printing in China, varying from the middle of the +6th to the end of the 9th century.[d] No

[a] Cf. Wang Yü (*1*), pp. 474–8.   [b] Cf. Stein (*4*), IV, pl. XCIV.
[c] A buff paper stencil coated with oil, collected by the late Berthold Laufer, is in the Field Museum of Natural History in Chicago.
[d] Some two dozen theories are summarised and discussed in Shimada Kan (*1*), Sun Yü-Hsiu (*1*), Pelliot (*41*), Carter (*1*), Li Shu-Hua (*11*), and Chang Hsiu-Min (*5*).

147

Fig. 1109. Paper stencil with perforated designs and its print on paper. British Library.

printed specimen dated earlier than the +8th century survives, but literary evidence indicates that printing may have started earlier than this. Actually, most of the opinions are based primarily on the interpretation in early literature of certain key terms which may have meant engraving or printing. These include words referring to methods of cutting, such as *khan*[1] (to cut), *kho*[2] (to engrave), *tiao*[3] (to carve in relief), and *lou*[4] (to incise); to methods of printing or publishing, such as *yin*[5] (to print or impress), *shua*[6] (to brush), and *hsing*[7] (to distribute); to the material of the block, such as *mu*[8] (wood) and *pan*[9] (wood block); or to certain kinds of wood used for preparing the blocks, such as *li*[10] (pear), *tsao*[11] (jujube), and *tzu*[12] (catalpa). Thus, such combinations of these words as *tiao pan* (to carve wood blocks), *pan yin* (to print from wood blocks), or *tzu hsing* (to cut catalpa and distribute) have been used in one way or another to mean cutting, or publishing. Some of the theories relating to +6th-century printing, based on unrelated terms, are certainly untenable, because the terms thus cited are either misinterpreted or misrepresented, or because the reliability of the sources appears to be questionable.

Four instances of gross misinterpretation relating to +6th-century printing are illustrated here as examples. Shimada Kan takes the term *shu pen*[13] in Yen's *Family Instructions*, written in +550–77, to imply the existence of printing; but the term clearly means 'book' or 'manuscript', not 'printing'. Again, the term *tiao chuan*[14] (to carve images and compose *sutras*) in a Buddhist work of +593 was mistakenly quoted as *tiao pan* (to carve wood blocks) in many secondary sources, which were in turn used by some early sinologists and in certain Western-language works, including the earlier editions of the *Encyclopaedia Britannica*. Then, a *dharani* from Tunhuang printed in +980 was thought to be a reprint of one originally printed in the Sui dynasty (+581–618) because the characters *ta sui chhiu*[15], representing the name of a *sutra*, were misread as referring to the Great Sui dynasty. Actually, the text of this *sutra* was not translated into Chinese until the later part of the +8th century. Finally, a fragment of paper with characters bearing a date equivalent to +594, discovered by Stein in Chinese Turkestan, was reported as a printed poster. But examination of this document by the Research Laboratory of the British Museum showed no indication of printing.[a] Thus none of these four items is acceptable as proof of printing in the +6th century.

As for theories about printing in the +7th century, the evidence is more convincing, though some points are controversial. One statement in a Thang work says the famous monk Hsüan-Tsang (+602–64), who travelled to India in 624–45, printed (*yin*) pictures of Samomta Bhadra on paper, perhaps after his return to China and before his death. This statement may be reliable in spite of the

---

[a] Cf. Carter (1), pp. 40–1.

[1] 刊	[2] 刻	[3] 雕	[4] 鏤	[5] 印
[6] 刷	[7] 行	[8] 木	[9] 板	[10] 棃
[11] 棗	[12] 梓	[13] 書本	[14] 雕撰	[15] 大隨求

fact that the dating of the book has an element of controversy.[a] Another text, this time by the pilgrim I-Ching (+634–713) reported in +692 that Indian Buddhists printed (*yin*) Buddhist images on silk and paper; however, there is a suspicion that the word *yin* may have meant impression with image blocks rather than real printing.[b] On the other hand, a more recent opinion suggests that monk Hui-Ching[1] (fl. +600–50), who arrived at Chhangan in 600, and defended Buddhism by saying that *sutras* were written and engraved (*shan kho*[2]) to insure their existence, was clearly implying that printing was used at the end of the Sui dynasty.[c] Since Buddhist texts preserved on stone had been a common practice, this word *kho* could refer to engraving on stone instead of on wood, and thus this statement may not in fact be taken as definitive evidence for printing at this time.

Finally, yet another document mentions that a women's code compiled by the Empress Chhang-sun was ordered to be carved on wood blocks for distribution (*tzu hsing*) at her death in +636, which may imply printing in that year.[d] This story is due to the Ming historian Shao Ching-Pang[4] (+1491–1565), but the passage containing the term *tzu hsing* does not appear in the two Thang standard histories or in other documents which include the same story. Thus certain evidence is lacking in support of this statement from a secondary source.[e]

As for +8th-century printing, several specimens are extant. The earliest of these is a *dharani sutra* scroll (Fig. 1110) which was discovered in +1966 in a stone stupa in the Buddhist temple Pulguk-sa, Kyongju, in southeast Korea. The scroll bears no date, but it includes certain special forms of characters created and used when Empress Wu (r. +680–704) was ruling in China. It is believed that this charm must have been printed no earlier than +704, when the translation of the *sutra* was finished, and no later than +751, when the building of the temple and stupa was completed.[f] Another piece of old printing, chapter 17 of the *Lotus sutra*, is said to have been found in Turfan and is preserved in Japan. It is printed on yellowish hemp paper with nineteen characters per line in the text, which contains also some of the peculiar characters commanded by Empress Wu.[g] If this is evidence of

[a] The story is included in Feng Chih, *Yün Hsien San Lu*[3], preface dated +926, but some critics have suspected it is a forgery because of an edition with a wrong date of +901 in the preface, in which he says the book was completed several years after his return home in +904. If the date +926 in the Sung edition is correct, the authenticity of the work should be no problem. Both Hsiang Ta (*13*), pp. 5–6, and Chang Hsiu-Min (*17*), p. 346, are inclined to believe that printing in the +7th century is highly possible.

[b] Both Pelliot (*41*), pp. 14–19, and Li Shu-Hua (*11*), pp. 82–4, believe that the use of paper for printing in India may have been introduced from China, since paper was rare in India until the Moslem dynasty in the +12th century.

[c] Chang Chih-Che (*1*), pp. 154–5; Yü Wei-Kang (*1*), pp. 231–4, says that the term *shan kho* derives from the *Chuang Tzu* and does not mean writing and engraving.

[d] See *Hung Chien Lu* (*SLHP*), ch. 46, pp. 3*a,b*; Chang Hsiu-Min (*5*), p. 59.

[e] Chang Hsiu-Min (*17*), p. 345, argues that the exclusion of the term from the standard histories does not necessarily prove that this story is not true, especially since the author is considered to be a reputable scholar, who must have based it on sources other than the standard history.

[f] Cf. Goodrich (*31, 32*); Ledyard (*2*); Yi Hong-jik (*1, 2*); and further discussion below, p. 322.

[g] Nagasawa Kikuya (*3*), pp. 5–6; this copy is now kept in the Calligraphy Museum, Tokyo.

[1] 慧淨   [2] 繕刻   [3] 雲仙散錄   [4] 邵經邦

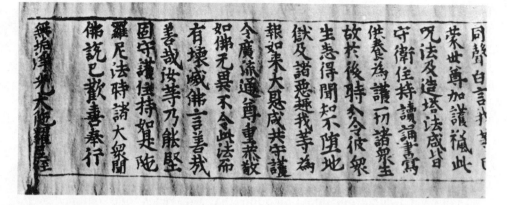

Fig. 1110.　The end section of a *dharani* charm scroll, printed in the early +8th century, found in Pulguk Temple in southeast Korea in +1966.

printing during her reign, it could be another specimen contemporary with the one found in Korea.

The specimen found in Korea predates the *dharani* charm from Japan printed *c.*+764–70, and which until recently was considered to be the world's oldest extant example of printing.[a] That there were a million copies of the charm in Japan is recorded in several contemporary documents, which also relate that in +764 a copy of one of the four versions of the *dharani* was placed in every one of a million tiny wooden pagodas (Fig. 1215) ordered by the Empress Shōtoko to be distributed and stored among ten different temples. Although the records do not say that the charm was printed,[b] not only do many printed specimens of it survive, but it seems certain that such a multitude of copies could not have been made without the aid of printing. The text is from the same *sutra* as the one recently discovered in Korea, but the latter consists only of the prayers in Chinese characters transliterated from the Sanskrit, while the scroll from Korea is much longer, including the story as well as the prayers of the *sutra*.

Since both of these examples were found outside China, it may, of course, be questioned whether they are, in fact, Chinese printing. While the *dharani* found in Japan is believed to have been produced there, as it is documented in Japanese records, the single piece of *dharani* found in Korea is very likely to have been printed in China. The fact that there were frequent pilgrimages of Korean monks and students to the Thang capital, that the Silla Kingdom in Korea was zealous in adopting Chinese culture and practices, together with both the presence of special writing forms of Empress Wu, and also the lack of any collateral evidence of early printing in Korea, suggests that this printed *sutra* originated in China and was

[a] For a fuller description of this charm, see Carter (*1*), ch. 7; Hunter (*9*), ch. 3; Nagasawa Kikuya (*3*), pp. 6–8; cf. below, pp. 336 ff.
[b] Chang Hsiu-Min (*5*), p. 134, questions this because of the absence of mention of printing in all the documents and the fact that no other printing earlier than +1172 in Japan is mentioned or has survived.

perhaps brought to Korea for cermonial use when the temple was built. In any case, the use of printing in China must have begun some time before the date of the examples found in Japan and Korea.[a]

The two charms just mentioned are both miniature examples of printing, unlike the usual size of a Chinese book. The first complete printed book is probably the famous *Diamond Sutra* of +868, discovered in Tunhuang by Stein during his second expedition in 1907.[b] This book, in roll form, is made of seven sheets of white paper pasted together to form a scroll with a total length of $17\frac{1}{2}$ feet. Each sheet is $2\frac{1}{2}$ feet long and $10\frac{1}{2}$ inches wide. The text is the complete work of *Chin Kang Po Jo Po Lo Mi Ching*[1], translated into Chinese from the Sanskrit *Vajracchedikâ Prajñâ Paramitâ* by Kumârjîva (b. +344) in the +4th century. Both the picture on the frontispiece (Fig. 1167) and the calligraphy in the text show a highly advanced technique in cutting and printing, more refined than those found in Japan and Korea, or in Europe of pre-Gutenberg date. At the end of the roll a colophon says: 'On the fifteenth day of the fourth moon of the ninth year of Hsien-thung (+868), Wang Chieh reverently made this for blessings to his parents, for universal distribution' (Fig. 1111). This is the earliest clearly dated printing in complete book format extant today.

Among printed materials of the Thang dynasty, several other examples may be mentioned. These include the printed versions of the *dharani* (Fig. 1112), of Buddhist verses, and the two oldest printed calendars, all discovered at Tunhuang.[c] One calendar, for the year +877 (Fig. 1113), is a fragment printed with minute drawings and diagrams, solar terms, and pictures of the animals corresponding to the twelve branches, very similar to those used on calendars even in modern times. The other, for +882, is printed with a line of very heavy characters as the heading, which reads: 'Family calendar of Fan Shang[2] of Chhengtu-fu in Hsi-chhuan,[3] province of Chhien-nan'. Apparently the private printing of family calendars was very popular in Szechuan and all along the Yangtze valley; indeed, a memorial of +835, submitted by Feng Su[4] (+767–836), a regional commandant of Szechuan, requested that the private woodblock printing of calendars be forbidden, because large numbers of unauthorised calendars were being printed and sold in markets before the Board of Astronomers had submitted the approved calendar for the new year to the emperor.[d]

Besides Buddhist *sutras* and calendars, many books on various subjects were also printed and sold in bookstores. A Thang official, Liu Phien[5], who was in Szechuan with the refugee emperor in +883, said in his family instructions that during his

---

[a] Despite the findings of earliest printing specimens outside China, there is little doubt that these printings followed the exact model and method of Chinese printing. As Goodrich (32), p. 378, says 'everything still points, in my opinion, to the beginnings of the invention in China and its spread outward from there'.
[b] Preserved in perfect condition at the British Museum; cf. Giles (17), pp. 1030–1; Carter (1), ch. 8.
[c] Giles (17), pp. 1033–4, 1036–7.
[d] See *Chhüan Thang Wen* (1818), ch. 624, pp. 14*b*–15*a*.

[1] 金剛般若波羅蜜經   [2] 樊賞   [3] 西川   [4] 馮宿   [5] 柳玭

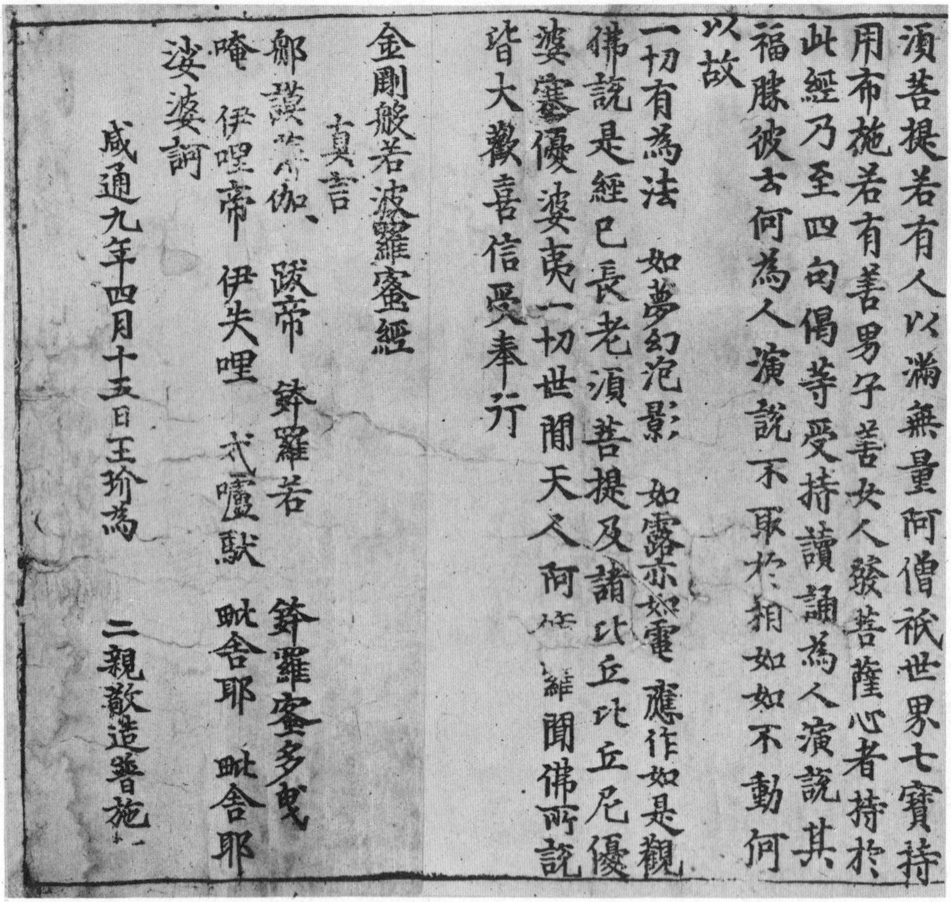
須菩提若有人以滿無量阿僧祇世界七寶持
用布施若有善男子善女人發菩薩心者持於
此經乃至四句偈等受持讀誦為人演說其
福勝彼云何為人演說不取於相如如不動何
以故
一切有為法 如夢幻泡影 如露亦如電 應作如是觀
佛說是經已長老須菩提及諸比丘比丘尼優
婆塞優婆夷一切世間天人阿修羅聞佛所說
皆大歡喜信受奉行
金剛般若波羅蜜經
真言
那謨薄伽跋帝 鉢羅若 鉢羅蜜多曳
唵 伊哩帝 伊失哩 戍嚧馱 毗舍耶 毗舍耶
莎婆訶
咸通九年四月十五日王玠為
二親敬造普施

Fig. 1111. Colophon at end of the *Diamond sutra* printed in +868, from Tunhuang. British Museum.

stay in Szechuan he saw a number of books on astrology, the divination of dreams, and geomancy, as well as dictionaries and other lexicographical works, being printed on paper from engraved wood blocks; but he commented that the ink was smeared and could not be read clearly.[a] Another Thang official, Ho-kan Chi[1] (fl. +847–59), a regional supervisory commandant in Chiangsi, is said to have spent much time in the study of Taoist alchemy and to have composed a biography of Liu Hung[2], several thousand copies of which were printed between +847 and +851 and distributed to interested alchemists.[b]

Among the books brought back by the Japanese monk Shūei[3] on his return from China in +865 were two rhymed dictionaries, *Thang Yün*[4] and *Yü Phien*[5],

[a] See *Ai Jih Chai Tshung Chhao* (*SSKTS*), ch. 1, p. 3*a*.
[b] See *Yün Chhi Yu I* (BH), ch. 10, pp. 6*a*,*b*.

1 紇干臮 2 劉弘 3 宗叡 4 唐韻 5 玉篇

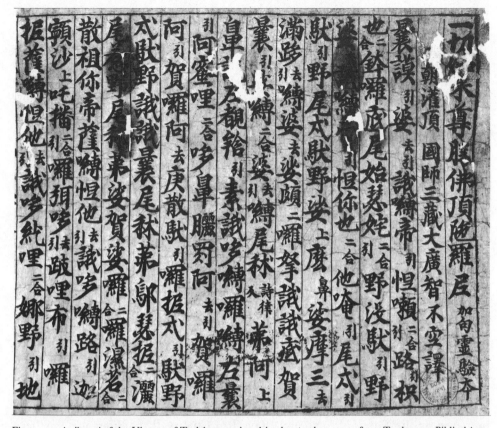

Fig. 1112. A *dharani* of the Victory of Tathâgata printed in the +9th century from Tunhuang. Bibliothèque Nationale, p. 4501.

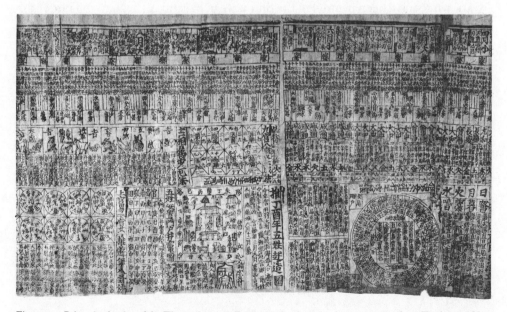

Fig. 1113. Printed calendar of the Thang dynasty. Fragment for the year *ting-yu* or +877 from Tunhuang. Very closely cut in horizontal strip with minute diagrams and drawings, including the 12 animals of the duodenary cycle. British Museum.

which were recorded as 'printed in Hsi-chhuan (modern Szechuan)', along with other titles apparently referring to manuscript calendars, medical prescriptions, and other secular works, besides Buddhist *sutras*.[a] A Thang official, Ssu-khung Thu[1] (+837–908), wrote, probably in +871–9, that the printed copy of the *vinaya* sutra in the Ching-ai Temple of Loyang, which had been destroyed, probably in +845, should be reprinted.[b] Therefore, a printed copy of the *sutra* must have existed before the date of destruction.

Also, a *dharani* charm printed by the Pien family in Chhengtu is reported to have been found within a hollow bracelet in a Thang tomb in Chhengtu in 1944. This piece, about a foot square, was printed on very thin but strong paper, which was probably made of mixed fibres of silk, mulberry bark, hemp, and *than*[3] wood (*Pteroceltis tartarinowii*, Maxim.) It bears a line of Chinese characters, from which a few are missing, indicating the place and name of the printer at the right, with Buddhist figures on the four sides and also at the centre, surrounded by the Sanskrit text forming a square in seventeen lines (Fig. 1114). The Chinese text says: 'This charm is printed for sale by Pien ... near ... Lung-chhih-fang, Chheng-tu hsien, Chheng-tu fu.' It has no date, but the tomb has been dated as c. +850–900 in the late Thang.[c] This adds another specimen of early printing preserved in China.

By the early part of the +10th century, under the Five Dynasties (+907–60), the application of printing seems to have been much wider both in subject-matter and in geographical distribution. Printed materials include, for the first time, the Taoist canon and Confucian classics, literary anthologies, historical criticism, and encyclopaedic works, besides the Buddhist *sutras* and calendars. Printing centres included Loyang and also Khaifeng in modern Honan; most of the works being prepared by the National Academy (Kuo Tzu Chien)[5] throughout the Northern dynasties (+907–60) were printed there. In the Southern Kingdoms, a few books are known to have been printed in the Shu state in Szechuan, in Nanking under the Southern Thang, and in Hangchow under the Wu-Yüeh state.

The first work of the Taoist canon to be printed was a study by the Taoist monk Tu Kuang-Thing[6] of the commentary on the *Lao Tzu* by Emperor Hsüan Tsung (r. +713–55), entitled *Tao Te Ching Kuang Sheng I*.[7] It was privately printed at the author's expense from some 460 blocks, in +913, twelve years after he had

---

[a] The catalogue is included in the *Taisho Daizokyo*, vol. 55 (Mokuroka-bu), pp. 1108–11. Another Japanese monk Ennin[2] (+793 to +864) mentions that he bought in China in +838 a copy of the *Vimalakirti-nirdesa-sutra* for 450 pieces of cash and saw in +839 1000 copies of the *Nirvana-sutra* at Mount Wu-Thai. The cheapness of the cost and numerous copies lead one to believe that they were printed editions; see Reischauer (4), pp. 48, 137; Goodrich (28), p. 38.

[b] The dates in connection with this statement are suggested by Hsiang Ta (*1*); 9, 12; see also Carter (1), p. 61; Li Shu-hua (*11*), p. 117–21.

[c] This printing can be dated no earlier than +757, since Chheng-tu was not called *fu*[4] until this date, or +841–6, when the coins from the I-chou mint found in the tomb were manufactured. See Feng Han-Chi (2) with picture, which is also reproduced in *Chung-Kuo Pan Kho Thu Lu*, plate 1.

[1] 司空圖   [2] 圓仁   [3] 檀   [4] 府   [5] 國子監
[6] 杜光庭   [7] 道德經廣聖義

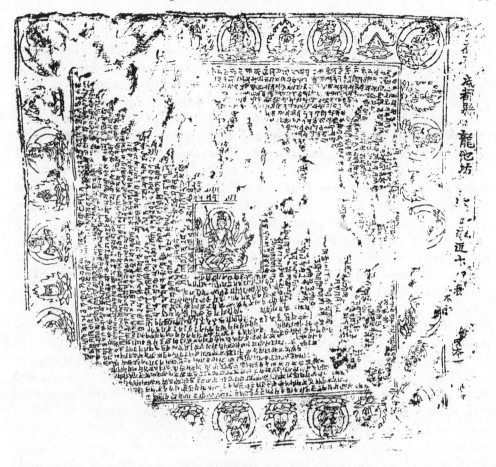

Fig. 1114. A *dharani* printed in Lung-Chhih-Fang, Chhengtu-fu, *c.* late +9th century, 31 × 34 cm., discovered in Chhengtu in 1944. From *Chung-Kuo Pan Kho Thu Lu*, 1961.

completed writing it.[a] Also printed in Shu was a collection of one thousand poems by the monk Kuan-hsiu[1] (+842–923), upon whom the King of the Shu state bestowed the honorific title Master Chhan-yüeh.[b] Publication of this book, *Chhan Yüeh Chi*[2], by a disciple of the author's in +923, marked the beginning of the printing of individual literary collections which has proliferated among Chinese publications ever since. In the Northern Dynasties, the Taoist canon was printed by a Taoist priest under the auspices of Emperor Kao-Tsu of the Chin in +940. A new preface was written by the scholar Ho Ning[3] (+898–955), who printed several hundred copies of his own poems and songs in 100 *chüan* for distribution.[c]

[a] Li Shu-Hua (*11*), pp. 142–3.
[b] Yeh Te-Hui (*2*), p. 22; the book is reprinted in the *SPTK* from a Sung manuscript copy.
[c] *Chiu Wu Tai Shih* (*ESSS/TW*), ch. 127, p. 7*a*.

[1] 貫休   [2] 禪月集   [3] 和凝

However, the most important printing of this period consisted of two separate publications of the Confucian classics. In order to standardise the text, these had been engraved on stone at least three times since the end of the +2nd century, long before they were ever printed. It was, it seems, the economy and wide distribution made possible by printing, as exemplified by the many religious and secular works available at this time, that inspired the prime minister Feng Tao[1] (+882-954) and his associate Li Yü[2] (d. +935) to undertake publication of the Confucian classics. Despite the disturbances of civil wars during this period, Feng Tao was able to remain in high position in the government through ten reigns of five different dynasties, and this enabled him to carry out this task. It began in +932 when the chief counsellor recommended the recruiting of learned scholars from the National Academy to make a collation of the text based on the version of the Stone Classics of the Thang dynasty, and of good calligraphers to transcribe the correct text on paper in standard style, and this to be cut on wood blocks. Five sheets of text were to be prepared each day.[a] The text included the eleven classics authorised in the Thang, plus two supplementary works on the forms of characters used in the Five Classics and Nine Classics. The task took some twenty-two years; the collection was completed and printed in 130 volumes in +953.[b] This was the first time that the Confucian classics were printed, and the beginning of official publications for sale by the National Academy.

Two years later, in +955, a lexicon of classical terms, *Ching Tien Shih Wen*[4], was entrusted to Thien Min[5] for printing. Four years later (959), the *Shang Shu Shih Wen*[6], a section of the above, was prepared for printing by Kuo Chung-Shu[7] (+918?-77). It was twice revised and reprinted, in 972 and in 999.[c]

At the time when the printing of the Confucian classics was completed in Khaifeng, in 953, another project for printing them was started in the state of Later Shu (modern Szechuan) under the private sponsorship of its minister Wu Chao-I[8] (d. 967). Wu first entered the Shu administration in 935 and was promoted to minister in 944. It was said that in his youth he was poor, and when he wanted to borrow certain books from friends, they showed reluctance to lend them; he thereupon made a vow that if he should be prosperous one day, he would print the work for scholars. It was when he became minister of Shu that he fulfilled his pledge. After the conquest of Shu by the Sung, all the powerful families which had served the Shu state were punished and their property confiscated, except for the Wu family. Emperor Tai-tsu (r. +960-76) was fond of books and had discovered

---

[a] *Wu Tai Hui Yao* (*TSHCC*), p. 96, says that the text was copied in good calligraphy, and that 'five sheets every day' were given to the cutters; while the *Tshe Fu Yüan Kuei* (1640 ed.), ch. 608, pp. 30 *b*-31 *a*, says 'every 500 sheets of paper', the exact meaning of which is not clear.

[b] See all related documents translated in Carter (1), pp. 70-2, 76-9.

[c] The format of the Confucian classics printed by the National Academy has been preserved through a facsimile reprint in the +12th century. A copy of the *Erh Ya*, bearing the name of the calligrapher, Li O,[9] survives in Japan and is reproduced in the *Ku I Tshung Shu*[10] (1884).

[1] 馮道   [2] 李愚   [3] 國子監   [4] 經典釋文   [5] 田敏
[6] 尚書釋文   [7] 郭忠恕   [8] 毋昭裔   [9] 李鶚   [10] 古逸叢書

Wu's name in his publications. He therefore ordered that all the wood blocks be returned to the Wu's, and Wu Chao-I's descendants were supported from the large profit made by printing them; moreover, Wu's son, Wu Shou-Su[1], served in high positions both under the Shu ruler and later at the Sung court.[a] Not much information exists about the actual printing of the Confucian classics by Wu Chao-I who is said also to have printed other works. These included the *Wen Hsüan*[2], *Chhu Hsüeh Chi*[3], and *Po Shih Liu Thieh*[4], the last two being collections of literary quotations used by students in preparing for the civil service examination. The story of Wu Chao-I is probably the first used by Confucian scholars to exemplify the rewarding of the virtue of printing books, and similar stories were repeated in many later works concerned with books and printing.[b]

Other works printed during this period include the first book on historiography: *Shih Thung*[5] by Liu Chih-Chi[6] (+661–721), and the first collection of regulated verses; *Yü Thai Hsin Yung*[10], compiled by Hsü Ling[11] (+507–82). None of the printed works mentioned in various sources now survive, except for a few Buddhist scrolls and pictures printed toward the end of the tenth century. Of these the best known are the invocation *sutras*, titled *Pao Chhieh Yin Tho-Lo-Ni*[12], printed by Chhien Chhu[13] (i.e. Chhien Hung-Chhu, +929–88), prince of the Wu-Yüeh state (modern Chekiang together with parts of Chiangsu and Fukien, with its capital of what is modern Hangchow). At least three versions printed with different dates are known to have survived. One, printed in +956, two and a half inches wide and about twenty inches long, was found in a pagoda in the Thien-ning Temple in Huchow in 1917. Its text consists of some 341 lines of characters, with eight or nine characters to each line, preceded by an illustration of human figures, while a colophon says: 'The generalissimo of the empire and prince of the Kingdom of Wu and Yüeh, Chhien Hung-Chhu, has printed the *Pao Chhieh Yin* sutra in 84,000 rolls, and presented them for safe-keeping in precious pagodas. Recorded in the third year of Hsien-Te, *ping chhen* (+956).'[c]

Another version, dated *i chhou* (+965), was contained in a red wooden box, 10cm long, within a gilded stupa when discovered in 1971 at Shao-hsing, Chekiang. This copy consists of lines of eleven to twelve characters each, preceded by a frontispiece and colophon similar to but not identical with those of the other

[a] The stories about Wu Chao-I seem to have been derived from his disciple Sun Feng-Chi[7] (not to be confused with Sun Feng-Chi[7] of S/Sung), who was first quoted by Chhin Tsai-Ssu[8] (+10th century) in his book *Chi I Lu*[9], which was quoted repeatedly by Ming and Chhing authors, with factual discrepancies; see Li Shu-Hua (*11*), pp. 145–50; Weng Thung-Wen (*3*), pp. 27–8.

[b] See Yeh Te-Hui (*2*), pp. 1–4.

[c] See Wang Kuo-Wei (*3*); Giles (*15*), pp. 513–15; and Sören Edgren (*1*), pp. 141 ff., with plates of the full text. Only two copies of this version are known. One copy, in the Royal Library of Sweden, was acquired from a private collector in New York. Another copy is reported to have been found under a Sung pagoda in 1971 in Wu-wei, Anhui, which was out of the jurisdiction of the Wu-Yüeh Kingdom, indicating the circulation beyond its border; see Chang Hsiu-Min (*14*), p. 74.

[1] 毋守素	[2] 文選	[3] 初學記	[4] 白氏六帖	[5] 史通
[6] 劉知幾	[7] 孫逢吉	[8] 秦再思	[9] 紀異錄	[10] 玉臺新詠
[11] 徐陵	[12] 寶篋印陀羅尼		[13] 錢俶(弘俶)	

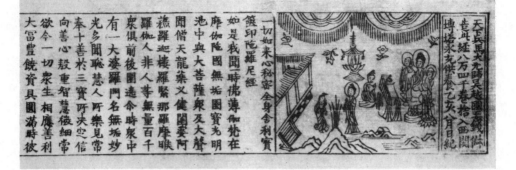

Fig. 1115. Invocation *sutra* printed by Prince Chhien Chhu of the Wu-Yüeh Kingdom, dated +975. Far Eastern Library, University of Chicago.

two. It is better cut with fine lines and printed on white rattan paper, while the other two are on yellowish paper.[a]

The third scroll, 6 feet 3·5 inches long and 1·2 inches wide, was found in 1924 when the Thunder Peak Pagoda in Hangchow collapsed during a storm. This text consists of 271 lines of ten or eleven characters each, preceded by a picture showing the consort of the prince with attendants making votive offerings (Fig. 1115). A colophon only slightly different from the others gives the name Chhien Chhu and date *i hai* (+975), which was three years before the kingdom was absorbed by the Sung empire.[b]

Chhien Chhu was not the only sponsor of printing at this time. A Buddhist monk named Yen-Shou[1] (+904–75), a chief priest at the famous Ling-Yin Temple[2] in Hangchow, printed probably more than a dozen titles of *sutras*, charms, and pictures of which over 400,000 copies are known to us, ranging from 20,000 of a Kuan-yin portrait on silk to 140,000 of a picture of a Maitreya pagoda, actually printed by him.[c] In addition, three versions of 84,000 copies each were printed by Chhien Chhu, making a total in this area alone within the short period of thirty-six years, from +939 to 975, that is extremely impressive. It was not only the most extensive printing known to us at this time, but also there is no doubt that it had a tremendous impact on printing during the early Sung dynasty. Some of the cutters and printers of this period may either have participated in some of the early Sung printing or trained apprentices for the profession, but whichever they did, the result was that Hangchow became the most prosperous printing centre for the next three to four hundred years.

[a] This new discovery is reported in Chang Hsiu-Min (*14*), p. 75.

[b] This version is more common; see descriptions by Wang Kuo-Wei (*4*); Chuang Yen (*1*); Carter (1), pp. 73, 80–1; Giles (15), pp. 513–15; Li Shu-Hua (*11*), pp. 150–5; within some of the bricks, a picture of a pagoda was accompanied by the charm; see Chang Hsiu-Min (*14*), p. 74.

[c] See the Sung edition of *Hsin Fu Chu*[3], written with commentary by Yen-Shou and printed in Hangchow in +1160; Chang Hsiu-Min (*14*), p. 75.

¹ 延壽   ² 靈隱寺   ³ 心賦注

Besides the printed scrolls from eastern China, a number of printed fragments and single sheets from this period have been found in Tunhuang.[a] Many of these are undated, though a few bear exact dates and the names of donor and block cutter, and duplicate copies of some of them seem to have been printed from the same block. Some have been coloured by hand on both sides.[b] Two, printed in +947, depict Kuan-yin or some other divinity on the upper half, with an inscription of praise in about 100 characters on the lower half (Fig. 1116). One print, from +950, is a fragment in eight sheets of the *Diamond sutra*, and this is probably one of the earliest specimens of paged paper books still surviving. An undated rhyme book, *Chhieh-yün*[1], also in paged format, is believed also to have been printed during this period.[c]

(3) INCUNABULA OF THE SUNG AND PRINTING UNDER FOUR EXTRANEOUS DYNASTIES

From its modest beginnings, printing became a fully developed and advanced art in the Sung dynasty (+960–1279). Techniques were improved, new devices introduced, and the scope of printing was widened further still. The methods spread not only to many neighbouring nations in the east, west, and south, which had been in contact with Chinese culture for many centuries, but for the first time also to several non-Chinese peoples in the north. From there, printing began to cross the Chinese border and move westwards. The excellent block printing of the Sung period became a model to be emulated by later printers, while the invention of movable type was one of the most important developments in history. This was the golden age of Chinese printing, and books printed at this time equal in importance the incunabula produced in Europe three or four centuries later.

At the beginning of the Sung dynasty, the Buddhists began the gigantic project of printing the *Tripitaka*, which was followed by the government-sponsored printing of the Confucian classics, the standard histories, and other literatures. The Taoists also began to print their canon, comparable in scope and quantity to the Buddhist collection. Many government offices, schools, monasteries, private families, and bookshops participated in the printing business; in fact publishing proliferated in almost every field of knowledge, extending from the canonical literature to include history, geography, philosophy, poetry and prose, novels and dramas, divination and occultism, and scientific and technical writings, especially on medicine. The centres of printing were in Khaifeng in the north and Hangchow in the south, the two capitals of the Sung, and Mei-shan in Szechuan, where literary tradition can be traced back to the Thang and Five Dynasties, was the

---

[a] Cf. Carter (1), pp. 57–8, 64–5.   [b] See p. 280, n. b.
[c] Several fragments of this dictionary from Tunhuang, including two printed editions, are kept in the Bibliothèque Nationale, Paris (Nos. P 2214 and P 5531).

[1] 切韻

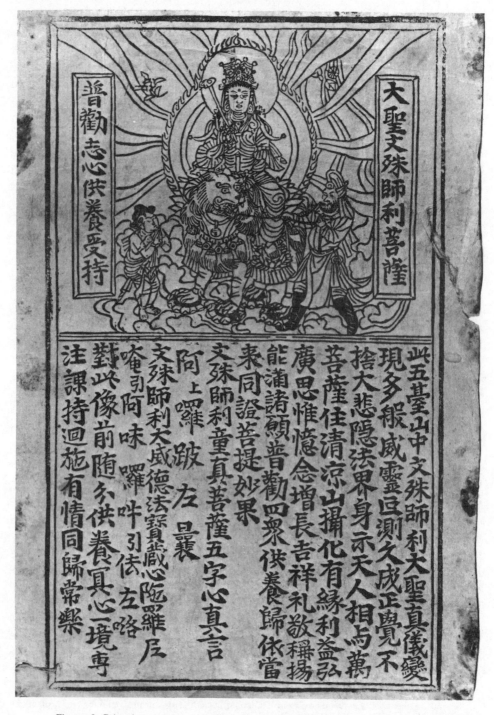

Fig. 1116. Printed prayer sheet on white buff paper, depicting Mañjuśrī riding on a lion with two attendants, c. +950. British Library.

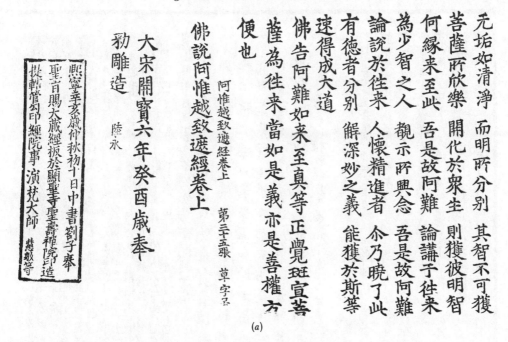

Fig. 1117. Early printing of the Buddhist *Tripitaka*. (*a*) Fragment of the Khai Pao edition, dated +973.

cultural centre of west China. Printing also flourished in Chien-yang, Fukien, one of the major centers of papermaking in south China.

Throughout the three centuries of the Sung, at least six different editions of the Buddhist *Tripitaka* had been printed,[a] and so this proved to be the most productive period for the printing of that comprehensive collection of Buddhist literature. The six editions were the Khai-pao[1] printed in I-chou (modern Chhengtu) in +971–83 (Fig. 1117a), the Chhung-ning[2] in Fuchow in +1080–1112, the Phi-lu[3] also in Fuchow in +1112–72, the Yüan-chüeh[4] in Hu-chou (in Chekiang) in +1132, the Tzu-fu[5] in An-chi (in Chekiang) in +1175, and the Chi-sha[6] edition in Phing-chiang (modern Soochow) in +1231–1321 (Fig. 1117b).[b] All but one of them consisted of from 5000 to almost 7000 *chüan* (rolls) bound in the continuously folded form known as *sutra* binding; the exception was the Khai-pao edition which was in roll form.[c] Supposing that an average of fifteen blocks were needed for one *chüan*, a

[a] Not counting the two editions produced under the Liao and Chin dynasties during the same period, and four more, including one in the Hsi-hsia script, under the Yüan dynasty, as described below.

[b] All the volumes of these editions have been lost except for the Chi-sha edition and a few fragments of other editions; for details of the various editions see Yeh Kung-Chho (*1*); Kenneth Chhen (8).

[c] A few volumes of the Khai-pao edition printed in Khai-feng (Honan) have been discovered, but it is not known whether this edition was printed in both Chhengtu and Khai-feng, or whether part of the blocks were transferred from Szechuan to Honan.

[1] 開寶　　　[2] 崇寧　　　[3] 毗盧　　　[4] 圓覺　　　[5] 資福
[6] 磧砂

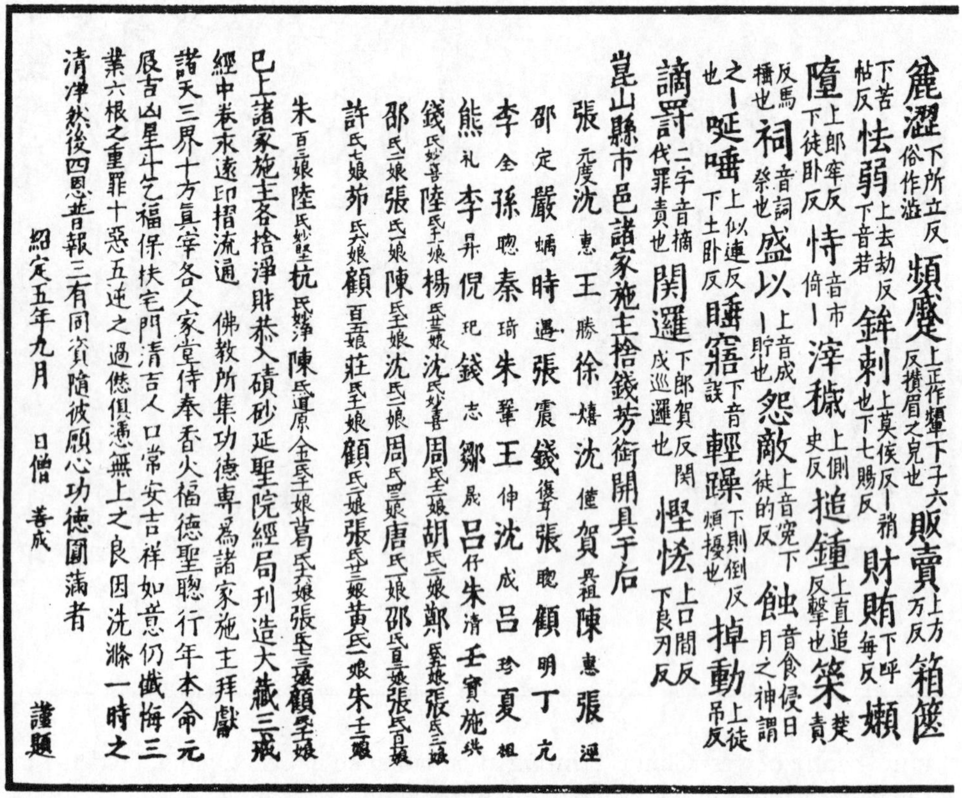
麁澀下所立反俗作澁 頻蹙上正作顰下子六反攢眉之皃也 販賣上方万反 箱篋
下苦帖反 怯弱上去劫反下音若 銼刺上莫侯反一䂓也下七賜反 財賄下呼每反 嬾
墮上郎牢反下徒卧反 恃音市倚一 滓穢上側史反 搥鍾上直追反擊也 榮楚責
反馬播也 祠音詞祭也 盛以上音成一貯也 怨敵上音宛下徒的反 蝕音食侵日月之神謂
之一蜒嚾上似連反下土卧反也 睡寤下音誤 輕躁下則倒反煩擾也 掉動上徒吊反
讁罸二字音摘伐罪責也 閑邏下郎賀反閑戍巡邏也 慳悋上口間反下良刃反
崑山縣市邑諸家施主捨錢芳銜開具于后
張元度 沈惠 王勝 徐熺 沈僅 賀昇祖 陳惠 張逕
邵定 嚴蝺 時遇 張震 錢復亨 張聰 顧明 丁亢
李全 孫聰 秦琦 朱羣 王仲 沈成 呂珍 夏祖
熊礼 李昇 倪玘 錢志 鄒晟 呂仔 朱清 壬實 施烘
錢氏妙音 陸氏十一娘 楊氏廿三娘 沈氏妙喜 周氏十一娘 胡氏二娘 鄭氏大娘 張氏三娘
邵氏二娘 張氏一娘 陳氏十娘 沈氏二娘 周氏四娘 唐氏一娘 邵氏百三娘 張氏百娘
許氏七娘 茆氏六娘 顧百五娘 莊氏十娘 顧氏二娘 張氏廿三娘 黃氏一娘 朱壬一娘
朱百三娘 陸氏妙堅 杭氏妙淨 陳氏道原 金氏十娘 葛氏六娘 張氏七三娘 顧氏十娘
已上諸家施主各捨淨財恭入磧砂延聖院經局刊造大藏三藏
經中卷永遠印摺流通 佛教所集功德専爲諸家施主拜獻
諸天三界十方眞宰各人家堂侍奉香火福德聖聰行年本命元
辰吉凶星斗乞福保扶宅門清吉人口常安吉祥如意仍懺悔三
業六根之重罪十惡五逆之過愆俱憑無上之良因洗滌一時之
清淨然後四恩普報三有同資隨彼願心功德圓滿者
紹定五年九月 日僧 善成 謹題

(*b*)

Fig. 1117. (*b*) Specimen of the Chi-Sha edition, dated +1232.

total of some 60,000 to 80,000 blocks must have been cut for each set.[a] This involved the training of calligraphers, cutters, and printers, as well as other skilled workers at a variety of different locations, and so contributed to the spread of these arts. Although there is no known record of the number of copies printed of each set, enough must have been prepared not only for deposit in the many monasteries in the empire, but also for distribution to other nations such as Tangut, Korea, Japan, and Vietnam, which sometimes requested more than one copy at one time.[b]

From the late 980s, about thirty years after the founding of the Sung, the printing of Confucian classics and standard histories by the National Academy began. This included the new commentaries on the twelve classics printed in +988–96 and the re-engraving in 1005 of the standard Feng Tao edition of the twelve classics, while with the addition of the *Meng Tzu* in 1011, the standard

[a] This estimate is based on the *Korean Tripitaka* of 6791 *chüan* printed in 1458 from 81, 258 blocks which are still preserved in the Haein-sa in Korea.

[b] The introduction of the *Tripitaka* and other books from China to other countries will be discussed in the section on the spread of Chinese printing; see below, pp. 319 ff.

number of the Thirteen Classics as we have them today was completed. The printing of the seventeen standard histories started in 994 and was completed in 1061,[a] taking two-thirds of a century of careful collation, copying, proof-reading, block-cutting, and printing. This was the first time that standard histories were printed collectively under imperial auspices. At about the same time, the printing was begun of several Taoist works, including the collection of the Taoist canon, the *Wan Shou Tao Tsang*[1], 5481 chüan in 540 cases, which was cut in Fukien and printed in 1116–17.[b] Included in the collection were two books of the Manichaean scriptures, which may have been first printed before 1000 or earlier.[c] Also, under official sponsorship, printing was carried out of several rhymed dictionaries, classified encyclopaedias of quotations, anthologies of literature, as well as other selected books.

Aside from the orthodox literature, a good number of scientific and technical works were also printed at the beginning of the Sung dynasty. These included the famous +6th-century work on agriculture and farming, the *Chhi Min Yao Shu*[2], printed in 1018, and a series of ten works on arithmetic, including the *Chou Pei Suan Ching*[3] (Fig. 1118) and the *Chiu Chang Suan Shu*[4], which were printed in 1074. Most important, however, was the extensive printing and distribution of medical books, including the famous Khai-pao edition of the herbal *Pen Tshao*[5] printed in 973, with a revised and enlarged edition in 974, and yet another edition with additional commentaries and illustrations in 1044. A collection of specialist studies was also printed; these included the *Chu Ping Yüan Hou Lun*[6] on the origin of diseases (+1027); the *Mo Ching*[7] on the principles of the pulse (+1068); the *Shang Han Lun*[8] on fevers caused by cold (+1065) and the *Thung Jen Chen Chiu Thu Ching*[9] on acupuncture and moxibustion in 1026. Furthermore, no fewer than ten different collections of medical prescriptions, including the most popular work *Thai Phing Sheng Hui Fang*[10], were printed and distributed before 1100. Besides these editions in large characters, the printing of a small-character edition of the prescriptions and other medical classics was ordered in 1088 by the Office of Medical Administration (Thai I Yüan[11]), to be sold at the cost of the paper, ink, and labour by local government offices in order to meet the demand of local physicians.[d]

After the fall of Khaifeng to the Jurchens in 1126, all the printing blocks in the National Academy were looted and removed to the north. However, when the Sung capital was resettled in Lin-an (modern Hangchow), many of the lost books were engraved anew, copying the original editions put out by the former National

---

[a] See the documented discussion of titles in Wang Kuo-Wei (5); a few remnants of the *Shih Chi*, *Han Shu*, and *Hou Han Shu* exist.
[b] Cf. Liu Tshun-Jen (2), p. 113; Chang Hsiu-Min (6), p. 15.
[c] Cf. Carter (1), pp. 93–4; Chavannes & Pelliot (1), pp. 300–2.
[d] See memorial quoted in Wang Kuo-Wei (5).

[1] 萬壽道藏　　[2] 齊民要術　　[3] 周髀算經　　[4] 九章算術　　[5] 本草
[6] 諸病源候論　[7] 脈經　　　[8] 傷寒論　　　[9] 銅人鍼灸圖經
[10] 太平聖惠方　[11] 太醫院

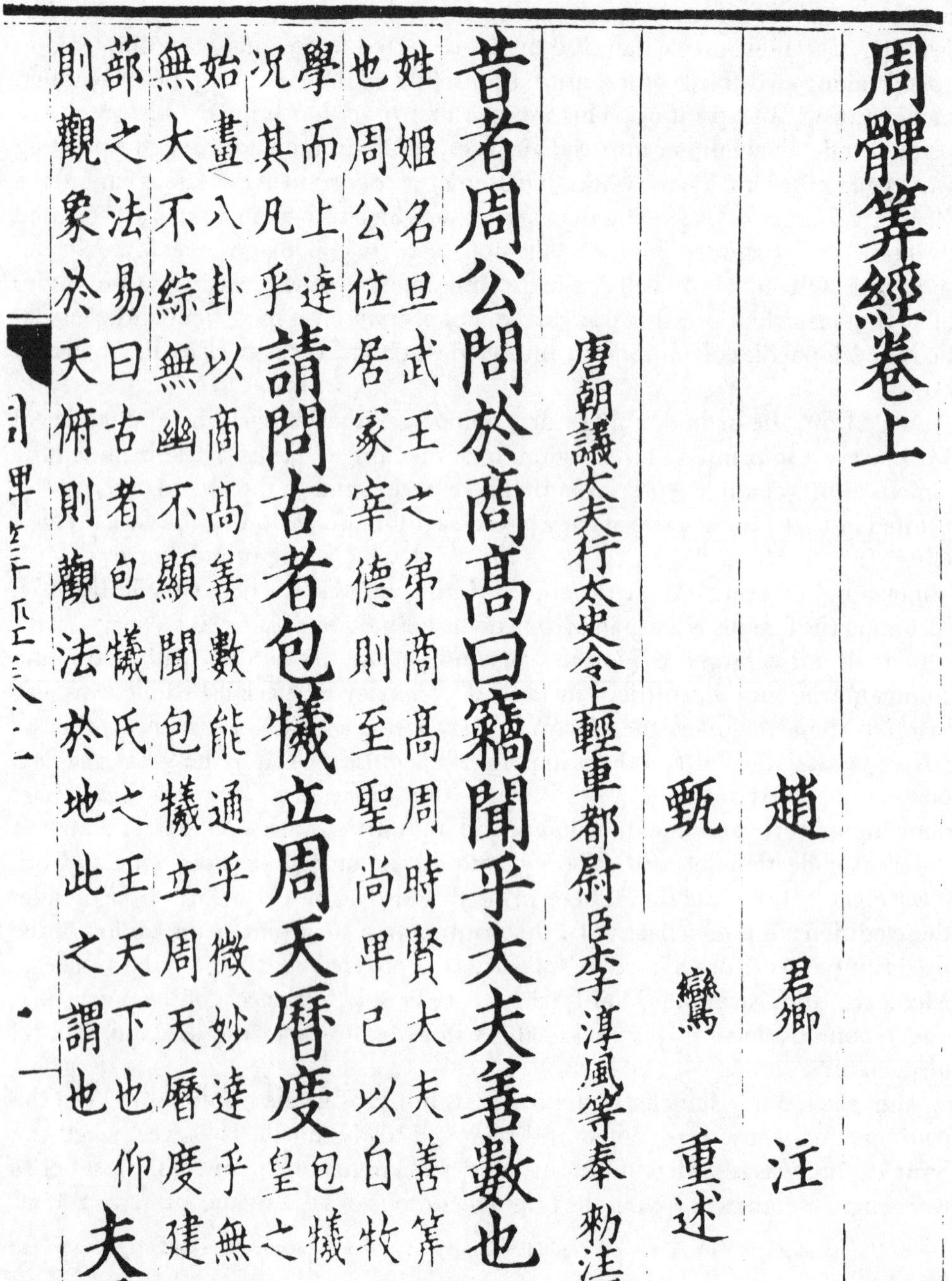

周髀算經卷上

趙君卿注

甄鸞重述

唐朝議大夫行太史令上輕車都尉臣李淳風等奉敕注釋

昔者周公問於商高曰竊聞乎大夫善數也

請問古者包犧立周天曆度

皇之包犧下一三

姓姬名旦武王之弟商高周時賢大夫善算者

也周公位居冢宰德則至聖尚甲已以白牧下

學而上達

況其凡乎

始畫八卦以商高善數能通乎微妙達乎無方一三

無大不綜無幽不顯聞包犧立周天曆度建章

部之法易曰古者包犧氏之王天下也仰

則觀象於天俯則觀法於地此之謂也

公周

善箸者周公

夫天

Fig. 1118. The Sung printing of the *Chou Pei Suan Ching*, Arithmetical Classic of the Gnomon and the Circular Path, written c. — 1st century.

Academy. Numerous works in all branches of knowledge were also printed by various local government agencies, private families, and bookshops in almost all parts of the empire, and of these three types of local government agencies were extremely active in printing and publishing.[a] First, various local government offices of different departments printed a great variety of histories, literary collections, and scientific and medical works, including the annalistic general history *Tzu Chih Thung Chien*[1] by Ssu-ma Kuang[2] (+1019–+86), printed by the Chekiang Office of Tea and Salt Revenue in 1133; and a collection of medical prescriptions, the *Thai Phing Sheng Hui Fang*, reprinted by the Fukien Office of Financial Administration in 1147. Second, the local governments of such prefectures and sub-prefectures as Chiang-ning and Phing-chiang in modern Chiangsu, Lin-an, Yü-yao, and Yen-chou in modern Chekiang, and Mei-shan in modern Szechuan, also played their part. The most famous of their publications were the Mei-shan edition of the seven standard histories of the Southern and Northern Dynasties printed in +1144 and the Phing-chiang-fu edition of the classical work on architecture, the *Ying Tsao Fa Shih*[3] by Li Chieh[4], printed in 1145.[b] Third, public and private schools, Confucian temples, and ancestral halls of various localities printed various kinds of books, including five medical works by the Huan Chhi Shu Yüan[5] in 1264 and the beginner's textbook *Thung Meng Hsün*[6] by the ancestral hall of the Lü family in Ching-hua in 1215. Such local and provincial agencies as these provided standard texts and any specially needed books as their responsibility.

Some of the most popular books printed and published in the Sung dynasty were associated with private families, of which the most outstanding was the Yü family of Chien-yang, Fukien, which had a tradition of continuous operation in the book business for over 500 years. As early as the +11th century the Yü family was already engaged in printing in Chien-yang, which became a famous centre of the book trade and printing from then on. Books engraved by them were still very popular in the +15th and 16th centuries, and a bookshop owned by one of the family was still operated at the original site in Chien-yang as late as the eighteenth century.[c] The most prominent printer of this family known to us in the Sung dynasty was Yü Jen-chung[7] (fl. +1130–93), who held a *chin-shih* degree and collected over 10,000 volumes; his studio was therefore called Wan Chüan Lou[8] (Tower of Ten Thousand Rolls). He printed many titles we still know, including the famous Nine Classics and Three Commentaries published at the end of the +12th century. Yet even more works were printed during the Yüan dynasty by one of his descendants, Yü Chih-an[9] (fl. +1300–45), whose printing firm was

---

[a] See details on titles and printing agencies given in Yeh Te-Hui (2), pp. 60–85.

[b] The book was first published in 1103 and reprinted in 1145. The original edition is no longer extant, except for the fragments of 1145 edition which were found in 1956. A reconstructed edition of the work based on manuscripts was published with colour illustrations in 1925.

[c] This unbroken tradition of the Yü family's interest in the book business prompted an inquiry by the Chhien-lung emperor in 1775; see the decree and report quoted in Yeh Te-Hui (2), pp. 42–3.

[1] 資治通鑑	[2] 司馬光	[3] 營造法式	[4] 李誠	[5] 環溪書院
[6] 童蒙訓	[7] 余仁仲	[8] 萬卷樓	[9] 余志安	

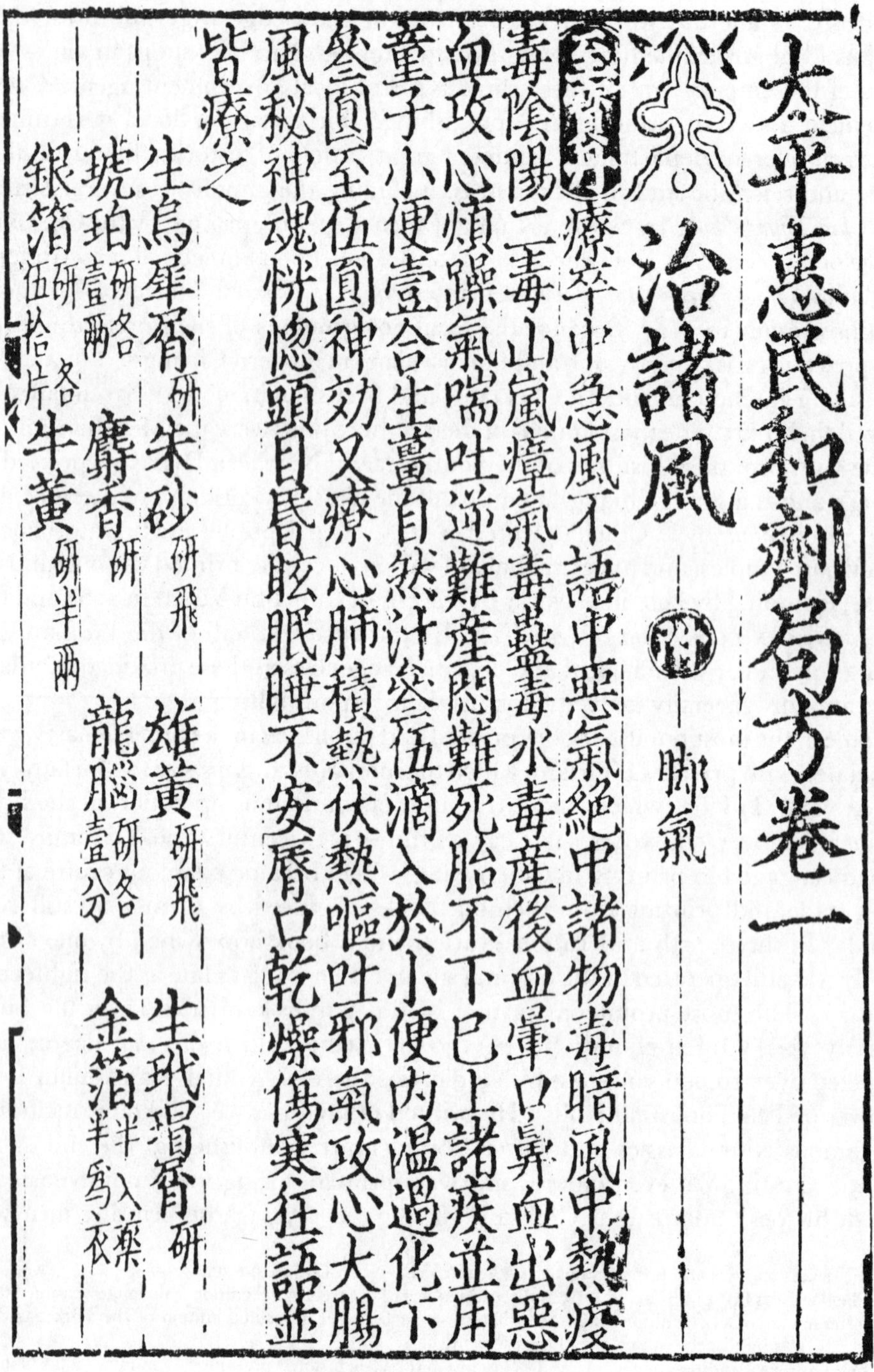

太平惠民和劑局方卷一

治諸風 〔附〕 脚氣

〔至寶丹〕療卒中急風不語中惡氣絶中諸物毒暗風中熱疫毒陰陽二毒山嵐瘴氣毒蠱毒水毒産後血暈口鼻血出惡血攻心煩躁氣喘吐逆難産悶難死胎不下已上諸疾並用童子小便壹合生薑自然汁叄伍滴入於小便内温過化下叄圓至伍圓神効又療心肺積熱伏熱嘔吐邪氣攻心大腸風秘神魂恍惚頭目昏眩眠睡不安唇口乾燥傷寒狂語並皆療之

生烏犀屑 研飛　朱砂 研飛　雄黃 研飛　生玳瑁屑 研

琥珀 研各壹兩　麝香 研　龍腦 研各壹分　生玳瑁屑 研

銀箔 研各拾伍片　牛黃 研半兩　金箔 半入藥　金箔 半為衣

Fig. 1119. Yüan edition of *Thai Phing Hui Min Ho Chi Chü Fang*, the Great Peace People's Welfare Pharmacies, written in +1151.

called Chhin Yu Thang[1] (Hall of Abundance through Diligence). Among his publications were the collection of medical prescriptions *Thai Phing Hui Min Ho Chi Chü Fang*[2], printed in 1304 (Fig. 1119), collected commentaries on poems of Li Po and Tu Fu (+1311–12), while many classics and histories came out between +1335 and +1345.[a]

Another noted Fukien printing house was the Shih Tshai Thang[3] (Hall of Colourful Generations) of Liao Ying-chung[4] (+1200?–75), a scholar-official from Shao-wu (in Fukien), who published a de luxe edition of the Nine Classics and Three Commentaries around +1270. This edition was noted for its careful collation, refined execution, its excellent ink and paper, and its rich decoration; unfortunately it was soon destroyed by the Mongol invasion of southern China. However, a facsimile of this edition was reprinted in about +1300, together with a manual of collation and printing, the *Chiu Ching San Chuan Yen Ko Li*[5] (Manual for the Transmission of the Nine Classics and Three Commentaries), which provided specifications for the selection of editions, the style of calligraphy, its emendation, pronunciation, punctuation, and collation.[b] This work has been praised as a classic model of textual criticism and printing ever since.

Competing with the many private printers of Fukien were those of the two Sung capitals, Khaifeng and Lin-an (modern Hangchow), where numerous bookshops flourished. The area around the Hsiang Kuo Ssu monastery in Khaifeng was the centre of a book market,[c] and the famous painting of the Spring Festival on the Pien River in Khaifeng, *Chhing Ming Shang Ho Thu*,[6] depicts one bookshop (Fig. 1120) together with many other shops along the river front. The prosperity of the book industry in Hangchow is testified by a cluster of bookshops in the city; at least a dozen bookshops with precise addresses can be identified from the colophons of the printed editions. One of them had two or three branches, while many bookshops specialised in publications on particular subjects.[d]

The most noted family printers in the Southern Sung capital was Chhen Chhi[10] (fl. +1167–1225), a poet and publisher, and his son, Chhen Chieh-Yüan[11] (fl. +1225–64), who stood first in the provincial examination. Together with other members of the family, they operated bookshops known as *shu pheng*[12], and printed no fewer than a hundred works, especially anthologies of poetry by almost all the

[a] For titles printed by the Yü family, see Yeh Te-Hui (2), pp. 43–7.
[b] The printing of this edition and authorship of the manual have traditionally been attributed to Yüeh Kho[7] (+1183–1242?), of Hsiang-thai (modern Thang-yin, Honan), a grandson of the Sung general Yüeh Fei[8] (+1103–42), but are now attributed to Yüeh Chün,[9] possibly a distant descendant of Yüeh Fei of I-hsin, Chiangsu; see Weng Thung-Wen (1), pp. 429–49; (1), pp. 199–204; Achilles Fang (3), pp. 65 ff. also notes in Hervouet (3), p. 53.
[c] See *Tung Ching Meng Hua Lu* (*TSHCC*), ch. 3, p. 19.
[d] See names and addresses of bookshops and their distribution in Hangchow in tables and maps in M. Finegan (1), pp. 374 ff.

[1] 勤有堂	[2] 太平惠民和劑局方	[3] 世綵堂	[4] 廖瑩中
[5] 九經三傳沿革例	[6] 清明上河圖	[7] 岳珂	[8] 岳飛
[9] 岳浚	[10] 陳起	[11] 陳解元	[12] 書棚

Fig. 1120. Detail of a bookstore of Northern sung appearing in a long scroll painting, *Spring Festival on the River*, attributed to Chang Tse-Tuan of early +12th century (enlarged from an old version, probably of the Yüan dynasty).

noted poets of the Thang and Sung dynasties.[a] Other family printers were the Yin of Lin-an, the Huang of Chien-an, the Liu of Ma-sha town in Chien-yang, the Juan of Min-hou (Fukien), as well as many others in Szechuan, Shansi, and the Chiang-Huai and Hu-Kuang regions (modern Chiangsu, Anhui, Chiangsi, and Hunan). Their publications cover a great variety of subjects, including classics, histories, poetry, individual literary collections, and medicine.

Contemporary with Sung China, four nomadic tribes—the Khitan, Tangut, Jurchen, and Mongol—established kingdoms along its northern border and gradually expanded into Chinese territory. Being unlettered and less civilised, they adopted Chinese culture and made use of printing as soon as they conquered and came to rule the Chinese. Early in the +10th century the Khitan kingdom or Liao dynasty (+907–1125) created a form of writing of some 3000 characters, based on the Chinese system, in which to express its own language, and many Chinese classics, histories, and medical works were translated into Khitan and printed, though their circulation was prohibited outside Khitan territory. None of this Liao printing survives, though there is a Sung reproduction of a Liao printed glossary in Chinese, the *Lung Khan Shou Chien*[1], with prefaces dated +997 and 1034. The most extensive Liao publication known to us is the Chhitan edition of the *Tripitaka* in Chinese, in some 6000 *chüan* in 579 cases, printed with Korean paper and ink in Peking in 1031–64; but nothing of it survives today.[b]

The Tangut or Hsi-hsia kingdom (+990–1227) in northwest Manchuria and Mongolia proclaimed itself an empire in 1031, with its capital in what is now Ning-hsia. A system of writing its language, based on Chinese and Khitan, was created in 1036, and many Chinese books were translated and then printed in this script. Gifts and exchanges of books were arranged with the Sung court from time to time; Buddhist *sutras* were donated no fewer than six times and some of them were translated and printed. After the Mongol conquest of Tangut and China, a Tangut edition of the *Tripitaka* in the Hsi-hsia script, in more than 3620 *chüan*, was printed in Hangchow and completed in 1302, and about a hundred copies were distributed to monasteries in the former Tangut region. Many fragments of books in Tangut and Chinese were discovered at the beginning of this century, including two editions of the *Diamond sutra* printed in 1016 and 1189, and two bilingual glossaries, the *Hsi-Hsia Tzu Shu Yün Thung*[2] (+1132), and the *Fan Han Ho Shih Chang Chung Chu*[3] (+1190). Apparently many books in their native tongue were also printed under the Tangut rulers.[c]

The Jurchen or Chin dynasty (+1114–1234) originated in Manchuria and

[a] For private printing in the Sung, see Yeh Te-Hui (2), pp. 42–59.

[b] See Chang Hsiu-Min (6), pp. 11–12; Kenneth Chhen (8), p. 212.

[c] Printed materials in Tangut were discovered in the ruined town of Khara-khoto by Kozlov in +1908 and by Stein in +1914, and also in Tunhuang, Turfan, and Ning-hsia; see K. T. Wu (7), pp. 451–3. Movable-type printings of c. +1139–94 were recently found and reported by Wang Ching-Ju (2) and Chang Ssu-Wen (1); also a collection of *Tangut Tripitaka* fragments in nine parts was published in Delhi in 1971; see Goodrich (29), pp. 64–5.

[1] 龍龕手鑑   [2] 西夏字書韻統   [3] 番漢合時掌中珠

Fig. 1121. An illustration of salt manufacture and transportation from the Chin edition of the *Pen Tshao*, printed in +1204. From East Asian History of Science Library, Cambridge.

occupied the northern part of China after the defeat of the Chhitan in 1125 and the Sung in 1126, when the capital, Khai-feng, was captured. Immediately, all the books and printing blocks in the National Academy were moved to the north, and a government printing office was established in Phing-yang[1] (in modern Shansi) four years later, while an Institute for the Promotion of Literature (Hung Wen Yüan[2]) was established in 1194. Many books of classics, history, philosophy, poetry, and science were then printed by various government and private agencies.[a] These included the famous Ta-kuan edition of the herbal *Ching Shih Cheng Lei Pen Tshao*[3], printed in Phing-yang in 1249 (Fig. 1121). Another edition of the *Tripitaka*, 7182 *chüan* in 682 cases, was printed on paper-mulberry paper in Chieh-chou, Shansi (+1148–73),[b] and competing with it was a most comprehensive collection of Taoist literature, entitled *Hsüan Tu Pao Tsang*[4], in 6455 *chüan* in 602 cases, which was printed in Khai-feng in +1188–91.[c] Many classical and philosophical books were translated into Jurchen, for which a script was created between +1123 and +1135, following the Chinese and Khitan model.

[a] Over thirty works in Chinese and fifteen in Jurchen are known to have been printed by the National Academy of the Chin dynasty, and eleven by private and commercial printers; see Yeh Te-Hui (*2*), pp. 89–90; K. T. Wu (*7*), p. 454; Chang Hsiu-Min (*6*), pp. 12–15.

[b] An incomplete set of this edition was discovered, in roll form, in 1934 at Chao-chheng, Shansi; a total of 4330 is kept in the Peking Library.

[c] Cf. Chang Hsiu-Min (*6*), pp. 14–15; Liu Tshun-Jen (*5*), p. 114.

[1] 平陽   [2] 宏文院   [3] 經史證類本草   [4] 玄都寶藏

The Mongols annexed the Jurchen kingdom in north China in 1235, and conquered the Southern Sung in 1280, thus unifying the empire under their Yüan dynasty (+1260–1368). Inheriting the legacy of Sung printing, the Yüan not only continued its tradition of fine workmanship, especially at the beginning of their rule, but also introduced some innovations. In addition to the National Academy, several other government agencies were established to edit and print books. These included the Institute of Compilation (Pien Hsiu So[1]) in Peking, the Institute of Literature (Ching Chi So[2]) in Phing-yang, Shansi, the Office for the Promotion of Literature (Hsing Wen Shu[3]), and the Bureau of Publications (Pi Shu Chien[4]), which last was reported in +1273 to have a staff of 106 members, including forty woodcutters, thirty-nine workmen, and sixteen printers.[a] A particularly interesting aspect of Yüan publication was the practice of joint enterprises in printing by local schools. Typical examples were the projects to print the seventeen standard histories jointly by nine circuit schools of the Chien-khang Region in +1305 (nine were actually completed), and the collection of eleven classics by schools of the Chiang-hsi Region at an earlier date.[b] A local history of Nanking, *Chin-Ling Hsin Chih*[5], in fifteen *chüan* and thirteen volumes, was co-operatively printed from 1217 blocks contributed by several local schools and government agencies. Funds were raised in the same way for the cutting of some 5688 blocks to print the encyclopaedia *Yü Hai*[6] and thirteen other works by Wang Ying-Lin[7] (+1223–96).[c] Remarks on the joint printing ventures appear in many books printed by similar co-operative efforts, some of which still survive.

The Mongols continued to print several more editions of the *Tripitaka*, including the Hung-fa[8] edition, 7182 *chüan*, printed in Peking in 1277–94, and the Phu-ning[9] edition, 6010 *chüan*, in Hangchow in 1278–94. The Chi-sha edition, 6362 *chüan*, was begun in Phing-chiang (Soochow) in 1231 under the Sung but not completed until 1322 at the time of the Yüan dynasty.[d] The Khitan edition in Hsi-hsia script mentioned earlier was completed in 1302, the Chin edition of the Taoist collection, the *Hsüan Tu Pao Tsang*, was reprinted in some 7800 *chüan* in 1237–44, but it was destroyed under the Mangu emperor in 1258, when Taoism was persecuted. Despite the imperial order, however, some of the blocks were saved by being hidden, and no fewer than six or seven sets of the blocks for printing this Taoist collection were still preserved in Taoist monasteries in north China around 1281.[e]

Because so many skilled cutters and printers had been trained in Chekiang and

[a] *Pi Shu Chien Chih*[10] (*KTHS*), ch. 7, p. 17*a*, *b*.
[b] K. T. Wu (7), pp. 464–9.
[c] *Ibid.* pp. 69–71; *Ssu-Ming Hsü Chih*[11] (1854), ch. 7, pp. 12*b*–14*a*.
[d] See Kenneth Chhen (7), pp. 213–14. Two nearly complete sets of the Chi-sha edition survive today; one discovered in 1931 has been reprinted in 1935 and the other, of over 5000 volumes, is in the possession of the Gest Oriental Library of Princeton University. Nearly 700 volumes are of the original Sung printing and the rest are of Yüan and later editions; see Hu Shih (12), pp. 113–41.
[e] Chang Hsiu-Min (6), p. 15; Liu Tshun-Jen (5), p. 115.

[1] 編修所	[2] 經籍所	[3] 興文署	[4] 祕書監	[5] 金陵新志
[6] 玉海	[7] 王應麟	[8] 弘法	[9] 普寧	[10] 祕書監志
[11] 四明續志				

Fukien under the Sung, the printing centres in these two regions continued to prosper under the Yüan and no fewer than 220 titles are known to have been printed by 107 family firms, mostly in the fields of classics, history, individual literary collections, dictionaries, encyclopaedias, and medicine (Fig. 1122).[a] Among the medical works, at least five editions of the herbal and more than twenty collections of prescriptions are known to have been printed by such families as the Kao, Liang, Liu, Ssu, Tshao, Tuan, and Hsü of Phing-yang (in Shansi), Chien-an (in Fukien), and other places.[b] Though they did not call themselves book dealers, the private printing of such technical works was, it seems, for profit.

Phing-yang in the north and Chien-an in the south were the two most flourishing commercial printing centres under the Yüan. In Chien-an alone there were forty-eight commercial firms which are known to have published during this period, the most active being the Jih Hsin Thang[1] of Liu Chin-Wen[2] and the Chhin Yu Thang of Yü Chih-An. Each of them printed nearly two score works, the Jih Hsin Thang publishing an average of one work each year from 1335 to 1357, and the Chhin Yu Thang one every two years in 1304-45.[c] Both the Chhin Yu Thang and the Tsung Wen Shu Thang[3] of Cheng Thien-Tse[4] flourished for many generations; the former operating from the 11th century and the latter from the 14th, and both continued active until the 16th century. While most publishers printed a variety of books, a few concentrated on certain subjects only. For instance, the Yüan Sha Shu Yüan[5] in Chien-yang published at least four large sets of encyclopaedias in 1315-25, and the Ku Lin Shu Thang[6] in Lu-ling (in modern Chiangsi), the Huo Chi Thang[7] of Yen-shan (in modern Hopei), and the Kuang Chhin Thang[8] of Chien-an, all specialised in printing medical books.[d] Also important in this period was the publication of popular stories and dramatic texts which flourished during the Yüan dynasty, though few of them exist now.[e]

(4) NEW DIMENSIONS OF MING PRINTING

Ming printing was distinguished by the extended scope of its subject-matter and by its technical innovations and artistic refinement. In contrast to that of previous periods, the printing under the Ming included not only the traditional works in classics, history, religion, and literary collections, but also such new subject-fields as popular novels, music, industrial arts, accounts of ocean voyages, shipbuilding, and scientific treatises from the West, which had never before been seen in print in China. Significant increases were also noted in printing of dramatic texts, medical

---

[a] Nagasawa Kikuya (8), pp. 35-46. K. T. Wu (7), pp. 493-4.
[b] See titles listed in Yeh Te-Hui (2), pp. 97-103.   [c] K. T. Wu (7), pp. 487-8.
[d] See titles listed in Wang Kuo-Wei (5), pp. 4521-2.
[e] A collection of thirty dramatic texts printed in Yüan was reprinted in Japan in 1914; an incomplete dramatic tale printed about +1300 was found by Kozlov in Khara-khoto.

[1] 日新堂   [2] 劉錦文   [3] 宗文書堂   [4] 鄭天澤   [5] 圓沙書院
[6] 古林書堂   [7] 活濟堂   [8] 廣勤堂

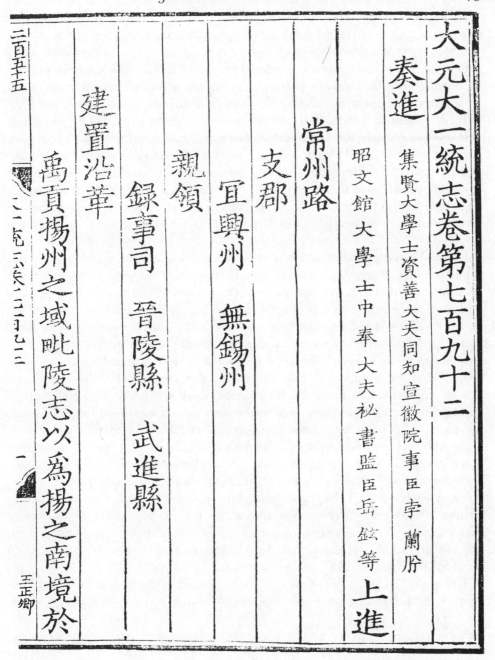

Fig. 1122. Universal Geography of the Yüan Empire, *Ta Yüan Ta I Thung Chih*, printed in Hangchow, +1347.

writings, records of foreign countries, especially of south and southeast Asia, local gazetteers, and such large compilations as collectanea and encyclopedias. In technique, the Ming printers introduced metal typography, improved the multi-colour process of block printing, refined the woodcut for book illustrations, and used xylography for facsimile reproductions of old editions.[a] In both content and technology, then, printing in the Ming was distinguished in many respects.

Its development during the dynasty can be divided into two distinctive periods with 1500 as roughly the dividing line. For more than a century before that date, Ming printing generally inherited the tradition of Yüan in both techniques and format, and the continuation of the civil service examinations resulted in the printing by the National Academy and other official agencies of Confucian classics, standard histories, and works of reference for reading by the candidates. Commercial printers also primarily busied themselves in printing such materials as textbooks.

Under the Ming, the fifteenth century witnessed political stabilisation, economic expansion, and increasing contacts with foreign countries, especially southwards over the sea. Numerous writings on the Western Oceans (*hsi yang*[1]), on navigation, and naval architecture were produced as a result of the increasing marine activity, and particularly the seven expeditions of Cheng Ho[2] that he made between 1405 and 1431. At the same time, great intellectual development was manifested by the enrichment of the imperial library, the establishment of numerous private collections, and by the compilation of the *Yung-Lo Ta Tien*,[3] an encyclopedia of scope and size unprecedented in history. This gigantic work consisted of 22,937 *chüan* of passages extracted from over 7000 titles from the classics, history, philosophy, literature, religion, drama, fiction, industrial arts, and agriculture, all of them grouped by subject and arranged according to the order of the rhymes. Compiled and copied by a task force of some 3000 scholars between +1403 and +1408, the work contained fifty million characters and was bound in yellow brocade in 11,095 volumes, each sixteen inches in height and ten inches in width.[b] The main text was written in black ink, with headings and sources in vermilion, on white paper ruled with red vertical lines. A plan for printing this work did not materialise, except perhaps for a small portion,[c] primarily, it seems, because the cost would have been prohibitive. Only one additional set was made by hand.[d]

The original copy of this work was first deposited in the imperial library, Wen Yüan Ko[5], in Nanking, and was removed to Peking in 1421. The library there,

[a] Cf. K. T. Wu (6); Chhien Tshun-Hsün (2).
[b] Cf. Kuo Po-Kung (1); Wu (5), pp. 167 ff.; Giles (1); Goodrich (26); these figures are based on the account given in the preface of the Peking reprint of 1960; different sources give varying figures.
[c] Paul Pelliot is reported to have been told by the official-collector Tuan-Fang[4] (+1861 to +1911) that he had seen some 100 titles of this work in printed form; see Goodrich (34), p. 18.
[d] The texts of 385 works in 4946 *chüan* were extracted for the *Ssu Khu Chhüan Shu*. The original set and part of the duplicate were destroyed at the end of the Ming and many of the remaining volumes were destroyed during the Boxer uprising in 1900. Only some 800 *chüan* are known to have survived today in various libraries throughout the world. A reprint of 730 *chüan* in 200 *ts'e* was made in Peking in 1960, and another of 742 *chüan* in 100 volumes in Taipei in 1962.

[1] 西洋   [2] 鄭和   [3] 永樂大典   [4] 端方   [5] 文淵閣

which inherited the books of the former imperial collections of the Sung, Chin, and Yüan dynasties, consisted in +1441 of 7350 titles in some 43,200 volumes (*tshe*) containing one million *chüan*. Thirty per cent of the materials were in print and seventy per cent in manuscript.[a]

The period after 1500 turned out to be the most productive era for the development of literature, art, and technology in China. The popular novels written in the colloquial language at this time set the standard style of Chinese traditional fiction for the following centuries, while after 1600, the development of fiction and dramatic texts, in turn, encouraged the refinement of woodcut book illustrations in popular literature. At the same time, illustrated works on the industrial arts, on the design of inkcakes, and manuals of painting and stationery, were also produced at a high standard of excellence. Multi-colour processes with wood blocks were developed from two to more than five colours in the printing of texts with commentaries, maps, letter papers, and other artistic works.[b]

The arrival of Jesuit missionaries in China at the end of the sixteenth century marked the beginning of the introduction of Western knowledge to Chinese intellectual circles. As a result, during the next two centuries, more than four hundred writings and translations added to Chinese scholarship in such new fields of knowledge as Christianity, Western humanities and institutions, and scientific literature. Among the earliest printed works from the West were Michele Ruggieri's catechism of Christianity, *Sheng Chiao Shih Lu*[1], printed in Canton in 1582; Matteo Ricci's map of the world, *Khun Yü Wan Kuo Chhüan Thu*[2] (1584) and his translation, in collaboration with Hsü Kuang-Chhi[3], of Clavius' *Euclidis Elementorum*, *Chi Ho Yüan Pen*[4] (1607). Many other works on mathematics, astronomy, physics, geology, biology, psychology, medicine, and world geography and history, besides those on the Christian religion, were included in the late Ming and early Chhing printing.[c]

Many official agencies in both the central and local governments engaged in printing various kinds of books for different readers. The printing facilities at the imperial palace, known in general as Nei Fu[5], were in charge of the Supervisorate of Ceremonies (Ssu Li Chien[6]), which was one of the twelve supervisory offices established early in the Ming. Three printing shops (Ching Chhang[7]), including those for the Confucian classics, Buddhist *sutras*, and Taoist canons, were operated under this agency. The shop for the Confucian classics printed many prestigious editions of the Five Classics and Four Books, and a collection of neo-Confucian philosophy called *Hsing Li Ta Chhüan Shu*[8], printed in 1415.[d] Although the physical

---

[a] Cf. K. T. Wu (5), p. 184.
[b] Book illustration and colour printing will be discussed more fully on pp. 262 ff. below.
[c] Cf. Pfister (1); Bernard-Maître (18), (19); Tsien (12), p. 306.
[d] For the Nei Fu editions, see titles in *Ku Chin Shu Kho*[9] by Chou Hung-Tsu[10]; for works compiled under imperial auspices, see Li Chin-Hua (1).

[1] 聖教實錄　　[2] 坤輿萬國全圖　　　　　　[3] 徐光啓　　[4] 幾何原本
[5] 內府　　　　[6] 司禮監　　[7] 經廠　　　　[8] 性理大全書　　[9] 古今書刻
[10] 周宏祖

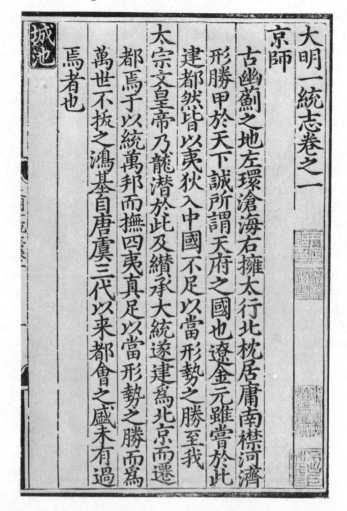

Fig. 1123. National Geography of the Ming Dynasty, *Ta Ming I Thung Chih*, printed by the Imperial Palace in +1461. Far Eastern Library, University of Chicago.

format of these editions was excellent, their scholarly value was not high because the printing agency in the imperial palace was under the charge of eunuchs who were not competent in scholarly matters. Imperial instructions and official documents and compilations were also printed by the imperial printing shop. For example, the official geography of the Ming empire, *Ta Ming I Thung Chih*[1], ninety *chüan*, came out in 1461 (Fig. 1123), the collected statutes of the Ming dynasty, *Ta Ming Hui Tien*[2] in 180 *chüan*, in 1511, and its revised edition, 228 *chüan*, in 1587. The official gazette, *Ti pao*[3], the predecessor of the *Peking Gazette*, was first transcribed by hand but, beginning in 1628, was printed with wooden movable type. It is said that

[1] 大明一統志   [2] 大明會典   [3] 邸報

more types were cut for the frequently used characters than for those that were less used.[a]

Various branches of the government such as the Board of Rites, Board of War, Board of Works, Censorate, Imperial Observatory, National Academy, and Bureau of Physicians are all known to have printed books. For instance, the Board of Rites printed a collection of documents concerning the bestowing of posthumous imperial honours, *Ta Li Chi I*[1], in four *chüan*, compiled under imperial auspices and printed in early 1526. The Board of War printed an illustrated work with maps on the defence organisation of the northern border regions, titled *Chiu Pien Thu Shuo*[2], presented to the throne in 1538. The Bureau of Physicians published several works on medicine, including an illustrated book on acupuncture and moxibustion, *Thung Jen Chen Chiu Thu Ching*[3], in three *chüan*, printed from a Sung text on stone, with illustrations from a bronze model, under imperial auspices, in 1443. But though most of the publications of government agencies related to their respective fields of administration, a fair number of them did not. For example, many works for enjoyment and amusement were printed by the Ming Censorate, the Tu Chha Yüan[4], and among some thirty of their titles there were two popular novels, the *San Kuo Chih Yen-I*[5] (Romance of Three Kingdoms) and the *Shui Hu Chüan*[6] (Water Margin), three works on the game of chess, and two on music and songs.[b]

The most productive among the government agencies was the National Academy, Kuo Tzu Chien[7] which, between its southern and northern branches, printed no fewer than three hundred works including classics, histories, local gazetteers, imperial documents, manuals of calligraphy, classified encyclopedias, as well as works on medicine, agriculture, and technology. The most notable of these were the *Thirteen Classics* and the *Twenty-one Standard Histories*, which were first printed from old blocks accumulated from previous dynasties during the previous seven centuries, and later from new blocks carved between 1530 and 1596. The teachers and students of the academy were responsible for the collation, revision, printing, and custody of blocks, and their names often appear on the blocks themselves.[c]

The distinguished contributions of Ming printing include the publications of various local officials such as governors, provincial judges, princes, and certain lower administrative units. Many of the books were printed by the offices of prefectures of almost all the provinces in the empire, including such border and interior regions as Kuangtung, Kuangsi, Yunnan, and Kueichow, where printing was scarcely known in previous dynasties, and especially significant was the compilation and publication of local gazetteers, which began to proliferate to cover all

[a] See *Thing-Lin Wen Chi* (*SPTK*), ch. 3, p. 15a.
[b] See titles listed in *Ku Chin Shu Kho* (*KKT*).
[c] Cf. an account of the printing of the National Academy at Nanking by Liu I-Cheng (1).

[1] 大禮集義　　[2] 九邊圖說　　[3] 銅人針灸圖經　　　　[4] 都察院
[5] 三國志演義　[6] 水滸傳　　[7] 國子監

provinces and numerous prefectures, sub-prefectures, and even towns and villages throughout the Ming empire.[a]

Among the many local official printers, most interesting were the various enfeoffed princes, who had the wealth, leisure, and the opportunity for book collecting and printing. Their libraries contained many rarities of Sung and Yüan editions bestowed upon them by the emperors, while quite a few were prominent authors, collectors, and printers of fine editions, and they constituted one of the distinguished features of Ming scholarship; more than thirty of them are known to have engaged in writing and printing. Not only were many of their own works printed, but also the writings of local scholars which were sometimes sponsored by these princely establishments, and more than 250 titles of these principality editions are known to have been printed; one Prince of Ning printed as many as 137 titles.[b] Their publications included works on medicine, longevity, meditation, amusement, music, games, instruction and conduct of princes, and textbooks for women, besides traditional subjects. The most notable among these publications included the collected works on music and acoustics, *Yüeh Lü Chhüan Shu*[2], in thirty-eight *chüan*, by the Prince of Cheng, Chu Tsai-Yü[3] (+1536–1611), printed about 1606 (Fig. 1124); a collection of medical prescriptions, *Hsin Khan Hsiu Chen Fang Ta Chhüan*[4], printed by the Prince of Chou, Chu Yu-Tun[5], in 1391 and again in 1505; a classified encyclopedia of quotations, *Chin Hsiu Wan Hua Ku*[6], printed by the Prince of Hui in 1533; a treatise on incense and perfumes, *Hsiang Phu*[7], and on tea, *Chha Phu*[8], both printed by the Prince of I in 1640; and a cookbook, *Yin Shan Cheng Yao*[9], reprinted by the Prince of Chhing. All the books they produced are known to have been well collated, based on the best editions, and printed with excellent workmanship on good paper,[c] while wooden movable type is also known to have been employed by the princes of Shu and I in the 16th century.[d]

Under the Ming, private printing was not common before 1500, but became very popular during the 16th and 17th centuries. Many of the private printers, including scholars, families, book collectors, local schools, and monasteries, were motivated by altruism toward the spread of literature, and did not act for profit or because of official obligations. Thus the privately printed editions were usually carefully collated and high in quality and workmanship.[e] Works by individual authors were normally published after their deaths and sponsored by their descendants, friends, or families, though in some cases works were printed during their lifetimes, a witness to their own sponsorship. The famous work on technology,

---

[a] See titles listed in *Ku Chin Shu Kho*; some 770 titles of Ming local gazetteers are recorded in the first edition of Chu Shih-Chia (*3*), as compared with twenty-eight for the Sung and eleven for the Yüan periods.
[b] *Ning Fan Shu Mu*[1], see the *SKCSTM*.
[c] See Chhang Pi-Te (*7*).
[d] Cf. Chang Hsiu-Min (*11*), p. 58.
[e] Cf. K. T. Wu (*5*), p. 231.

[1] 寧藩書目   [2] 樂律全書   [3] 朱載堉   [4] 新刊袖珍方大全
[5] 朱有燉   [6] 錦繡萬花谷   [7] 香譜   [8] 茶譜   [9] 飲膳正要

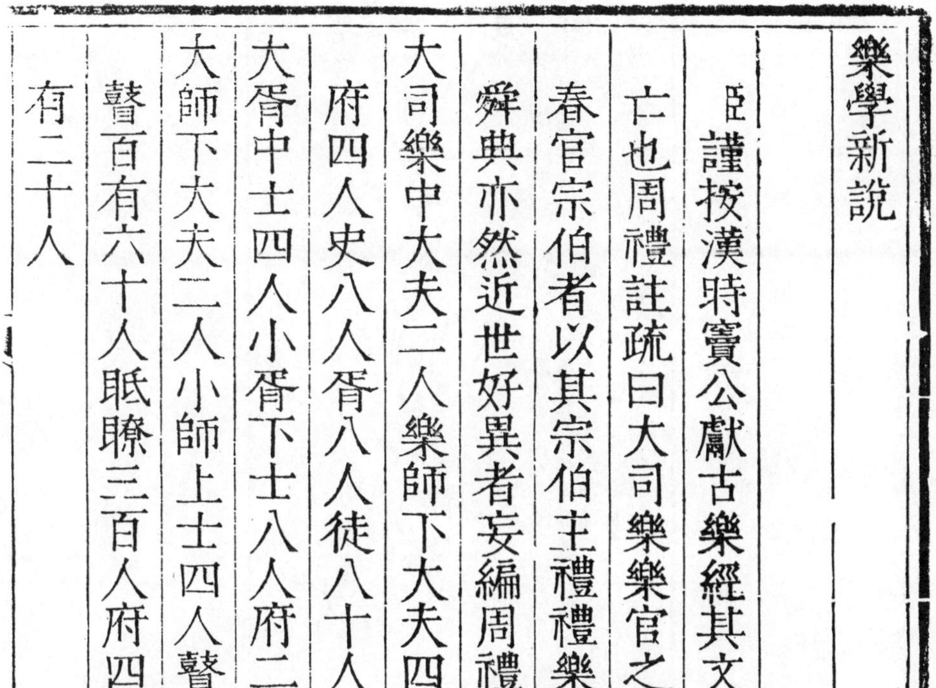

樂學新說

鄭世子臣載堉謹撰

臣謹按漢時竇公獻古樂經其文與大司樂同然則樂經未嘗

亡也周禮註疏曰大司樂樂官之長掌教六樂六舞等事而在

春官宗伯者以其宗伯主禮禮樂相將是故列職於此臣考諸

舜典亦然近世好異者妄編周禮改屬地官司徒誤矣

大司樂中大夫二人樂師下大夫四人上士八人下士十有六人

府四人史八人胥八人徒八十人

大胥中士四人小胥下士八人府二人史四人徒四十人

大師下大夫二人小師上士四人瞽矇上瞽四十人中瞽百人下

瞽百有六十人眡瞭三百人府四人史八人胥十有二人徒百

有二十人

Fig. 1124. Collected works on music and acoustics, *Yo Lü Chhüan Shu*, printed by Chu Tsai-Yü, prince of the Ming dynasty, *c.* +1606. Far Eastern Library, University of Chicago.

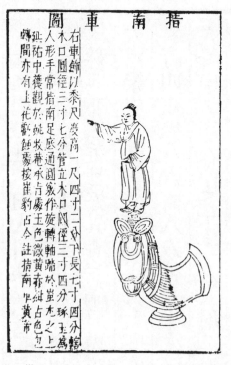

Fig. 1125. South-pointer as illustrated in the Ming encyclopaedia, *San Tshai Thu Hui*, + 1620 ed. Far Eastern Library, University of Chicago.

Thien Kung Khai Wu[1] by Sung Ying-Hsing[2], was printed by the author in 1637, the illustrated encyclopedia, *San Tshai Thu Hui*[3], (Illustrations of the Three Powers), in 106 chüan, compiled by Wang Chhi[4] (*chin shih*, 1565) of Shanghai, in collaboration with his son Wang Ssu-I[5], was printed by his friends in 1609 (Fig. 1125). In the case of the noted scholar-official Hsü Kuang-Chhi[6] (+ 1562–1633), also of Shanghai, his illustrated encyclopedia on agriculture, the *Nung Cheng Chhüan Shu*[7], in sixty *chüan*, was edited after the author's death by one of his disciples, Chhen Tzu-Lung[8], and printed at the Phing Lu Thang[9] in 1640.

Most of the reprints of earlier works or fine editions were usually made by book collectors who, with careful collation, reproduced rare or unique editions from their own libraries, primarily for the diffusion or preservation of the original works for the scholarly world, or as a hobby to spend the wealth accumulated from their businesses. One of the many Ming collector-printers was An Kuo[10] (1481–1534), of Wuhsi, who printed no fewer than two dozen titles known to us, one half of them by metal movable type.[a] The most famous of these men was Mao Chin[11] (+ 1599–1659), of Chhang-shu in Chiangsu, who printed more than 600 works on

[a] See list of works printed with metal movable type under Ming in Chhien Tshun-Hsün (2), pp. 14–15.

[1] 天工開物   [2] 宋應星   [3] 三才圖會   [4] 王圻   [5] 王思義
[6] 徐光啓   [7] 農政全書   [8] 陳子龍   [9] 平露堂   [10] 安國
[11] 毛晉

a variety of subjects, especially many multi-volume works of classics, histories, literary collections, and *tshung shu*. One record shows that he used 11,846 wood blocks for the Thirteen Classics, 22,293 blocks for the Seventeen Standard Histories (Fig. 1126), and 16,637 blocks for the collection *Ching Tai Pi Shu*[1], which consists of 140 titles.[a] At one time, during its early stage, Mao employed some twenty cutters and printers in his workshop and accumulated as many as 100,000 blocks for the printing of various works in his studio Chi Ku Ko[2], a name for both his private library and his printing shop.[b] His work had a great impact on printing in the early Chhing period.

Many private schools printed textbooks for their students and other titles of scholarly importance. The Chhung Cheng Shu Yüan[3] of Kuangtung printed the collected commentaries of the Four Books in 1535 and part of the Standard Histories in 1537, while the famous Tung Lin Shu Yüan[4] and other private academies printed many individual literary collections. Certain officials also printed books in their private capacity. Local scholars serving in the capital usually printed a special kind of gift book, the *shu pha pen*[5], to be presented as a souvenir to their colleagues on their return to the capital from their native provinces. Because the contents of such editions were not carefully collated, they were usually not considered of scholarly value.

As for religious works, at least three and perhaps four editions of the Buddhist *Tripitaka* and one edition of the Taoist canons were printed under the Ming. The most famous of these is the southern edition of the *Tripitaka*, *Nan Tsang*,[6] including 1610 works in 6331 *chüan*, printed in Nanking in 1372, and the northern edition, *Pei Tsang*[7], including 1615 works in 6361 *chüan*, produced in Peking in 1420. Both were printed under imperial auspices and bound in the folded format. A third edition of the *Tripitaka*, known as the *Ching-shan Tsang*[8], was printed in Wu-thai[9] and Ching-shan as well as several other places in Chiangsu and Chekiang between 1589 and 1677; it was the first *Tripitaka* to be bound with thread stitching in the flat style. Another edition is said to have been printed in Hangchow in the Chia-ching period (+1522–66), but it may not have been completed, and no such work is now known to exist.[c] As for the Taoist collection, the *Tao Tsang*[10], the original set included 5305 *chüan*, compiled under imperial auspices, was completed in 1445 and a sequel in 180 *chüan*, in 1607. The two series were later printed together and distributed to Taoist temples as an imperial favour.[d]

[a] Numbers of folios of these volumes are given in Wang Ming-Sheng (+1722–97), *I Shu Pien* (1841), ch. 14, p. 14*b*; these figures equal the number of blocks.
[b] Quoted in Yang Shao-Ho's annotated catalogue; see K. T. Wu (5), p. 245.
[c] Cf. Yeh Kung-Chho (1).
[d] Cf. Liu Tshun-Jen (5), p. 104: Wieger (6) lists 1464 writings, based on a Chinese catalogue compiled by Pai Yün-Chi[11] of the 17th century. The complete set, now kept in the White Cloud Temple in Peking, was reprinted by the Commercial Press in Shanghai in 1923.

[1] 津逮秘書   [2] 汲古閣   [3] 崇正書院   [4] 東林書院   [5] 書帕本
[6] 南藏   [7] 北藏   [8] 徑山藏   [9] 五臺   [10] 道藏
[11] 白雲霽

史記集解序

裴駰

班固有言曰司馬遷據左氏國語采世本戰國策述楚漢春秋接
其後事訖于天漢其言秦漢詳矣至於采經撫傳分散數家之事
甚多疏略或有抵捂亦其所涉獵者廣博貫穿經傳馳騁古今上
下數千載閒斯巳勤矣又其是非頗謬於聖人論大道則先黃老
而後六經序游俠則退處士而進姦雄述貨殖則崇勢利而羞賤
貧此其所蔽也然自劉向揚雄博極群書皆稱遷有良史之才服
其善序事理辯而不華質而不俚其文直其事核不虛美不隱惡
故謂之實錄駰以為固之所言世稱其當雖時有紕繆實勒成一
家總其大較信命世之宏才也考挍此書文句不同有多有少莫
辯其實而世之惑者定彼從此是非相貿眞偽舛雜故中散大夫

Fig. 1126. One of the Seventeen Standard Histories, *Shih Chi*, printed in the craft style of calligraphy by the Chi Ku Ko of Mao Chin, *c.* +17th century. Far Eastern Library, University of Chicago.

Ming commercial printers inherited the tradition of the previous dynasties of the Sung and Yüan with publishing centres in Fukien, Chekiang, and Szechuan. Such bookshops as Chhin Yu Thang[1], which had operated in Chien-yang since the Sung, and the Shen Tu Chai[2], had a history of over a hundred years. The latter, owned by Liu Hung[3], printed numerous titles of histories, literary collections, encyclopedias, and medical works. Especially notable were such large multi-volume sets as the *Wen Hsien Thung Khao*[4], an encyclopedia of institutions in 348 *chüan*; the *Shan Thang Chhün Shu Khao So*[5], an encyclopedia of quotations in 212 *chüan*; and the *Ta Ming I Thung Chih*[6], a national geography of the Ming dynasty, in ninety *chüan*. As might be expected, certain printers specialised in medical books. The Chung Te Thang[7] of Hsiung Chung-Li[8] of Ao-feng printed at least eight medical classics, including a collection of pediatric prescriptions, *Hsiao Erh Fang Chüeh*[9], in ten *chüan* (1440); prescriptions for smallpox, *Tseng Cheng Chhen Shih Hsiao Erh Tou Chen Fang Lun*[10], in two *chüan* (1448), a complete work on surgery, *Wai Kho Pei Yao*[11], in three chüan (1568); and a supplement to the gynaecological treatise, *Hsin Pien Fu Jen Liang Fang Pu I Ta Chhüan*[12], in twenty-four *chüan*, also printed in 1568.[a]

After 1500 Huchow in Chekiang and She-hsien in Anhui were among the best printing centres, and from the late 16th and early 17th centuries, many skilful cutters moved to the area south of the Yangtze, where such cities as Nanking, Soochow, Chhangshu, and Wuhsi became very prosperous in printing and book production.

Generally speaking, Ming printing, especially during the later part of the period, significantly influenced the format of Chinese books in the next four or five hundred years. Ming books printed before 1500 inherited the traditional format of the Yüan (Fig. 1123); their calligraphy was in the soft style, with a black folding line in the block, and the volumes were bound in a wrapped back binding. After 1500, the Sung traditional format was generally followed, where the calligraphy was more rigid and straight, lacking free and swift movement (Fig. 1124). The folding line was white or blank, and names of calligraphers and cutters and numbers of characters appeared on the blocks, similar to Sung practice. From the middle of the 16th century, calligraphy became more stereotyped in style (Fig. 1126), and this has remained the standard form of Chinese printing to the present time. Unbleached yellow bamboo paper was generally used in early Ming, bleached white paper around 1500, and yellow paper again toward the end of the 16th century, at which time binding also underwent a transformation, from the wrapped back style to the stitched binding which remains in use.[b]

---

[a] See list in Yeh Te-Hui (2), pp. 127–42.
[b] See further discussion of book formats and bindings on pp. 222 ff. below.

[1] 勤有堂   [2] 愼獨齋   [3] 劉宏   [4] 文獻通考
[5] 山堂羣書考索   [6] 大明一統志   [7] 種德堂   [8] 熊忠立
[9] 小兒方訣   [10] 增證陳氏小兒痘疹方論   [11] 外科備要
[12] 新編婦人良方補遺大全

(5) PROSPERITY AND DECLINE OF TRADITIONAL PRINTING IN THE CHHING PERIOD

China under the Chhing dynasty (+1644–1912) inherited a great cultural tradition and, although under the alien Manchu rule, in general enjoyed a period of intellectual development and prosperity. Activities in literature, classical research, and the compilation, collecting, and production of books and documents were especially remarkable and abundant. During the first half of the dynasty, from the latter part of the 17th and through most of the 18th century, government leadership in academic pursuits resulted in widespread printing. Numerous distinguished scholars and officials were recruited by the imperial court to engage in the compilation of books, and publication of the results of their research followed naturally. Economic development also gave rise to a group of book collectors who had the financial resources to reprint the rare editions and manuscripts which came into their possession.

The gradual decline of the empire during the latter part of the dynasty, that is from the beginning of the 19th century, was accompanied by a general recession of activities everywhere and a lack of resistance to Western influence. Although the printing industry continued, it generally degenerated in quality if not in quantity in both official and private sectors, and it was during this period that traditional printing proved insufficient to meet modern needs and gradually gave way to the new technology of the West.

Since the imperial palace was active in printing and compilation, Peking naturally acquired central importance in publishing. Printing and publishing also flourished in places like Nanking, Soochow, Hangchow, and Yangchow. While these centres emerged, Fukien was no longer as influential and its editions were less widely circulated than before. Publishing in battle-stricken Szechuan was also declining. Throughout the Chhing dynasty, Chekiang and Chiangsu remained centres of book-collecting and publishing because of their favourable geographic locations, natural resources, and commercial prosperity. As time went on, Hunan and Hupei came on the scene, while Shanghai, the chief entrance for Western influence, became the main publishing city at the turn of the 19th century.[a]

Not unlike previous dynasties, the Chhing saw active imperial patronage of compilation and publication, but with even greater vigour. The Manchu rulers actually started printing before establishing control over China proper, though their early imprints consisted mainly of Manchu translations of Chinese works. After their ascent to the throne, the publishing policy common to all previous dynasties was continued, the body chiefly responsible for central government printing being the Imperial Printing Office at the Wu Ying Palace (Wu Ying Tien

---

[a] Cf. Yeh Te-Hui (2), pp. 253–4; see also distribution of eighty bookshops which produced popular fiction during the Ming-Chhing period in Liu Tshun-jen (4), pp. 36–44.

Hsiu Shu Chhu[1]), a division of the Imperial Household (Nei Wu Fu[2]) and located within the imperial compound. The publications of this office were generally known as Palace editions (*tien pen*[3]) and were noted for careful collation, good paper and ink, elegant calligraphy, and excellent execution and binding (Fig. 1127).

The palace publications included a variety of books and documents produced by different methods. Manuscripts copied by hand included imperial admonitions, authentic records, and such huge collections as the *Ssu Khu Chhüan Shu*[4] (Complete Library of Four Treasures).[a] Block-printed editions included the collected administrative regulations, penal codes, and records of imperial birthdays, tours, and military campaigns, as well as all works on classics, philosophical and literary writings, dictionaries (Fig. 1128), and a miniature 'sleeve edition' (*hsiu chen pen*[5]) of selected titles. Also, bronze and wooden movable type was used for printing such large sets as encyclopaedias, collectanea, and scientific compendia. At least 382 individual titles are registered as having been printed in one way or another at the Imperial Printing Office at the Wu Ying Palace between 1644 and 1805.[b]

Among the huge sets was the gigantic compilation *Thu Shu Chi Chheng*[6] (Grand Encyclopaedia of Ancient and Modern Knowledge) in 5020 volumes, printed with bronze movable type in +1728 (Fig. 1147).[c] It consists of 6109 topical headings grouped into thirty-two sections under six major divisions: Heaven, Earth, Man, Science, Literature, and Government. Direct quotations with occasional illustrations are further classified by types of materials under each heading, with a total of over 100 million words in the work.[d] Another large set was the *Wu Ying Tien Chü Chen Pan Tshung Shu*[7] (Collection of Assembled Gem Editions of the Wu Ying Palace), which included 138 titles printed with wooden movable type from 1773 to around 1800.[e] One Buddhist collection, *Lung Tsang*[8] (The Dragon *Tripitaka*), was also printed in Peking. Among some fifteen *tripitakas* printed in China up to the end of the Chhing dynasty, this set of 1662 works in 7168 *chüan* was perhaps the largest in number of works included and the quickest in printing, as it was completed in the remarkably short time of three years from 1735 to 1738.[f]

As in previous dynasties, central government printing in the Chhing was followed by local agencies, and the central government actually encouraged such activity. In 1776, for example, copies of the Palace edition of the *Wu Ying Tien Chü Chen Pan Tshung Shu* were given to southeastern provinces with permission to

[a] This manuscript library consisting of 3511 titles in 36,275 volumes was compiled under imperial auspices from 1773 to 1782. Four original copies were deposited in the imperial palace and Yüan Ming Yüan in Peking, and in Mukden and Jehol. Three additional copies were made in 1790 for Hangchow, Chenkiang, and Yangchow.
[b] See tabulation in Shaw (1), p. 20.
[c] See also discussion on pp. 215 ff. below.
[d] For its history, nature, and scope, see Introduction in Giles (2).
[e] Cf. Rudolph (14), pp. 323–4.
[f] Two sets comparable to the size of *Lung Tsang* were the one printed in Peking in 1177–1194 and another in Ching-shan in 1589–1677, each of which consisted of 1654 works and 7182 *chüan* and 6956 *chüan* respectively.

[1] 武英殿修書處   [2] 內務府   [3] 殿本   [4] 四庫全書
[5] 袖珍本   [6] 圖書集成   [7] 武英殿聚珍板叢書   [8] 龍藏

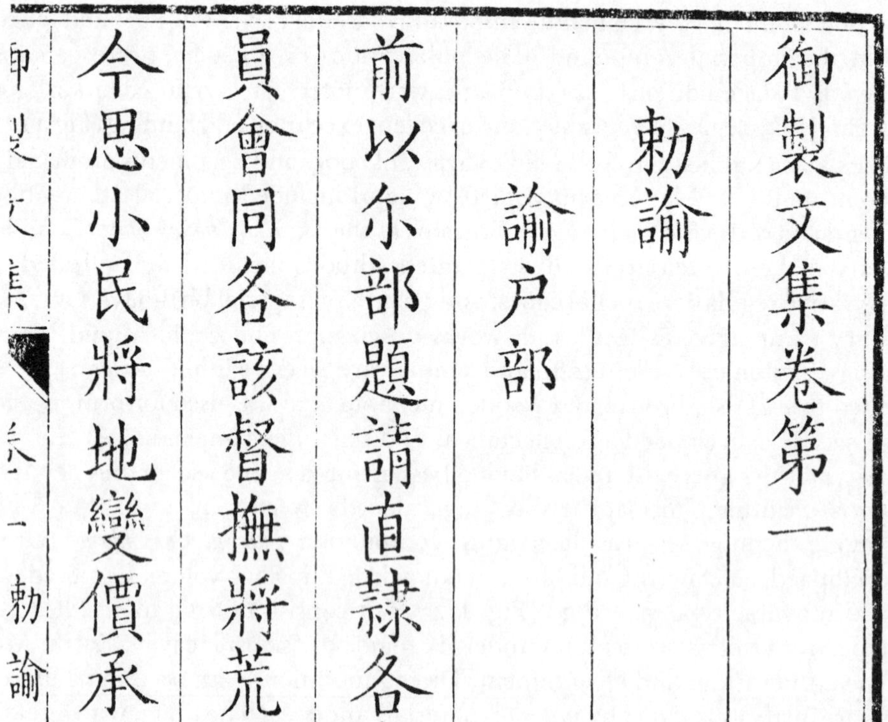

御製文集卷第一

勅諭

諭戶部

前以尔部題請直隷各省廢藩田產差部

員會同各該督撫將荒熟田地酌量變價

今思小民將地變價承買之後復徵錢糧

Fig. 1127. Imperial writings of Emperor Khang-Hsi printed by the Wu Ying Tien palace printing shop in +1771.
From *Yü Chih Wen Chi*. Far Eastern Library, University of Chicago.

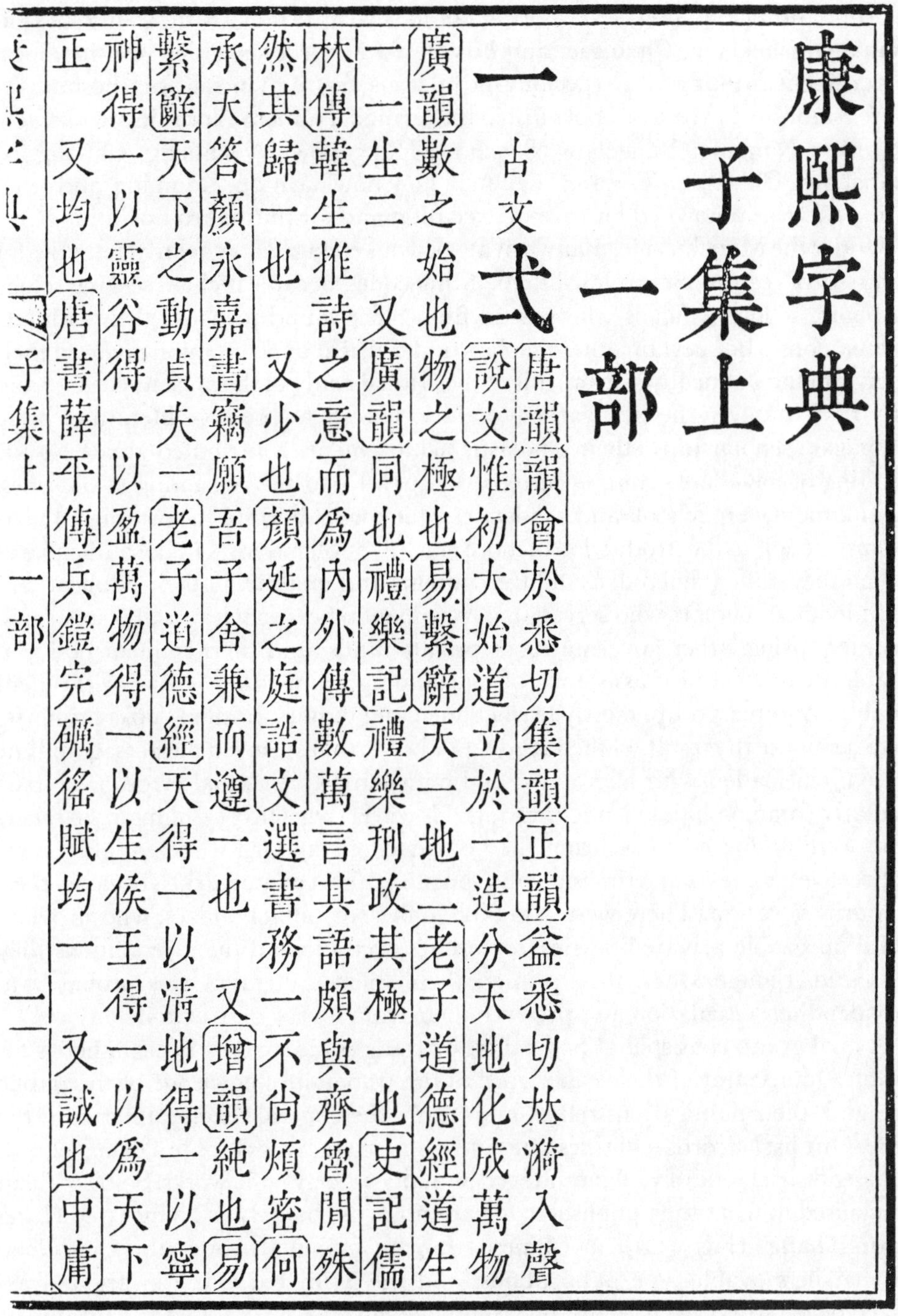

康熙字典
子集上
一部

一　古文　弌

〔唐韻〕〔韻會〕於悉切〔集韻〕〔正韻〕益悉切，竝音漪入聲。數之始也。〔易·繫辭〕天一地二。〔老子·道德經〕道生一，一生二。又〔廣韻〕同也。〔禮記〕禮樂刑政，其極一也。〔史記·儒林傳〕韓生推詩之意而爲內外傳，數萬言，其語頗與齊魯閒殊，然其歸一也。又少也。〔顏延之·庭誥文〕選書務一，不尚煩密。又〔增韻〕純也。〔易·繫辭〕天下之動，貞夫一者也。〔老子·道德經〕天得一以清，地得一以寧，神得一以靈，谷得一以盈，萬物得一以生，侯王得一以爲天下貞。承天答顏永嘉書，竊願吾子舍兼而遵一也。又正也。又均也。〔唐書·薛平傳〕兵鎧完礪，徭賦均一。又誠也。〔中庸〕

子集上　一部　一
二

Fig. 1128. Original block edition of the most popular dictionary, *Khang-Hsi Tzu Tien*, with some 49,000 characters, printed *c.* +1716. Far Eastern Library, University of Chicago.

reprint, and before long, several local editions were printed with wood blocks at Nanking, Chekiang, Chiangsu, and Fukien. After the Thai-Phing uprising, local government printing was especially prosperous and provincial printing bureaus were established to restore books lost in the turmoil. Numerous printing works were set up in Nanking, Yangchow, Soochow, Hangchow, Wuchhang, Chhangsha, Nanchang, Chhengtu, Tsinan, Thaiyuan, Foochow, Canton, Kunming, and other cities; all were renowned for careful execution and for mutual co-operation.[a]

Under the Manchu rule, many private persons engaged in scholarly activities for a variety of reasons. Some, loyal to the Ming court, became recluses seeking refuge in books, scholar-officials who retired from office found in books their ultimate companions while certain other scholars under threat of persecution diverted their energy to the politically safe pursuit of the critical study of the ancient Classics and history. Whatever the motivation—vanity, a wish to preserve literature, or to propagate scholarship—their scholastic achievements contributed greatly to the printing of new books and, even more frequently, to the reprinting of old ones. Again, there were schools and academies which went beyond the printing of texts for instruction to the production of more general publications, but the audiences at which they aimed differed, as did the contents and qualities of their publications.

Individual scholars who served in government offices sometimes also sponsored printing, using either government or private funds, and their eminent positions enable them to secure assistance of outstanding scholars for works of excellent quality. Another group, mostly learned men and writers, printed books primarily to disseminate the results of their study. But books that survived the passage of time were often loaded with alterations and omissions, and textual errors were particularly common in late Ming imprints. However, a group of Chhing researchers who were interested in securing the best texts of standard works, applied their entire lives to textual criticism, and printing the collated texts became their favourite vocation. There were also bibliophiles, or bibliomaniacs, who strove to build up sizable private libraries, sometimes even idolised the rare editions they possessed, though when they printed their choicest items it was probably to broaden their circulation and preserve them from loss or destruction.

A final group consisted of booksellers or trade printers; they dealt in books for profit. On account of their experience of the trade and knowledge of the public demand, their publications reflect more truly the general reading interest of the time. One list[b] records 246 titles put out by sixty-two bookshops in Peking's Liu-li chhang district, one of which produced as many as forty-nine works. Some of them specialised in what they published; for example, the San Huai Thang[1] and Hung Yüan Thang[2] (Fig. 1129) in Chinese-Manchu bilingual works; the Chü Chen Thang[3] in movable-type publications, and the Tsun Ku Chai[4] in art and archaeological works. At least one bookshop, the Sui Ya Chai,[5] edited and printed its

---

　　[a] Cf. Ching Yü (*1*), pp. 342–3.
　　[b] This list covers imprints from mid-Chhing to early Republic; see Sun Tien-Chhi (*1*), pp. 127–56.

　　[1] 三槐堂　　　　[2] 鴻遠堂　　　　[3] 聚珍堂　　　　[4] 尊古齋　　　　[5] 邃雅齋

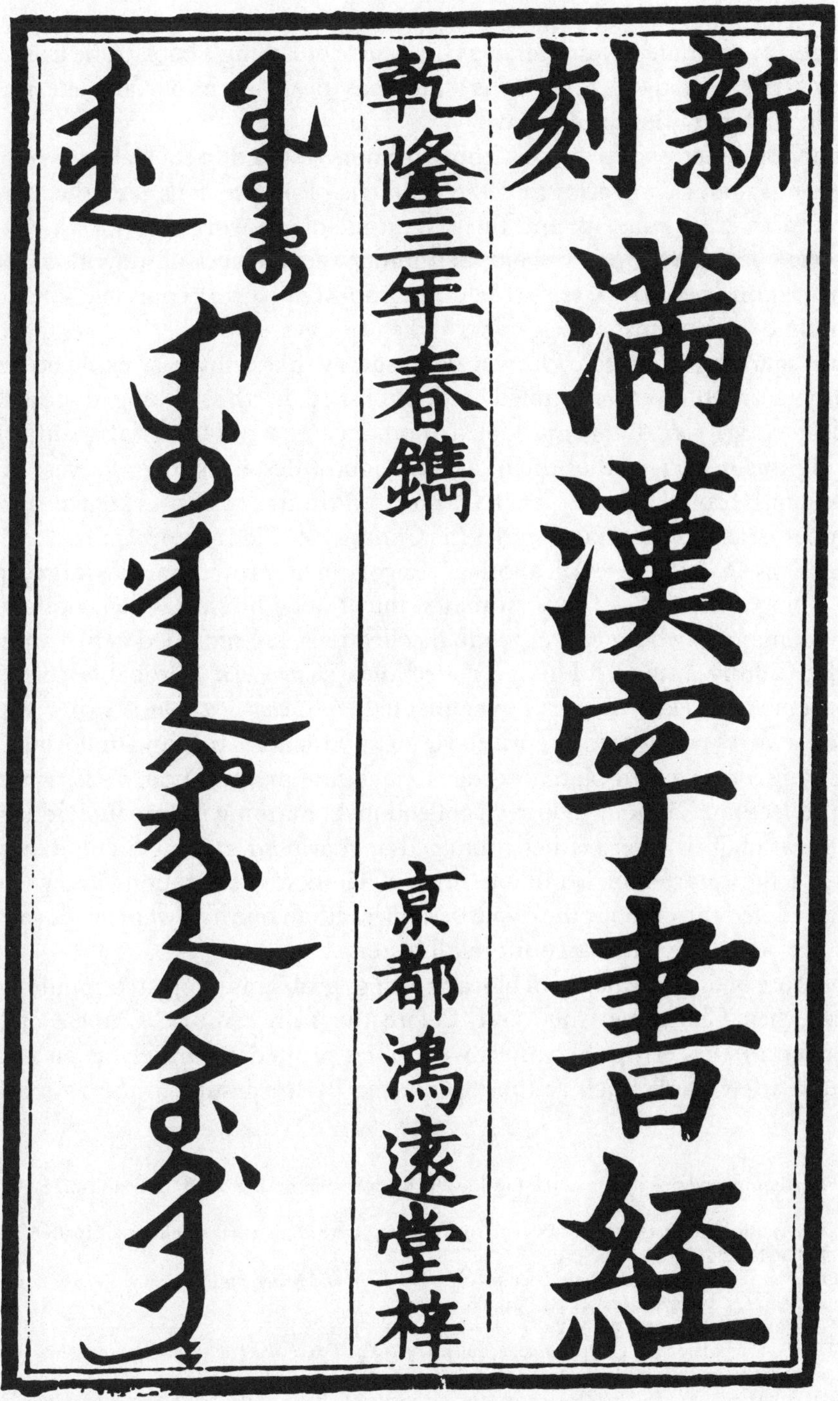

新刻
滿漢字書經
乾隆三年春鐫
京都鴻遠堂梓

Fig. 1129. Chinese-Manchu bi-lingual text of the *Shu Ching* printed by the Hung Yüan Thang in Peking in +1738. Far Eastern Library, University of Chicago.

own collections. The Liu Li Chhang district, where over 300 bookshops were in business at one time or another, was the centre of Peking's book trade from the late Khang-Hsi period (+1662–1722) and has been an important site for many intellectual activities until today.[a]

The combined efforts of the Chhing printers resulted in such a surge of printing in several major categories that the products of no previous period can be compared with it for quantity and the magnitude of the works produced. First, local histories (*ti fang chih*[1]) were compiled under the auspices of individual local administrations of provinces, prefectures, sub-prefectures, counties, villages, and sometimes mountains, passes, rivers, dykes, bridges, salt wells, temples, academies, tombs, gardens or guilds. Of over 7000 such works known to exist, no less than eighty per cent were compiled and published in the Chhing dynasty.[b] Clan registers (*tsung phu*[2] or *chia phu*[3]) form another category of materials printed mainly during this period, and of them, at least 4000 titles are known to exist in public collections throughout the world. Of them, the source of 1550 is known, and of this number 1214 were compiled under the Chhing (see Fig. 1130).[c] Individual literary collections (*pieh chi*[4]) are still another category of literature that was largely printed in Chhing times; indeed, it is estimated that 14,000 literary writers of the Chhing period, many of whom had individual collections, are recorded in five anthologies of the Chhing authors.[d] Finally, collectanea (*tshung shu*[5]) probably compose the most extensive class of works ever printed. A *tshung shu* consists of a variety of literary works published or reprinted under one general title and uniformly bound, following a prescribed plan, in order to facilitate preservation, wider circulation, and collecting. Of some 3000 such collections containing 70,000 individual works, the great majority were either produced or reprinted in the Chhing dynasty.[e] In fact, of the quarter of a million titles of Chinese publications known to have accumulated throughout the dynasties, no less than one half were produced during this period, the greatest amount in all history.[f]

Because of the prevalence of block printing, problems in printed communication arose when China met the West. Before the 19th century, Chinese characters included in Western publications were often printed separately as an appendix, though a few works such as those produced by the Jesuits in the 16th and 17th

---

[a] For a history and description of the Liu-Li-Chhang district in Peking, see Sun Tien-Chhi (*1*), Wang Yeh-Chhiu (*1*).

[b] Of 5832 titles listed in Chu Shih-Chia (*3*), 1935 ed., 4655 were printed in the Chhing; the 1958 ed. added 1581 titles, but no breakdown is given.

[c] Only fourteen are known to have been produced in Yüan and Ming, and 322 in the Republican period; see Taga Akigorō (*1*), pp. 220–49; Leslie (*1*), p. 86.

[d] For sources of the estimate, see Yang Chia-lo (*2*), pp. 25–6.

[e] The most complete catalogue of the collectanea, *Chung Kuo Tshung Shu Tsung Lu*, vol. 1, registers 2797 works by subject classification; certain classes are arranged by periods.

[f] It is estimated that 253,435 titles are registered in various dynastic and other bibliographies from Han to the 1930's; 126,649 were produced under the Chhing; see Yang Chia-Lo (*2*), p. 27.

¹ 地方志　　　² 宗譜　　　³ 家譜　　　⁴ 別集　　　⁵ 叢書

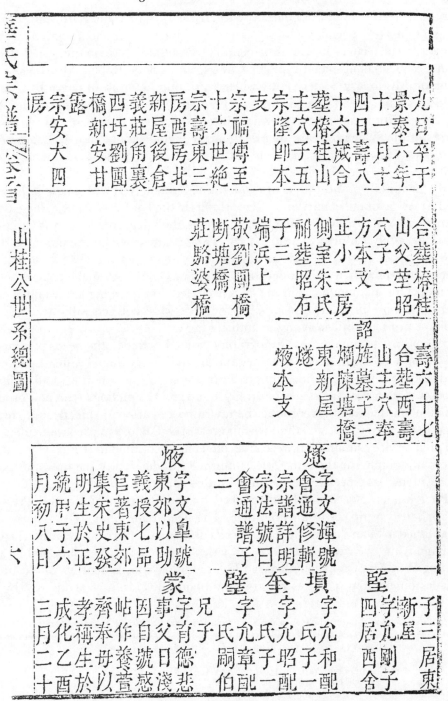

Fig. 1130. Genealogical record of the Hua clan, printed with wooden movable type in +1872, listing names of Ming printers with Hua Sui in the fourth column and his nephew Hua Chien in the fifth column. From *Hua Shih Tsung Phu*, 1872 ed., Columbia University Library.

centuries were printed from blocks.[a] As early as 1555 or 1570, European printers began to make experiments to accommodate Chinese characters printed on their presses,[b] while later, in the 19th century, Protestant missionaries tried to cut punches for making metal type for Chinese characters. Robert Morrison of the London Missionary Society established a printing house in Malacca in 1814 to print his Chinese-English dictionary (Fig. 1131) and a translation of the *New Testament* with metal type cut by a Chinese engraver, Tshai Kao[1], and his assistants. In the following year, the first monthly periodical in Chinese, *Chha Shih Su Mei Yüeh Thung Chi Chuan*,[2] was started. This enterprise soon found many followers.[c]

By the middle of the 19th century, fonts of Chinese type were made in Europe and America for missionary and other printers in the Far East.[d] At first, these types were used primarily for printing bilingual texts, though gradually they found their place in purely Chinese printing. A Mr Tang created two fonts containing over 150,000 types cast in moulds at Canton in +1850,[e] and nine years later, the electrotype process was introduced by William Gamble to make a large set of Chinese type in Shanghai. But in Chinese circles block printing and wooden type were still more popular, and modern methods were not generally accepted by Chinese printers until early in the 20th century.

In contrast to typography, lithography catered better to the needs of Chinese books and successfully affected literary and artistic life in China. Printers in the past had usually reproduced fine editions in facsimile by the laborious and difficult process of re-engraving, so they naturally found great attraction in this new process which permitted them to reproduce handwriting and art work directly, or to make exact replicas of treasured editions with great speed. Books to prepare readers for the civil service examination were also conveniently printed in reduced format by photolithography in the late 19th century. Reproduction of pictures and book illustrations was especially successful with this new method (Fig. 1132).

Photolithography was first introduced by the Catholic Thu-se-wei Press in Shanghai to print Christian literature. The Tien Shih Chai[3] (Fig. 1133), also established in Shanghai in 1874, produced editions with such small characters that a magnifying glass was sometimes provided along with a purchase, while the Thung Wen Shu Chü,[4] founded in 1881, published mostly reprints of old titles. Besides monochrome pictures, multi-colour photolithography was employed by Fu

[a] A catechism and a Latin-Chinese vocabulary prepared by Father Michele Ruggiero were printed from blocks in Macao in +1585; a German chronology printed in Berlin in 1696 and Baeyer's *Museum Sinicum* printed in 1730 were also from blocks; Cf. Lach (5), II 3, pp. 486–97; Hirth (28), p. 165.
[b] Lach (5), I 1, pp. 679–80; II 3, p. 527.
[c] For the development of typography in modern China, see K. T. Wu (8).
[d] Ching Yü (1), 345–55; Hirth (28), p. 166.
[e] Method of making type and a sample page are given in *Chinese Repository*, 19 (1850), pp. 247–9.
[f] Walter Medhurst (1) says that he began to use lithography in Batavia in 1828 or 1829. In a dozen years from 1823, he printed thirty works in Chinese, nineteen with blocks and eleven with lithography.

[1] 蔡高   [2] 察世俗每月統紀傳   [3] 點石齋   [4] 同文書局

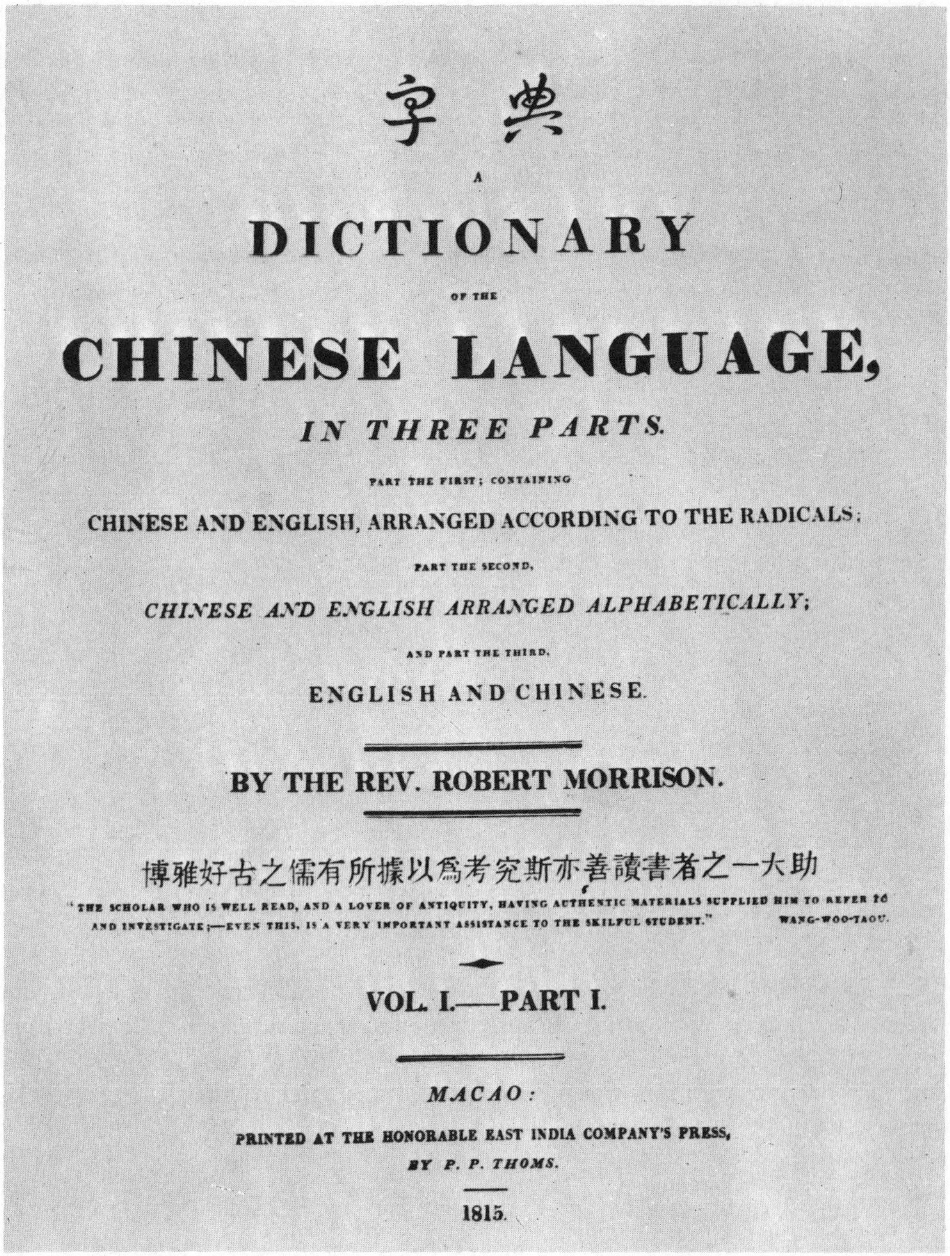

字典

A

DICTIONARY

OF THE

CHINESE LANGUAGE,

IN THREE PARTS.

PART THE FIRST; CONTAINING

CHINESE AND ENGLISH, ARRANGED ACCORDING TO THE RADICALS;

PART THE SECOND,

CHINESE AND ENGLISH ARRANGED ALPHABETICALLY;

AND PART THE THIRD,

ENGLISH AND CHINESE.

BY THE REV. ROBERT MORRISON.

博雅好古之儒有所據以爲考究斯亦善讀書者之一大助

"THE SCHOLAR WHO IS WELL READ, AND A LOVER OF ANTIQUITY, HAVING AUTHENTIC MATERIALS SUPPLIED HIM TO REFER TO
AND INVESTIGATE;—EVEN THIS, IS A VERY IMPORTANT ASSISTANCE TO THE SKILFUL STUDENT."   WANG-WOO-TAOU.

VOL. I.——PART I.

MACAO:

PRINTED AT THE HONORABLE EAST INDIA COMPANY'S PRESS,

BY P. P. THOMS.

1815.

Fig. 1131. First Chinese-English dictionary printed with metal type in Macao, 1815.
University of Chicago Library.

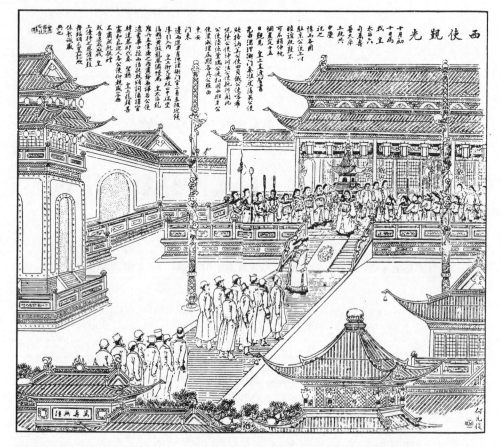

Fig. 1132. Foreign envoys in Peking paying tribute to Empress Dowager Tzhu-Hsi on her sixtieth birthday, received by the Emperor Kuang-Hsü at the imperial court in +1894. Drawing by Ho Yüan-Chün and printed by lithography.

Wen Ko,[1] Tsao Wen Shu Chü,[2] and the Commercial Press, all in Shanghai, whence the process of photolithography spread to many other cities.[a] Founded in 1897, the Commercial Press grew into the largest printing house in the Far East, and has been a landmark for the intellectual and educational development of modern China.[b]

(f) TECHNIQUES AND PROCEDURES OF CHINESE PRINTING

UNLIKE papermaking, the technical procedures of printing have scarcely been documented in Chinese literature. No information on how printing blocks were made nor on how many copies were printed from each block in earlier times is

[a] See Ching Yü (1), pp. 356–7; Hirth (28), pp. 169–71.
[b] For its history and development, see Florence Chien (1).

[1] 富文閣   [2] 藻文書局

Fig. 1133. A lithographic printing shop, Tien Shih Chai, operated in Shanghai after +1874. Drawing by Wu Yu-Ju., c. +1884.

available, except for occasional remarks made by a few foreign observers and writers.[a] On the other hand, some of the movable type methods were recorded. Details of materials, tools, and the methods used for preparing block printing can only be deduced from an interpretation of related terms, by examination of printed editions, and from the oral testimony of a few surviving craftsmen or observers of the work of carving and printing since its gradual disappearance in the early part of the 20th century. Only recently an article describing the methods of block printing gave some details of the technical procedures,[b] yet even so, many questions still cannot be fully answered.

While traditional terms for printing such as *tsai li*[1] (to spoil pear wood), *chhien tsao*[2] (to incise jujube wood), or *fu tzu*[3] (to send for engraving on catalpa wood) refer to the kinds of materials used, such modern terms as *yin shua*[4] (to print and brush) or *huo tzu*[5] (movable characters) are concerned with the methods used, and some of

[a] Brief reports about methods of Chinese printing were given by Rashīd al-Dīn in c. 1300, Matteo Ricci in c. 1600, and Isaiah Thomas in 1810; others mention Chinese printing but not its technical procedures; see below, pp. 306 ff.

[b] See an article written in 1947 by Lu Chhien (1), based on the information supplied to him by a surviving block printer.

[1] 災梨   [2] 鋟棗   [3] 付梓   [4] 印刷   [5] 活字

these are often found in printers' colophons or elsewhere in books. Based on available information and interviews with block cutters, the following accounts are accompanied with figures to illustrate how traditional printing was done.[a]

(1) MATERIALS, TOOLS, AND PREPARATION FOR PRINTING FROM WOODBLOCKS

Woods most commonly chosen for printing blocks included pear, jujube, catalpa, and sometimes apple, apricot, or the wood of other fruit trees with similar qualities. Boxwood, gingko, and Chinese honey locust were also used. Pear (*li*, *Pyrus sinensis*, Ldl.) has a smooth and even texture with a medium hardness that is ideal for carving in any direction. Jujube (*tsao*, *Zizyphus vulgaris*, Lam.) has a straight, even grain with fine pores and is harder than pear. While its fruit bears a superficial resemblance to a date, the jujube is unrelated to the palm family and does not resemble it in any way. Catalpa (*tzu*, *Lindera tsu-mu*, Hemsl.) is a hard wood with a straight but coarse grain. It is commonly considered to be the best material for coffins. Boxwood (*huang yang*,[4] *Buxus sempervirens*, L.) is the softest and is often used for carving regular text; gingko (*yin hsing*,[5] *Gingko biloba*, L.) is very absorbent, and the Chinese honey locust (*tsao chia*[6] or *pai thao*,[7] *Gleditsia sinensis*, Lam.) is a very hard wood, useful for cutting delicate lines in illustrations.

These woods from deciduous trees were chosen, it seems, because they were both abundant as well as suitable. The wood of coniferous trees, while soft and straight-grained, is impregnated with resin which would probably affect the evenness of the ink coating and so is unsuitable for printing blocks; such wood is also valuable for other uses. For printing paper money and articles in large quantities, copper plates and other materials were sometimes used.[b]

There are two methods of cutting wood into blocks for carving. One is to cut with the grain, making a block with a straight or an irregular grain. The other is to cut across the grain. For best results in carving and inking, Chinese craftsmen usually chose the former method, preferring blocks with a straight and close grain. This not only made possible a larger area for the text, but also avoided having to use the heart of the wood. The wood had, of course, to be free of knots and spots, since these would interfere with both carving and printing. After the blocks have been cut, they are usually soaked in water for about a month before use but, if they are needed for immediate use, they can be boiled instead; they are then left to dry in a

---

[a] The writer is grateful to Mr Ku Thing-Lung,[1] Director of the Shanghai Library, for his supply of information, photographs, and drawing concerning the procedures of block printing and for his arrangement of an interview in September 1979 with the staff of the Shanghai Shu Hua She[2] and To Yün Hsien[3] printing works, where block books and colour prints are still made.

[b] A block for printing deposit slips of a Chinese bank in the 19th century was composed of fourteen pieces of water-buffalo horn; see illustrations in Britton (3), pp. 103–5.

[1] 顧廷龍   [2] 上海書畫社   [3] 朶雲軒   [4] 黃楊   [5] 銀杏
[6] 皂筴   [7] 白桃

shaded place before being planed on both sides. Vegetable oil may be spread over the block surface, which is then polished with the stems of polishing grass (*chi chi tshao*,[1] *Achnatherum*). The size of a block depends upon that of the sheet to be printed; normally it is rectangular, averaging twelve inches wide, eight inches high, and half an inch thick. Both sides are usually carved to enable the printing of two pages, or one leaf, on each side (Fig. 1134).

In the preparation for engraving and printing, the manuscript is transcribed onto thin sheets of paper by a professional calligrapher. To do this a blank sheet is ruled into columns and spaces with a centre line in each column known as a variegated space (*hua ko*[2]); this is used as a guide for writing the characters within each space in a balanced arrangement. The paper is waxed lightly and smoothed with a stone burnisher to make the surface easier to write on with a brush. The transcript is placed, written side down, on a block over which a thin layer of paste has been evenly spread (Fig. 1136a). The back of the paper is then rubbed with a flat palm-fibre brush (Fig. 1137a) so that a clear impression of the inked area is transferred to the block. When the paper has dried, its upper layer is rubbed away with finger tips and brush to expose a fine mirror image of the characters or designs which have been applied to the block, looking as if they had been inscribed directly on it. The block is then ready for carving.

The engraving of the woodblocks requires a set of sharp-edged tools of different shapes (Fig. 1135).[a] The cutting knife (*kho tzu tao*[3]), is a steel graver with a sharp blade, and is the most important tool for cutting the main lines along the edges of inked areas. A double-edged chisel (*tsan*[4]) is then applied to cut away surfaces of the block not covered with ink. Next a gouge (*chho*[5]) scoops out the space from the surface, sometimes leaving a groove in the columns and, finally, a pick with two sharp ends (*liang thou mang*[6]) is used for work too fine for other tools.

The general practice in engraving is to leave all black lines in the mirror image in relief. A cut is first made narrowly bordering each character; this process is called *fa tao*,[7] or starting cut (Fig. 1136b). The knife is held like a dagger in the right hand, steadied and guided by the middle finger of the left hand, and is usually drawn towards, not pushed away from, the cutter. This ensures that a close cut is made along the very edge of the black line. All the vertical lines are cut first, in one direction, and then the block is turned around for the cutting of the horizontal or slanting strokes as well as the dots. When this has been done, the blank space between the outer and inner lines is cut away in a step known as *thiao tao*,[8] or close cut leaving the characters with a relief of about one-eighth an inch. The gauge with a semi-circular edge is used to cut away all the blank surfaces, known as *ta khung*,[9] or chiselling blanks. (Fig. 1136c), and a wooden hammer (*pao tzu*[10]) (Fig. 1135f) is

---

[a] A list of printing and engraving tools exhibited at the First International Library Exhibition in Rome, 1929, is given in the *Bulletin of the Library Association of China*, 4:5 (3 April 1929), pp. 12–23.

[1] 芨芨草   [2] 花格   [3] 刻字刀   [4] 鏨   [5] 鑿
[6] 兩頭忙   [7] 發刀   [8] 挑刀   [9] 打空   [10] 拍子

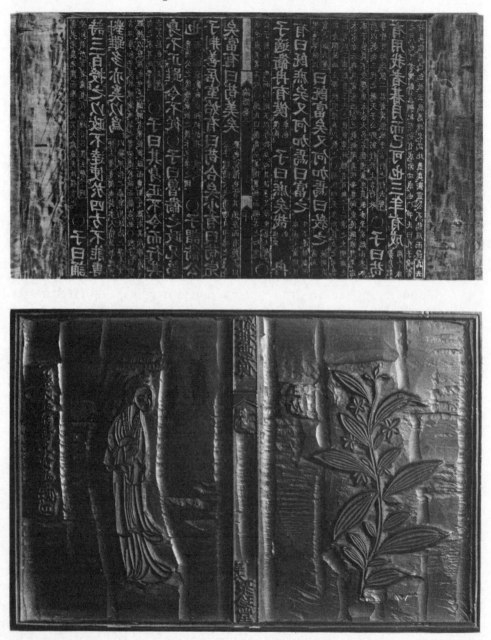

Fig. 1134. Carved woodblocks ready for printing. Above is the text and commentaries of the *Confucian Analects* with reverse characters in relief; below is an illustration of the *Dream of Red Chamber*, both were carved on two sides. Far Eastern Library, University of Chicago.

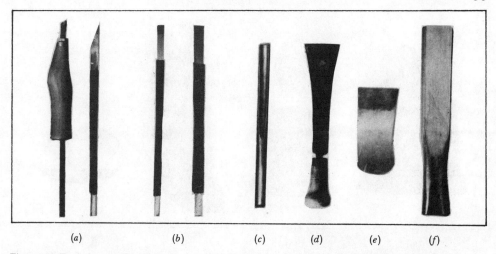

Fig. 1135. Tools for woodblock carving. (a) Cutting knives. (b) Double-edged chisels. (c) Gouge with semi-circular edge. (d) Flat edge chisel. (e) Scraper. (f) Wood mallet. Original photos from Jung Pao Chai, Peking and To Yün Hsien, Shanghai.

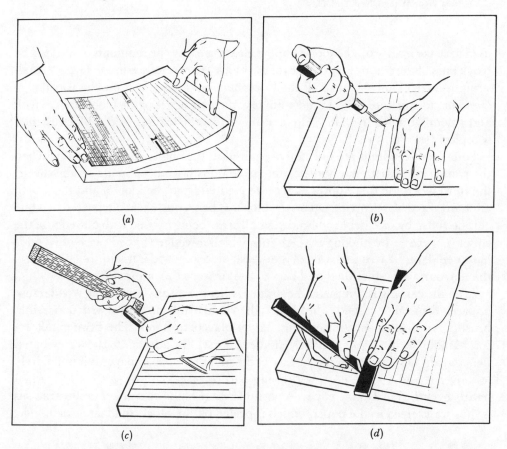

Fig. 1136. Preparation of block for printing. (a) Transferring text to block. (b) Starting cut. (c) Chiseling blanks. (d) Cutting lines. Drawings by staff of Shanghai Library.

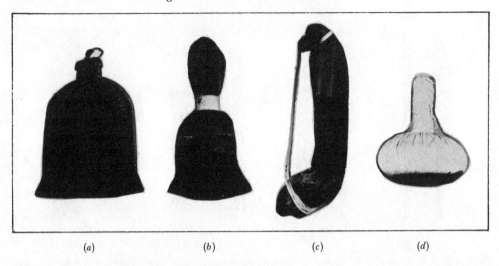

(a)   (b)   (c)   (d)

Fig. 1137. Brushes and accessories for printing and rubbing. (a) Flat brush for transferring text to block or stone. (b) Round brush for inking. (c) Long brush for taking impression. (d) Stuffed pad for inked rubbing. Original photos from Weng Wango, New York City.

used to strike lightly on the tools to aid in clearing away the remnants of wood. The black lines of the columns and edges of the block are carefully trimmed with a small straight-edged knife to make all the lines sharp and clear; this process is called *la hsien*,[1] or cutting lines (Fig. 1136d). Finally, the four edges of the block are sawn off and smoothed with a plane. The carving and preparation of the block is then complete (see Fig. 1134).

Four proof-readings are normally required in the engraving process: one when the transcript is written, another after corrections have been made on the sheets, the third when the first sample sheets are printed from the blocks, and the fourth after any repairs are carried out on the block. When a mistake is discovered or a line chipped off, a block can be repaired; a small error being excised with one edge of the chisel (*ting chho*[2]) by making a notch into which a wedge-shaped piece of wood is hammered, but if area is involved, a suitable piece of wood is inlaid. In either case, the new surface is smoothed and carved as if it were the original.

After the carving is complete, the surface is cleaned of any remaining wood refuse or paper tissue and washed. The block is then held firmly on a table with paper, ink, brushes, and other accessories (Fig. 1137) placed to hand. The printer takes a round inking brush (*yüan mo shua*[3]), made of horsehair, dips it into the water-based ink, and applies it to the raised surface of the block. A sheet of paper is immediately laid over it and a long, narrow rubbing pad (*chhang shua*[4], or *pa tzu*[5]) is brushed lightly over the back of the paper. A positive image of the characters or illustrations is thus transferred to the paper, which is peeled off the block and laid aside to dry.

---

[1] 拉綫   [2] 釘鑿   [3] 圓墨刷   [4] 長刷   [5] 耙子

The process is repeated until the necessary number of copies is obtained.[a] Sample copies were sometimes made in red or blue, but final copies were always printed in black, and it is said that a skilled printer could print as many as 1500 to 2000 double-page sheets in a day.[b] Fifteen thousand prints can be taken from the original block, and another 10,000 after slight touching up. Blocks can be stored and used again and again when additional copies are needed.

(2) VARIETIES AND METHODS OF MOVABLE TYPE PRINTING

Movable type printing was invented as an alternative to the cumbersome process of block printing. For economy and efficiency, experiments were naturally made from time to time with movable type, even though it was not entirely satisfactory for printing Chinese. The principle of assembling individual characters to compose a piece of text can be traced back many centuries before Christ, as inscriptions on bronze vessels, pottery objects and cast metal seals have made evident,[c] but the use of movable type for printing was not begun until the middle of the +11th century.

(i) *Earthenware types*

At the present time, the only known authoritative account of the invention of movable type by a commoner named Pi Sheng[1] (*c.* +990–1051) is the contemporary record of Shen Kua (+1031–95), which says:

> During the reign of Chhing-li [+1041–48] Pi Sheng, a man of unofficial position, made moveable type. His method was as follows: he took sticky clay and cut in it characters as thin as the edge of a coin. Each character formed, as it were, a single type. He baked them in the fire to make them hard. He had previously prepared an iron plate and he had covered his plate with a mixture of pine resin, wax, and paper ashes. When he wished to print, he took an iron frame and set it on the iron plate. In this he placed the types, set close together. When the frame was full, the whole made one solid block of type. He then placed it near the fire to warm it. When the paste [at the back] was slightly melted, he took a smooth board and pressed it over the surface, so that the block of type became as even as a whetstone.
>
> If one were to print only two or three copies, this method would be neither simple nor easy. But for printing hundreds or thousands of copies, it was marvelously quick. As a rule he kept two formes going. While the impression was being made from the one forme, the type was being put in place on the other. When the printing of the one forme was finished, the other was then ready. In this way the two formes alternated and the printing was done with great rapidity.
>
> For each character there were several types, and for certain common characters there were twenty or more types each, in order to be prepared for the repetition of characters on

---

[a] It is hard to estimate how many copies were printed for an edition in average demand; Lu Chhien (*1*), p. 632, says that thirty copies were normally made for the first impression and additional copies made when needed.

[b] Matteo Ricci reported 1500 sheets per day; see Gallagher (1), p. 21. But John F. Davis (1) said 2000 and others estimate 6000–8000 per 10-hour day.

[c] See detailed discussion of the prehistory of printing on pp. 132 ff. above.

[1] 畢昇

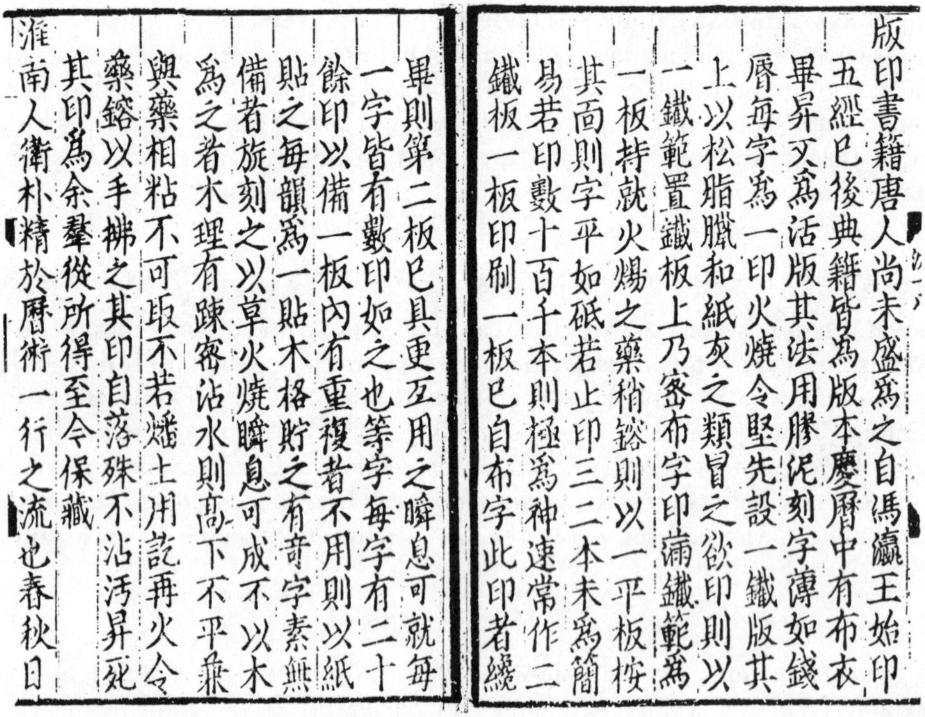

版印書籍唐人尚未盛為之自馮瀛王始印
五經已後典籍皆為版本慶曆中有布衣
畢昇又為活版其法用膠泥刻字薄如錢
唇每字為一印火燒令堅先設一鐵版其
上以松脂臘和紙灰之類冒之欲印則以
一鐵範置鐵版上乃密布字印滿鐵範為
一板持就火煬之藥稍鎔則以一平板按
其面則字平如砥若止印三二本未為簡
易若印數十百千本則極為神速常作二
鐵板一板印刷一板已自布字此印者纔
畢則第二板已具更互用之瞬息可就每
一字皆有數印如之也等字每字有二十
餘印以備一板內有重複者不用則以紙
貼之每韻為一貼木格貯之有奇字素無
備若旋刻之以草火燒瞬息可成不以木
為之者木理有踈密沾水則高下不平兼
與藥相粘不可取不若燔土用訖再火令
藥鎔以手拂之其印自落殊不沾污昇死
其印為余羣從所得至今保藏
淮南人衛朴精於曆術一行之流也春秋日

Fig. 1138. Earliest extant edition of the *Meng Chhi Pi Than*, printed in the +14th century. The passage rearranged into one double-leaf above records the first use of earthenware movable type printing by Pi Shêng in the middle of the +11th century. Copy preserved at the National Library of China.

the same page. When the characters were not in use, he had them arranged with paper labels, one label for words of each rhyme-group, and kept them in wooden cases. If any rare character appeared that had not been prepared in advance, it was cut as needed and baked with a fire of straw. In a moment it was finished.

The reason why he did not use wood is because the tissue of wood is sometimes coarse and sometimes fine, and wood also absorbs moisture, so that the forme when set up would be uneven. Also the wood would have stuck in the paste and could not readily have been pulled out. So it was better to use burnt earthenware. When the printing was finished, the forme was again brought near the fire to allow the paste to melt, and then cleansed with the hand, so that the types fell off of themselves and were not in the least soiled.

When Pi Sheng died, his font of type passed into the possession of my nephews, and up to this time it has been kept as a precious possession.[a]

The account, concise as it is, contains full technical details of type-making, type-setting, printing, and the breaking up of type, discussing also its advantages, and the disadvantages of using unsuitable materials (Fig. 1138). Unfortunately, we know of no other information about Pi Sheng or books printed with his movable

[a] *Meng Chhi Pi Than* (*TSHCC*), ch. 18, p. 117; cf. tr. Carter (1), pp. 212–13. Part of the original text was cited in *Huang Chhao Shih Shih Lei Yüan*[1] (*SFSTK*), compiled by Chiang Shao-Yü in 1145.

[1] 皇朝事實類苑

type.[a] Although the process went into eclipse after its inception, it was a complete invention and fully four hundred years ahead of Gutenberg.[b]

In the six hundred years following Pi Sheng, two occasions of the use of earthenware type are recorded. Yao Shu[1] (1201–78), a councillor of Kublai Khan, persuaded his disciple, Yang Ku,[2] to print philological primers and works by Sung neo-Confucianists with 'movable type of Shen Kua'.[c] Wang Chen[3] (fl. 1290–1333), before describing his wooden movable type in *Nung Shu*, mentioned an alternative method by others of baking earthenware type together with an earthenware frame and filling it so as to make whole blocks.[d] Both of these statements are rather obscure, but at least they indicate that earthenware type was probably employed again during the middle of the 13th century.

There is no evidence that earthenware movable type was used in the Ming period, and not until the mid-Chhing dynasty was its use as well as another ceramic type introduced.[e] In 1718, Hsü Chih-Ting[4] (*chü jen* 1723, studio name Chen Ho Chai[5]), a scholar in Thai-an, Shantung, developed a printing process with porcelain type (*tzhu pan*[6]) and printed at least two books known to us, *Chou I Shuo Lüeh*,[7] a commentary on the *Book of Changes*, printed in 1719 (Fig. 1139), and *Hao-An Hsien Hua*,[8] miscellaneous notes compiled by Chang Erh-Chhi,[9] and printed about 1730.[f] Another record says that a certain scholar in Thai-an printed with earthenware movable type in 1718 to 1719, and though the name of the printer is not mentioned, it is most likely to have been Hsü Chih-Ting.[g]

Once again Shen Kua's record inspired scholars to try the use of movable type. Chai Chin-Sheng[10] (b. 1784), a teacher in Ching-hsien, Anhwei, spent thirty years in making a font of earthenware movable type, utilising every available hand in his family. By +1844 he had made over 100,000 sets in five sizes, and it was with these that he printed at least three books. The first of these was his own collected poems, entitled *Ni Pan Shih Yin Chhu Pien*[11] (First Experimental Edition with Earthenware Type), in which five poems concern his own work of writing, editing, cutting, setting, and printing (Fig. 1140a), and he was certainly the earliest and perhaps the only author-printer in China known to us. Chai also used his type to print 400

[a] It has been suggested that Pi Sheng died in the reign of Huang-yu (+1049–53), only a few years after this experiment, so was unable to pass on this ingenious device to a fellow printer. This would explain the eclipse of the process and the scarcity of information about it. It has further been deduced that Shen Kua's relatives collected the font in Hangchow, the leading centre of Sung printing and the logical place for movable type printing to be invented; see Hu Tao-Ching (4), pp. 61–2.
[b] Johann Gutenberg (1400? to 1468) printed the 42-line Bible in c. 1455.
[c] See *Mu An Chi* (*SPTK*), ch. 15, p. 4b.
[d] See *Nung Shu* (Peking, 1956), p. 538.
[e] Wang Shih-Chen (1634–1711), *Chhih Pei Ou Than* (1701 ed.), ch. 23, p. 7a, says a certain official in Jao-chou, surnamed Chai[3] and a *chin shih* of I-tu, completed an edition of *I Ching* with porcelain after making several experiments, but he does not say whether it was printed with porcelain plates or porcelain type.
[f] See Chu Chia-Lien (1), pp. 61–2.
[g] See Chang Hsiu-Min (7), p. 31.

[1] 姚樞　　[2] 楊古　　[3] 王禎　　[4] 徐志定　　[5] 眞合齋
[6] 磁版　　[7] 周易說略　　[8] 蒿盦閒話　　[9] 張爾歧　　[10] 翟金生
[11] 泥版試印初編

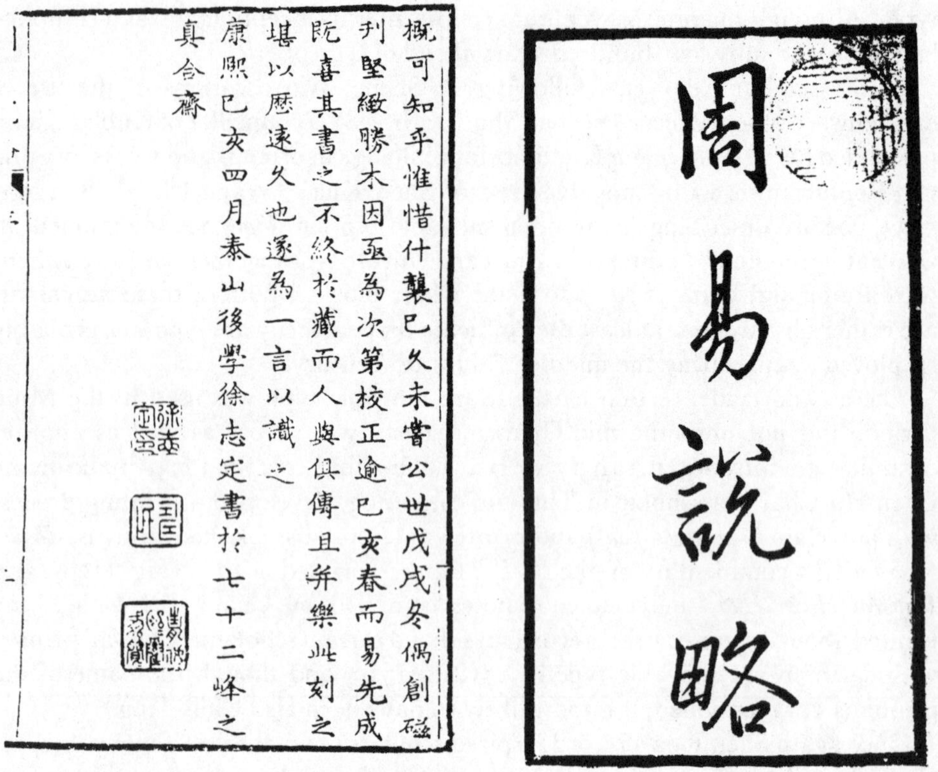

Fig. 1139. *Chou I Shuo Lüeh* printed with porcelain type, *c.* +1719. Courtesy of the National Library of China.

copies of poetry by one of his friends in 1847–8, and his family register in 1857 (Fig. 1140b);[a] a set of the types, clay moulds, and blanks used for his printing has recently been found in Hui-chou, Anhui. These types have been found to agree exactly with the style of calligraphy in the text of the printed editions (Fig. 1141).[b]

Apart from the above occasions, earthenware movable-type printing is known to have been employed also in Chhangchow and by Wu-hsi of Chiangsu and I-huang of Chiangsi during the Chhing dynasty.[c] The earthenware type printers of Chhangchow were noted for their method of typesetting, in which they spread a layer of clay in the frame and arranged the type in it, the clay holding the type firmly. Books printed in this way were so esteemed for their typographic quality that printing orders were sent to Chhangchow from distant parts of the country. Wooden type was invariably engraved individually, and earthenware and porcelain type seems to have been made in the same way, although the earthenware

[a] See an article on Chai Chin-Sheng and his printing in Chang Hsiu-Min (7), pp. 30–2.
[b] This font, which includes types in four different sizes measuring 9 mm. to 4 mm. in length, 8.5 to 3.5 mm. in width, and all 12 mm. in height, was discovered in 1962 and is kept in the Institute of the History of Natural Science, Academia Sinica, Peking; see report of Chang Ping-Lun (1), pp. 90–2.
[c] Cf. Chang Hsiu-Min (5), p. 79; Hu Tao-Ching (4), p. 63.

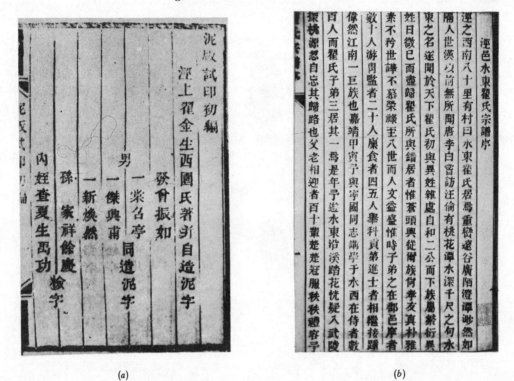

(a)   *(b)*

Fig. 1140. Printing with earthenware type of the +19th century. (a) First experimental edition, *Ni Pan Shih Yin Chhu Pien*, c. +1844. (b) Clan record of the Chai family, c. +1857. National Library of China.

type of Chai Chin-Sheng is said to have been cast with moulds before being baked to harden.[a]

Some scholars have doubted the feasibility of making movable type of earthenware, but the extant copies of books printed by Hsü Chih-Ting and Chai Chin-Sheng are sufficient to prove the existence of such type. In fact, porcelain movable type was considered even 'tougher and more durable than wood',[b] while earthenware movable type were 'as hard as stone and horn', and had a further advantage over wooden type in that the latter would become 'swollen and muddled' after printing about two hundred copies.[c]

(ii) *Wooden type*

Wooden movable type had been tried by or before Pi Sheng's time, but it was discarded because the material was found unsuitable for printing. The technical

[a] See Chang Hsiu-Min (7), p. 31. However, it is uncertain whether the moulds were made of copper or clay, as those discovered in Hui-chou in 1962 are said to have been made of earthenware; see Chang Ping-Lun (1), p. 92.
[b] See Hsü Chih-Ting's preface to *Chou I Shuo Lüeh*, cited in Chu Chia-Lien (1), p. 61.
[c] See preface to *Ni Pan Shih Yin Chhu Pien*, cited in Chang Hsiu-Min (7), p. 32.

Fig. 1141. Earthenware types of Chai Chin-Sheng, *c.* +1844, discovered in 1962 in Hui-chou, Anhui province. Above are four different sizes of the type and below are the printed characters from the large size of the type. Courtesy of Mr Chang Ping-Lun of the Institute of History of Science, Academia Sinica, Peking.

difficulties must have been overcome later as wood was used again after a lapse of some three centuries. The credit for the first practical wooden movable type must be given to Wang Chen[1] (fl. +1290–1333), whose treatise on agriculture, *Nung Shu*,[2] contains the first detailed record of its use. Wang Chen was a magistrate of Ching-te in Anhuei in 1295–1300, and wooden type was made in the two years 1297 and 1298 of his administration. He first summarised his process as follows:

Now, however, there is another method that is both more exact and more convenient. A compositor's forme is made of wood, strips of bamboo are used to mark the lines and a block is engraved with characters. The block is then cut into squares with a small fine saw till each character forms a separate piece. These separate characters are finished off with a knife on

[1] 王禎   [2] 農書

Fig. 1142. Drawing of Wang Chen's wooden movable-type printing process, *c.* +1300, showing, at right, typesetting with characters in compartments arranged by rhymes and, on the left, brushing on the back of paper from the type frames. From Liu Kuo-Chün (*1*), 1955 ed.

all four sides, and compared and tested till they are exactly the same height and size. Then the types are placed in the columns [of the forme] and bamboo strips which have been prepared are pressed in between them. After the types have all been set in the forme, the spaces are filled in with wooden plugs, so that the type is perfectly firm and will not move. When the type is absolutely firm, the ink is smeared on and printing begins.[a]

He then proceeded to describe each step in greater detail, covering type cutting and finishing, making the type case and revolving table, and typesetting and printing. First, all characters were divided according to the five tones and the rhyme sections according to the official book of rhymes. A calligrapher wrote the characters for the types, which were then pasted on the wooden blocks to be engraved, with spaces left between them for sawing. For common words, a large number of types were made and in all, more than thirty thousand types were needed. After engraving the characters on the wooden block, each single character was cut out with a small fine-toothed saw, and all were finished off with a small knife for exact uniformity in size and height. The types were then arranged in wooden cases, held in rows with bamboo strips, and kept in place with wooden plugs. They were arranged upon a revolving table according to the five tones and rhymes, with large characters used for labels.

The revolving table, about seven feet in diameter, was supported by a central leg about three feet high (Fig. 1142). Upon the table was a round bamboo frame in which the movable types were kept, each section being numbered from top to

<hr>

[a] *Nung Shu*, p. 538; tr. Carter (1), pp. 213–17.

bottom. Two such tables were provided: one for the types from the official book of rhymes and the other for the special types of most-used characters. The typesetter could then select types as needed, and put them back into their proper rhyme-compartments after they had been used. A special list of characters was prepared according to the book of rhymes, and the characters were all numbered. A man holding the list called for the types by number, while another took them from the compartments on the table and placed them directly in the forme. Unusual characters that were not found in the book of rhymes were made by the wood-cutter as needed.

Wang Chen's method of printing included estimating the dimensions of the book to be printed and making a wooden edging about the four sides of the block. The right-hand edge was left open until the forme was full, then the edging was mounted and fastened tightly with wooden plugs. The types within the columns of the forme must be so fixed that they were exactly even and correct, with bamboo slivers of various sizes wedged in. When the types stood absolutely even and firm, printing began. The inking of the forme was done with a brush moving vertically down the columns, never across. In taking an impression on paper the columns had likewise to be rubbed with the brush from the top down.

The incentive underlying Wang Chen's enterprise was his voluminous work on agriculture, *Nung Shu*, although that work finally came out in block-printed form.[a] However, his set of wooden movable type was used to print a local gazetteer of Ching-te, Anhui, which is said to have contained 60,000 characters; one hundred copies of this are claimed to have been printed in less than a month in 1298.[b] The chief contributions of Wang Chen to Chinese printing were that he improved the speed of typesetting by simple mechanical devices, and that he left us a record of the systematic arrangement of wooden movable types, although none of the books printed with his type survive.

About two decades after Wang Chen's enterprise, Ma Chheng-Te,[1] a magistrate of Feng-hua, Chekiang, printed a classical commentary, *Ta Hsüeh Yen I*,[2] in twenty *tshe*, and other books in 1322 with his set of 100,000 movable characters.[c] Although the material used for his type is not mentioned in the record, it is assumed that it was wood.[d]

We have little record of Ming government printing with wooden type, but the device was fairly well received by local feudatory princes and other printers. At least two princes printed several movable-type editions with large and small sizes of characters, the prince of Shu[3] bringing out a literary collection of the Sung poet Su Chhe[4] in Chhengtu in 1541, and the prince of I[5] a book against superstition by a

[a] Wang Chen moved to Yung-feng in Chiangsi in 1300 and his book was completed and printed there with wood-blocks in 1313. See his postscript to *Nung Shu*, p. 540.
[b] *Ibid*.   [c] Chang Hsiu-Min (*11*), p. 57.
[d] Since movable-type editions printed with bronze, earthenware, or enamelware types were usually so indicated; those without such indication were probably wooden-type editions.

[1] 馬稱德   [2] 大學衍義   [3] 蜀   [4] 蘇轍   [5] 益

Yüan author, with a supplement, in 1541; the latter also printed a rhyming dictionary earlier. Other users of wooden type included local academies, local government offices, families, and commercial printers in Nanking, Soochow, Chhangchow, Hangchow, Wenchow, and Foochow, as well as in Szechuan and Yunnan.[a] Subject coverage widened in Ming wooden-type editions, which included novels, art, science and technology, and especially family registers and the official gazettes.[b]

Wooden type was used on a much wider scale in the Chhing. Not only was it employed officially by the imperial court, but it was also widespread among private printers. Sometimes, too a font of movable type was made as an investment, which could later be pawned, sold, or presented as a gift. A plan for printing a collection of Confucian classics (*ju-tsang*[1]) using wooden type—a project comparable to printing the vast Buddhist *Tripitaka* or the Taoist canon—was suggested but not realised.[c] When, however, the *Ssu Khu Chhüan Shu*[2] was compiled, it was planned to print the lost books recovered from the *Yung-Lo Ta Tien*[3], but the bronze type previously made was found to have been destroyed. On the suggestion of Chin Chien[4] (d. 1794), an official in charge of printing at the Wu-ying Palace, the emperor in 1733 ordered the making of 253,000 wooden types, a great task, yet completed in only one year.

Using this font, a series called the *Wu Ying Tien Chü Chen Pan Tshung Shu*,[5] 138 titles in more than 2300 *chüan*, was printed. About five to twenty copies of each title were printed on white paper to be deposited in the palaces, and about 300 copies on bamboo paper for sale and distribution in the provinces.[d] Chin Chien then wrote a manual titled *Wu Ying Tien Chü Chen Pan Chheng Shih*[6] (Imperial Printing Office Manual for Movable Type), summarising his experience in this printing project. This illustrated account includes nineteen sections on such steps as making the type body, cutting the type, making type cases, forme trays, strips in variable thickness, blanks, centre columns, sorting trays, page and column rule formes, setting the text, making ready, proofing, printing, distributing the type, and a schedule for rotation (see Fig. 1143).

Chin Chien's method began with cutting pieces of jujube wood into boards from which rectangular strips were prepared, dried in the open air, and planed smooth. Then they were cut into blocks for the type. These blocks were planed in grooves precisely cut in pieces of hardwood until they were just level with the lip of the groove, so that the proper breadth, thickness, and length were obtained. Thus each large type block was 0·28 inch thick, 0·3 inch wide, and 0·7 inch long. For small type the dimensions for length and thickness were the same, but the breadth was

---

[a] Chang Hsiu-Min (*11*), pp. 58–60.
[b] *Thing Lin Wen Chi (SPTK)*, ch. 3, p. 15a; see also pp. 172 ff. above.
[c] See Chang Hsiu-Min (*12*), p. 60.
[d] Many of the copies of this collection now extant are woodblock reprints by various local government offices.

[1] 儒藏　　　　　　[2] 四庫全書　　　[3] 永樂大典　　　[4] 金簡
[5] 武英殿聚珍版叢書　　　　　[6] 程式

(a)

(b)

(c)

(d)

Fig. 1143. Making and setting wooden movable type at the Imperial Printing Office of the Chhing court, *c.*
+1733. (*a*) Making type blanks. (*b*) Carving types. (*c*) Making sorting trays. (*d*) Setting types. From *Wu-Ying-
Tien Chü Chen Pan Chheng Shih, c.* +1733.

only 0·2 inch. After planing, the blocks were tested individually by being passed through two rectangular copper tubes whose inside dimensions corresponded to the sizes of the large and small type blocks.

Characters to be cut on these blocks were written on thin paper, cut out one by one, and pasted upside down on the block face. The blocks were then locked tightly together in a wooden 'bed' and engraved, so that it was just like cutting characters on a solid slab of wood. The engraved type was stored in type cases arranged according to the order given in the Khang-Hsi dictionary. To set text, lists of desired characters were compiled, and the proper number of types for each character were taken from the type cases and placed in the sorting trays. Next, following the text, the corresponding characters, strips, and blanks were set in the forme trays, which were then labelled with the name of the book and chapter and page numbers for easy identification.

In the meantime, ruled sheets were printed off with blocks that had been cut with the column rules. With this done, the text formes were proofed one by one and printed within the previously printed ruled lines on the sheets. Thus the printing of the book was complete.[a]

Printing with wooden types differed somewhat in the two processes recorded in great detail by Wang Chen and Chin Chien. First of all, in making the types Wang carved the characters on wooden blocks and then sawed them apart, while Chin first prepared type bodies before the characters were cut individually to make them into types. Second, in setting type, Wang employed a method whereby the type came to the workers, while Chin's workers went to the type. Third, Wang's frame was added after the type was set, while Chin printed the ruled sheets and the text separately on the same paper.

Wooden movable type was also used during the Chhing by local governments and local academies, and even more widely by private printers. One interesting feature of this printing from wooden type was the production of clan records or family registers by travelling printers, who carried the facilities to the customers and were most active in the Chiangsu and Chekiang regions. After the autumn harvests they travelled in groups of five to ten, bringing with them some 20,000 types (including large and small sets) made of pear wood, and with cooperative effort, were able to print one family register in anything from one to six months. For characters which they did not carry and where pear wood was not available, they made new types of earthenware. As a side line, they also printed works other than family registers.[b]

(iii) *Bronze and other metal type*

Printing with bronze movable type began to be widely applied in China in the late 15th century. The enterprise was sponsored by wealthy families in southern

[a] Cf. translation of the manual and an article by Rudolph (14, 15).
[b] Chang Hsiu-Min (12), pp. 61–4.

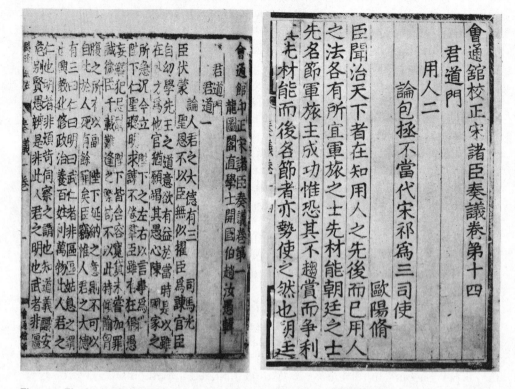

Fig. 1144. First book printed with metal type in China, *Chu Chhen Tsou I*, by Hua Sui in +1490 in large and small character editions. National Central Library, Thaipei.

Chiangsu, a region of affluence, and was continued by commercial printers in Fukien. Most notable among the Ming bronze-type printers were the Hua and An families, both of Wu-hsi, Chiangsu, who like many who accumulated sizable fortunes, strove to establish reputations by the sponsorship of printing. Hua Sui[1] (1439–1513), who did not engage in scholarly pursuits until the age of fifty but then became so interested in books that he pursued his passion to the detriment of his family's finances, embarked on the first bronze-type printing in his clan in 1490. All books printed by him contain the signature Hui Thung Kuan[2] (Studio of Mastery and Comprehension), by which he meant that he had mastered the process of using movable type in printing. At least fifteen titles in over 1000 *chüan* are known to have been printed with bronze type at Hui Thung Kuan in about twenty years.[a]

His first experiment, on a collection of memorials, *Chu Chhen Tsou I*[3] was printed in two editions, using large type for one and small type for the other (Fig. 1144). Hua Chheng[4] (1438–1514), a distant relative of Hua Sui, was an antiquarian and

[a] Phan Thien-Chen (1) argues that the movable type of the Hua family was made of tin with bronze frames and was not the bronze type it is generally considered to have been.

[1] 華燧   [2] 會通館   [3] 諸臣奏議   [4] 華珵

a book-collector who named his studio Shang Ku Chai[1] (Studio for Esteeming Antiquities), and many of the rare books he obtained appeared in printed form within a short time, as the result of his using bronze type. These included one of the earliest collections, *Pai Chhuan Hsüeh Hai*,[2] in 160 *chüan*, printed in +1501. Hua Chien[3] (fl. 1513–16), a nephew of Hua Sui, was another bronze-type printer of the Hua family. He used the signature Lan Hsüeh Thang[4] (Hall of Orchid and Snow), and his printing included the Thang encyclopaedia, *I Wen Lei Chü*,[5] in 100 *chüan*, (1515). Some other relations simply signed 'the Hua family', with no personal name, and altogether, about two dozen titles in over 1500 *chüan* are known to have been printed with bronze movable type by family members in less than three decades between 1490 and 1516.[a]

An Kuo[6] (+1481–1534), also of Wu-hsi, a wealthy merchant with a vast financial empire, continued the enterprise of printing with bronze type. He is known to have issued at least ten titles from 1516 until the time of his death; all bear his studio name, Kuei Pho Kuan[7] (House of the Cassia Slope). His books included local histories, works on water conservancy, literary collections, and two encyclopaedias which are noted both for their craftsmanship and their careful collation.[b]

Bronze-type publications printed in Chhangchow, Soochow, and Nanking apparently belong to the same category as those of the Hua and An families; the sponsors were described as enterprising people or big families. On the other hand, bronze-type books produced by Fukien printers in Chih-chheng (i.e. Chien-ning) and Chien-yang were commercial in nature. Sometimes a font was jointly owned by more than one printer, and a distinguished example produced from such cooperation is an edition of *Mo Tzu*,[8] which was excellently printed in blue ink from bronze type (Fig. 1145).[c] Some fifty titles of poetry by individual Thang poets were also very nicely printed (Fig. 1146) in this way.

Since An Kuo participated in bronze-type printing in the same district at a later date than the Hua family,[d] it seems that he was inspired by the Huas. But where did the bronze movable type of the Hua family come from? It was said that Ming use of movable type in southern Chiangsu was inspired by Shen Kua's description in *Meng Chhi Pi Than*.[e] But given the marked technical differences between earthenware and bronze types, such a transition involved many technical problems of engraving, casting, type-setting, inking, and printing, and so a creative mind was required for such an innovation. The first Ming printer to use bronze type, Hua Sui, appears

[a] See bibliography in Chhien Tshun-Hsün (*2*), pp. 11–14, and his article on Hua Sui in Goodrich (*30*), pp. 647–9.
[b] See Chhien Tshun-Hsün (*2*), pp. 5–6, 14–15; Goodrich (*30*), pp. 9–12.
[c] *Chung Kuo Pan Kho Thu Lu*, vol. i, p. 101.
[d] Some scholars mistakenly believed that the Hua family followed the An family in movable-type printing, but the reverse is true; see Chhien Tshun-Hsün (*2*), p. 6.
[e] *Wei Nan Wen Chi* (*SPTK*), postscript by Chu Yün-Ming.

[1] 尚古齋   [2] 百川學海   [3] 華堅   [4] 蘭雪堂   [5] 藝文類聚
[6] 安國   [7] 桂坡舘   [8] 墨子

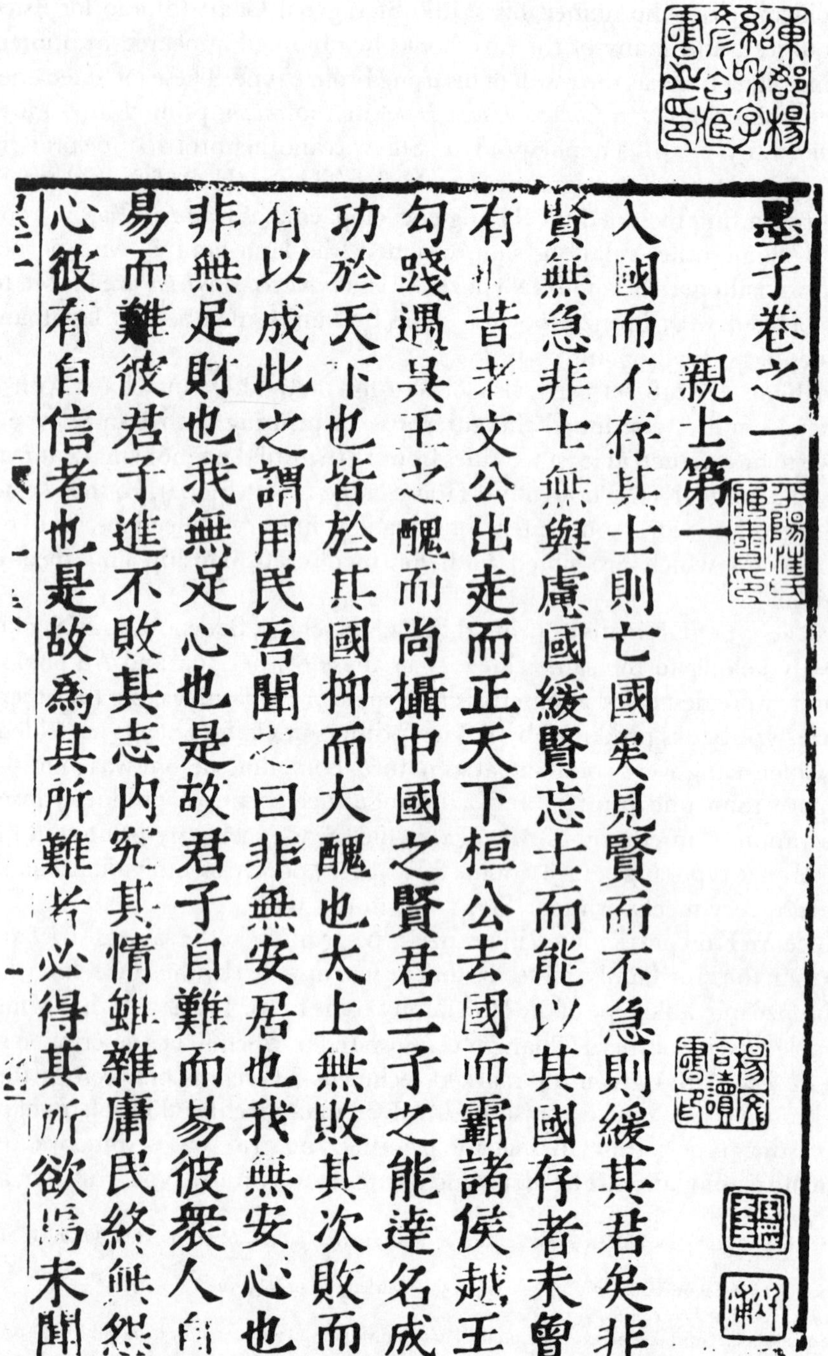

墨子卷之一

親士第一

入國而不存其士則亡國矣見賢而不急則緩其君矣非
賢無急非士無與慮國緩賢忘士而能以其國存者未曾
有也昔者文公出走而正天下桓公去國而霸諸侯越王
勾踐遇吳王之醜而尚攝中國之賢君三子之能達名成
功於天下也皆於其國抑而大醜也太上無敗其次敗而
有以成此之謂用民吾聞之曰非無安居也我無安心也
非無足財也我無足心也是故君子自難而易彼眾人自
易而難彼君子進不敗其志內究其情雖雜庸民終無怨
心彼有自信者也是故為其所難者必得其所欲焉未聞

Fig. 1145. A specimen of the *Mo Tzu* printed with bronze movable type in blue ink in +1552. National Library of China, Peking.

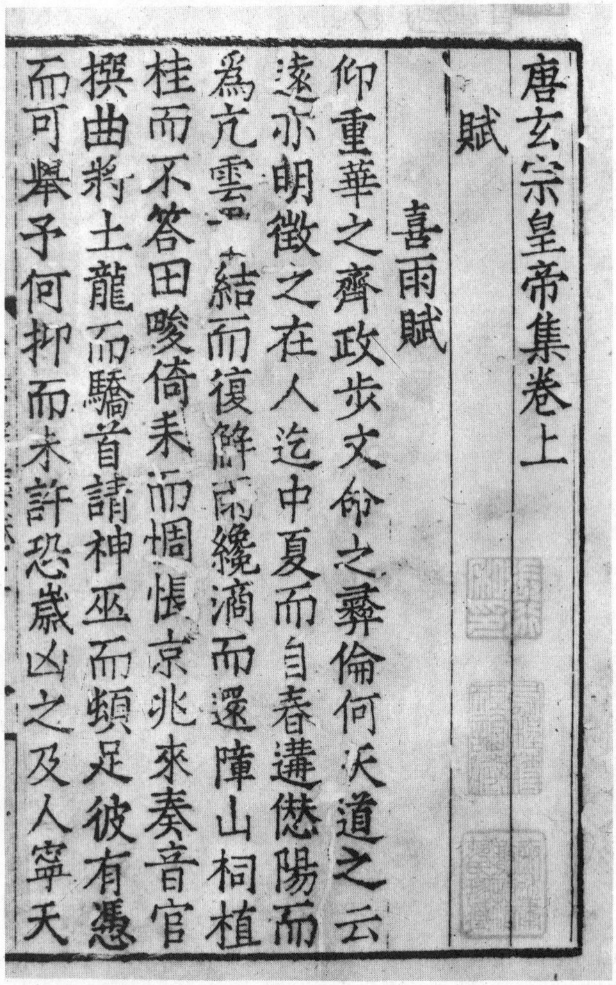
唐玄宗皇帝集卷上
賦
喜雨賦
仰重華之齊政步文命之彝倫何沃道之云
遠亦明徵之在人迄中夏而自春遘愆陽而
爲亢雲結而復解雨纔滴而還障山椆植
桂而不答田畯倚耒而惆悵京兆來奏音官
撰曲將土龍而驕首請神巫而頻足彼有憑
而可舉予何抑而未許恐歲凶之及人寧天

Fig. 1146. Bronze type printing of the +16th century, *Thang Hsüan Tsung Huang Ti Chi*, a collection of imperial writings by Thang Emperor Hsüan-Tsung (r. +713–55). National Central Library, Thaipei.

to have been a pedantic scholar of some wealth who claimed only 'mastery and comprehension' of this method; his hobby of studying and printing began only after middle age, and nowhere in the prefaces of his publications did he claim that he invented bronze type. Though Korean influence has been suggested, no concrete evidence can be found to support this theory.[a]

Since the practicability of metallic movable type had been amply demonstrated by Ming printers, large-scale production of some 250,000 bronze characters was undertaken by the imperial court early in the Chhing dynasty, ostensibly to print the Grand Encyclopaedia *Thu Shu Chi Chheng*[1] around 1726 (Fig. 1147), although

[a] Cf. Chhien Tshun-Hsün (*2*), pp. 7–8.

[1] 圖書集成

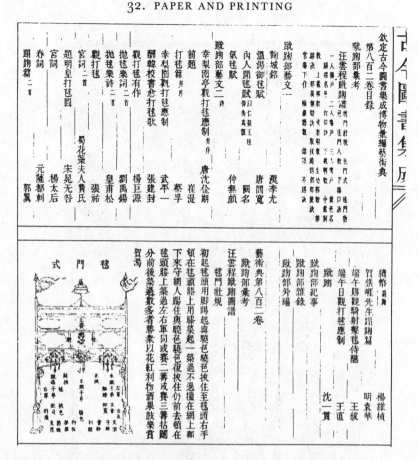

Fig. 1147. The Grand Encyclopaedia printed with bronze movable type in +1726. The pages show the chapter on football with a diagram of the goal cited from a Ming work. From *Thu Shu Chi Chheng* reprint, 1934.

still earlier Chhing attempts have been recorded. Sixty-six copies of the encyclopaedia, each comprising 5020 volumes, were printed,[a] the font of bronze type then being deposited in the Wu-Ying palace but later, in 1744, it was melted down to make coins.[b]

Apart from the imperial palace, bronze type was employed by a number of private printers. These included the Chhui-Li-Ko[1] in Chhang-shu who printed a literary collection in 1686; a Manchu general, Wu-Lung-A,[2] who printed a collection of imperial edicts in Taiwan around 1807; Lin Chhun-Chhi[3] in Foochow in 1846–53; Wu Chung-Chün[4] in Hangchow in 1852; a Chhangchow printer who printed a Hsü family register in 1858; and officials of the Thai-Phing Kingdom in

---

[a] See Chang Hsiu-Min (*10*), pp. 49–50.
[b] The font was depleted by theft by the officials in charge. In order to cover up their crime, in 1744 when Peking experienced a coin shortage these officials suggested melting the type; *Ta Chhing Hui Tien Shih Li* (1899 ed.), ch. 1119, p. 1*b*.
[c] See Chang Hsiu-Min (*10*), pp. 50–3.

[1] 吹藜閣　　　[2] 武隆阿　　　[3] 林春祺　　　[4] 吳鍾駿

1862. Especially interesting was a font made by Lin Chhun-Chhi (b. 1808), who spent twenty-one years from 1825 to 1846 in cutting some 400,000 individual characters in large and small sizes for a font known as Fu Thien Shu Hai;[1] this cost him over 200,000 silver taels. In regular style calligraphy and well executed, it was used for a number of books on phonology, medicine, and military strategy, and it was possibly used, too, in 1852 by Wu Chung-Chün of Hangchow to print two other works.

Because of its low melting point, tin is a good material for casting movable type in matrices. In fact, tin was used for that purpose as early as the late +13th century. Wang Chen said:

> In more recent times, type has also been made of tin by casting. It is strung on an iron wire, and thus made fast in the columns of the form, in order to print books with it. But none of this type took ink readily, and it made untidy printing in most cases. For that reason they were not used long.[a]

Thus the reason for rejecting tin for type was its incompatibility with Chinese ink. This deficiency was apparently overcome by the middle of the 19th century, when a fount of tin type was known to have been successfully made by a Mr Tang in Kuangtung (Fig. 1148).[b]

Among the chief materials for modern movable type, lead was never extensively used by traditional Chinese printers. Lu Shen[2] (+1477–1544) vaguely reported that printers in Chhangchow in the early 16th century made movable type in bronze and lead (thung chhien tzu[3]),[c] but this term here may mean an alloy of bronze and lead rather than two separate metal types. The material for the 'bronze' movable type was, in all probability, an alloy because pure copper is too soft to be serviceable. Copper must be combined with tin or lead to increase its hardness, as was done in the manufacture of all ancient bronze weapons and vessels.

A question that remains unanswered is whether metal movable type for traditional printing in China was cast from moulds or engraved individually. There are no surviving specimens for examination, nor any existing detailed records to answer the question; our conjectures have to depend upon comparing the style of writing and calligraphy of different editions, and on our interpretation of such literary records as are still in existence. Examination of Ming imprints of bronze movable type reveals that they are markedly irregular in the shaping of characters and writing styles. The characters are not rounded but wedge-shaped, and even on the same leaf, the writing style of the same character is still not uniform; it is as if they were individually carved and not cast from matrices. However, this alone cannot be considered definitive evidence for engraving, because it is not necessary

---

[a] *Nung Shu* (Peking, 1956), p. 538.
[b] See *Chinese Repository*, vol. 19 (1850), pp. 247–9; Hirth (28), pp. 166–7.
[c] *Chin Thai Chi Wen* (*TSHCC*), p. 7.

[1] 福田書海   [2] 陸深   [3] 銅鉛字

馬潘莫謝鄭吳錢陸容林石葉盧徐聶金邱雞梁蕭楊蔡黎張孫
彭孔李招鄧徐馮金黃侯江蕭尹麥侯詹崔王游桂杜賴何廖
黃江尹麥馮賴楊鄒張鄒孫梁招詹洪羅金侯嚴杜何廖黎鄭李
孔謝莫鄒雞嚴蕭潘容侯石馬徐林盧侯金邱鄧李崔王游桂
蔡賴徐黎侯金蕭何杜廖潘孫王葉莫謝鄭麥江馮孔桂王彭
楊楊賴徐黎侯金蕭張梁容侯黃錢馬盧金蕭游彭招徐鄧洪王
尹陸吳林蕭石侯張梁容侯黃錢馬盧金蕭游彭招徐鄧洪王
盧石楊蔡葉金邱梁孫賴黎廖何杜林徐金馬容潘錢莫侯陸
尹彭招孔麥桂江游崔王李侯徐鄧詹洪馮黃羅金吳謝鄭陸蕭
黃招孔桂崔李鄧蕭盧孫詹蕭侯游葉馬梁徐尹金鄭江蕭彭蕭
楊容賴蔡邱黎潘聶張杜金石莫杜麥蕭徐陸侯金謝吳廖何羅

Fig. 1148. Tin type by Mr Tang of Kuangtung, +1850. From *Chinese Repository*, vol. 19, p. 248.

to have only one mould for each character, and the types may have been retouched individually after being cast.[a]

However, to exploit fully the advantages of movable-type printing, casting seems the logical method, because it is much harder to cut characters in bronze than in wood, and since each font includes tens or hundreds of thousands of

[a] The Korean bronze movable types were all cast from matrices yet the shapes of their characters are often irregular and the strokes not uniform, showing that this problem was not peculiar to Ming types alone.

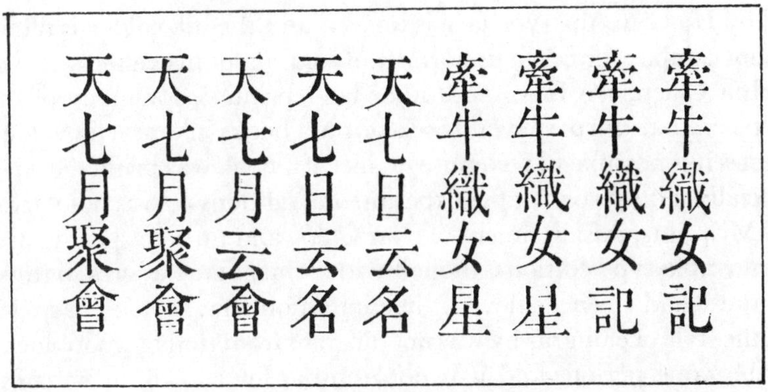

天天天天天牽牽牽
七七七七七牛牛牛
月月月日日織織織
聚聚云云云女女女
會會會名名星星記
　　　　　牛牛牛
　　　　　織織織
　　　　　女女女
　　　　　星記記

Fig. 1149. Variation of same characters appearing on a page in the *Thu Shu Chi Chheng*, showing the types might have been cut by hand instead of cast from matrix.

characters, to cut them individually by hand would be against the very economic principle which dictated the use of movable type. Yet things do not always happen according to principle, and the engraving of individual type is still a possibility. Indeed, in the case of the bronze type for the Grand Encyclopaedia *Thu Shu Chi Chheng*, the evidence seems to favour engraving. Certainly, the wage for engraving on bronze is said in an official record to be many times that for carving characters on wood,[a] implying that bronze is much harder to cut, and after examining the copies, experts have concluded that it seems practically certain that the types were cut, not cast in matrices (Fig. 1149).[b]

Because a set of metal type might comprise two to four hundred thousand characters, we can readily comprehend the magnitude of the work and cost involved in printing with movable type. Very often two sets of type, large and small, had to be made to print, for example, text and commentaries; this was the case with the bronze type of both the Hua and An families and with that of Lin Chhun-Chhi. The tin types of Fo-shan even included three sets.

Printing with movable type involved, in addition to applying ink to the plate and then rubbing paper on it as was the case in block-printing, there was also the labour of assembling and setting the type and redistributing it after printing. Such steps were, it seems, carried out by division of labour, for some of the metal movable-type editions give lists of workers testifying to this. For example, the bronze-type edition of the Sung encyclopaedia *Thai Phing Yü Lan*, in 1000 *chüan*, printed in Chien-yang in 1574, indicates that two persons were in charge of typesetting and another two of the printing.[c] Since metal is impervious, there were obvious difficulties in obtaining even inking and uniform printing required continuous efforts to improve techniques. In the first few Ming bronze-type editions

---

[a] Cf. *Ta Chhing Hui Tien Shih Li*, ch. 1119, pp. 1*a*–2*b*.

[b] Giles (2), p. xvii, says that Alfred Pollard of the British Museum and Émile Blochet of the Bibliothèque Nationale at Paris were both inclined to believe that the type was not cast with the font.

[c] See item 7 in Chhien Tshun-Hsün (2), p. 15.

produced by Hua Sui, the type face is uneven and the ink colour is often cloudy. They cannot compare either in uniformity of columns or in evenness of ink with the work of Hua Chien, An Kuo, and other later printers, for many of these later bronze-type editions surpass even block-printed books in aesthetic value.[a]

Sometimes it is possible to recognise at once if a book was printed from movable type, especially if it was one of the experimental editions or was not very carefully executed. Misprints, misalignment of characters, and uneven spacing are distinct marks of movable-type editions, though with prints made after a high typographical standard had been achieved, differentiation may not be easy. Generally speaking, the style of calligraphy was not different from that of woodblock printing prints of the same period, and it is not unusual for a book to be considered a movable-type print by one scholar, and a block printing by another. Likewise, it is difficult, if not impossible, to know whether a book was printed from bronze or from wooden type. The printer's statement in the book, if any, seems to be the only reliable evidence.

(iv) *Disadvantages of movable type in Chinese printing*

Until the advent of modern typography, woodblock printing had always been the principal vehicle of traditional Chinese printing, and it is natural to ask why printing with movable type, although invented as early as the mid-eleventh century, was not more widely used in China. The most important and obvious reason is, of course, the nature of written Chinese. It is composed of thousands of ideograms which are needed in any extensive writing, and since several types are needed for each character, and for the commoner ones twenty or more, a fount of at least 200,000 Chinese types is not unusual.[b] The contrast with an alphabetical language becomes clear when it is realised that a complete fount containing upper- and lower-case letters, numerals and other signs, consists of no more than a hundred different symbols. So it seems that the need for such great numbers in an ideographic language reduced the practicability of movable-type printing in China.

Another significant factor has been indicated in quoting Shen Kua, who said that for only two or three copies the movable-type method would be neither simple nor easy, though for printing hundreds or thousands it was marvellously quick. The technique of inking and rubbing is only a minor part of the whole process of using movable type, while the major use of labour is in assembling the type and, after use, distributing it for future service.[c] Thus movable-type printing is desirable

---

[a] For example, the bronze-type edition of *Mo Tzu* in blue ink, and the several collected works of Thang authors printed in Chih-chheng (Chien-ning).

[b] More than 200,000 bronze characters were made by the imperial palace printing shop of the Chhing dynasty for printing the *Thu Shu Chi Chheng* around 1725 and over 250,000 wooden characters for printing the *Wu Ying Tien Chü Chen Pan Tshung Shu* in 1733. Some 400,000 bronze characters were also made by a private printer in the early 19th century.

[c] Reprinting of movable-type editions was not easy. Many books are known to have been printed first from movable type and later from blocks. Apparently it would not have been economical to reset the types for another edition.

only for large-quantity production, because only then is the average time for each copy reduced to a practical and economic level.[a]

Unlike plates of set type, printing blocks can be preserved indefinitely and used over and over again, with only occasional retouching or repairing.[b] Block printing and movable-type printing therefore serve different needs: the former, recurrent demands for small quantities over relatively long periods; the latter, large quantities at one printing. The former was precisely the pattern of book demand and supply in traditional Chinese society; therefore movable type could not replace the printing block. Printers in old China made tens of copies at a time, and stored the printing blocks, which could be taken out at any later date for additional copies. Thus they avoided the unnecessary holding of printed books in stock and of tying up capital. Block printing was therefore predominant in traditional Chinese publication.

As far as capital investment was concerned, movable-type printing posed much greater financial burdens on printers. Costs of paper and ink were relatively constant, but for the movable-type itself a tremendous initial investment was needed for making the vast number of characters needed, and compared very unfavourably with the small cost of wood blocks and of the labour of engraving them. In the long run, the fact that movable types could be re-used was an advantage, but very few printers could afford such a long-term investment, while the fact that block-engravers were plentiful and inexpensive made printers the more reluctant to change a well-established process.

Furthermore, scholars required that the printed page be free of textual errors and that the calligraphy be artistic. Movable type, especially in the early stages of its development, did not always fulfil these requirements, while printing from wood-blocks made possible a great variety of typographical effects, and lent a distinction and an individuality to the printed page which fonts of uniform type could not equal. Moreover, the rigidity of the one-piece block made for a better appearance of the printed page than did movable type, and when, as sometimes, the text was carved directly from the author's copy, errors which occur with typesetting and proof-reading were eliminated.

It has been mentioned that metal type did not hold Chinese watery inks well. This was also true of earthenware and porcelain type, which had the additional disadvantage that uneven changes in size sometimes occurred during the necessary baking process, resulting in uneven matching of the type. All these factors contributed to an aesthetic inferiority which prevented movable-type printing from becoming popular.

[a] For instance, in the 1574 edition of the Sung encyclopaedia *Thai Phing Yü Lan* we find the statement, 'Over one hundred copies were printed with bronze movable type.' An edition of 400 copies was printed by Chai Chin-Sheng with earthenware type in 1847.

[b] There are numerous records in Chinese documents of the transmission of printing blocks from generation to generation. Some editions, known as *san chhao pen*,[1] were printed with blocks cut in Sung, repaired in Yüan, and re-used in Ming, through three dynasties.

[1] 三朝本

From the technological point of view, the production of a hundred movable types was much more difficult than engraving a printing block with a hundred characters. Grouping the types into retrievable order posed another problem, and to deal with it very skilled labour, usually involving considerable linguistic knowledge, had to be employed. The collective effect of all these factors therefore produced a situation very unfavourable for the development of movable-type printing in the very culture where it was invented.[a]

(3) FORMAT AND BINDING OF CHINESE BOOKS

(i) *Signs, columns, and scripts*

Chinese books in the traditional style have always been printed on one side of paper. Each leaf of the paper is folded double at the centre of the sheet, making a double-leaf page, and each part of the leaf consists of special signs or lines, the name of which would help in explaining their nature and functions on the page (see Fig. 1150). The printed portion of the leaf, which is the actual size of the block, is called the block face (*pan mien*[1]) and the centre fold the heart of the block (*pan hsin*[2]). At the centre of the leaf there may appear such signs as the elephant trunk (*hsiang pi*[3])—a light or heavy line used to mark the centre for folding—and the fish tail (*yü wei*[4])—a pair of sharp-angled spots at the upper and lower parts of the centre used to indicate the level for folding. A running title, leaf number, and sometimes the number and heading of the chapter, or the number of characters on the leaf and the name of the cutter, may be given in a narrow column at the fold. A square sign with the chapter number occasionally appears in some of the Sung editions on the upper left side of the margin, called the book ear (*shu erh*[5]), serving as a thumb index for the book, and is especially useful for the butterfly binding.[b]

The upper and lower margins of the leaf are called respectively the book eyebrow (*shu mei*[6]) or heavenly head (*thien thou*[7]), which is usually wider than the lower margin or the earthly foot (*ti chüeh*[8]). The page on each side of the leaf is ruled into columns and spaces (*hang ko*[9]) with border lines (*chieh*[10]) to divide individual columns and marginal lines (*pien lan*[11]), single or double, on four sides. The characters of the text are usually arranged in one vertical line within a column, and notes or commentaries in smaller characters in two lines. Each page may contain from five to ten columns, with from ten to thirty characters in each column. This basic format of the printed sheet and some of the bibliographical terms used for

---

[a] See Poon Ming-Sun (2), pp. 185–7.
[b] See discussion of the terms for Chinese books in Yeh Te-Hui (3), pp. 27–8; Li Wen-Chhi (2), pp. 17 ff.

[1] 版面   [2] 版心   [3] 象鼻   [4] 魚尾   [5] 書耳
[6] 書眉   [7] 天頭   [8] 地脚   [9] 行格   [10] 界
[11] 邊欄

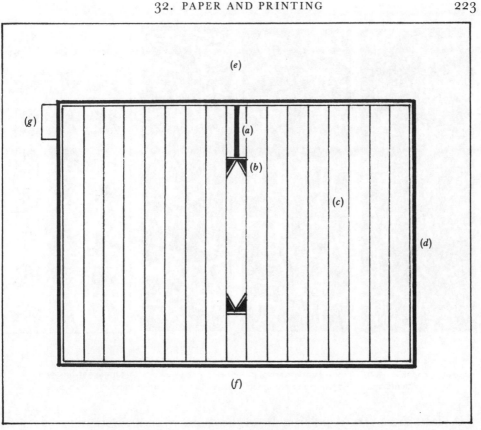

Fig. 1150. Typical format of a printed leaf, showing (a) elephant trunk, (b) fish tail, (c) borderline, (d) marginal line, (e) book eyebrow or heavenly head, (f) earthly foot, and (g) book ear.

referring to the traditional format of the Chinese book, have been continued into modern times.[a]

The most important part of the book is certainly the text, which is normally printed in different styles of the standard script (*khai shu*[1]). Since books printed in different periods and in various locations show variation of calligraphic styles, this has become one of the special features not only for judging a book's artistic qualities but also for dating the printed editions.[b] The different styles of Chinese calligraphy are primarily derived from models created by prominent calligraphers at successive periods, but the standard script, which was developed from the clerical script (*li shu*[2]) in or around the Later Han, and stabilised into the standardised form in the Thang dynasty, has been used for printing ever since its original invention.[c]

[a] See the translation of the bibliographic terminology in Tsien (7), pp. 1–18.
[b] See the criteria for dating the Sung editions in Poon (2), pp. 204 ff.
[c] See samples of various calligraphic styles of the masters in Chiang Yee (1); Léon Chang (1); Fu Shen (1); and Tseng Yu-Ho Ecke (1).

[1] 楷書   [2] 隸書

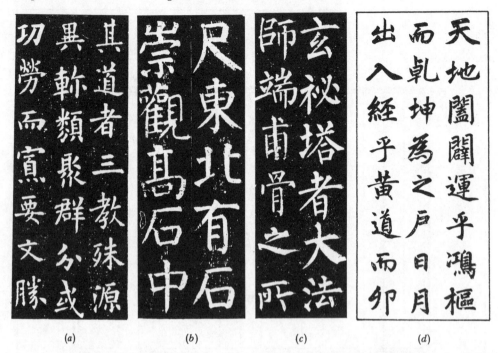

Fig. 1151. Popular styles of Chinese standard script by influential calligraphers. (*a*) Ouyang Hsün, +631. (*b*) Yen Chen-Chhing, +771. (*c*) Liu Kung-Chhüan, +841. (*d*) Chao Mêng-Fu, *c.* +1302.

At least three very popular styles of the standard script were adopted by the block printers from the Sung period. These included those of Ou-yang Hsün[1] (+557–641), Yen Chen-Chhing[2] (+709–85), and Liu Kung-Chhüan[3] (+778–865) (Fig. 1151). The Ou-yang style is very well balanced in composition with elegant, slender, and even lines. The Yen style, on the other hand, is muscular, rigid, and broad with thick and heavy strokes. The Liu style is something of a compromise between the two with neat and forceful strokes in elongated or perpendicular shapes. Generally speaking, the Northern Sung editions follow closely the Yen style (Fig. 1152) and those of Southern Sung imitated the Ou-yang style if produced in the Hangchow area and the Yen and Liu styles if produced in Fukien. While Yen and Liu styles were also prevalent in Szechuan, they show some influence of the 'slender gold style' (*shou chin thi*[4]) of Emperor Kao-tsung (r. +1127–62).

The early Yüan editions continued the Southern Sung tradition of Yen-Liu style, but later shifted to that of the contemporary calligrapher Chao Meng-Fu[5] (+1254–1322), whose style is particularly soft, feminine, and charming (Fig. 1153). The earlier Ming editions continued the Sung and Yüan tradition of the Chao and Ou-yang styles, but from the middle of the 16th century, the style of

[1] 歐陽詢　　[2] 顏眞卿　　[3] 柳公權　　[4] 瘦金體　　[5] 趙孟頫

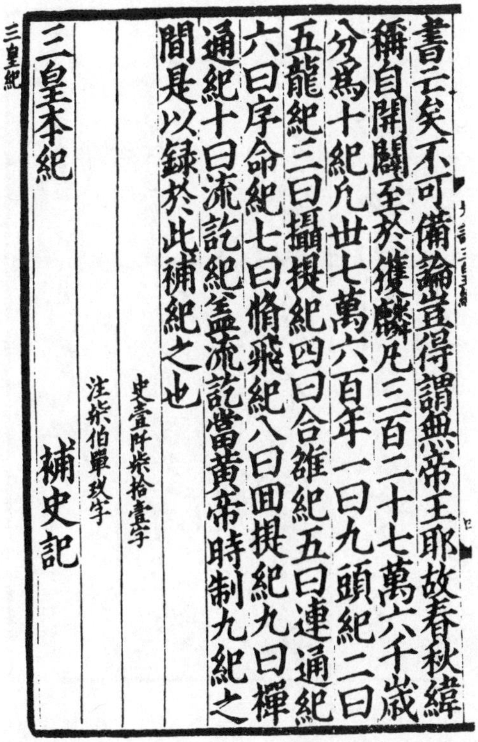

三皇紀

書云矣不可備論豈得謂無帝王耶故春秋緯
稱自開闢至於獲麟凡三百二十七萬六千歲
分爲十紀凡世七萬六百年一曰九頭紀二曰
五龍紀三曰攝提紀四曰合雒紀五曰連通紀
六曰序命紀七曰循飛紀八曰回提紀九曰禪
通紀十曰流訖紀蓋流訖當黃帝時制九紀之
間是以錄於此補紀之也
史壹阡柒拾壹字
注柒伯單玖字
三皇本紀　補史記

Fig. 1152. Northern Sung printing of *Shih Chi* with Yen style of calligraphy.

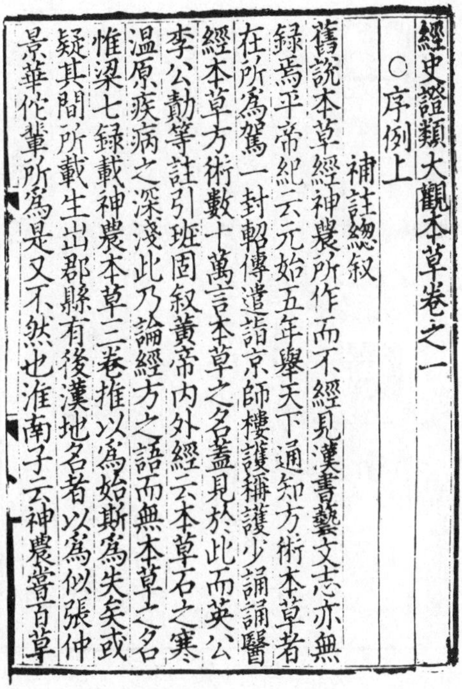

經史證類大觀本草卷之一
○序例上
補註總敘
舊說本草經神農所作而不經見漢書藝文志亦無
錄焉平帝紀云元始五年舉天下通知方術本草者
在所爲駕一封軺傳遣詣京師樓護稱護少誦醫
經本草方術數十萬言本草之名蓋見於此而英公
李公勣等註引班固敘黃帝内外經云本草石之寒
溫原疾病之深淺此乃論經方之語而無本草之名
惟梁七錄載神農本草三卷推以爲始斯爲失矣或
疑其間所載生出郡縣有後漢地名者以爲似張仲
景華佗輩所爲是又不然也淮南子云神農嘗百草

Fig. 1153. Yüan printing of *Chêng Lei Pên Tshao* in the Chao calligraphic style printed by the Tshung Wên Academy in +1302.

characters gradually changed to the craftsmen script (*chiang thi tzu*[1]), or so-called Sung style (*Sung thi tzu*[2]), which is more rigid and square in construction with heavy lines for vertical strokes, lighter lines for horizontal strokes, and a heavy tail at the end of the strokes (Fig. 1154). This stereotyped form has been followed by printers ever since, though with slight variations from time to time. The modern metal types have adopted this style for all printed matter, and two other styles called imitated Sung script (*fang Sung thi*[3]) and regular script (*cheng khai*[4]) are used primarily for headings and other special purposes (Fig. 1155).

In some cases, the text was not prepared by the craftsmen, but was written by the author himself, using specially appointed calligraphers, or by a member of the family whose handwriting was considered exceptional.[a] Even when the text was copied in the printing style by professionals the preface of the book was often made in an extraordinary fine calligraphy either by the author of the preface or by a noted calligrapher on his behalf (Fig. 1154b). In a visual sense, these fine

[a] See calligraphy in the Sung, Yuan, and Ming editions discussed in Yeh Te-Hui (*2*).

[1] 匠體字　[2] 宋體字　[3] 倣宋體　[4] 正楷

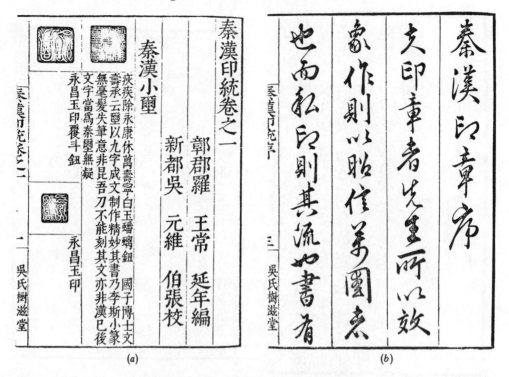

(a)　　　　　*(b)*

Fig. 1154. Ming printing, *c.* +1606, with (*a*) text in craftsman style calligraphy and seal illustration in red; and (*b*) preface in free handwriting style. Far Eastern Library, University of Chicago.

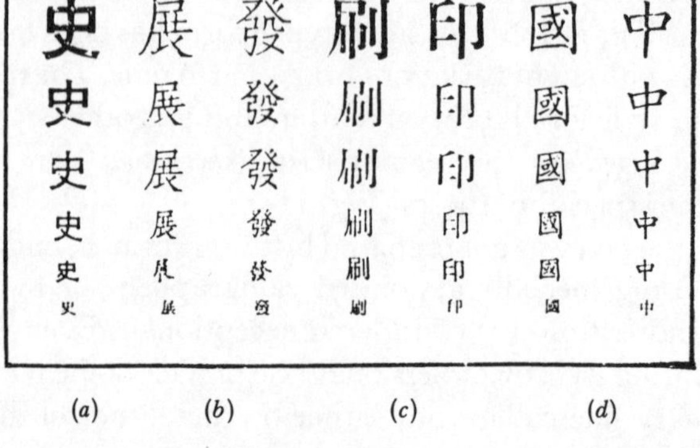

史	展	發	刷	印	國	中
史	展	發	刷	印	國	中
史	展	發	刷	印	國	中
史	展	發	刷	印	國	中
史	展	發	刷	印	國	中
史	展	發	刷	印	國	中

(a)　　　*(b)*　　　*(c)*　　　*(d)*

Fig. 1155. Printing styles of modern metal type in six different sizes. (*a*) Square for headings. (*b*) Imitated Sung style. (*c*) Old Sung or printing style. (*d*) Standard or regular style. From Shih Mei-Chhen (*1*).

calligraphic specimens not only serve as examples of Chinese art, but frequently also represent the most aesthetic part of the book.

Another special feature of the printed style is that of the taboo characters (*hui tzu*[1]) in the personal names of reigning emperors and sometimes their ancestors, which were not permitted to be used in print. The problem posed by this prohibition was generally dealt with either by omitting certain strokes of a taboo character or by substituting a homonym or synonym for it.[a] Similar taboos were sometimes observed by the printers with regard to the personal names of their own ancestors.[b] The appearance or absence of taboo characters in a text is another criterion used for dating printed editions.

(ii) *Evolution of the format of Chinese books*

Close to the time when the printed sheet was first being adopted for books, there was a gradual change in the way books were bound. Up until the +9th century, in the middle of the Thang, the units of material that made up Chinese books were bound into long continuous sections. Bamboo and wooden tablets and silk and paper rolls exemplify this style of binding. Even the pleated book (*che pen*[2]), which came into use just prior to the appearance of individual printed sheets, remained one extended piece of material. But as soon as printing began to be mentioned in literature, the new format of the folded leaves also appeared; it was more compact and made the different parts of the text more immediately accessible than had been possible with the rolls. Folded leaves developed in various stages from the 'butterfly' binding of the Sung,[c] through the wrapped back binding of Yüan and Ming times, to the stitched binding of the Ming and Chhing periods.[d] Then in the early 20th century the stitched binding began to be superseded by Western-style binding as the modern printing press came into general use in China.

Originally, bamboo and wooden tablets were fastened together with silk and hemp cords; a series of tablets so fastened was then either rolled up on itself or had succeeding tablets folded against each other like accordion pleats.[e] To make such a book the cord used to bind the tablets was doubled or sometimes tripled and the first tablet was placed in the bend. The cord was knotted to fit tightly in a notch on the edge of the tablet to prevent the cord's movement along the tablet when the book was read. The notch made in a second tablet was placed against the same knot and one open strand of the cord went under, the other over, the tablet. The two

---

[a] Sometimes a note is given underneath the substituted characters, indicating that the original character is the emperor's name, in order to preserve textual accuracy.

[b] See the list of such taboo characters from the Sung to the end of the Chhing in Chhü Wan-Li and Chhang Pi-Te (*1*), pp. 109 ff., also Chhen Yüan (*4*).

[c] See below, pp. 230 ff.

[d] For a general history of the development of the Chinese traditional bookbinding, see Ma Heng (*1*), Li Wen-Chhi (*1*), Li Yao-Nan (*1*), and Martinique (*1*), (*2*).

[e] For the system of binding and sealing bamboo and wooden tablets, see Tsien (*2*), pp. 111–13.

[1] 諱字   [2] 摺本

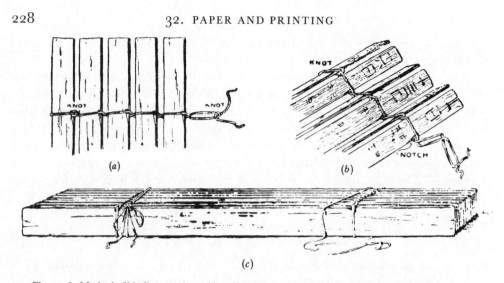

Fig. 1156. Method of binding wooden tablets, showing (a) fascicles open, (b) manner of binding, and (c) complete fascicle closed. From Aurel Stein (6).

strands were then half twisted around each other. The third tablet was set into this twist and the cords were pulled over and under this tablet and again half twisted, this continuing until the last tablet was tied into place with a final knot. Enough of the cord was left free, so that it could be used to tie the entire book together when it was rolled or closed by folding (Fig. 1156).[a] When the bound documents were dispatched, they were sealed in silk or cloth bags of various colours, which indicated the methods of delivery.[b]

Rolls of silk and paper gradually replaced the books of bamboo and wooden tablets, silk rolls being known to have existed from the -7th to the $+5$th century, while paper rolls came into use during the early centuries of this millennium. The length of a silk roll depended upon the length of text, the silk being cut at the end of the composition, but if a text required more than forty feet of silk,[c] more pieces were sewn on. Paper sheets, made in two-foot lengths,[d] were pasted together until the necessary length was attained, and this is the only difference between the two. The other components of silk and paper rolls were identical because paper rolls, which came to be substituted for the more expensive silk books, continued to be made in the same way. For example, both kinds of rolls were fastened at the end to wooden rollers that had their tips made of precious materials like porcelain, ivory, tortoise shell, coral, gold, and red sandalwood.[e] And at the outer, unattached end of the roll, there was an extension of silk gauze, brocade, or paper that protected the outermost section of text from damage. To this extension a ribbon was attached

---

[a] See description of the procedures of binding in Stein (6), pp. 251–3; Tsien (2), pp. 111–13.
[b] For sealing and dispatching of documents, see Lao Kan (8), vol. 1, p. 75b.
[c] The standard size of plain silk in the Han and Chin periods was 2 feet 2 inches in width and 40 feet in length.
[d] Ancient papers from Tunhuang measure in most cases about 24 cm. in width and 48 cm. in length for each sheet.
[e] Cf. Ma Heng (1), p. 207.

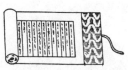

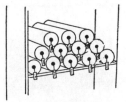

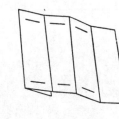

(a)   (b)   (c)

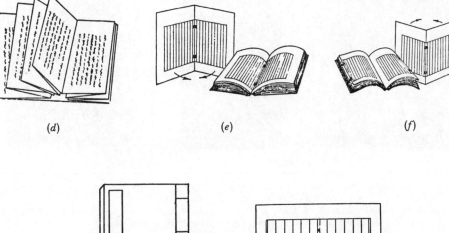

(d)   (e)   (f)

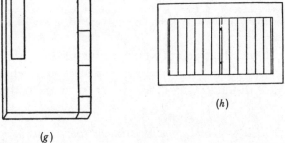

(h)

(g)

Fig. 1157. Evolution of the format and binding of Chinese traditional books. (a) Paper roll with silk extension and ribbon at the head and roller at the end. (b) Identification labels attached to rolls at the end. (c) *Sutra* binding with pleated leaves. (d) Pleated leaves with continuous pages pasted together. (e) Butterfly binding with leaves folded down the centre. (f) Wrapped-back binding with two edges of the leaves pasted on the spine and covered with stiff paper cover on the back. (g) Thread-stitched binding. (h) A double-leaf with printed columns. Adapted from Liu Kuo-Chün (1), (2).

with which the roll was tightened and fastened (Fig. 1157a). The colour of this ribbon was used sometimes to denote the class of literature to which the text belonged, and this form of colour coding was used also on the labels fixed to the end of the roller to identify the particular work in the roll (Fig. 1157b). The rolls were protected by wrappers, called book cloth (*shu i*[1] or *chih*[2]), made of silk or

¹ 書衣   ² 帙

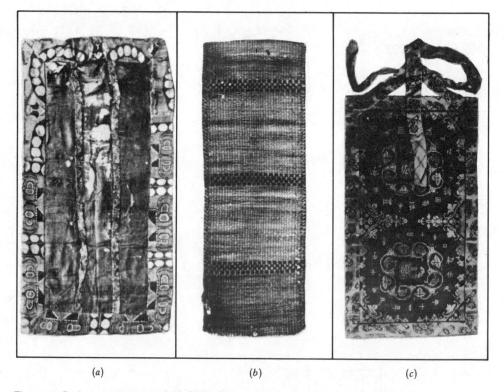

(a)   (b)   (c)

Fig. 1158. Book wrappers made of silk fabric for wrapping paper rolls of the Thang dynasty. (a), (b) From Tunhuang at Louvre Museum, Paris. (c) Preserved at Shōsōin, Japan.

bamboo matting that had borders of white or coloured silk (Fig. 1158), thin gauze, or some other cloth. A protective wrapper held about ten rolls[a] which were placed on shelves with identifying labels attached to one end of each roller.[b]

(iii) *Folded, wrapped back, and stitched binding*

A transitional format of bookbinding is associated closely with Buddhism. The Buddhist sutras came to China from India on the long, narrow, single leaves of palm-leaf books. This format is believed to have suggested the *sutra* binding (*ching che chuang*[1]), in which the long continuous span of joined paper sheets that made the roll book was folded over into accordion pleat-like leaves (Fig. 1157d).[c] These accordion-fold books were easier to handle than rolls when the Buddhist clergy and

---

[a] The word *i*,[2] sometimes meaning 'ten', is derived from the system of book wrapping.
[b] Cf. Tsien (2), pp. 155–6.
[c] Cf. Li Yao-Nan (*1*), p. 212; Chhang Pi-Te (*3*), p. 3, believed that separate flat sheets of paper, known as leaves (*yeh*[3]), existed before they were bound together and that the use of word *yeh* to denote a leaf or page of a book was derived from this meaning.

    [1] 經摺裝   [2] 帙   [3] 葉

laity read and reread a *sutra* without interruption.[a] Another style of binding was called 'whirlwind binding' (*hsüan feng chuang*[1]), in which separate sheets of paper written on both sides overlapped and were pasted along the right edge of each sheet on a blank piece of paper. When rolled or opened up, the overlapping sheets with openings towards one direction looked like a whirlwind current moving forward.

The increasing use of reference books and textbooks, which were easily worn out along the folded lines, together with the new means of rapid reproduction of books by printing, spurred the creation of the binding style known as 'butterfly binding' (*hu tieh chuang*[2]) in the 9th and 10th centuries (Fig. 1157e). Here, the large printed leaves were folded down the centre and gathered into a pile, and a stiff paper cover was pasted against the spine which was made up by the folded centres of the leaves. When the book was opened, the leaves suggested the wings of a butterfly. The spine itself was relatively safe from damage, and if the outer edges on the three sides away from it suffered injury, the parts affected were trimmed away without any loss of the text.[b] Some time in the 13th or 14th century, during the Yüan dynasty, the format changed again. Although, as with the butterfly binding, this new style had a pasted spine and a stiff cover, the two edges of the leaves in the volume were pasted into the spine and the folded centers were brought to the mouth of the book. However, volumes made with this kind of 'wrapped back binding' (*pao pei chuang*[3]) presented a problem when set on shelves with the spine above and the mouth resting on the shelf as the butterfly binding had been; they frequently and readily split into two half-leaves at centre folds of the leaves. To prevent this the volume was laid on its side on the shelf, thereby rendering the stiff book covers unnecessary (Fig. 1157f). Another innovation was the use of paper twists, that is squares of paper twisted into long threads that were passed through two holes pierced near the spine of the fascicle and pasted there. These twists helped to prevent the pasted spine from breaking away from the cover.[c]

The wrapped back binding was difficult to repair properly once the fascicle had come apart, and it was particularly difficult to replace the paper twists without running the risk of doing more damage. As a result, silk or cotton thread was stab-stitched at the spine to reinforce the twists and finally, some time in the 16th or 17th century,[d] pasting the covers to the spine was replaced by the stitched thread binding format (*hsien chuang*[4]) (Fig. 1157g). This format was very durable compared with the butterfly and wrapped back styles of binding, and was quick and simple to do. The printed sheets were folded singly and gathered into a fascicle rather than into signatures, and the next step, the alignment of the sheets, was almost always done using folding signs at the center or the bottom frame line of the sheets. With the fascicle aligned, the flyleaf and sometimes the inner half of the double-leaved

---

[a] Cf. Liu Kuo-Chün (2), p. 44.   [b] Cf. Li Wen-Chhi (1), p. 545.
[c] Cf. illustrations in Liu Ping (1), p. 38; Martinique (2), thesis, pp. 54–5.
[d] Li Yao-Nan (1), p. 216.

[1] 旋風裝   [2] 蝴蝶裝   [3] 包背裝   [4] 綫裝

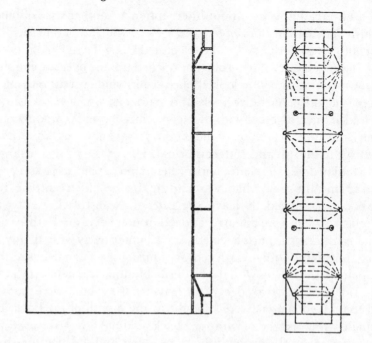

Fig. 1159. Thread-stitched binding and diagram showing steps of stitching. From Nordstrand (1).

cover paper was attached to the book with two paper twists. Next, the edges (except the edge at the book mouth where the leaves were folded) were trimmed with a knife and polished with a pumice stone, and it was at this time that the holes for the thread were pierced. These holes were placed farther from the spine than the holes into which the paper twists had been inserted so that the area encircled by the thread would include the twists. Usually only four holes were pierced, being placed at points where the thread would minimise any stress placed on the spine when the fascicle was opened, though two more holes might be pierced at the corners of oversized books to counteract the extra strain put there by the weight and size of the volume (Fig. 1159).[a]

There were many variations of the order in which a volume was stitched. Usually the thread was passed into one of the outer holes, brought back over the spine into the same hole to form a spine ring, and was then passed over the top or bottom edge of the book and back into the hole, closing another ring. All the other rings were made similarly until the thread returned to the first hole where it was tied and the knot pulled out of sight[b] into the middle of the fascicle. The upper and lower corners on the back of the book were sometimes reinforced by wrapping with brocade to prevent them from being damaged easily.

[a] See description and illustration in Martinique (1), thesis, pp. 228–30.
[b] See step-by-step procedures and illustrations in Nordstrand (1), pp. 106 ff.

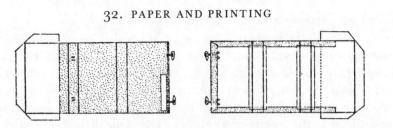

Fig. 1160. Protective covers for traditional thread-stitched books made of cloth and paper. From Nordstrand (1).

(iv) *Protective cases and covers*

A suitable number of fascicles of the same work were placed together in a detached protective case (*han*[1]) made exclusively for them out of wood or paper-board. Such cases began to be used in the +9th century, during the Thang, and coincided with the first use of flat sheets.[a] They were made in various designs, but all covered the sides, mouths, and spines of the fascicles, and very often the tops and bottoms too. The measurements of the height, width, and depth of the set of fascicles to be encased together were used to make the boards for the case, then linen, brocade, or some other fabric was covered with paste and the boards were placed down on it so that, after the paste had dried, the fabric would fold over into the shape of the protective case (Fig. 1160).[b] Wooden chests specially made to hold books were fashioned from wood like Chiangsi cedar (*Phoebe nanmu* or *machilus nanmu*) that was decay resistant and insect repellent. In Southern China wooden boards called pressing boards (*chia pan*[2]) were used as book covers, and tightened to the fascicles by cloth straps. The woods preferred for making these boards were the catalpa, Chinese rosewood, and jujube.[c]

Unlike its counterpart in the West, the stitch-bound Chinese book could be separated from its protective case. It therefore was much lighter and free from the strains that had to be sustained by Western books hanging by a cumbrous mechanism of tapes, thread, and mull to their heavy protective covers. Moreover, the compactness of the Western book together with its heavy paper, increased the strain on its covers considerably. Yet by trading off compactness and permanent union to its protective shields, the Chinese book with its simple and easily repaired binding and its leaves printed on one side only, was superior in its ability to remain in a good state to books bound in the Western style.[d]

(4) DEVELOPMENT OF INKMAKING IN CHINA

Chinese ink underwent a long process of development, commencing perhaps as early as three millennia ago, and played an important role, together with paper

[a] Chhang Pi-Te (*3*), p. 4. [b] Nordstrand (1), pp. 109–12.
[c] Cf. Yeh Te-Hui (*4*), p. 46; tr. Achilles Fang (1), p. 142–3.
[d] For the advantages of the Chinese traditional bookbinding, cf. Martinique (2), thesis, p. 227.

[1] 函 [2] 夾板

and printing, in the progress of Chinese civilisation. The prominence of ink in Chinese culture is evident not only in its extensive use in writing but also in the fact that it has been featured as an object of art and a subject of scholarship; while few names of Chinese papermakers and printers are known to us today, hundreds of inkmakers are extensively recorded in literature, including numerous works devoted exclusively to the study and lore of ink. The value placed on it is evident both in China, where a prized piece might literally be worth its weight in gold, and outside China, where it was borrowed or imitated in both the eastern and western hemispheres.

The earliest samples of Chinese ink are to be found in traces of writings and drawings in black or colour on bone, stone, clay, bamboo, wood, silk, and paper documents dating from the -14th to the $+4$th centuries. A few specimens of solid ink dating from the -3rd to the $+3$rd or $+4$th century have recently been discovered, and a number of artifacts from later times still survive. Since they have not been exhaustively analysed, documentary sources are needed for the study of their nature, composition, and manufacture, but unfortunately, ancient literature is not very informative for the period preceding the Han dynasty.

From the time of the Han until the Sung dynasty, most ink appears to have been made from a combination of pine soot, glue, and miscellaneous additives. Lacquer might have been used to write short inscriptions on certain hard-surfaced objects prior to this time, but it was not used for general writing. 'Stone ink', possibly a form of graphite, was also in use at this time, probably to a much more limited extent. From the Sung period on, lampblack, made from animal, vegetable, or mineral oils, was often substituted for pine soot, but the pigments, binding agents, and additives, remained much the same in spite of the passage of time, though their proportions tended to vary with individual inkmakers.

Ink was probably first decorated with designs and calligraphy after it came to be manufactured in a prismatic shape with flat surfaces, a development which may have occurred before the Thang dynasty. Such decoration of ink evolved into an elaborate enterprise, and large sets of ink-sticks, decorated with associated designs, were produced for a collector's market. Eventually, connoisseurs of ink in China probably attached as much importance to the decoration of the ink as to its writing qualities.

(i) *Role of ink in Chinese and other cultures*

Writing in China was something more than just a functional means of recording. From times at least as early as the Han dynasty, calligraphy was considered a major art form, and eventually all objects associated with it came to share in the general aesthetic of writing and were themselves elevated to the status of forms of art. Consequently, paper, ink, the writing brush, and the inkstone (Fig. 1161), the basic equipment used in writing known as the 'four treatures of the scholar's studio' (*wen*

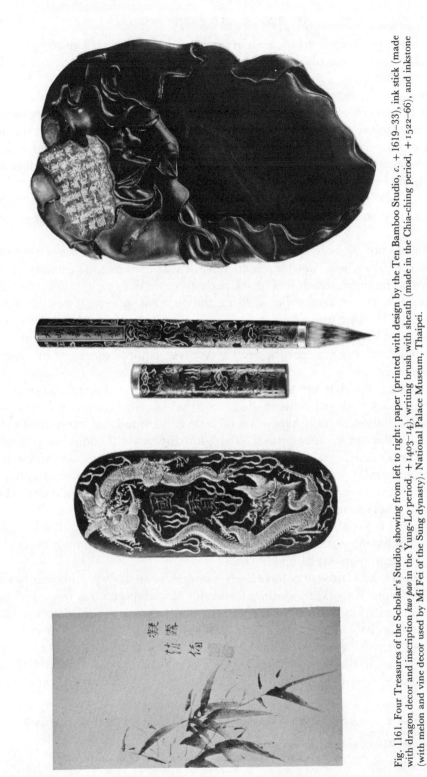

Fig. 1161. Four Treasures of the Scholar's Studio, showing from left to right: paper (printed with design by the Ten Bamboo Studio, c. +1619–33), ink stick (made with dragon decor and inscription *kuo pao* in the Yung-Lo period, +1403–14), writing brush with sheath (made in the Chia-ching period, +1522–66), and inkstone (with melon and vine decor used by Mi Fei of the Sung dynasty). National Palace Museum, Thaipei.

fang ssu pao[1]), were collected by connoisseurs as *objets d'art* as well as being used in the production of works of art. In addition to its use in writing, ink was also used in painting, printing, as a cosmetic and even in medicine.

Chinese ink is generally kept in a solid form until shortly before use, when a small amount is converted into a liquid by grinding the ink on an inkstone moistened with water. The solid form was conducive to lengthy preservation since it was not subject to evaporation. It also facilitated the development of ink as an artistic medium, since designs could easily be incorporated in the moulds used for forming the solid ink or applied directly to the surface. Permanence and lustre are other important qualities of Chinese ink, and these are clearly apparent in documents and paintings more than a thousand years old that have survived until the present day.

China was not the only ancient civilisation to use ink, for there is evidence that Egyptian ink may be somewhat older, since it was used at least as early as the invention of papyrus, which took place perhaps more than 2500 years before the Christian era. It was kept in liquid form, the pigment forming it possibly being produced by burning animal bones, and it may have been used in Western Asia after − 1100.[a] Several references in both the Old and New Testaments to ink or to writing in a book suggest that the Jews may have acquired the use of ink during their stay in Egypt.[b]

The Greeks used different kinds of ink for writing on parchment and papyrus; they were black with pigments made from dried wine-lees or burnt ivory. Like Chinese ink, it was generally kept in a solid form and ground and mixed with water prior to use. Roman ink was similar, although it appears that additional pigments were utilised, made from half-charred human bones from graves, certain kinds of earth or minerals, resin or pitch soot, sepia (from cuttlefish), and lampblack. Lampblack and sepia inks were used for writing on papyrus, and could be washed away with a wet sponge, a feature that distinguishes them from Chinese inks, which are permanent. Another ink, made from galls caused by the ova of gall wasps, was used on parchment. Arabian ink is believed to have been derived from that of the Greeks, its main pigment being pine soot.[c]

The use of ink in India may have been retarded by an early prohibition against recording religious works in writing, for in the Mahabharata it is remarked that anyone who sells, forges, or writes the Veda is condemned to hell. The earliest use of ink in India probably occurred with the development of prepared cotton-stuffs and birch bark as writing materials, and first reported by Greek writers in the −4th century. Later, in the +1st century another Greek writer notes that 'Indian ink' (*Indikon milan*) was exported from the Indian port of Barbarikon, and Pliny

---

[a] For the use of ink in Egypt and Western Asia, see Wiborg (1), pp. 70, 137; Breasted (2), pp. 230 ff.
[b] Cf. Wiborg (1), p. 71.
[c] For Greek, Roman, and Arabian inks, see Wiborg (1), pp. 71–6.

[1] 文房四寶

compared this to some of the best ink made in Rome in his day.[a] However, it is possible that this Indian ink in fact originated in China, as suggested by Berthold Laufer,[b] since many other valuable articles of trade, such as silk, regularly made their way from China to Europe by way of India from as early as the Former Han period.

Japanese and Korean inks, made either from pine soot or lampblack, were apparently borrowed from those of China, since the manufacturing processes are almost the same, while inkmaking, papermaking, and brush-making were crafts that foreign students at the Thang court were required to learn. Tibet also appears to have learned the use of ink from the Chinese, although their ink was kept in a liquid rather than a solid form.

Without samples of all these inks, it is difficult to make meaningful comparisons between them and Chinese ink, but we know that the special qualities of Chinese ink caused it to be actively sought after and imitated in many areas of the world, including Europe. Louis LeComte in the 17th century said of Chinese ink that 'it is most excellent; and they have hitherto vainly tried in France to imitate it'.[c] Du Halde also wrote in 1735 that 'the Europeans have endeavored to counterfeit this ink, but without success'.[d] In commenting on the general characteristics of Chinese ink which have probably been responsible for its popularity, Laufer said:

It produces, first of all, a deep and true black; and second, it is permanent, unchangeable in color, and almost indestructible. Chinese written documents may be soaked in water for several weeks without washing out.... In documents written as far back as the Han dynasty ... the ink is as bright and well preserved as though it had been applied but yesterday. The same holds good of the productions of the printer's art. Books of the Yüan, Ming, and Ch'ing dynasties have come down to us with paper and type in a perfect state of composition.[e]

These distinguishing qualities were due, of course, to various ingredients in the composition and to the elaborate methods used in the manufacture of Chinese ink, which will be discussed below.

(ii) Origin and early specimens of Chinese ink

Traditionally, the invention of ink in China has been attributed to the calligrapher Wei Tan[1] (+179–253) early in the +3rd century. Archaeological and literary evidence, however, attests to the widespread use of various kinds of ink, or pigments which functioned like ink, well before this time. The early symbols and signs appearing on painted pottery found in Pan-pho, Shensi, indicate the use of red and

[a] Wiborg (1), pp. 62–5.
[c] LeComte (1), Eng. ed. 1697, p. 192.
[e] Wiborg (1), pp. 41–2.
[b] Wiborg (1), p. 2.
[d] Du Halde (1), Eng. ed. 1736–41, vol. I, p. 370.

[1] 韋誕

black pigments as early as the neolithic period,[a] and a considerable number of late Shang oracle bones bear traces of red and black pigments used in conjunction with characters either before or after they were incised into the surface. The red pigment has been identified as cinnabar and the black pigment as a carbonaceous material variously identified as ink or dried blood.[b] Characters written in black fluid have also been found on the surfaces of stone objects, jade, and pottery of the Shang period.

The earliest form of the character for ink (mo^1) was used in Western Chou bronze inscriptions, and apparently refers to a punishment of blackening or tattooing the face rather than to use of a writing fluid.[c] The earliest textual reference to *mo* as a fluid used in writing appears in the *Chuang Tzu* of the Warring States period, which mentions that when Prince Yüan of Sung expressed a desire to have his picture painted, all the court scribes stood up and started 'licking their brushes and mixing their ink'.[d] Two later works refer to the use of ink as a writing medium in the Spring and Autumn period. A minister of the state of Chin is quoted as remarking to his master: 'I wish I could be your critical subordinate, handling tablets with brush and ink and watching over you to record whatever faults you have'.[e] And Duke Huan of Ch'i 'asked the officials to record his orders on a wooden board with brush and ink'.[f] All three references imply that brushes were used to apply ink to a writing or painting surface, while the *Chuang Tzu* passage implies that ink was kept in a solid form prior to use.

Archaeological excavations have turned up numerous documents of various kinds from the Spring and Autumn, Warring States, and Chhin periods written in ink on precious stones and on bamboo or wooden tablets. Since the 1950s, archaeology has also yielded several artifacts of Chinese ink. The oldest is a small piece of ink found in a group of twelve Chhin graves dating back to the −3rd century, in the late Warring States or Chhin dynastic period, excavated at Shui-hu-ti, Yün-meng hsien, Hupei in late 1975 and early 1976. It is reported that this ink (M4:12) is cylindrical in shape and of pure black colour, with a diameter of 2.1 cm. and height of 1.2 cm. In the same grave there was also found an inkstone and a small piece of stone apparently used for grinding on the inkstone, for both items bear traces of grinding and remnants of ink.[g]

Again, five specimens of ancient ink were discovered in several Later Han graves located at Liu-chai-chhü, Shan-hsien, Honan province in 1965. In the site report, three of these specimens (8:60, 37:45, and 102:9) are said to be relatively well preserved and are described as cylindrical in shape; they were formed by moulding

[a] Cf. *Hsi-an Pan-pho* (Peking, 1963), p. 156.
[b] Cf. Britton (2), pp. 1–3; Benedetti-Pichler (1), pp. 149–52.
[c] Cf. Karlgren (1), no. 904*b*.   [d] *Chuang Tzu* (*SPTK*), ch. 7, p. 36*b*.
[e] See *Han Shih Wai Chuan* (*SPTK*), ch. 7, pp. 6*a*,*b*.
[f] See *Kuan Tzu* (*SPTK*), ch. 9, p. 1*b*.
[g] See excavation report of the Shui-hu-ti site in *WWTK*, 1976, no. 9, p. 53, and illustration in plate 7, fig. 5.

1 墨

with the hands, and either one or both ends had been used for grinding. Their sizes vary from 1·5 cm. to 2·4 cm. in diameter and from 1·8 cm. to 3·3 cm. in height; one of them has a wood base.[a] Two specimens of ink were also found in two Chin period graves, M2 and M3, located at Lao-hu-shan, Nanking in 1958. According to the site report, the M2 specimen, which is bar-shaped, is approximately 6 cm. in length and 2·5 cm. in width, but the M3 specimen has crumbled, and consequently its measurements are not given. Both these specimens were analysed by Chhiu Chia-Khuei[1] of Nanking University, who concluded that the M2 specimen was not ink but a kind of earth with organic material present in it. His analysis of the M3 specimen yielded markedly different results:

The ink is black in colour and light in weight. There are small yellow grains in it which appear to be loess impurities. When this sample is placed under the microscope and compared with contemporary ink, the particles seen are very similar. They form clusters and are combustible when heat is applied, leaving behind a small amount of ash. It is evident that this sample is also very comparable to contemporary ink in this respect. Thus I conclude that this is ink.[b]

On the basis of this study, it has been conjectured that the M2 specimen was similar in composition to the 'stone ink' mentioned in early records and that the M3 specimen was representative of a more advanced ink.[c]

(iii) *Pigmentation and composition of Chinese ink*

From the +13th century, scholars have suggested that Chinese ink was first made from lacquer, then minerals, and finally pine soot and lampblack. There has been a controversy among modern scholars, however, about whether Chinese ink was ever made from lacquer. The earliest surviving literary references to lacquer writing, appearing in the *Hou Han Shu* and *Chin Shu*, compiled in the +5th and +6th centuries, are made in connection with texts that were probably written on bamboo or wooden tablets prior to the Han dynasty. The meaning of the term lacquer writing (*chhi shu*[2]) in these references, however, has been frequently questioned and the subject of different interpretations.[d]

From recent discoveries, it seems entirely possible that hard, non-absorbent surfaces may have necessitated the use of a more adhesive ink than was used eventually on silk and paper, and that lacquer may have suited this requirement. The text of a collection of bamboo tablet documents discovered in one of the previously mentioned Chhin graves found at Shui-hu-ti, indicates that lacquer and cinnabar were used in writing the names of official units on the surfaces of government tools, armour, and weapons when these did not lend themselves to

[a] See report in *KKHP*, 1965, no. 1, p. 160.
[b] See *KKTH*, 1959, no. 6, p. 295.   [c] *Ibid.*
[d] See controversial opinions on lacquer writing in Tsien (2), pp. 168 ff.

[1] 裘家奎   [2] 漆書

incising.[a] These tablets also cite regulations for lacquer orchards and testing the quality of lacquer.[b] The references indicate that lacquer was used at an early date for writing on certain kinds of materials such as metals, which do not absorb watery ink. But it seems certain that lacquer was not a major vehicle for writing, since there is no archaeological evidence that it was used on more conventional hard-surfaced writing media, such as bamboo or wooden tablets. However, lacquer was possibly present as a minor ingredient in some inks.

Silk was used as a writing medium at least as early as the −5th century, and silk documents, dating from the Warring States through to the Han period, have been found at many sites in China and Central Asia.[c] Paper was also used for writing beginning in the Later Han dynasty. Specimens of writing with black ink from the +2nd century have been found at Chü-yen, Tun-huang, Lou-lan, and other sites,[d] but since the ink on these early silk and paper documents has never been chemically analysed, it is difficult to say what its composition is.

Pine soot, traditionally the favourite pigment in ink, was used in ink manufacture in the time of Wei Tan, as is attested in a poem by Tshao Chih[2] (+192−224).[e] An inkmaking formula appearing in a +5th-century work and attributed to Wei Tan calls for the use of fine and pure soot, pounded and strained to remove any adhering vegetable substance.[f] Although the source of this soot is not indicated, it seems likely, in view of the fact that there is a procedure to remove 'adhering vegetable substance', that it was made from wood, perhaps pine. Recent studies carried out with a scanning electron microscope have shown that the sizes of carbon particles found in 14th-century Chinese ink made from pine soot are remarkably small and uniform, superior in these respects to a sample of modern ink also made with soot.[g]

The method of making ink from pine soot is given by the Ming author Sung Ying-Hsing (c. +1600−60) as follows:

Ordinary ink is made from pine wood after all the resin has been eliminated. The least amount of resin left in the wood will result in a non-free-flowing quality in the ink produced. To get rid of the resin, a small hole is cut near the root of the tree, into which a lamp is placed and allowed to burn slowly. The resin in the entire tree will gather at the warm spot and flow out.

For making pine wood soot, the tree is felled and sawn into pieces. A rounded chamber of bamboo is built, resembling in appearance the curved rain-shield on small boats and constructed in sections; it has a total length of more than 100 feet. The external and internal surfaces of this chamber and the connecting joints are all securely pasted with paper and

---

[a] See *Shui-hu-ti Chhin Mu Chu Chien* (Peking, 1978), pp. 71–2, 121–2, 138.
[b] In the quality test, the process used is designated by the term *yin shui*.[1] The meaning of this term is not clear, but the general idea seems to have been that the quality of lacquer varied in inverse proportion to the amount of water (*shui*) needed for the test.
[c] Cf. Tsien (2), pp. 116–17.   [d] See above, pp. 41 ff.
[e] See Tsien (2), p. 166.   [f] *Chhi Min Yao Shu* (*TSHCC*), ch. 9, p. 231.
[g] See J. Winter (1), pp. 209, 213–14, 219.

[1] 飲水   [2] 曹植

Fig. 1162. Inkmaking with pine soot by gathering resin from a pine tree and scraping out the soot from the end of the cover. From *Thien Kung Khai Wu, c.* +1637.

matting, but small holes are made at certain intervals for the emission of smoke. The floor of the chamber is constructed of brick and mud with channels for the smoke built in.

After the pine wood has burned for several days, the chamber is allowed to cool and workers will now enter and scrape out the soot [Fig. 1162]. That which is obtained from the last one or two sections of the chamber is of 'pure' quality and is used as raw material for the manufacture of the best ink. The soot obtained from the middle sections is of 'mixed' quality and is used in ordinary ink. That from the first one or two sections, however, is scraped and sold only as low-grade soot; it is further pounded and ground by printers for printing books. In addition, lacquer workers and plasterers also use the coarse grade as black paint.[a]

Although pine soot probably remained the most popular pigment used in making ink, it was soon rivalled from the Sung dynasty onwards by lampblack made from combustion, in lamps with wicks, of animal, vegetable, and mineral oils such as fish oil, rapeseed oil, bean oil, hemp oil, sesame oil, tung oil, and petroleum. In Ming times, it is said that nine-tenths of all ink was made from pine soot and one-tenth from oil lampblack.[b] In 1738 Du Halde described a lampblack inkmaking operation as follows:

[a] See *Thien Kung Khai Wu (KHCP)*, pp. 276–7; tr. Sun & Sun (1), pp. 286–7.
[b] See *Thien Kung Khai Wu*, p. 276; tr. Sun & Sun (1), p. 286. It notes that one catty of oil, after burning, will yield more than one ounce of fine quality lampblack.

They put five or six lighted wicks into a vessel full of oil, and lay upon this vessel an iron cover, made in the shape of a funnel, which must be set at a certain distance, so as to receive all the smoke. When it has received enough, they take it off, and with a goose feather gently brush the bottom, letting the soot fall upon a dry sheet of strong paper. It is this that makes their fine and shining ink. The best oil also gives a lustre to the black, and by consequence makes the ink more esteemed and dearer. The lampblack which is not fetched off with the feather, and which sticks very fast to the cover, is coarser, and they use it to make an ordinary sort of ink, after they have scraped it off into a dish.[a]

Another kind of ink mentioned in early Chinese sources is 'stone ink' (shih mo[1]). This appears to have been a mineral substance of some sort which was either used as found or was prepared by grinding. It was possibly a form of coal, petroleum, or graphite, for the discovery sites specified in early records are all located in areas where graphite is produced at the present time.[b]

Pine soot and lampblack consist principally of carbon, which, in its free state, does not combine readily with other materials; consequently, the use of carbon in ink necessitates the use also of an agent that will bind the carbon pigment to the writing surface. Binding agents also play another role in Chinese ink in holding the carbon particles together in the solid form.

The binding agents used in Chinese ink were traditionally glues made from a variety of animal remains, including raw hides or leather, muscles, bones, shells, horns, fish skin, fish scales, and fish maw;[c] the quality of the water used was also important. After one of these substances was boiled, the resulting hot viscous fluid was strained through a silk gauze or cotton filter to remove lumps and then allowed to condense into solid form until needed for use.[d] The solid glue was then dissolved before use in inkmaking with solvents such as the juice of the bark of the chhin[2] tree (Fraxinus bungeana, D.C. var. Pubinervis wg.).[e] The ratio of glue to pigment probably varied with the nature of the materials used and the stickiness of the ink desired, the latter probably being dictated by the quality of the writing surface to be employed. We know, for instance, that equal weights of lampblack and glue are specified in an inkmaking formula contained in an encyclopedia compiled in the early 18th century.[f]

In addition to the essential pigments and binding agents, other materials were often added, especially in periods prior to the Ming dynasty, to improve consistency, colour, and aroma. As many as 1100 miscellaneous additives were sometimes used.[g] These included egg whites, gambage, raw lacquer sap, soaptree pods, and croton seeds to improve consistency; cinnabar, chhin tree bark, purple herb, madder root, yellow reed, black beans, copper vitriol, gall nuts, ti yü[3] (Sanguisorba

[a] Du Halde (1), vol. 1, p. 371.   [b] Cf. Tsien (2), pp. 71–2.
[c] See Kecskes (1), p. 55.   [d] Cf. Tsien (2), p. 167.
[e] See Chhi Min Yao Shu (TSHCC), ch. 9, p. 231.
[f] See Ko Chih Ching Yüan (1735 ed.), ch. 37, pp. 26–7.
[g] See H. Franke (28), p. 59.

[1] 石墨   [2] 梣   [3] 地芋

officinalis), curled pine (*Selaginella involvens*), walnut, peony rind, pig and carp galls, pearls, tonka beans, pomegranate skins, and vermilion to improve colour and gloss; and cloves, sandalwood, sweet pine (*Nardostackys jatamansi D.C.*), camphor, and musk to improve scent.[a]

(iv) *Technical processes of inkmaking*

It seems likely that inkmaking formulae were usually kept secret to guard against competition; consequently, the formulae which were recorded and survive to the present day probably represent only a very small fraction of those actually used. Although the ingredients used in making any ink are generally not very numerous, the exact composition, preparation, and quantity of each ingredient were subject to considerable variation. According to early works on inkmaking by such authors as Li Hsiao-Mei[1] (fl. +1095), Chhao Kuan-Chih[2] (c. +1100), and Shen Chi-Sun[3] (fl. +1598),[b] the steps involved in making ink consist of gathering soot or lampblack, straining and then mixing with pre-dissolved glue and miscellaneous additives, kneading, pounding, steaming, moulding, covering with ashes, drying, waxing, storing, and testing (Fig. 1163).

The earliest known formula for inkmaking in China, often attributed to Wei Tan (+179–253), appears in a work on agriculture and manufacture written by Chia Ssu-Hsieh in the +5th century.

Fine and pure soot is to be pounded and strained into a jar through a sieve of thin silk. This process is to free the soot of any adhering vegetable substance so that it becomes like fine sand or dust. It is very light in weight, and great care should be taken to prevent it from being scattered around by not exposing it to the air after straining. To make one catty of ink, five ounces of the best glue must be dissolved in the juice of the bark of the *chhin* tree which is called *fan-chi*[4] wood in the southern part of the Yangtze Valley. The juice of this bark is green in colour; it dissolves the glue and improves the colour of the ink.

Add five egg whites, one ounce of crushed pearl, and the same amount of musk, after they have been separately treated and well strained. All these ingredients are mixed in an iron mortar; a paste, preferably dry rather than damp, is obtained after pounding thirty thousand times, or pounding more for better quality.

The best time for mixing ink is before the second and after the ninth month in a year. It will decay and produce a bad odour if the weather is too warm, or will be hard to dry and melt if too cold, which causes breakage when exposed to air. The weight of each piece of ink cake should not exceed two or three ounces. The secret of an ink is as described; keeping the pieces small rather than large.[c]

All the main ingredients used in manufacturing ink in later times, even in the present day, are to be found in this early formula: a pigment (soot), a binding agent

---

[a] Cf. Kecskes (1), p. 59.
[b] Cf. German tr. of their works in H. Franke (28), pp. 33 ff.
[c] *Chhi Min Yao Shu* (*TSHCC*), ch. 9, p. 231; tr. Tsien (1), pp. 166–7.

[1] 李孝美   [2] 晁貫之   [3] 沈繼孫   [4] 樊雞

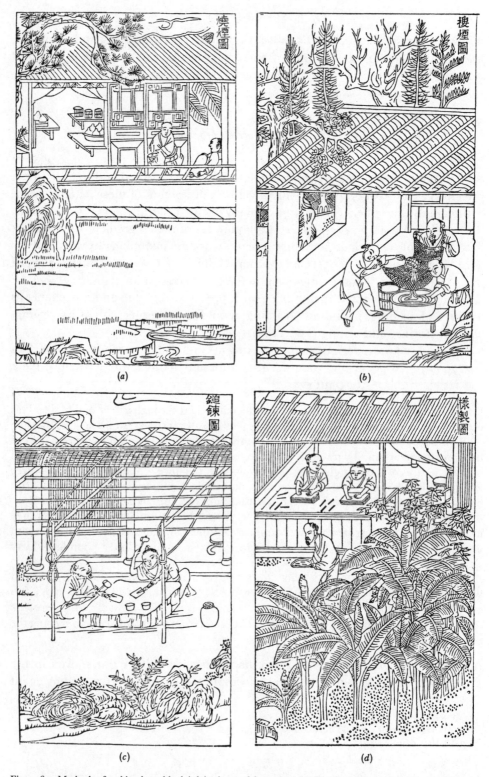

Fig. 1163. Methods of making lampblack ink in the +16th century. (a) Burning oil in enclosed room with paper curtain. (b) Gathering and sieving the lampblack. (c) Pounding and beating the ink paste. (d) Moulding the paste into different shapes. From *Mo Fa Chi Yao*, c. +1598.

(glue), and miscellaneous additives (*chhin* tree bark, egg whites, cinnabar, and musk).

Another formula of Master Chi[1] of the Liang dynasty (+502–77) specified that:

Two ounces of pine soot are added with small amounts of cloves, musk, and dried lacquer and mixed with glue to form a stick, which is then dried over the fire. The ink will be ready for use in a month. The colour turns purplish when the purple herb [*tzu tshao*,[2] *Lithospernum erythrorhizon*] is put in and bluish if the *chhin* bark powder is added. Both colours are pleasant.[a]

Still another formula is ascribed to Li Thing-Kuei of the +10th century in the Southern Thang period:

Wash, clean and shred three ounces of cow horn, soaked in 10 catties of water for seven days. Boil three honey locust pods [*tsao chiao*,[3] *Gleditschia sinensis*] for one day to get three catties of juice, soak in the juice one ounce each of gardenia kernel [*chih tzu*,[4] *Gardenia florida*], the bark of the amur cork tree [*huang po*,[5] *Phellodendron amurense*], bark of the ash [*chhin phi*,[6] *Fraxinus bungeana*] and sappan wood [*su mu*,[7] *Casalpinia sappan*], half an ounce of white sandalwood, and one piece of sour pomegranate skin, for another three days. Bubble up the mixture to get one catty of juice and mix the juice with two and a half ounces of fish glue; soak overnight. Cook again, add a little green vitriol, and it will be ready to mix with one catty of sieved soot.[b]

A later formula attributed to Shen Chi-Sun (fl. +1398) used lampblack in place of soot for pigmentation: ten ounces of tung oil lampblack mixed with four and one-half ounces of cowhide glue, one-half ounce of fish glue, and one-half ounce each of the bark of the ash and of sappanwood.[c] The formulae of the Ming and Chhing period were probably somewhat simpler on the whole than those given here, for it appears that inkmakers of these periods did not use the many additives as freely as before, because they were considered to reduce the quality of the ink. Economic considerations may also have been a factor.

The qualities sought in Chinese ink are often reflected in remarks made concerning the ink of noted inkmakers. The ink of Wei Tan (+179–253), the earliest inkmaker of fame, was described by Hsiao Tzu-Liang,[8] a prince of the Southern Chhi dynasty, as so black that 'each drop looks like lacquer'.[d] The ink of Chang Yung,[9] an inkmaker during the period of the Northern and Southern Dynasties, was also compared with lacquer.[e] From the late Thang period on, the names of numerous distinguished inkmakers have been recorded in history.[f] The most famous is perhaps Li Thing-Kuei[10] (fl. +950–80) a member of a prominent family

[a] *Wen Fang Ssu Phu* (*TSHCC*), p. 67.   [b] *Mo Phu Fa Shih* (*MSTS*), pp. 172–3.
[c] *Mo Fa Chi Yao* (*TSHCC*), p. 22.
[d] See *Ko Chih Ching Yüan* (1735 ed.), ch. 37, p. 21a.
[e] *Lao Hsüeh An Pi Chi* (*HCTŶ*), p. 4.
[f] Lu Yu (c. +1330) in his *Mo Shih* and Ma San-Heng (fl. +1637) in his *Mo Chih* list a total of 448 inkmakers from ancient times to +1637.

[1] 冀公   [2] 紫草   [3] 皂角   [4] 栀子   [5] 黃蘗
[6] 秦皮   [7] 蘇木   [8] 蕭子良   [9] 張永   [10] 李廷珪

of inkmakers. After his father, Li Chhao[1] (fl. +907–36), also a noted inkmaker originally from a Hsi[2] family at I-shui, Hopei, moved to She-hsien, Anhui, Li Thing-Kuei served as an official in charge of inkmaking at the Southern Thang court of Prince Li Yü[3] (+937–78), who granted the royal surname of Li to Thing-Kuei in honour of his distinguished services. The ink made by him and his father is especially famous for its qualities of hardness and insolubility in water.[a]

In the Sung period, Chang Yü[4] (fl. +1068–85) was known for an ink made from oil lampblack, musk, camphor, and gold flakes, and which appeared in the form of small round coin shapes.[b] Phan Ku[5] (fl. +1086) was noted for making ink with very small amounts of glue, only five to ten ounces per pound of soot, and for pounding his ink dough ten thousand times. Ink made by the Sung Emperor Hui-tsung (r. +1101–25) was also highly sought after; the ingredients included a special su-ho[6] (*Liquidambar orientalis*) resin. Chu Wan-Chhu[7] (fl. +1328–30) of the Yüan period is remembered for using only soot from pine trees.[c]

The inkmakers of the Ming and Ch'ing periods were known primarily for the forms and styles of their inks, often produced in elaborate sets, although quality of composition was no doubt still important. Wu Shu-Ta[8] was famous for his ink of lacquer-like blackness and stone-like hardness; its ingredients included tung oil, glue, powdered gold, and musk, and the dough was beaten ten thousand times.[d] Hu Khai-Wen[9] in the early 19th century was noted for an ink made from lard, lampblack, antler glue, and twelve miscellaneous additives, including pearl powder and musk.[e]

Although Chinese ink was generally produced in a solid form, some liquid ink was also made. It was sometimes prepared in quantity and stored in bamboo tubes by persons who wished to avoid the labour of mixing their own ink prior to use. Special liquid inks were also made for commercial applications, like printing, where the volume of ink required tended to prohibit grinding each time it was to be used. Printing ink, for instance, was first prepared as a paste made of coarse soot taken from the far end of a smoke chamber and mixed with glue and wine, and then preserved in jars or vats for later use. It had to be kept for three or four summers, for its bad odour to disappear, and in fact the longer the period of preservation, the better it became; printing done using freshly prepared ink was easily smeared. When needed, water was added to the paste and it was mixed thoroughly and strained through a sieve made of hair from horse's tails.[f]

The best ink for printing in red was a mixture of vermilion and red lead boiled in water with the mucilaginous root of a plant called po chi[10] (*Bletilla striata*). Next best

[a] See *Mo Shih* (*TSHCC*), pp. 10 ff.   [b] *Mo Chih* (*TSHCC*), p. 2.
[c] See inkmakers cited in Kecskes (1), pp. 27–9.   [d] Cf. Mu Hsiao-Thien (1), p. 28.
[e] Cf. Kecskes (1), p. 42.
[f] See the description of Chinese printing ink in Lu Chhien (1), pp. 632–3, which is probably the only account of its preparation in Chinese literature.

[1] 李超   [2] 奚   [3] 李煜   [4] 張遇   [5] 潘谷
[6] 蘇合   [7] 朱萬初   [8] 吳叔大   [9] 胡開文   [10] 白芨

was the liquid obtained from boiling the red-stem amaranth (*hsien tshai*,[1] *Amaranthus tricolor*, L.), but this easily turns purple and does not give as fresh a colour as the vermilion and red lead mixture. Blue ink was made from indigo (*tien*,[2] or *mu lan*,[3] *Indigofera tinctoria*), a Chinese native blue dye with a permanent colour used for dyeing textiles. Prussian blue is not suitable for printing, as the colour runs when paper is wet.[a]

Invisible ink was already known to the Chinese perhaps no later than the +12th century. A story of the early Southern Sung says that the son of a military official Wang Shu[4] was deprived of his title because he had spread scandal about Chhin Kuei[5] (+1090–1155), but during his banishment met a magician who could write invisible characters with a liquid on paper. When it was treated with water the characters appeared. So for fun Wang's son wrote the four characters 'death to Chhin Kuei' and applied water to test the technique. The magician then went away intending to show the paper to the government, and was only prevented by being bribed with much money.[b] Although the process was called magic, the characters were apparently written with chemicals, perhaps alum, on paper; they appeared when treated with some kind of solution.

Ink was also commonly used in medicine as early as the +10th century. Ink mixed with wine was given to the daughter-in-law of the Sung prince to relieve bleeding stemming from childbirth,[c] and Li Shih-Chen[6] mentioned in his *materia medica* a number of prescriptions in which ink was administered. Ink made from fine pine soot, roasted, ground, and mixed with water, vinegar, and other ingredients such as turnip, onion, foxglove juice, bile, wine, and dried ginger, was also used as a cure for bleeding following childbirth, dysentery, ulcers and sores, nose-bleeds, swelling, and eye irritations, among other disorders.[d] Depending on the nature of the disorder, the ink mixture was taken either orally or applied externally. Its curative effects were due, it was said, to its alkali nature absorbing acid humours and sweetening the acrimony of the blood; it was also claimed that the glue from animal skins was a supreme remedy for a haemorrhage.[e] Li also reminded readers that ink made of lampblack from other materials such as oil, petroleum, or straw should not be used for medical purposes.

(v) *Art and connoisseurship of Chinese ink*

An early shape of Chinese ink in solid form can be discerned in a Han tomb mural discovered in the vicinity of Wang-tu, Hopei, in 1953. The painting depicts a

---

[a] Cf. Lu Chhien (*1*), p. 633.
[b] See an anonymous work about the Southern Sung capital written *c.* +1270, *Tung Nan Chi Wen* (*SSKTS*), ch. 1, pp. 7*a–b*.
[c] See *Mo Shih* (*TSHCC*), pp. 16–17.   [d] *Pen Tshao Kang Mu* (Peking, 1975), pp. 446–9.
[e] See Du Halde (*1*), Eng. ed., vol. 1, p. 372. A similar practice prevailed in the West; Francesco Carletti, *Ragionamenti sopra le cose da lui vedute ne' suoi viaggi*, vol. 1, p. 84, notes that a wound caused by an injurious insect in Peru, where the author visited in +1595, was healed by the application of a little ink; see Wiborg (*1*), p. 48.

[1] 莧菜   [2] 靛   [3] 木藍   [4] 王庶   [5] 秦檜   [6] 李時珍

scholar seated on a low platform and beside him an inkstick standing on a round three-legged inkstone and a cup, presumably filled with water, for grinding the ink. The inkstick is standing on end and this suggests that it is probably paraboloidal in shape.[a] Literary sources of the Han and Chin periods generally refer to ink in units of *wan*[1] or *mei*,[2] but they contain no full descriptions of the actual shapes of these units, though *wan* generally refers to something round in shape and *mei* to something flat and thin. There are references to ink in a *lo*[3] (conch) shape used after the Chin dynasty, but the use of this word as a unit defies precise definition.[b]

Chinese ink is generally said to have been first manufactured in a prismatic shape during the Thang period.[c] The discovery of a bar-shaped piece of ink in a Chin grave may indicate that this shape was actually developed somewhat earlier, for another bar-shaped specimen attributed to the Thang period was discovered by Aurel Stein in Chinese Turkestan.[d]

Prismatic shapes, of course, feature flat surfaces. The development of such ink surfaces may have been due to their capacity for facilitating design, which became increasingly prevalent as ink was transformed from a simple object of utility to an *objet d'art*. The earliest known decorative elements used on ink surfaces, dating from the Thang period, consisted of propitious animals, such as the dragon and the carp, as well as calligraphy. During the Ming and Chhing periods, many inksticks were decorated with a pictorial design on one side and calligraphy on the other. The pictorial designs, often symbolic in nature, included dragons, lions, carp, deer, pine trees, cranes, tortoises, gourds, plum flowers, pomegranates, bamboo shoots, landscapes, scenes from everyday life, inventions, religious personages and symbols, as well as others. The inscriptions, which are sometimes gilded, include details of manufacture, explanations of the pictorial design on the reverse side, moralisations, religious sayings, auspicious phrases, poems, and examples of calligraphy. Ink was also made in a variety of special shapes, often in imitation of different artistic objects such as jade pendants, bronze mirrors, and ancient knife-shaped coins.

The various shapes and designs of the inkstick were conditioned by the construction and engraving of the mould, which was made of either copper or wood. Copper moulds produced sharp and clear images of the design but were hard to engrave. Wood was easier to carve but sometimes showed its grain on the surface of the ink. The designs into which the ink paste was pressed were cut intaglio into the mould, resulting in their appearance in relief on the surface of the inkstick. Shen Chi-Sun (fl. +1398) provides in his work an illustration of a six-piece mould for the six sides of an inkstick, in addition to a base into which the six-piece mould would be assembled and fitted (Fig. 1164).[e]

---

[a] See *Wang-Tu Han Mu Pi Hua*, pp. 13–14; also Tsien (2), p. 169, Plate XXVII.
[b] Cf. Wiborg (2), pp. 22–3.   [c] *Ibid.*   [d] See Stein (4), I, p. 316.
[e] See illustration and explanation of the ink mould in *Mo Fa Chi Yao* (*TSHCC*), pp. 64–5.

[1] 丸   [2] 枚   [3] 螺

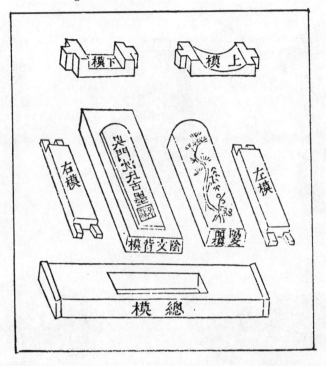

Fig. 1164. Mould for inkmaking. The six-piece set with designs and inscriptions was assembled and fitted into the base. From *Mo Fa Chi Yao, c.* +1598.

Decorated sets of inksticks were also quite popular in the Chhing period and are still sold today. Usually a set was organised around a common theme, such as different kinds of animals, the eight trigrams, views of the imperial palace, landscapes of scenic splendours, etc. Each set was usually encased in a specially made ornate box which could be opened to show off its contents to best advantage. The largest set of inkcakes ever made in China was perhaps a group of sixty-four pieces entitled 'Pictorial Inks Commemorating the Gardens' (Fig. 1165) made by the Chien Ku Chai[1] by imperial order of the Chia-chhing emperor (r. +1796–1821).[a]

Ink was probably collected in China almost as soon as it was noticed that two different specimens could differ widely in quality, but extant records do not reveal much about the art of collecting before the +10th century, in the Southern Thhang and Sung periods. The poet Su Shih[2] (+1036–1101) was an avid ink collector who amassed a collection of five hundred pieces, and his contemporary, Lü Hsing-Fu,[3] was also a noted collector. The imperial collections of the Ming and Chhing dynasties featured numerous inks which still survive.

Many catalogues of ink collections have been published since the late 16th

[a] The set is kept in the collection of the Metropolitan Museum of Art in New York City; see Wang Chi-Chen (2), p. 130.

[1] 鑑古齋   [2] 蘇軾   [3] 呂行甫

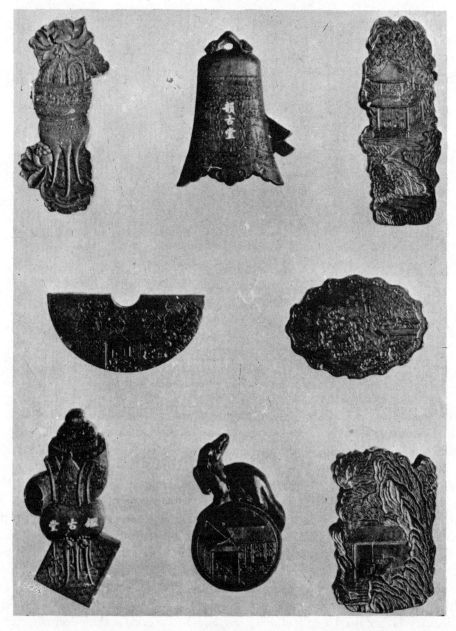

Fig. 1165. Pictorial ink-cakes in various shapes and designs, depicting views of imperial palaces or villas in and around Peking, c. +1800. Metropolitan Museum of Art, New York City.

century by inkmakers, ink dealers, and ink collectors, primarily for appreciation and connoisseurship of the artistic aspects of ink tablets. The earliest and most influential examples are two collections of ink designs reproduced by woodcuts. One titled *Fang Shih Mo Phu*[1] (Fang's Album of Ink Designs) by Fang Yü-Lu[2] (*c.* +1580), containing more than 380 illustrations, arranged by the form and subject-matter of the designs under six categories, as well as a number of laudatory essays, was published in She-hsien, Anhui in +1588. Eighteen years later, his professional competitor Chheng Ta-Yüeh[3] (fl. 1541–1616) published another collection called *Chheng Shih Mo Yüan*[4] (Chheng's Album of Ink Designs), which contains some 500 designs printed in colour together with essays, poems, eulogies, and testimonials from his friends.[a] The two works are similar in nature and content and many of their designs are even identical, but the latter surpassed the former not only in the number of illustrations it provided but also in artistic excellence; furthermore, it included some special features such as the Western alphabets and biblical pictures copied from European engravings given to Chheng by Matteo Ricci (1552–1610) in +1606 (Fig. 1166).[b] It is perhaps the first Chinese book which includes illustrations of an occidental origin.

Another kind of ink catalogue, produced by ink dealers, includes among other things the prices at which the items featured were apparently offered for sale. One early example is the *Mo Shih*[5] (History of Ink) by Chheng I[6] (fl. 1662–1722), an inkshop owner from She-hsien, Anhui, who listed ink titles, kinds of materials, weights, and prices, together with eulogies of inks written by his friends.[c] A third category of ink catalogues is represented by those of private collections. Two early examples are the *Hsüeh Thang Mo Phin*[7] (Ink Collection of the Snow Pavilion) by Chang Jen-Hsi[8] published in 1670, and the *Man Thang Mo Phin*[9] (Ink Collections of the Boundless Pavilion) by Sung Lo[10] (1634–1713), published in 1684.[d] Both list the names of inkmakers, ink titles, designs, dates of manufacture, forms, number of pieces, and weights, all of them standard items described in such catalogues. Interest in ink collecting and connoisseurship has continued up to the present time, and an album of rubbings and descriptions of eighty-three old specimens of Ming-Chhing ink kept in four private collections in Peking, titled *Ssu Chia Tshang Mo Thu Lu*[11] (Illustrated Catalogue of Four Ink Collections), was published in about 1956.[e]

---

[a] For stories of the two rival inkmakers, see Wang Chi-Chen (2), pp. 126 ff.; K. T. Wu (6), pp. 204 ff.; and his two articles in Goodrich & Fang Chao-Ying (1), pp. 212–15, 438–9.

[b] A postscript to the illustrations of four Biblical stories, dated 6 January 1606, is possibly in the handwriting of Ricci.

[c] See Shu-Chhao (1), pp. 72–3.   [d] Cf. Kecskes (1), pp. 81–2.

[e] The four collectors include a chemist, a noted calligrapher, and two other scholars, who wrote the descriptions in their own handwriting; see bibliography under Yeh Kung-Chho (2).

[1] 方氏墨譜   [2] 方于魯   [3] 程大約   [4] 程氏墨苑   [5] 墨史
[6] 程義   [7] 雪堂墨品   [8] 張仁熙   [9] 漫堂墨品   [10] 宋犖
[11] 四家藏墨圖錄

Fig. 1166. Ink design by Chheng Ta-Yüeh with Biblical story provided by Matteo Ricci, *c.* +1606. From *Chheng Shih Mo Yüan*, +1606 ed.

(g) AESTHETIC ASPECTS OF CHINESE PRINTING

BLOCK PRINTING not only involves various procedures of a technical nature, but also consists of many elements that are of artistic significance, while the text itself can display different styles of writing and so represents a piece of calligraphy which can be read for its aesthetic appreciation. Illustrations using woodcuts and other methods are forms of graphic art; they supplement and adorn the text, aid interpretation, assist the memory, and can also provide additional understanding to supplement the written word. Without illustrations, the text may even lack a sustaining interest or, in some cases, be quite unintelligible. Book illustration is nearly as old as the earliest books, but its practical development began with the application of printing.

In the course of its development, the woodcut became a highly sophisticated art not only in the way it depicted the subject-matter, but also in the techniques and styles of its design and engraving, which depend entirely on the skill and vision of individual artists and craftsmen. This is especially true in the case of Chinese colour prints, which require the exactness of the originals in line, in colour, in gradation, and even in the texture and expression of the brushwork. It is perhaps the only kind of graphic art which depends completely on the mutual artistic understanding of the designer, cutter, and printer. Consequently, the woodcut and the subject of book illustration serve a double purpose in their utilitarian and aesthetic qualities. Their contents may represent the thoughts, ideas, events, and personalities of a period. The picture may be at the same time a work of art and the only surviving evidence of an element of the culture of its time.

(1) BEGINNINGS OF CHINESE GRAPHIC ART IN PRINT

Pictorial representation in Chinese documents can be traced back to the beginnings of writing itself. The pictographic nature of Chinese characters indicates the use in ancient times of drawings as a means of communication; the majority of the Shang characters were, in fact, pictograms or combinations of pictograms to indicate ideas. Symbols of birds or beasts were used in ancient seal carvings, and either decorative bird signs were appended to ordinary characters, or individual strokes of a character were written with a motif of bird feathers.[a] Scenes of battle, hunting, and daily life were cast on bronzes or rendered on clay or on lacquered objects, and in particular illustrations on silk cloth were appended to books of bamboo tablets since the narrow tablets themselves were not suitable for drawings. Pictorial representations on stone were very common before any recorded use of wood for carving, and the techniques of relief or intaglio stone carving, the use of decorative designs and the line structure, may have exerted some direct influence on the woodcuts of later times. The close relationship between calligraphy and painting also may have influenced the trends of illustrated books, since a picture sometimes needs written description to tell the otherwise unintelligible story of the painting—the very reverse of the dependence of text on pictures for clarification.

The earliest woodcut illustration in a printed book known to us is that in the *Diamond sutra* of +868, discovered in Tunhuang at the turn of the 20th century. It has a frontispiece (*fei hua*[1]) at the beginning of the roll depicting a scene with the Buddha sitting in the centre, in discourse with his disciple Subhūti who kneels on the ground, and attended by divine beings, monks, and officials in Chinese attire (Fig. 1167).[b] The picture is carefully executed, displays complicated details, lifelike facial expressions, delicate lines to the costumes, and decorative effects in the background, all of which show the artistic and technical maturity of woodcuts at that time. There is no question that the art of woodcut illustration had developed much earlier than the production of this picture, though no other printed illustrated specimens of the +9th century or earlier are extant; however, quite a few survive from the +10th century. These include a number of Buddhist pictures from Tunhuang, each on a single sheet with the illustration above and the text below, some undated and others dated.[c] Buddhist images also appear on individual blocks

---

[a] Cf. the development of decorative inscriptions described in Tsien (2), pp. 24–5, 46–7, 54–5.
[b] Reproduced in Stein (4), IV, plate C.
[c] Cf. dated documents of +947–71 mentioned in Carter (1), pp. 64–5, note 12.

[1] 扉畫

Fig. 1167. Frontispiece of the *Diamond sutra* printed in +868 from Tunhuang, depicting the story of Buddha in discourse with his disciple Subhūti and surrounded by attendants and divine beings. British Library.

printed together in red or black; as well as drawings on calendars; and several other printed texts with illustrations.[a]

While the specimens mentioned are all from the northwestern region, woodcut pictures are also known to have been produced in the southeastern part of the country. The most prominent of these are the three different versions of the invocation sutra, *Pao Chhieh Yin Tho Lo Ni Ching*,[1] printed by Prince Chhien Chhu[2] (+929–88) of the Wu-Yüeh Kingdom, dated 956, 965, and 975.[b] The frontispieces of the three versions are similar though slightly different in design, depicting the prince's consort, née Huang, kneeling before an altar and praying for blessings (Fig. 1115). The technique of engraving is not so refined as in the frontispiece of

---

[a] For the general history of Chinese woodcuts, see Kuo Wei-Chhü (*1*), Wang Po-Min (*1*), and Josef Hejzlar (*1*); for facsimile reproductions of Chinese woodcuts, see Aoyama Arata (*1*), Cheng Chen-To (*6*), *Chung-Kuo Pan Kho Thu Lu* (Peking, 1961), v. 7, and Higuchi Hiroshi (*1*).

[b] Cf. above, pp. 157 ff.

[1] 寶篋印陀羅尼經   [2] 錢俶

the *Diamond Sutra*, but its appearance at the beginning of the *sutra* shows that 84,000 copies of each version were illustrated. Indeed, it seems that all the Buddhist works were printed in great quantities, as the Buddhist religious outlook required, and it is recorded that 140,000 copies of a picture of a Maitreya pagoda were printed by the monk Yen-Shou[1] (+904–75); in addition, 20,000 copies of a Kuan-Yin portrait were printed on silk, and 70,000 copies of the *Fa Chieh Hsin Thu*[2].[a] Although none of these woodcuts survive, the large quantities indicate a significant printing power at this early stage.

Further advances in the art of the woodcut were made during the Sung, Chin, and Yüan periods. Not only were standards of artistic and technical skills improved, but the scope of illustrations was extended from the religious to such secular fields as art, archaeology, scientific works, and Confucian classics, subject-matter being widened to include designs, landscapes, portraits, pictures of daily happenings and amusements, all reflecting the Chinese life of the time.[b] Religious pictures continued, of course, and among the Buddhist *sutras*, a few surviving items include the drawing of a Bodhisattva with eight arms in the *Ta Sui chhiu*[3] *dharani* printed in +980 and found in Tunhuang; a picture of Maitreya seated on a lotus throne under a canopy, painted by an academician artist, Kao Wen-chin,[4] in Yüeh-chou (Shao-hsing, Chekiang), and printed in +984 (Fig. 1168);[c] a life of Wen-Shu or Manjusri, God of Wisdom, printed during the Southern Sung by the Chia family in Lin-an (Hangchow), and frontispieces to different editions of individual works of the *Tripitaka* depicting the Buddha and his disciples; these were printed from +971 onward. The most unusual illustration in the Buddhist texts is a set of four landscape woodcuts from chapter 13 of the *Yü Chih Pi Tsang Chhüan*,[5] one of the imperial prefaces to the Khai-Pao illustrated edition of the *Tripitaka*, printed in +984–91.[d] Showing an affinity to Sung landscape paintings, the woodcut prints are excellently composed and executed with meticulous care (Fig. 1169).

Confucian classics began to be printed in the +10th century, but illustrations were not included until the +12th, when a special edition having illustrations with the text below them, known as *tsuan thu hu chu*,[6] was printed for the use of students preparing for the civil service examinations. Noted works of this type included the *Liu Ching Thu*,[7] an illustrated book on 309 objects which are mentioned in the Six Classics; this was printed in Fukien in +1166, with at least three other editions known to have been printed in the Sung period;[e] the *San Li Thu*,[8] a book on rituals containing illustrations of altars, insignia, costumes, and other ceremonial articles (+1175); and the *Erh Ya Thu*,[9] an illustrated lexicon of classical

---

[a] Chang Hsiu-Min (*14*), p. 76.
[b] For a full discussion of Sung illustration, see K. T. Wu (*9*), pp. 173 ff.
[c] Cf. *Artibus Asiae*, vol. 19, no. 1; Chi Shu-Ying (*3*), pp. 29–30.
[d] See an analytical study of this landscape woodcut by Max Loehr (*1*).
[e] One woodblock edition followed rubbings from a stone carved at Hsing-chou; two others mentioned in the Sung dynastic history are no longer extant.

[1] 延壽   [2] 法界心圖   [3] 大隨求   [4] 高文進   [5] 御製祕藏詮
[6] 纂圖互註   [7] 六經圖   [8] 三禮圖   [9] 爾雅圖

Fig. 1168. Woodcut pictures of Northern Sung, depicting Maitreya seated on the lotus throne, painted by court
artist Kao Wen-Chin and printed in +984. From *Artibus Asiae*, vol. 19, no. 1.

257

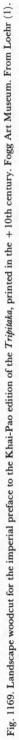

Fig. 1169. Landscape woodcut for the imperial preface to the Khai-Pao edition of the *Tripitaka*, printed in the + 10th century. Fogg Art Museum. From Loehr (1).

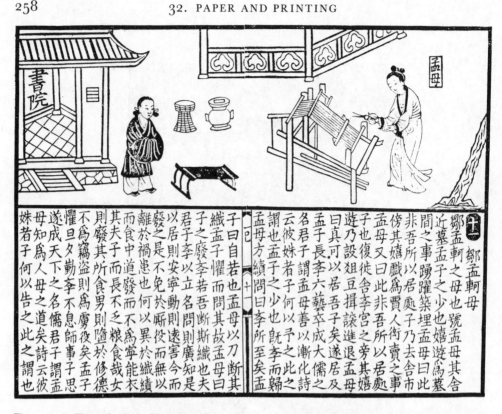

Fig. 1170. An illustration in the *Biographies of Famous Women*, c. +13th century, depicting the story of Mencius who was admonished by his mother to continue study, for stopping study would be like cutting threads from weaving. From *Lieh Nü Chüan*, *KSTK* ed.

terms on various objects and activities. Historical events were another popular subject for illustration, and one of the best works of the kind from this early period is a collection of 123 biographies of famous women, the *Lieh Nü Chuan*,[1] with original illustrations attributed to the noted painter Ku Khai-Chih[2] of the +4th century; it was first printed c. +1063, and reprinted by the Chhin Yu Thang of the Yü family in the latter part of the +13th century (Fig. 1170).

Before the use of photography, all kinds of objects were the subjects of woodcuts. The best-known works of this type include three archaeological catalogues of the Sung period: the *Khao Ku Thu*,[3] which describes bronzes in the imperial and private collections, printed c. +1092; the *Hsü Khao Ku Thu*,[4] which was a supplement to it; and the *Hsüan-Ho Po Ku Thu*,[5] dealing with some 600 bronzes compiled during the Hsüan-ho period (+1119-25). These are important and have frequently been cited as a reliable secondary source for the study of Chinese ancient bronzes.[a]

[a] Cf. Robert Poor (1), p. 33.

[1] 列女傳   [2] 顧愷之   [3] 考古圖   [4] 續考古圖   [5] 宣和博古圖

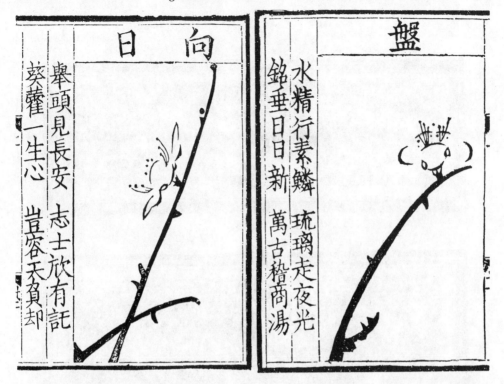

日　向

舉頭見長安
葵藿一生心
志士欣有託
豈容天負卻

盤

水精行素鱗
銘垂日日新
琉璃走夜光
萬古稽商湯

Fig. 1171. An album of flowering plum printed in the Sung dynasty, showing various stages of the blooming plant with caption on the top and a poem at the side for each of the 100 drawings. From *Mei Hua Hsi Shen Phu*, 1261 ed., preserved at Shanghai Museum.

Another notable example was an album on plant life, the *Mei Hua Hsi Shen Phu*[1] (Fig. 1171) containing 100 excellent drawings of the plum in different stages of blossoming; it was printed in 1238 and reprinted in 1261. An album on farm life, *Keng Chih Thu*,[2] includes twenty-one scenes of tilling and rice cultivation, and twenty-four on sericulture, spinning, and weaving; it was perhaps printed first in 1145 and then again in 1237, based on a stone carving of 1210.

In scientific and technical works, illustrations have proved to be even more essential for understanding and interpretation. One outstanding illustrated work on architectural design, *Ying Tsao Fa Shih*,[3] was first published in +1103 as a guide to public construction (Fig. 1172), while a book on astronomy describing the armillary sphere, *Hsin I Hsiang Fa Yao*,[4] contains sixty drawings of the instrument; it was printed in Chhü-chou, Chekiang, in +1127. Two medical works also are well illustrated, one on acupuncture, the *Thung Jen Chen Chiu Ching*,[5] first printed in +1026, the year in which two brass anatomical figures were made by imperial order, and the book was probably the first to contain illustrations of the human anatomy. The other is the celebrated work on *materia medica*, the *Ching Shih Cheng Lei*

¹ 梅花喜神譜   ² 耕織圖   ³ 營造法式   ⁴ 新儀象法要   ⁵ 銅人針灸經

Fig. 1172. Interior decorative designs as illustrated in an architectual work of the Sung dynasty, *Ying Tsao Fa Shih*, printed in the early +12th century. From a facsimile reprint ed., 1925.

Pen Tshao,[1] which includes pictures and descriptions of various kinds of medicines from the mineral, vegetable, and animal kingdoms; first printed in +1108, there were numerous reprints throughout the centuries. Illustrated books were also published on divination, calendars, and other subjects of popular interest, including paper money.[a]

In spite of the alien rule of the Jurchens and Mongols, most printing activities continued, and while the Sung capital, Khaifeng, declined as a printing centre, and Mei-shan in Szechuan was destroyed during the Mongol invasion, Phing-yang (in modern Shansi) developed into one of the major centres of printing from the +12th century onwards. A few surviving examples illustrate the style and format of the woodcuts of this time, the earliest known work from this period being the Thien Ning Ssu[2] edition of the *Tripitaka*, printed in 1148–73, in which a frontispiece precedes each *chüan*. Another example is the revised version of the *materia medica*, *Chhung Hsiu Cheng-Ho Ching Shih Cheng Lei Pei Yung Pen Tshao*,[3] which was cut and printed by Chang Tshun-Hui[4] of Phing-yang in 1249 (Fig. 1121).[b] Another interesting item from this period is a large woodcut on a single sheet, a picture of four beauties, discovered by a Russian expedition in Karakhoto in 1909.[c] It depicts four women famous in Chinese history, and is entitled *Sui Chhao Yao Thiao Chheng Chhing Kuo Chih Fang Jung*[5] (Beautiful Ladies of Successive Dynasties who overturned Empires). It was carved and printed in the Chi family shop in Phing-yang, probably in the +12th century, and its layout, and its cutting and printing techniques represent the height of perfection and harmony. The picture is believed to be one of a pair used most likely for decoration in a private house.

During the Yüan period, woodcuts increased in number in such popular works as textbooks, fiction, and drama. They followed the established format of the earlier period, with about one third of a page devoted to illustrations and two thirds underneath to the text. This arrangement, indicating the function of illustration as a visual aid and an adornment to the text, continued until the 16th century. Illustrated books of this time, known as *chhüan hsiang*,[6] or fully illustrated [editions], were carved by a group of cutters and published mostly by bookshops in Chien-yang. The *Hsin Khan Chhüan Hsiang Chheng Chai Hsiao Ching Chih Chieh*,[7] the newly cut and fully-illustrated edition of the *Book of Filial Piety* with commentaries, was published in Chien-yang in +1308,[d] together with the *Ta Hsüeh*[8] and *Chung Yung*,[9] forming a series of textbooks for beginners. Many historical romances were published in the same manner, such as the *Chhüan Hsiang San Kuo Chih Phing Hua*[10] (Fig.

---

[a] See illustrations of early paper money on p. 97 ff. above.
[b] This edition in thirty *chüan* was reprinted in Peking in 1957.
[c] This picture was published first in *Geibon*, 1916, 7 (no. 2), p. 119, with comments by Ueda Juzō; cf. also Nawa Toshisada (*1*).
[d] Reprinted by Lai Hsün Ko in Peking in 1938.

[1] 經史證類本草   [2] 天寧寺   [3] 重修政和經史證類備用本草
[4] 張存惠   [5] 隨朝窈窕呈傾國之芳容   [6] 全相   [7] 新刊全相成齋孝經直解
[8] 大學   [9] 中庸   [10] 全相三國志平話

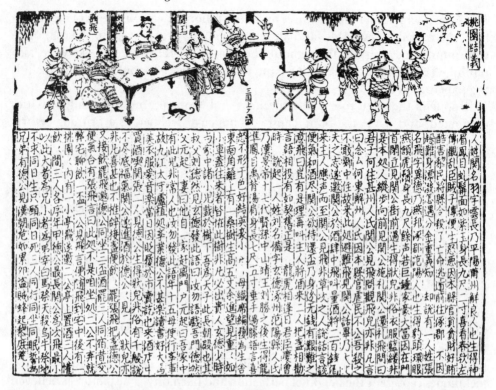

Fig. 1173. Illustration of the *Romance of the Three Kingdoms* with pictures above and text below as the typical format of early illustrated editions. The page above depicts the famous story of Liu Pei, Kuan Yü, and Chang Fei, the three heroes of the +3rd century in the popular novel *San Kuo Chih Yen I*, swearing the oath of brotherhood at Peach Garden. From the *Chhüan Hsiang San Kuo Chih Phing Hua*, facsimile reprint, 1976.

1173), the fully illustrated edition of the *Romance of the Three Kingdoms*, which was one of five such stories published as a series during this period.[a] All the illustrations are of a somewhat simple and pithy style, and this had some influence on early Ming illustrations.

(2) WOODCUTS IN THE MING AND CHHING PERIODS

During the Ming dynasty, especially in the latter part of the 16th and early part of the 17th century, woodcuts formed the greater part of book illustrations and reached their highest degree of excellence in Chinese history. In both quantity and quality, they not only surpassed anything in the past but have never since been equalled. Thousands of such illustrations survive, covering a greater variety of subject matter and representing different schools of format and style, using greatly refined techniques and a highly sophisticated polychrome process.[b] All this was

[a] A reprint of the collection, *Chhüan Hsiang Phing Hua Wu Chung*,[1] was published in Shanghai in 1955.
[b] See separate discussion on pp. 282 ff. below.

[1] 全相平話五種

Fig. 1174. First illustrated edition of the Ming drama *Romance of West Chamber*, printed in +1498, depicting the meeting of scholar Chang and his lover, Ying Ying, and her maid, Hung Niang (on the right) and Hung Niang reporting to Ying Ying's mother (on the left). From *Chung-Kuo Pan Hua Hsüan*, facsimile recut by Jung Pao Chai, Peking in 1958.

accomplished with little official support but was primarily the achievement of private and commercial agencies located in such centres as Nanking, Hui-chou, Hangchow, and Chien-yang in southeast and south China, as well as at Peking in the north. This resulted from the political and economic stability of the country during most of the period, and from the rise of a new audience who were seeking reading materials for pleasure rather than for purely scholarly purposes or religious enthusiasm as in previous times.[a]

The books most frequently illustrated were fiction, drama, poetry and art albums, scientific works, and primers,[b] as well as historical, geographical, and biographical writings. As might be expected, the greatest number of woodcuts were produced for popular literature, and almost every edition of novels, short stories, and dramas carried pictures to illustrate the story; these ranged from a few to as many as forty or fifty in one book, or even over a hundred in some cases. After the first illustrated edition of the famous drama, *Hsi Hsiang Chi*[1], or Romance of the Western Chamber, was published in 1498 (Fig. 1174), no fewer than ten others followed up to the end of the Ming. The earliest edition includes 150 illustrated themes some of which contain as many as eight pictures in sequence for a single theme. If connected, these would make a scroll some two or three feet long. The printer's colophon of this edition says: 'This large-character edition offers a combination of narrative and pictures, so that one may amuse his mind when he is staying in a hotel, travelling in a boat, wandering around, or sitting idle.'[c] Again, in a collection of some 300 classical dramatic texts, over seventy per cent are illustrated,

[a] For a complete study of Ming illustrations, see Shih Hsio-Yen (1).
[b] See the 15th-century *Illustrated Chinese Primer: Hsin Pien Tui Hsiang Ssu Yen*,[2] with introduction and notes by L. Carrington Goodrich (Hong Kong, 1967; reprinted 1976). The primer, which includes 306 drawings to illustrate 388 characters, was originally published in 1436 or earlier.
[c] See the facsimile reprint by the Commercial Press, Peking, 1955.

[1] 西廂記   [2] 新編對相四言

together containing 3800 woodcuts for Ming works.[a] The Fu Chhun Thang bookshop of the Thang family in Nanking alone printed almost 200 illustrations in some ten works of the *chhuan chhi*[1] stories,[b] while such famous novels as the *Hsi Yu Chi*,[2] *San Kuo Chih Yen I*,[3] and *Chin Phing Mei Tzhu Hua*[4] all include illustrations of the story; in addition, the *Shui Hu Chuan*[5] is known to have had at least seven different editions printed with illustrations during the Ming period.

Also illustrated were such collections of poetry as the *Pai Yung Thu Phu*[6], printed in 1597, with illustrations of 100 poems; and the *Thang Shih Hua Phu*[7] which is a combination of poetry and painting printed in 1600. The woodcut is especially suitable for reproduction of works of art and over a dozen such albums were published under the Ming, including the *Kao Sung Hua Phu*,[8] a painting manual of plants and birds by Kao Sung, (1550–4), and the *Ku Shih Hua Phu*,[9] a collection of paintings by famous artists of successive dynasties copied by a court painter, Ku Ping[10] (fl. 1599–1603), that was printed in 1603 (Fig. 1175). Many biographical, historical, and geographical works were also well illustrated. For example, the *Lieh Nü Chuan*[11] is known to have had some half-dozen illustrated editions published between 1587 and 1644; the *Chuang Yüan Thu Khao*,[13] printed in 1607 and 1609, portrays twenty-nine candidates who passed at the top of the civil service exam-inations in 1436–1521, and the *Hsi Hu Yu Lan Chih*[14] with scenes of the West Lake in Hangchow, came out in 1547. Scientific and technical works were especially well illustrated to help explain the text, and included the famous work on agricul-ture and technology, *Thien Kung Khai Wu*[15] (Fig. 1071), printed in about 1637; the *Nung Cheng Chhüan Shu*,[16] a comprehensive treatise on agriculture by Hsü Kuang-Chhi (1562–1633), that appeared in 1639; the *Wu Ching Tsung Yao*,[17] a collection of military classics printed in Chien-an in 1506–21; the *San Tshai Thu Hui*[18] (Fig. 1125), an illustrated encyclopaedia devoted to pictures, maps, charts, and tables (1609); and three editions of the book of *materia medica*, the *Pen Tshao Kang Mu*,[19] printed in 1596, 1603, and 1640.[c]

Among the few known artist-designers of the Ming woodcuts, Chhen Hung-shou[20] (*tzu* Lao-Lien[21], 1599–1652) is the most noted for his creativity in showing the individual personality in human figures. An accomplished figure painter before he earned a living as an illustrator, he is known to have designed five books, the *Chiu Ko Thu*,[22] the *Nine Songs*, composed by Chhü Yüan (*c.* −343 to −277), and printed

---

[a] Based on the titles in the *Ku Pen Hsi Chhü Tshung Khan*,[12] series 1–3, reprinted in Peking in 1954–7, which includes 212 Ming works with illustrations.

[b] See Kuo Wei-Chhü (*1*), pp. 79–80.

[c] Reproductions in facsimile of Ming illustrations are found in Cheng Chen-To (*6*), vols. 2–16, 19–22; *Chung Kuo Pan Kho Thu Lu*, vol. 7; Chhang Pi-Te (*5*); Aoyama Arata (*1*); Nagasawa Kikuya (*9*); Higuchi Hiroshi (*1*), and Tschichold (*3*).

[1] 傳奇	[2] 西遊記	[3] 三國志演義	[4] 金瓶梅詞話	[5] 水滸傳
[6] 百詠圖譜	[7] 唐詩畫譜	[8] 高松畫譜	[9] 顧氏畫譜	[10] 顧炳
[11] 列女傳	[12] 古本戲曲叢刊		[13] 狀元圖考	[14] 西湖遊覽志
[15] 天工開物	[16] 農政全書	[17] 武經總要	[18] 三才圖會	[19] 本草綱目
[20] 陳洪綬	[21] 老蓮	[22] 九歌圖		

Fig. 1175. A landscape painting by the Yüan artist Mi Fu as copied by the Ming court painter Ku Ping in his album and printed in +1603. From *Ku Shih Hua Phu*, facsimile reprint, 1931.

in 1638; the *Hsi Hsiang Chi*[1] and the *Yüan Yang Tsung*,[2] both of which came out in 1638; the *Shui Hu Yeh Tzu*[3] (Fig. 1176), which portrays forty characters of the novel *Water Margin* for playing cards and was designed in 1640; and the *Po Ku Yeh Tzu*,[4] first published in 1653, which includes forty-eight figure designs of historical personages, that were also used for playing cards.[a] His designs for all these works consist of single figures against blank surfaces without any background, showing each personality as described in the literature, and the bold and sharp lines of the bodies and clothing are especially remarkable. The first two works were cut by members of the Huang clan of Hsin-an, and display clear and delicate lines.

Most of the cutters are unknown to us, but the names of a few appear on the wood blocks which they carved. As the profession was highly technical, transmitted perhaps only through tradition, they usually came from certain families or clans, spread sometimes over several generations, and sometimes moving from one place to another. Best known among them are the cutters of the families of Huang, Wang, and Liu of Hsin-an (Hui-chou or She-hsien in modern Anhui), where the best ink and paper were produced. Especially significant was the Huang family, of which more than one hundred members are known to have cut wood blocks, including thirty-one who produced the majority of all the known book illustrations during the Ming period. Their cradle of activities was the village Chhiu-chhuan[5] in Hsin-an; later they migrated to Nanking, Soochow, Hangchow, Peking, and elsewhere as their profession demanded. Because of the special delicacy of their style, which was characterised by fine, soft lines, their works are generally spoken of as the Hui school. Especially noted were Huang Lin[6] (b. 1564), one of the earliest cutters of the family, who produced the famous multi-colour manual of ink-sticks, *Chheng Shih Mo Yüan*[7] (Fig. 1166); Huang Te-Shih[8] (1560–1605), who contributed to the cutting of three archaeological catalogues; Huang Te-Hsin[9] (1574–1658), who cut the collection of Yüan drama, and whose five sons were all cutters; Huang I-Khai[10] (1580–1622) and his brother Huang Ying-Kuang[11] (b. 1592), who migrated to Hangchow and together produced the largest number of illustrated books of popular literature, including the *Chin Phing Mei Tzhu Hua*[12] (Fig. 1177) and several editions of the *Hsi Hsiang Chi*. A total of some fifty titles are known to have been credited to members of the Huang family during the last seven decades of the Ming dynasty.[b]

Generally speaking, woodcuts under the Ming developed gradually to a standard of artistic excellence and maturity. At the beginning of the period, both the technique and the subject matter continued the Sung-Yuan tradition, with cruder lines and composition, used mostly for religious and scholarly literature. Towards the end of the 15th and through the 16th century, the demand for illustrations for

[a] See Huang Yung-Chhüan (*1*).   [b] Chang Hsiu-Min (*13*), pp. 61–5.

[1] 西廂記   [2] 鴛鴦塚   [3] 水滸葉子   [4] 博古葉子   [5] 虬川
[6] 黃鏻   [7] 程氏墨苑   [8] 黃德時   [9] 黃德新   [10] 黃一楷
[11] 黃應光   [12] 金瓶梅詞話

Fig. 1176. Woodcut design for playing cards by Chhen Hung-Shou, + 1640, portraying the hero Sung Chiang in the *Water Margin*. From *Shui Hu Yeh Tzu*, facsimile reprint, Shanghai, 1979.

Fig. 1117. Ladies playing on the rope swing, one of the 100 illustrations of the novel *Chin Phing Mei*, or *Golden Lotus*, cut by members of the Huang family from Hsin-An, Anhui, who resided in Hangchow during their career years. From *Chin Phing Mei Tzhu Hua*, Chhung-Chen ed., *c.* +1628–44, facs. reprint.

popular literature, art albums, and pictures for amusement prompted some complexity and sophistication in design. The four decades from the beginning of the 17th century through to the end of the dynasty in 1644 produced the greatest number of woodcuts, introducing new techniques which led to the greater refinement of the art, with delicate lines, detailed design and composition, and exquisite execution. It was the golden age of woodcuts and book illustration in Chinese graphic history.

Under the Chhing dynasty, the woodcut in book illustration showed less creativity and prosperity than in the Ming. In one sense, the decline resulted from the suppression of certain categories of fiction and drama which had inspired the greatest part of the Ming illustrations. On the other hand, a new horizon of official patronage had developed, and also a new interest in popular woodcuts of New Year pictures.[a] Peking became the centre of official printing, while private and commercial printers continued to produce in other cities in both the north and the south.

During the first part of the Chhing, over a dozen illustrated books were produced at the imperial printing office, Wu Ying Tien, designed and cut by court artists and expert craftsmen. Many of these works were produced to record imperial ceremonies, journeys, military campaigns, or imperial establishments; the records were accompanied by poems composed by the emperors.[b] One of such earlier works was the *Wan Shou Sheng Tien Thu*,[1] which depicts various scenes at the celebration of Emperor Khang-Hsi's sixtieth birthday in 1713, when a procession extended some six miles in Peking. It was painted by the court artist Wang Yüan-chhi[2] (1642–1715) on silk, and later reproduced by the most skilful Chu Kuei[3] in 148 woodcuts which, if joined in a scroll, would be 166 feet long. A similar work recording the events of the celebration of Emperor Chhien-Lung's eightieth birthday in 1791 was printed in 1796, but it was not so well executed as the earlier one. Another pictorial record, known as the *Han Hsün Sheng Tien*[4] (Fig. 1178), was made during the emperor's inspection tours to four provinces in 1751–65. It depicts the most beautiful landscapes along several thousand miles of the journeys.[c]

Imperial gardens, establishments, and collections of ritual articles were other themes for illustrated works, which always included poems composed by the emperors. These included the *Pi Shu Shan Chuang Shih Thu*,[5] depicting scenes of the imperial summer palace in Jehol, printed in 1712, and the *Yüan Ming Yüan Ssu Shih Ching Shih Thu*[6] (Fig. 1179), consisting of forty pictures of the summer palace in a suburb of Peking (1745). Another illustrated work is the *Huang Chhao Li Chhi Thu Shih*,[7] which portrays sacrificial vessels, robes, musical instruments, astronomical apparatus, weapons, insignia, etc., used in the imperial rites of the reigning

---

[a] See separate discussion on pp. 287 ff. below.
[b] Wang Po-Min (1), pp. 139–48; Kuo Wei-Chhü (1), pp. 133–44; Higuchi Hiroshi (1), pp. 24–8.
[c] See an excellent reproduction of forty-eight colour woodcuts of various local sceneries from the 1765 edition and sixteen black-and-white pictures from the 1784 edition in Fuchs (10).

[1] 萬壽盛典圖   [2] 王原祁   [3] 朱圭   [4] 南巡盛典   [5] 避暑山莊詩圖
[6] 圓明圓四十景詩圖   [7] 皇朝禮器圖式

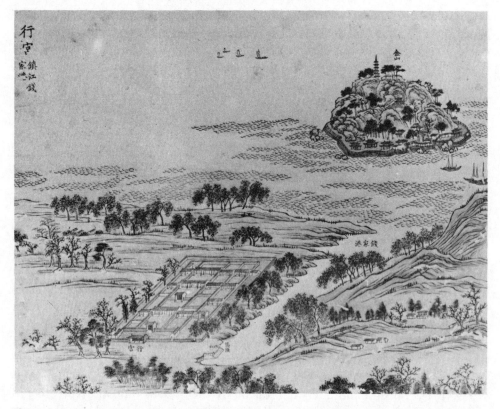

Fig. 1178. Temporary palace of Emperor Chhien-Lung near the Golden Mountain at Chenchiang, Chiangsu during his inspection tour to the South, printed in colour in +1766. From *Nan Hsün Shêng Tien*, reproduced in Fuchs (10).

dynasty; this was printed in 1759 and revised in 1766. Portraits of meritorious persons were preserved in such works as the *Phing Ting Thai Wan San Shih Erh Kung Chhen Thu Hsiang*,[1] which contains some thirty-two portraits of military heroes of the conquest of Taiwan in 1683. Non-Chinese persons are portrayed in the *Huang Chhing Chih Kung Thu*,[2] (1751) and depicts some 600 figures and their costumes from the native tribes of southwest China to the peoples of distant European countries, based on reports and observations by Chinese envoys who had been sent abroad.

Among noted works on agriculture and other subjects, the most famous is probably the *Yü Chih Keng Chih Thu*,[3] which was based on the Sung version but includes twenty-three pictures each of tilling and weaving, designed by the court painter Chiao Ping-Chen,[4] with one additional poem for each subject on the top margin of the picture by the Emperor Khang-Hsi (Fig. 1180) and others. It was printed in colour in 1696, carved later on stone, and reproduced in woodcuts by Chu Kuei in 1712. Other similar works include the *Mien Hua Thu*[5] on the culture

[1] 平定台灣三十二功臣圖象   [2] 皇清職貢圖   [3] 御製耕織圖   [4] 焦秉貞
[5] 棉花圖

Fig. 1179. Part of the scenery of the imperial garden, Yüan Ming Yüan, printed in +1745. From *Yüan Ming Yüan Shih Shih Ching Shih Thu*, reproduced in Cheng Chen-To (*1*).

and processing of cotton, printed in 1765; the *Shou Shih Thung Khao*[1] (1742), primarily based on an earlier work on agriculture by Hsü Kuang-chhi; the *Wu Ying Tien Chü Chen Pan Chheng Shih*[2] (1776), a manual for the various stages of making movable types and printing (Fig. 1143); and the grand encyclopaedia, *Thu Shu Chi Chheng*,[3] which includes thousands of illustrations in its various sections, even though it was printed with bronze movable type in 1728 (Fig. 1147).

At this time, Western art was influencing that of the Chinese court due to the presence of Roman Catholic priests who were serving in various capacities at the imperial palace. Thus when the *Phing Ting I-Li Hui Pu Chan Thu*,[4] with paintings of sixteen scenes from battles and memorable events in the conquests of Ili and

[1] 授時通考   [2] 武英殿聚珍版程式   [3] 圖書集成
[4] 平定伊犁回部戰圖

Fig. 1180. An album on tilling and weaving, redrawn under imperial suspices in +1712, here depicting rice planting with poem from original Sung edition in the upper right corner and an additional poem by Emperor Khang-Hsi in his own calligraphy in the top margin. From *Yü Chih Keng Chih Thu*, +1712 ed., British Library.

Chinese Turkestan, was completed in 1766, it was sent to Paris by imperial order to be engraved on thirty-six copper plates, which were completed in 1774.[a]

The official interest in printing book illustrations and the demand for excellence promoted production by private and commercial printers. Although the pre-

[a] Cf. Pelliot (63), pp. 183 ff.; Ishida Mikinosuke (1).

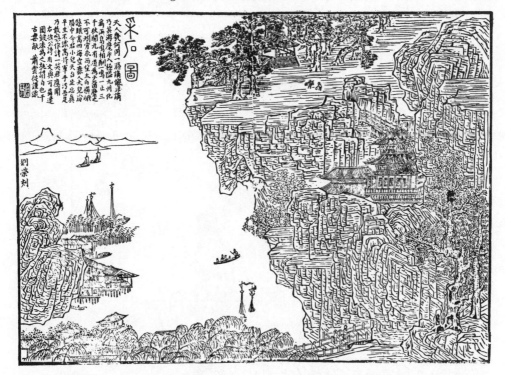

Fig. 1181. Scenic mountains and rivers of Thai-Phing, Anhui, painted by Hsiao Yün-Tshung and cut by Liu Yung in +1648. From *Chung-Kuo Pan Kho Thu Lu*, pl. 711.

dominance of many famous Ming family publishers had gradually disappeared, a few noted designers collaborating with skilful cutters were able to contribute to give an excellence to early Chhing woodcuts, especially those illustrating landscapes and human figures. Most noted among these artists was Hsiao Yün-Tshung[1] (1596–1673), whose design of pictures was most outstanding. He painted forty-three scenes of the landscape of Thai-phing (in modern Anhui), *Thai-Phing Shan Shui Thu Hua*[2] (Fig. 1181), at the request of an official of that place. These were engraved on wood by a number of skilful cutters in 1648 to accompany poems and essays, and consisted of panoramic views of the scenic mountains and rivers, all with very elaborate and delicate carving.[a] Another of his works is an illustration with figures of Chhü Yüan's *Elegy on Encountering Sorrows*, *Li Sao Thu*[3], printed in 1645 (Fig. 1182). Inspired by the expressive style of Chhen Hung-Shou and cut by some of the same cutters as the previous work, it shows both vivid facial expressions and dynamic flowing lines of garments.

Other important illustrations with figures include the *Ling-Yen-Ko Kung Chhen*

---

[a] Reproduced in Cheng Chen-To (*1*), vol. 16.

[1] 蕭雲從   [2] 太平山水圖畫   [3] 離騷圖

Fig. 1182. Chhü Yüan, the famous minister of the Chhu state and author of the *Li Sao*, *c.* −4th century, encountered a fisherman at the bank of Milo River before he drowned himself there, as illustrated in the *Li Sao Thu*, printed in +1645.

Thu Hsiang,[1] printed in 1668, which portrays twenty-four famous officials, scholars, poets, and artists of successive dynasties; and the *Wu Shuang Phu*,[2] printed in 1690, an album of forty unique personalities of Chinese history. Both of these works and two of the palace editions, the *Keng Chih Thu* and *Pi Shu Shan Chuang Shih Thu*, were cut by Chu Kuei, who was probably the most distinguished craftsmen of the early Chhing and who contributed so much to the excellence of the work of this period. Another important book is the *Wan Hsiao Thang Hua Chuan*,[3] printed in 1743, illustrating 120 persons noted in history from Han to Ming times. The most distinguished work of multi-colour woodcuts was the painting manual of the Mustard Seed Garden, which will be discussed in the following section.

During the second part of the Chhing period, after 1800, nothing particularly significant appeared, except for some illustrated works of a practical nature, archaeological repertories, local scenery, fiction, and other miscellaneous materials. However, among scientific and technical works, there was an illustrated book on botany, *Chih Wu Ming Shih Thu Khao*[4], printed in 1848 after the death of its author, Wu Chhi-Chün[5] (1789–1847); this included drawings of some 1714 specimens of plants, flowers, and fruits, based on previous records as well as the author's own collections and observations during his journeys throughout the country. Another illustrated book worthy of note is the *Ho Kung Chhi Chü Thu Shuo*,[6] on water conservancy, printed in 1836; this recorded the tools and materials for dyke building and river management. Of archaeological works, there were such illustrated catalogues as the *Chi-Ku-Chai Chung Ting I Chhi Khuan Shih*[7] on ancient bronze inscriptions, printed in 1804; and the *Ku Yü Thu Khao*[8] on ancient jade (1889).

A number of novels and short stories produced in the Chhing period were also illustrated. Especially notable is the *Dream of the Red Chamber*, of which the earliest edition, of 1791, includes some twenty or more pictures of heroes and heroines of the story, though the best is probably the *Hung Lou Meng Thu Yung*[9] (Fig. 1183), with fifty portraits of characters painted by Kai-Chhi[10] (1774–1829); it is elegantly designed and well executed. Another noted figure painter who should be mentioned is Jen Hsiung[11] (*tzu* Wei-Chhang,[12] *c.* 1815–57); he produced three biographical works on knights-errant, scholars, and hermits, printed in 1856–8, and designed a set of wine-game cards, printed in 1854. All his work is vivid and forceful, displaying the influence of the Ming artist Chhen Hung-Shou.

Towards the end of the Chhing dynasty Western printing techniques and facilities were introduced and gradually replaced the time-honoured art of woodcut illustration. Quite a few pictorial works, especially those from the Tien Shih Chai,[13] were printed by lithography and other photo-mechanical processes. However, the art of the woodcut has survived to modern times. Certainly, there is a

---

[1] 凌煙閣功臣圖像   [2] 無雙譜   [3] 晚笑堂畫傳   [4] 植物名實圖考
[5] 吳其濬   [6] 河工器具圖說   [7] 積古齋鐘鼎彝器欵式
[8] 古玉圖考   [9] 紅樓夢圖詠   [10] 改琦   [11] 任熊   [12] 渭長
[13] 點石齋

Fig. 1183. *Dream of Red Chamber* heroine Tai Yü drawn by Kai Chhi, *c.* +1884.
From *Hung Lou Meng Thu Yung*.

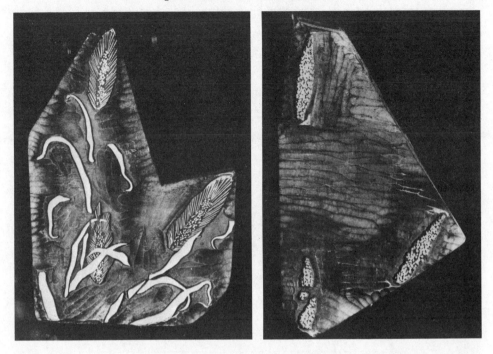

Fig. 1184. Two woodblocks bearing wheat design from a set of four, each used with a different coloured ink in printing stationery. Dard Hunter Paper Museum.

school influenced by Western techniques and styles of print-making, but a traditional school still continues and has been revitalised, especially by a demand for multi-colour woodblock printing for the reproduction of works of art, stationery, and New Year pictures.

(3) DEVELOPMENT OF MULTI-COLOUR PRINTING

Chinese multi-colour woodblock printing, known as *thao pan*[1] (set of blocks) or *tou pan*[2] (assembled blocks), was produced by a set of separate blocks (Fig. 1184), each of which was registered in position and printed in succession on the paper using a water-based ink in different colours. The number of blocks in a set varies from a few to several dozens or more, depending upon the variety of colours and tones printed (Fig. 1185). This polychrome process was used for printing text with punctuation and commentaries, for cartographic works, paper money, book illustration, ornamental letter-papers, New Year pictures, and works on painting, calligraphy, and the decorative arts.

Reproduction of works of art is usually done with the same kind of ink, colour, and paper as those used for the original. To some extent, the exactness of wood-

[1] 套板   [2] 餖板

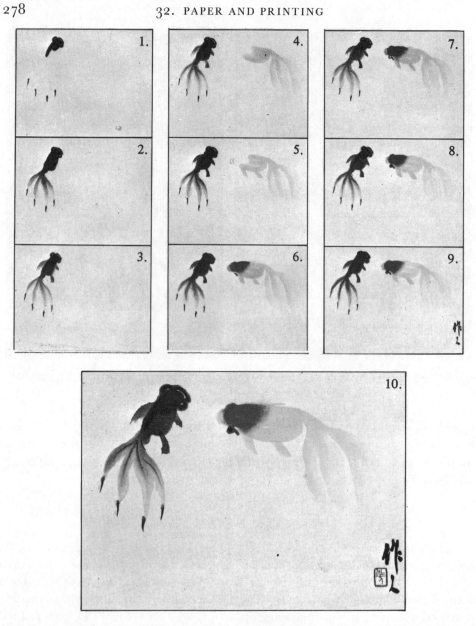

Fig. 1185. Ten steps for making a multicolour blockprint as exemplified by the picture of goldfish. Courtesy of Jung Pao Chai, Peking.

block prints cannot be equalled even by the modern photo-mechanical processes, because in photo-engraving, the fine reticulated lines do not express the exact texture and spirit of the original brushwork. Moreover, offset prints cannot reproduce the rich gradations and tones of Chinese ink painting, while the oil-based ink is unable to produce the same effect as the water colours of the original; in addition a background shadow produced by the photographic method is often observed.

The colour prints from woodblocks require considerable skill and expertise to

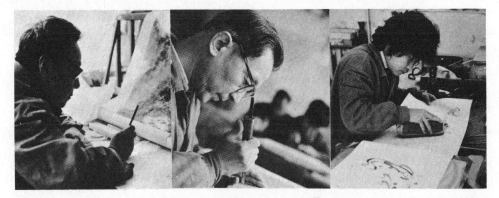

Fig. 1186. Processes of tracing, cutting, and printing for making colour blockprints at Jung Pao Chai, Peking.

master the various steps in designing, engraving, registering, and printing. Although no record is left to tell how the old colour prints were made, the process is believed to have been very similar to that used today.[a] For reproduction of colour painting, the first step is to study and analyse the colours used in the original. Separate outline copies of each basic colour are then traced on thin transparent paper, which is then stuck upside down on the smooth surface of the woodblock using a rice paste. After drying, the back layer of the paper is rubbed off and the block is ready for cutting.[b] As the lines and coloured areas must reproduce the original work exactly, the original is always kept beside the cutter and the printer (Fig. 1186).

The paper used for Chinese colour prints is usually the whitish, smooth, absorbent *Hsüan-chih* which is used by all Chinese artists for painting and calligraphy. The inks are the same water colours, most of which are earth pigments mixed with peach-tree resin or hide glue and water. These are mixed as they were for the originals so they produce the exact colours after drying. The worktable is made of two wooden boards placed to leave a slit between them (Fig. 1187). On the left side the engraved block is firmly fixed with pitch or wax to the table, with brushes and inks to hand. On the right side, sheets of paper are firmly held together under a clamp. When the block is inked, the printer must see that no colour runs beyond its proper boundaries. A sheet of paper is then laid upon the inked block and softly brushed over. Different pressure is applied to different parts of the block, depending upon the expression and texture needed for each stroke. Sometimes certain colours have to be printed first and dried before others are applied, and sometimes later printing must be done while the earlier colours are still wet. Gradation is achieved by applying varying degrees of colour from light to dark repeatedly from the same block, either by causing the ink to run on the block with a special brush, or by wiping away the ink at the desired place. In this way, an exact copy is produced

[a] The following description is based primarily on interviews with craftsmen of the Jung Pao Chai in Peking and To Yün Hsien in Shanghai in 1979; see also Yeh Sheng-Thao (1), pp. 27–8; Tschichold (3), pp. 41–4.
[b] Cf. above, pp. 197 ff.

Fig. 1187. Work table used for multicolour block printing at Jung Pao Chai, Peking.

which sometimes cannot be distinguished from the original. As a noted typographer has said: 'There is hardly another graphic art in the world that depends so entirely on the artistic sympathy and understanding of the printer as does the Chinese colour print.'[a]

This colour process apparently evolved from one in which prints with black outlines were coloured at first by hand and later by applying various colours to different parts of the same block. When separate blocks were used for different colours with gradations of tone, the technique became a highly sophisticated and refined art of printing. Several of the earliest examples of prints coloured by hand survive. A picture of Kuan-yin dated +947 and a few similar sheets with six colours were found in Tunhuang,[b] and paper money issued in +1107 was printed with legends in black, a circle design in vermilion, and 'blue face' in indigo, as a precaution against counterfeiting.[c]

One of the earliest examples of colour printing extant is perhaps a single sheet of woodcut recently discovered in Sian. It depicts the legend of Tungfang Shuo[1] (b. −106), a humorous official at the Han court who is said to have stolen the peaches of immortality from the Queen Mother of the West. The picture is attributed to the Thang artist Wu Tao-Tzu[2] (d. +792) and printed in black, grey, and green with a seal in red, possibly by a commercial printer at Phing-yang under the Jurchens in the early 12th century (Fig. 1188).[d] This piece is believed to have been used for

[a] Tschichold (3), p. 41.
[b] Specimens of early coloured prints are kept in the British Museum and in Musée Guimet, Paris.
[c] See description in *Shu Chung Kuang Chi* (*SKCS*), ch. 67, pp. 18*a*–23*b*.
[d] This print was found, together with a rubbing of calligraphy, some fragments of Jurchen documents, and fifty-eight iron and copper coins of the Sung and Chin period of which the latest dated to +1158, inside the cavity of a pillar for the stone tablet at the Forest of Steles (Pei Lin[3]), Sian in 1973 when the tablet was repaired; see a report in *WWTK*, 1979 (no. 5), pp. 3–4, plate 2.

[1] 東方朔   [2] 吳道子   [3] 碑林

Fig. 1188. Earliest extant multicolour printing, *c.* early + 12th century, depicting the legend of Tungfang Shuo stealing the peaches of immortality from the Queen Mother of the West. From *WWTK*, 1979, no. 5.

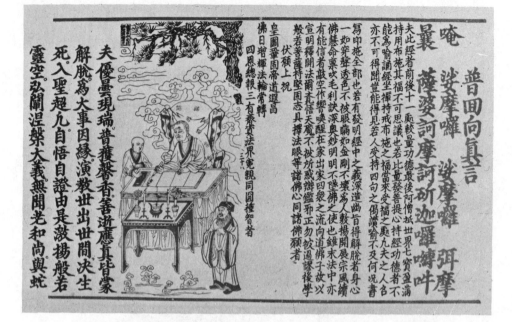

Fig. 1189. *Diamond sutra* printed in colour, +1340, with prayer in big characters, illustration in red and text in black. National Central Library, Thaipei.

house decoration or as one of the New Year pictures with a theme popular then.

Texts printed with colour commentaries date back to the early part of the 14th century, although the actual beginning must have been earlier.[a] One of the surviving examples is an edition of the *Diamond Sutra*, printed at the Tzu-fu Temple of Chung-hsin Circuit in 1340, using black for text and red for prayers and a picture of the *ling tzu* plant (Fig. 1189).

The multi-colour technique was further developed toward the end of the 16th century or around the turn of the 17th, when the Min and Ling families and other printers of the Chiangsu-Chekiang region printed hundreds of classical works, illustrated novels, dramatic texts, and medical writings in two to five colours. The most noted among them was Min Chhi-Chi[1]; he and others printed no less than one hundred titles of such a kind of work during this time.[b] One edition of the *Shih Shuo Hsin Yü*[2], a collection of short stories compiled in the +5th century, was printed in 1581 by Ling Ying-Chhu[3] in blue, red, and yellow. In the next century, in 1606, an album of ink-cake designs, *Chheng Shih Mo Yüan*[4] by the famous ink-maker

[a] A number of manuscript books with punctuation and comments in colour are listed in the bibliographies of the Three Kingdoms and Sui periods in the +3rd to early +7th century.

[b] See facsimile reprint of the *Hsi Hsiang Chi*[5], printed in five colours by Min Chhi-Chi in 1640, by Museum für Ostasiatische Kunst, Köln, 1977 with text by Edith Dittrich (1); Thao Hsiang (1) lists 110 titles in 132 works printed by the Min and other families in c. 1600–40.

[1] 閔齊伋   [2] 世說新語   [3] 凌瀛初   [4] 程氏墨苑   [5] 西廂記

Chheng Ta-Yüeh[1] (1541–1616) of Hui-chou, was printed in five colours. For the first time in Chinese woodcuts it incorporated some Western engraved designs from the Bible, and a romanised form of Chinese apparently supplied by Matteo Ricci (Fig. 1166).[a] At about the same time, numerous erotic picture albums in multi-colour were also printed, including the popular *Feng Liu Chüeh Chhang Thu*[2], pictures of the gay life and excessive pleasure, made by the famous cutter Huang I-Min[3] of Hsin-an, and printed with five colours in 1606.[b] The earliest extant atlas printed in colours is probably the *Chin Ku Yü Ti Thu*,[4] of 1643, a collection of sixty maps showing geographical areas in contemporary and ancient times. Place names and boundaries of the Ming period are shown in black, while those of antiquity, and the annotations, appear in red.[c] All these works, however, were printed with solid colours without gradations.

This polychrome technique was advanced to the highest degree of perfection in the first part of the 17th century, when many collections of painting, calligraphy, and ornamental letter-papers were produced with even more sophisticated processes. Unlike the previous illustrations with black outlines, the new technique was characterised by applying colours to the blocks without outlines but with a variety of graded tones. This development reached its peak with the *Shih Chu Chai Shu Hua Phu*,[5] a manual of calligraphy and painting from the Ten Bamboo Studio (Fig. 1190), and the *Shih Chu Chai Chien Phu*,[6] a collection of ornamental letter-paper designs from the same studio (Fig. 1191) both printed in five colours with graded tones by Hu Cheng-Yen in Nanking. The first of these two monumental works appeared in 1619–33, and the second in 1644 or 1645.

Hu Cheng-Yen[7] (*c.* 1582–1672) was a native of Hsiu-ning (Hui-chou) in modern Anhui, a centre for fine paper, ink, and printing for many centuries. He later lived in Nanking, to which many prominent cutters and printers moved from the not far distant Hsiu-ning. Hu was a scholar and physician by profession, but became a seal carver, painter, calligrapher, letter-paper designer, and printer of artistic talent, and his collections included a number of paintings and examples of calligraphy from his own hand as well as by some thirty other contemporary artists.[d]

His painting manual consisted of about 180 pictorial prints and 140 poems in calligraphy, which were grouped under eight categories, including birds, fruits, orchids, bamboo, plum blossoms, rocks, paintings in circular fan shape, and miscellaneous paintings and calligraphy. Each subject category contained about forty examples of painting and calligraphy, each printed on one double-page leaf.

---

[a] K. T. Wu (6), pp. 204–6; Pelliot (28), p. 1.
[b] This and other erotic albums are described in van Gulik (11).
[c] A few copies of this atlas are known to have survived in the Peking Library, the Bibliothèque Nationale, and the Library of Congress.
[d] Hsiang Ta (6), pp. 39–42.

[1] 程大約   [2] 風流絕暢圖   [3] 黃一民   [4] 今古輿地圖   [5] 十竹齋書畫譜
[6] 箋譜   [7] 胡正言

Fig. 1190. A multicolour blockprint of the Ten Bamboo Studio by Hu Cheng-Yen printed in *c.* +1627. From *Shih-Chu-Chai Shu Hua Phu*, reproduced by Tschichold (3).

The categorisation and selection of the pictures showed for the first time in Chinese art a systematic approach to the study of painting and calligraphy.

Before the complete edition of the *Shih Chu Chai Shu Hua Phu* was issued in 1633, a few advance editions are known to have been printed. These included one printed in 1622, of seventeen pictures of bamboos; one of unknown date on the four noble plants (plum blossom, orchid, bamboo, and chrysanthemum), with eight pictures; and one on birds printed in 1627, with ten pictures. The *Shu Hua Phu* was apparently a collection of previous works plus new ones printed in or before 1633. The earlier editions of this work are extremely rare.[a]

Hu's collection on ornamental letter papers, the *Shih Chu Chai Chien Phu*, consists of designs of rocks, ritual vessels, landscapes, human figures, and plants and

---

[a] An excellent facsimile reproduction of the first edition of this manual has been made with colour offset by Tschichold (3).

Fig. 1191. Letter-paper design from the Ten Bamboo Studio, printed with colours and gradation.
From *Shih Chu Chai Chien Phu*, facsimile ed. by Jung Pao Chai, 1934.

flowers. Some of these designs were made, with or without colour, from embossed blocks, another Chinese printing invention.[a] The process, known as *kung pan*[1] (embossed blocks) or *kung hua*[2] (embossed designs), employed either simple pressure on paper to produce the effect of relief from an engraved block, or a true embossing through pressure on paper placed between negative and positive blocks. One expert believes it was a process of true embossing with a negative cut from short-grained wood;[b] it is not true, however, that this technique was Hu's creation or that his collection of writing-paper designs was the first of its kind. At least two other collections of such stationery are known to have been published around or before this time. One is the *Lo Hsüan Pien Ku Chien Phu*[3] (Collection of Letter Papers with Antique and New Designs from the Wisteria Pavilion), compiled by Wu Fa-Hsiang[4] (*hao* Lo-hsüan, b. 1578), also from Nanking, and printed in 1626,[c] nineteen years earlier than the one from the Ten Bamboo Studio. The other is the *Yin Shih Chien Phu*[5] (Collection of letter papers from the Yin family), which includes embossed designs and is believed to have been printed at about the same time.[d]

The manual from the Ten Bamboo Studio was instrumental in the publication of later works of colour prints. Especially noted is the *Chieh Tzu Yüan Hua Chuan*[6] (Painting Manual of the Mustard Seed Garden) (Fig. 1192), which was even more influential than its predecessor and has enjoyed prestige as a model for beginners in brush work for the last three centuries. The Mustard Seed Garden was built in Nanking by Li Yü[7] (1611–80?), a playwright and prolific writer on many subjects, who printed a number of his own books and others under this name. Although he wrote a preface to the first series of this manual in 1679, he apparently was a sponsor but not the author. It is generally agreed that the first three series of this work, parts 2 and 3 of which were published in 1701, were prepared by his son-in-law, Shen Yin-Yu,[8] and illustrated by Wang Kai[9] and his two brothers Wang Shih[10] and Wang Nieh;[11] the fourth series was added by others in 1818, This work is a step-by-step instruction book on how to do paintings of landscapes (series 1); of plum blossom, orchid, bamboo, and chrysanthemum (series 2); of birds, insects, and flowers (series 3); and of human figures (series 4). This work has been widely

[a] Embossed paper first appeared in Europe after the middle of the 18th century, in Germany, and was patented in England in 1796; see Stoff (1).

[b] Tschichold (3), p. 33.

[c] The authorship of this work was mistakenly ascribed to Weng Sung-Nien 1647–1723, whose *hao* was also Lo-hsüan, when the second volume of an incomplete set was reproduced in the *Zuhon Sōkan*, compiled by Omura Seigai and printed in Tokyo in 1923. The first volume of the complete set discovered in Shanghai in 1964 contains a preface revealing the correct compiler and dating. The subject-matter and embossed designs of this work are similar to those of the Ten Bamboo Studio, but it uses outlines for colour and less gradation of tones; see *WWTK*, 1964, no. 7, pp. 7–9; also pp. 262 ff. above.

[d] This work survives in Japan, and two designs, one of which is embossed, are illustrated in Aoyama Arata (1), plate 10; also Nagasawa Kikuya (3), plate 102.

[1] 拱板   [2] 拱花   [3] 蘿軒變古箋譜   [4] 吳發祥
[5] 殷氏箋譜   [6] 芥子園畫傳   [7] 李漁   [8] 沈因友   [9] 王概
[10] 王蓍   [11] 王臬

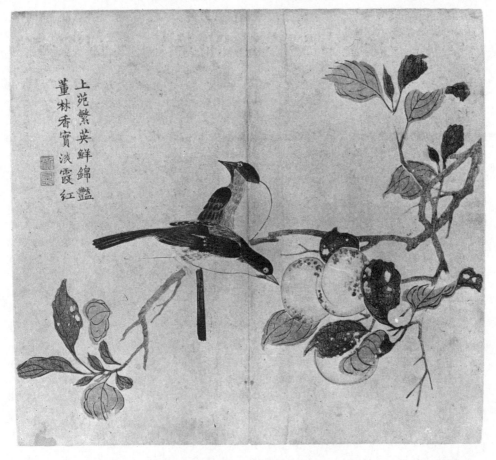

Fig. 1192. Birds on fruit tree as illustrated in the *Painting Manual of the Mustard Seed Garden*, printed in multicolour, c. +18th century. From Strehlneek (1).

circulated in China and Japan, with numerous editions reprinted in both countries and translation into different languages.[a]

(4) POPULARITY OF NEW YEAR PICTURES

No other significant works of colour prints are known to have been produced since the publication of the painting manual from the Mustard Seed Garden, except perhaps for some minor collections of ornamental letter-paper designs and for the numerous New Year pictures known as *nien hua*.[5] The New Year festival was the

---

[a] Over two score reprints were produced in China and Japan between the first publication in 1679 and 1937. For editions see A. K. Chiu (1), pp. 55–69; for translations see Petrucci (1) and Sze Mai-Mai (1); for reproduction of pictures see Tschichold (5).

[1] 年畫

most important event of the year, and was celebrated throughout the country as an extended holiday when people conclude the old year and look forward with new hopes and expectations for the coming one. Chinese in all walks of life liked to decorate their houses with pictures in bright colours and gay in content, either hand-coloured over black outlines or, more often, printed by a multi-colour process. The origin of *nien hua* may be traced back to the use of illustrated seasonal calendars, door gods to protect the house, and certain family scenes in paintings, all of which appeared in the Thang or earlier. Motifs and content continued to develop through the centuries, and this folk art increased in popularity when the technique of multi-colour woodcuts was widely applied to printing in the latter part of the Ming dynasty.[a]

New Year pictures fall into a wide variety of categories. Most popular are those reflecting human desires for happiness, prosperity, and longevity, symbolised by such figures as the bat, fish, peach, lotus, pomegranate, and peony. Gods of folk religion, heroes and heroines, landscapes and scenery, family life and children, farming and weaving, and many other topics were represented. One of the earliest extant examples is a delicately printed picture of Shou Hsing[1] (God of Longevity) (Fig. 1193) made in Soochow in 1597.[b] Soochow, a commercial and industrial city in the southeast of China that was most prosperous during the early Chhing period, was one of two major centres for the production and distribution of New Year pictures. Here, in a street named Thao Hua Wu[2], where a villa of this name had been built by an artist in 1505, a scenic area was formed during the following centuries, and this contained over fifty printing shops by the middle of the 19th century. Such pictures as the Wan Nien Chhiao,[3] or Everlasting Bridge (Fig. 1194), and the Chhang-men[4] City Gate, both printed in 1740, reflected the prosperous scenes of Soochow during this time.[c]

Another major centre was located in Yang-liu-chhing,[5] a town near Tientsin, where many workshops were established from the end of the 16th century on. It had become the largest production and distribution centre by the early part of the 17th century, when annual production reached twenty million copies. One single shop with several hundred cutters and printers produced over a million pictures a year, and over sixty workshops were still operating in the middle of the 19th century.[d] Two editions were made, one in the spring and autumn of each year. The spring edition was produced at great leisure, with more variety in designs and colours and in more refined style. The autumn edition, prepared in haste to meet

[a] For the history of the development see A-Ying (2); Kuo Wei-Chhü (1), pp. 182–217; Josef Hejzlar (1), pp. 48–51; Pommeranz (1).
[b] Reproduced in Aoyama Arata (1), plate 6.
[c] Most of the pictures from Soochow survive in Japan, and forty-seven selected samples are reproduced in Aoyama Arata (1), plates 15–67; see also A-Ying (2).
[d] A-Ying (2), p. 272.

[1] 壽星   [2] 桃花塢   [3] 萬年橋   [4] 閶門   [5] 楊柳青

Fig. 1193. One of the earliest extant New Year pictures, depicting the God of Longevity, printed in
Soochow in +1597. From Aoyama Arata (1), pl. 6.

Fig. 1194. Everlasting Bridge in Soochow as theme of a New Year picture printed in Soochow in +1740.
From Aoyama Arata (1), pl. 16.

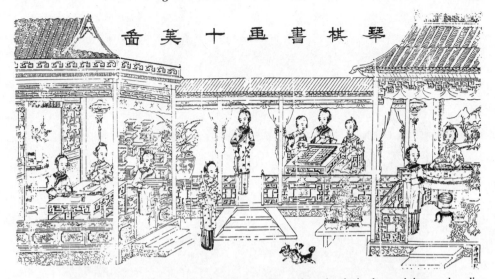

Fig. 1195. A New Year picture from Yang-Liu-Chhing depicting ten beauties playing lute and chess, and reading and painting, printed by the Tai Lien Tseng Picture Shop established from the +18th century. From Chang Ying-Hsüeh (1).

the immediate needs of customers, was more robust in style.[a] With the increasing demand for such pictures, many sub-centres were developed throughout the country, especially Yang-chia-wu in Shantung, Paoting in Hopei, Chu-hsien-chen in Honan, Yangchow in Chiangsu, and Fo-shan in Kuangtung. They became satellites of the major centres in the north and south, following the general style and pattern of their respective parent schools.

Generally speaking, the northern school inherited the techniques of woodblock printing of Phing-yang and Peking and was generally influenced by the style and subject-matter of traditional Chinese painting, especially those of the court painters. These themes put more emphasis on folk gods, women and children, scenes from the stage, and historical romances (Fig. 1195). The style of cutting was sharp and hard, and the work was more decorative with rich colours. On the other hand, the Soochow school followed the techniques of colour prints developed in the neighbouring areas during the Ming dynasty. The designs were more delicate and elegant, with subtler execution, rich decoration, and warm tones. The themes included such familiar scenes as children playing with toys or New Year decorations often seen in scroll painting. Some Western influence in both technique and content is noticeable in the pictures produced in the Chhing period. Both perspective and chiaroscuro are used, and such new scenes as a European opera house (Fig. 1196) or a locomotive were introduced as decorative art in Chinese homes. As a whole, both techniques and contents reflect the life styles and tastes of the people in the northern and southern parts of the country.

[a] See selected samples in Chang Ying-Hsüeh (1).

Fig. 1196. European opera house as theme of a New Year picture, c. +18th century. from Aoyama Arata (1), pl. 27.

(h) SPREAD OF PAPER AND PRINTING TO THE WEST

(1) DEVELOPMENT OF WESTERN KNOWLEDGE OF CHINESE PAPER

PAPER was introduced to Europe in the middle of the +10th century, manufactured there from the +12th century, and used for printing from the middle of the +15th century.[a] It was a rag paper, which was assumed at the time that it was invented by Arabs or by Europeans.[b] The fact of its true origin in China and gradual spread to Europe was not well established until around the turn of the present century. This slow recognition was due partly to the indirect transmission of the invention from China to the West through an intermediary, and partly to confusion about the nature of the material. The relationship between true paper and papyrus has been questioned from time to time, and belief in the Western origin of paper was discarded only recently.[c]

That the Chinese used paper was unknown to Europeans for three centuries after it had already been introduced to the West. Only from the beginning of the 13th century, when European travellers visited the Eastern parts of the world, did they observe the use of paper money in China, although their primary interest was in money and not in paper. The earliest report of paper money was made by William Ruysbroeck, a missionary who was sent by the king of France to the Mongol capital, Karakorum, in 1253–4. After his return to France, he mentioned in 1255 that the Chinese used a common money made of pieces of cotton paper for their business transactions;[d] the use of paper as a medium of exchange was probably not known in Europe prior to this. Ruysbroeck's report was soon adopted by Roger Bacon, who mentioned in his *Opus majus* (c. 1266) 'a card of mulberry tree on which are stamped certain lines.'[e] However, a more detailed and direct observation about paper money was made by Marco Polo during his travels to the East from 1275 to 1295. He described very briefly the use of mulberry tree bark for papermaking, and went into great detail on the processes of making paper money and the systems of circulating it, using it in transactions, and replacing it when worn out.[f]

Other pre-Renaissance writers of the 14th and early 15th centuries who gave similar accounts of the use of paper money in China were the prince of Armenia, Hayton (1307), the archbishop of Soltania (c. 1330), the Dominican John de Cora (c. 1330), the Franciscan friar Oderic of Pordenone (c. 1331), the Florentine merchant Francisco Balducci Pegolotti (1310–40), and the Venetian emissary Josafat Barbaro (1436).[g] All these early reports primarily expressed surprise that

---

[a] The earliest dates for the use of paper in Europe are given as 950 in Spain and 1102 in Sicily, and for its manufacture 1150 in Xátiva, Spain, and 1276 in Fabriano, Italy; see Hunter (9), pp. 470–4; also pp. 298 ff. below.

[b] See C. G. Gilroy (1), p. 404; Hoernle (1), pp. 663–4.

[c] For the confusion between paper and papyrus and for the theories on the Western origin of paper, see Tsien (2), pp. 140–2.

[d] Cf. Carter (1), p. 115, n. 21; Lach (5), I, p. 34.

[e] See R. Bacon (1), tr. B Burke (Oxford, 1928), I, p. 387.

[f] For the translation of a long chapter on paper money, see Yule (1), I, pp. 423–6.

[g] For discussions and quotations of these writers, see Carter (1), p. 115, n. 21; Lach (5), I, pp. 40–6.

the cheapest stuff could be substituted for the most valuable material, but not much information about paper itself or its origin in China was revealed by these writers. Thus the Renaissance historian Polydore Vergilius (d. 1555) mentioned paper made of linen cloth in his work on inventions, first published in Venice in 1499, but he did not say who invented it or where.[a]

During the latter part of the 16th century, when paper was already popular in Europe, some of the travellers who wrote exclusively on China no longer mentioned paper money, but discussed the materials for making paper and the great variety of its uses. Gaspar de Cruz, a Portuguese Dominican who visited China briefly in 1556 and published a book on the country in 1569, said that Chinese paper was made from the bark of trees, canes, silken rags, and also rags of any quality.[b] He mentioned, too, the uses of paper for different occasions—for sealing and authentication, when, for instance, a paper on which the authority's signature was written was glued across both the doors or the gates of a city. Again, he described its use at festivals, when doors were framed by triumphal paper arches, and scaffolds were erected and decorated in paper representing figures, statues, and pictures, all very well made, painted, and lighted with candles and lanterns. The use of paper at funeral ceremonies also came in for comment, and he explained how paper pictures of men and women were hung on cords to help send the dead to Heaven, while paper with printed images and cut in various shapes was burnt as offerings to the gods.[c]

Another missionary, Martin de Rada, a Spanish Augustinian friar who travelled to China twice, in 1575 and 1578, reported that Chinese paper was made from the inside pith of canes. 'It is very thin, and you cannot easily write on both sides of the paper, as the ink runs through.'[d] He also reported the ceremonial uses of paper and the burning of paper money as offerings to the dead. Rada's information was generally adopted by Juan González de Mendoza, a Spanish Augustinian monk, in his best-selling book on China published in 1585, and he added that 'they have abundance of paper and it is verie good cheape'.[e]

Not until the 17th and 18th centuries was more detailed information about Chinese paper and its early invention in China reported. At the beginning of the 17th century Matteo Ricci, the leader of the Jesuit missionaries to China, wrote that the use of paper was much more common in China than elsewhere, and its methods of production more diversified. He said that Chinese paper was thinner than any made in Europe and that made of cotton fibres was as white as the best paper found in the West.[f] The date of its invention was given by another Jesuit, Alvare de Semedo, who wrote around 1640: 'It is now 1800 years since they have had the invention of paper, of which is there so many sorts and in so great plenty, that I am persuaded, that, in this, China exceeds the whole world; and is exceeded

[a] See Polydore Vergil (1), *De Rerum Inventoribus*, Eng. tr. by Thomas Langley (New York, 1868), p. 67.
[b] See tr. Boxer (1), p. 120.   [c] *Ibid.* pp. 97, 101, 143, 147, 216.
[d] See tr. Boxer (1), pp. 295, 306.   [e] Cf. Mendoza (1), ed. Staunton, I, p. 123.
[f] Cf. P. M. d'Elia (2), I, p. 25; also *Journal of Matteo Ricci*, tr. Gallagher (1), p. 16.

by none in the goodness thereof.' He added that paper 'is made of a certain tree, which is called in India, *Bambù*, and in China, *Cio* [*chu*[1]], the art in making of it is like ours; but the best and whitest is made of cotton-cloth.'[a] Semedo's observation of bamboo paper was contemporary with Sung Ying-Hsing's work on Chinese technology, published in 1637,[b] which includes a whole chapter on the making of bamboo paper. His tracing of the origin of paper to the −2nd century is certainly interesting, because no other writer had ever said that the history of paper could be traced back to that early date; only very recently was this made clear when Western Han paper specimens from close to that date were found.

Although the Chinese invention of paper had been reported by Jesuit writers, the fact was apparently still unknown to the European scholarly community in the 17th century. A book on inventions, *Mundus mirabilis* (Wonderful world), by Everhard Happelius, published in Ulm in 1689, still said that the inventor of paper was unknown but was deserving of the highest honours.

The story of Tshai Lun may have been known to the Jesuit missionaries because it was popular in China and Tshai had become a legendary figure worshipped in many public places throughout the country, but not until 1753 was a summary of his biography published. Then, it appeared in Jean Du Halde's multi-volume history of China, originally published in Paris. The story says:

A great mandarin of the palace, whose name was *Tshai Lun*, invented a better sort of paper under the reign of *Ho Ti*, which was called *Tshai hou chih*,[2] paper of the Lord *Tshai*. This mandarin made use of the bark of different trees, and of old worn-out pieces of silk and hempen cloth, by constant boiling of which matter he brought it to a liquid consistence, and reduced it to a sort of thin paste, of which he made different sorts of paper; he also made some from knots of silk, which they called flaxen paper. Soon after the industry of the Chinese brought these discoveries to perfection, they found out the secret of polishing the paper, and giving it a lustre.[c]

Reports from missionaries to China in the 18th century contain little information on paper, except for one observation that Chinese paper was made from the bark of *chhu-kou*[3] [perhaps paper mulberry], which produced fine, white, fibrous, and silky fibres, and it was recommended for introduction to France.[d] Up to the early second half of the 19th century, all information on Chinese paper was based primarily on Du Halde, whose account seems to have been accepted as the authority on the history of Chinese papermaking.[e] It was not until the discoveries toward the end of the 19th and early in this century of paper specimens in Egypt, Tunhuang, and

---

[a] See Semedo (1), p. 34.
[b] *Thien Kung Khai Wu*; see discussion on pp. 68 ff. above.
[c] Du Halde (1), Eng. ed., II, pp. 417–18.
[d] See *Mémoires concernant l'histoire, les sciences, les arts, les moeurs, les usages, etc. des Chinois par les missionnaires de Pékin* ... (15 vols, Paris, Nyon, 1776–91), vol. II, p. 295.
[e] See citation in De Vinne (1), p. 133, n. 1.

[1] 竹   [2] 蔡侯紙   [3] 楮、構

Chinese Turkestan, and their scientific examination, that the fact of the Chinese invention of paper early in the Christian era and its step-by-step migration westward to Europe was firmly established.[a]

(2) TRAVEL OF PAPER WESTWARDS

Only after paper was perfected as a writing material and became used in daily life in China, did it spread in all directions throughout the world. Its introduction to other nations occurred in two stages: first by the arrival of paper and paper products, later by the adoption of papermaking methods by that nation. From the available evidence, it seems that at least one to two centuries were required to develop local manufacture after paper products were first introduced. In its westward migration, for example, paper was introduced to the Arab world no later than the 7th century, but its manufacture there was not begun until the 8th century; it reached Europe in the 10th century, but paper mills were not established there until the 12th.

It has often been said that the Chinese kept the secret of their knowledge of papermaking until a few papermakers were captured by Arabs in the 8th century,[b] but this is certainly not true.[c] That the westward movement of papermaking was slow was due primarily to China's geographical and cultural separation rather than to secretiveness, for papermaking was learned by China's immediate neighbours as soon as they began to have contact with Chinese culture. The introduction of paper to Korea and Japan in the northeast and to Indo-China in the southeast was early,[d] even though its migration to the west over the old silk road was slow and gradual. As archaeological evidence shows, the closer a country lay to China proper, the earlier is paper to be found there.

The westward migration of paper started with Eastern Turkestan, where it crossed the Chinese border from Tunhuang, perhaps in the 3rd century. In the Loulan region, paper fragments of the 3rd century were found by Sven Hedin and Aurel Stein, and in the Turfan and Kao-chhang area, paper of the 4th and 5th centuries was discovered by Prussian and Japanese expeditions early in the 20th century, and by Chinese excavations in more recent years. In the Khotan area, paper manuscripts in Chinese, Tibetan, Sanskrit, and the ancient Khotan languages, dated as 8th century, were found, also by Stein.[e] While some of the paper documents may have been brought from China itself to this region, there is evidence that paper was manufactured locally. Among the documents found in

---

[a]   Cf. Hoernle (1), pp. 663 ff.; also discussions below.

[b]   Cf. Hunter (9), p. 60.

[c]   The theory of Chinese secrecy about papermaking must have been based on the fact that it was a secret early in Europe, where paper-mill owners sometimes required an oath of loyalty of factory workers to guard the secrets of the craft against possible competitors, or applied for a patent for monopoly of raw materials as well as the manufacture of paper; see Hunter (9), pp. 233–4, and discussions on pp. 302 ff. below.

[d]   Cf. below, pp. 319 ff.

[e]   For discoveries of early paper specimens, see Conrady (5), pp. 93, 99, 101; Schindler (4), p. 225; Stein (11), I, pp. 135, 271; and a summary in Tsien (1), pp. 142–5.

Turfan in 1972, one dated to 620 bears the name of a papermaker, *chih shih* Wei Hsien Nu[1], along with names of administrative officers of Kao-chhang. Again, another piece has a message about sending prisoners to work in paper factories, which must therefore have been operated locally. After studies by Chinese scientists of a score of paper documents discovered in recent years, it is believed that some papers in this region were made locally no later than the beginning of the 5th century.[a] As to the Tibetan manuscripts mentioned earlier, it was learned that the raw fibre is not native to Sinkiang and may have been imported from Tibet.[b]

Paper probably moved farther westwards to the Arab world before the 7th century. Trade and other contacts between Arabs and Chinese furnished opportunities for the Arabs to know paper quite early, and such Arabic words as *kāghid* for paper and its equivalent *qirṭās*, which is found in the Koran, are believed to be of Chinese origin.[c] As early as 650 Chinese paper was imported to Samarkand, but it was a rare article used exclusively for important documents,[d] and it is generally believed that its manufacture in the Arab world was not begun until the middle of the 8th century. It is also said that in the battle on the banks of the Talas River in 751, when the allied Turkic-Tibetan forces routed the Chinese army of Kao Hsien-Chih[3] and captured the prisoners, among them were various craftsmen, including papermakers, who were taken to Samarkand to start paper manufacture.[e] Abundant crops of hemp and flax and the water supply from irrigation canals provided the natural resources for the paper industry at Samarkand, and manufacture grew; not only was the local demand filled, but 'paper of Samarkand' became an important article of commerce.[f]

From Samarkand the paper industry soon passed to Baghdad, where a second paper mill was established by Chinese workmen around 794. As well as a religious and cultural centre of Islam (Fig. 1197), Baghdad was then one of the richest cities of the world, and from this time, paper replaced parchment as the major writing material; the Arabian supply of the European market continued until the 15th century.[g] Another papermaking centre in Western Asia was established at Damascus, which supplied paper known in Europe as *charta damascena*, as well as products of its other handicrafts, for many centuries. Another Syrian town,

[a] Phan Chi-Hsing (*10*), pp. 137–8, 188.
[b] Stein (11), I, p. 426.
[c] Hirth said the Arabic word *kāghid* for paper can be traced to the Chinese term *ku-chih*,[2] paper-mulberry paper. S. Mahdihassan (49), pp. 148 ff, says *kagaz* and *qirṭās* are synonymous, meaning paper primarily and document secondarily, and that *qirtas* represents an earlier borrowing.
[d] Cf. Laufer (1), p. 559.
[e] While the Arabic source says that paper was brought to Samarkand by Chinese prisoners, the Chinese history records the battle without mention of papermakers as prisoners. In an account by Tu Huan,[4] who was one of the prisoners and who returned to China in 762, he mentioned several names of weavers, gold- and silversmiths, and painters among prisoners, but no papermakers; see his *Ching Hsing Chi*[5] (*HLW*), p. 2*b*; tr. Pelliot (32).
[f] Cf. H. Beveridge (1), pp. 160–4; also Carter (1), p. 134, citing the statement by the Arabian writer Tha' ālibī of the 11th century.
[g] Yao Tshung-Wu (1), p. 82.

[1] 紙師隗顯奴　[2] 穀紙　[3] 高仙芝　[4] 杜環　[5] 經行記

Fig. 1197. Drawing of a +13th-century Arabic library at Hulwan near Baghdad in Iraq.

Bambyx, also was known for its paper, which was mistakenly thought to have been made of cotton or bombycina.[a]

Paper migrated from Asia to Africa in the 9th century, and gradually replaced papyrus as the major writing medium. The content of the Rainer collection in Vienna, which includes some 12,500 documents in papyrus and paper, indicates that all documents before +800 were written on papyrus; after that, the later the date the more paper was used.[b] Towards the end of the 9th century, paper was evidently more popular than papyrus, and was also used for wrapping; rags became treasured as the raw material. Toward the middle of the 10th century, paper entirely displaced papyrus as writing material, as in China it had replaced bamboo and wood since the +3rd century. The northwest coast of Africa became familiar with paper probably in the 9th or 10th century, following the Arab conquest of Morocco, where the capital, Fez, became a centre of papermaking. But Fez lay in the strategic area that marked the struggle between the Arabs and the Spanish, and it was from this region that paper was introduced to Europe.[c]

Paper could have entered Europe by two different routes: one through Spain, the other by way of Italy. Documentary evidence shows that Spain was the first European country to have it for writing as well as to develop a flourishing paper industry. With the Arab conquest of the Iberian peninsula, paper appeared in

[a] The name *charta bambycina*, paper of Bambyx, was corrupted to *charta bombycina*, paper of cotton. In 1887 this description was proved incorrect by J. Karabacek and J. Wiesner through scientific analyses.
[b] Cf. Carter (1), pp. 135–6.
[c] Cf. Blum (1), pp. 24 ff.; Yao Tshung-Wu (1), p. 84.

Spain no later than the 10th century, and a manuscript of the 10th century found in Santo Domingo is said to be one of the earliest examples of paper there; it is made of heavy, long-fibred linen rags and sized with starch, and thus similar to Arabian papers. The Moors introduced its manufacture there, probably early in the 12th century. One old manuscript dating from 1129 was written on paper as well as parchment and the paper is believed to have been either imported to or made in Spain.[a] The first Spanish paper mills were established in the city of Xátiva, which was famous for its flax, and an Arab traveller wrote in 1150 that paper manufactured there was better than any from elsewhere in the civilised world; it was sent to both East and West.[b] The early mills were operated by Arabs, but after the Christian conquest by local people, and the first paper factory run by Christians was built in 1157 in Vidalon near the French border. Many Spanish Jews were also skilled in this craft, and after the conquest of the kingdom of Valencia, Jewish papermakers continued to work there, though a tax was levied on their product.[c]

Paper entered Italy not from other European nations but from the Arab world, perhaps from Damascus by way of Constantinople and Sicily. Appearing as early as the 12th century, several old Italian manuscripts still exist which show the early use of paper in Italy, while it is known to have been prohibited for official use in 1221; indeed, a legal ruling in Sicily stated that documents written on paper would have no authority.[d] All paper referred to must have been imported, since none was manufactured locally until more than a century later.

The earliest paper mill known to have been established in Italy is the one at Fabriano in 1268–76, which still continues in operation today. Originally it was a most important source of fine rag paper, and several innovations began there. Its pulp was made of short fibres thoroughly ground with metal beaters, the paper was sized with an animal glue, and watermarks with crosses and circles were introduced in 1282.[e] All these factors contributed to the success of the Fabriano paper, and were soon adopted by other European papermakers, especially at paper mills established in other cities in Italy, which included Bologna (1293), Cividale, Padua, and Genoa. As a result by the time the 14th century had dawned, Italian paper surpassed, in production and quality, that from Spain and Damascus.

In France, paper was probably introduced from neighbouring towns in Spain, for there was a close affinity between the two countries. Spanish paper was used in France at the beginning of the 13th century, but French papermaking started in the 14th, for a mill is known to have been established near Troyes in 1348, and others were set up at Essonnes, Saint-Pierre, Saint-Cloud, and Toiles between 1354 and 1388.[f] However, a legend relates that Jean Montgolfier was captured by Turks during the second Crusade and put to work at a paper mill, from which he escaped and returned to Europe in 1157. His grandsons, it was said, established several

[a] Cf. Carter (1), p. 139, n. 11.
[b] *Ibid.* p. 136, citing a statement by Al-Idrīsī.
[c] Cf. Blum (1), pp. 28–9.
[d] *Ibid.* pp. 22–3.
[e] Blum (1), pp. 32; Hunter (9), pp. 301–7.
[f] Blum (1), pp. 32–3.

Fig. 1198. Papermill established by Ulman Stromer at Nuremberg, c. +1390.

paper mills at Ambert in the Auvergne, and certainly this became an important centre for papermaking in the middle of the 14th century.[a]

Germany used paper early in the 13th century, mostly imported from Italy, and manufacture was not begun there until the end of the 14th century, when a mill was established by Ulman Stromer (Fig. 1198), who, apparently, had learned the trade from Italians, two of whom collaborated with him in setting up his mill in Nuremberg in 1390.[b] Stromer used the letter S as the watermark of his products, and it was his mill which, in 1391, suffered the first labour strike in the paper industry.[c] It was around this time that the demand for paper increased due to woodblock printing being introduced to Nuremberg, and rose more rapidly still after the introduction of typography in the middle of the next century (Fig. 1199).

In the Netherlands, paper is known to have been used in 1322; the oldest paper found in Dutch archives is dated to 1346 and has been preserved at the Hague.[d] A paper mill is said to have existed in 1428, but the industry was not well established until 1586, when two noted papermakers were authorised to manufacture the

[a] Hunter (9), p. 473; Kagitci (1), pp. 7–8.
[b] Blum (1), p. 33.   [c] Hunter (9), p. 234.   [d] Ibid. p. 474.

Fig. 1199. Earliest picture of papermaking in Europe. This woodcut by Jost Amman printed in Frankfurt in
+ 1568 shows tools and processes remarkably similar to those used by early Chinese papermakers, cf. Figs. 1071–2.
From Hunter (5).

product near Dordrecht. The Eighty-Years War of 1568–1648 resulted in the
migration of many craftsmen to Amsterdam, which had become an international
trade centre by the close of the 16th century, and improved paper production after
the important invention in + 1680 of the Hollander beater for the maceration of

raw materials.[a] Switzerland was content to import most of its paper from Italy and France until the middle of the 15th century, but then the proceedings of Church councils required a large amount of paper for record-keeping, and a paper mill was established in Basel in 1433; many others then followed in the same area, which became a papermaking centre.

In England, paper is known to have been used for written transactions at the beginning of the 14th century, much later than on the continent. It must have been imported, perhaps from Spain, for as late as 1476, the famous early printer William Caxton at first used only papers imported from the Low Countries.[b] However, a paper mill was established before 1495 by John Tate in Hertfordshire, and another by Thomas Thirlby at Fen Ditton in 1557, though the best known of the early mills was the one set up in 1588 at Dartford in Kent, by Sir John Spilman. Spilman was a jeweller to Queen Elizabeth, and he managed to obtain a patent in 1589 that gave him a monopoly throughout the kingdom for collecting all kinds of rags for making white writing paper.[c] Towards the end of the 17th century, some one hundred paper mills were operating in England.[d]

Paper did not reach certain other parts of Europe until the latter part of the 15th century. The first mill in Poland was established in Crakow in 1491, with others in Wilno (1522) and in Warsaw (1534).[e] Paper may have reached Russia early, but the first mill there was not established until 1576, and workmen were recruited from Germany when an extensive mill was set up in 1712. By 1801, there were some twenty-three mills operating in the Russian Empire.[f]

Paper reached the New World probably in company with the early explorers in the late 15th or early 16th century. 'Paper books doubled together in folds like Spanish cloth' were mentioned by Juan de Grijalva who arrived in San Juan de Ulna in + 1518.[g] These were probably a kind of quasi-paper made by beating fig or mulberry tree bark that had been used by the Mayas and Aztecs for writing.[h] The manufacture of true paper was introduced to America by European papermakers in the latter part of the 16th century, when a twenty-year concession was granted by the Spanish court in + 1575 to two papermakers to 'manufacture paper in New Spain'. In + 1580 they set up a mill at Culhuacán near Mexico City,[i] the first to be built on the American continent.

Paper used in colonial America north of Mexico was imported from Europe, mostly from the continent, before being locally manufactured in the late 17th century, when the first mill in northern North America was built in 1690 near Germantown, Philadelphia, by a German immigrant, William Rittenhouse, who had learned the craft in his native country. Only two years after his arrival in Philadelphia Rittenhouse, with a group of others in the German settlement, started paper manufacture. At the beginning of the 18th century two other mills were

  ^a Cf. Hunter (9), p. 483.   ^b Ibid. p. 476.   ^c Ibid. p. 480.
  ^d Ibid. p. 484.   ^e Ibid. p. 477.   ^f Ibid. pp. 479, 485.
  ^g See the expedition of Juan de Grijalva in Diaz del Castéllo (1), Eng. tr. A. P. Mandslay, 1, p. 162.
  ^h See discussion of huun and amatl in Hunter (9), pp. 25–9.
  ^i Ibid. p. 479.

established in Pennsylvania: one in 1710 by William de Wees, a relative of Rittenhouse who probably learned the craft at his mill; another, the Ivy Mill, in 1729 by Thomas Willcox, an English immigrant, at Chester Creek near Philadelphia.[a] Many workers from Willcox's mill later established their own mills for the manufacture of paper in neighbouring areas.

The paper produced at the Ivy Mill supplied the growing printing and publishing activities in Pennsylvania and New York. One of those much involved with it was Benjamin Franklin, who though primarily a printer was also interested in the development of the American paper industry and in improving papermaking methods. He presented an essay on this subject, criticising the European method of making large sheets of paper by pasting small sheets together and burnishing the joints with an agate or flint. He described the Chinese manner of making sheets as large as twenty feet by six feet, by two workmen, who dried these upon the flat, inclined sides of a heated kiln, making a remarkably smooth surface. After a detailed description of the Chinese method, Franklin concluded: 'Thus the great sheet is obtained, smooth and sized, and a number of the European operations saved.'[b]

In Canada, paper was imported primarily from the United States and Europe, before its first paper mill was established in 1803 at St Andrews, Quebec, by Walter Ware from Massachusetts. A little later, another was built in 1819 by R. A. Holland at Bedford Basin, near Halifax,[c] the increasing need for large quantities of paper for printing newspapers probably being the incentive for this local production of paper. This was true, too, in Australia, where the first paper mill was established near Melbourne in 1868.[d] By this time, papermaking had completed its journey from China to every corner of the world.

(3) INTRODUCTION OF PRINTING TO THE WEST

The travel of paper from China westwards to Europe by way of the Arab world can be traced step by step, but the spread of printing, on the other hand, is not so clear. Such information available indicates that it might have taken the same route to the West overland by way of the silk road or by sea, though at a much later date than paper. Printing appeared in Central and Western Asia as well as in Africa before it was known in Europe, while printed matter, including playing cards, printed textiles, woodcuts, and books printed from woodblocks, is known to have existed in Europe before Gutenberg. Although no direct relationship has yet been established between European typography and Chinese printing, a number of theories in favour of the Chinese origin of the European techniques have been advanced. Some of them are based on early references, others on the circumstantial evidence

---

[a] See Hunter (9), pp. 274–6.
[b] 'Description of the process to be observed in making large sheets of paper in the Chinese manner, with one smooth surface,' in *Transactions of the American Philosophical Society* (Philadelphia, 1793), pp. 8–10.
[c] Hunter (9), pp. 526, 539.   [d] Hunter (9), p. 568.

that close contact between the East and West, especially during the Mongol conquest, provided a Chinese background for the European invention of typography.

In a way similar to the migration of paper to the West, printing probably first crossed the northwestern border of China to reach Eastern Turkestan. This region, known as Turfan, was occupied by Turkic people from the +6th century and came under Chinese domination a hundred years later. In the middle of the +8th century, however, the area was conquered by the Uighurs, a Turkic tribe, which established an empire that lasted almost five hundred years, until they submitted to the Mongols at the beginning of the 13th century. During the Uighur period, Turfan was a place where many religions and cultures mingled, as discoveries by Prussian, Japanese and Chinese expeditions during this century of documents in seventeen different languages, and other cultural relics testify.[a] The documents found are mostly religious texts and commercial papers, including many examples of blockprinting in Uighur, Chinese, Sanskrit, Tangut, Tibetan, and Mongol, and correspond to the languages found in the Tunhuang documents.

The Uighur prints are all translations of Buddhist works in the Sogdian alphabet with occasional introductory matter by Uighur scholars. What is so interesting is that some of the books have titles and page numbers in Chinese characters (Fig. 1200), indicating that the blocks must have been carved or printed by Chinese craftsmen who used the characters for identification in handling and binding. The Chinese books also are Buddhist sutras printed in large characters and bound mostly in the folded format, with some in rolls as was the fashion in China. A Sanskrit *sutra* in Lantsa script, probably from the +13th century, has been found too; it is in the *pothi* form with two long, narrow sheets pasted together. Also included are Tibetan charms contained in hollow clay Buddhist figures, Mongol prints in the 'Phags-pa script, woodcut pictures, and some materials in the Tangut language. The Tangut people established an empire from the +11th through to the early +13th century in northwestern China, bordering Turfan, and used both blocks and movable type extensively for printing.[b]

Both movable-type prints in Uighur and block-printing from the Turfan area have also been discovered on other occasions. The Chinese expeditions in 1928–30 found three additional printed fragments of Buddhist texts in Chinese, two of which are written on the back of the paper in the Uighur language and bear a Chinese seal in red.[c] Also a font of several hundred wooden type for the Uighur language (Fig. 1201), dating to about +1300, was discovered in Tunhuang.[d] This shows that

---

[a] The items found by Albert Grünwedel and Albert von LeCoq of the Prussian group in the Turfan basin in 1902–7 were kept in the Museum für Völkerkunde, Berlin, and are said to have been partly destroyed during World War II. For a more detailed description of documents from the Turfan region, see von LeCoq (1), p. 62; Carter (1), pp. 141–6; Huang Wen-Pi (2).

[b] For the Tangut printing, see Goodrich (29), pp. 64–5; also pp. 169 ff. above.

[c] The Chinese findings are included in a report by Huang Wen-Pi (2).

[d] The Uighur type were found by Paul Pelliot in 1907; cf. Carter (1), pp. 146–7, 218. A recent report says the set can no longer be located.

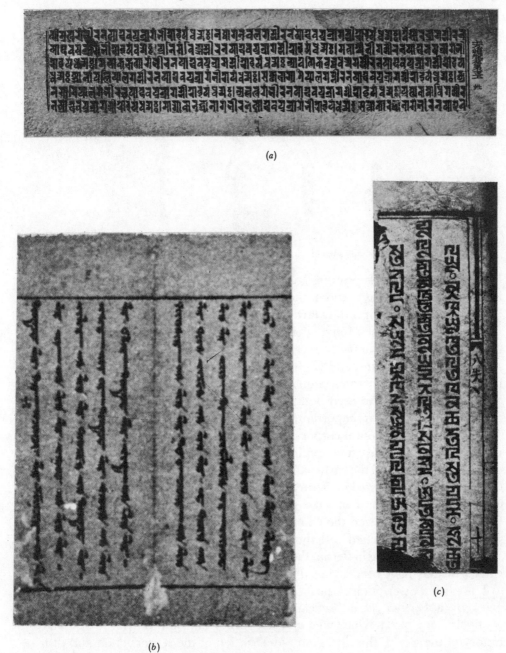

Fig. 1200. Printing in non-chinese languages, *c.* +1300, found in Turfan, bears features in Chinese. (*a*) Buddhist text in Sanskrit but titles and collation in Chinese characters on the right side. (*b*) Buddhist *sutra* in Uigur with Chinese page number on the left margin. (*c*) A *sutra* in 'Phags-pa' script with pagination in Chinese characters at the centre folding line. Museum für Völkerkunde, Berlin.

Fig. 1201. Wooden types and impression of Uighur scripts, *c.* +1300, found at Tunhuang. From Carter (1).

conversion from block printing to movable type is natural for an alphabetic language such as Uighur, and not only European languages.

The introduction of printing farther to the west was probably accomplished by the Uighurs during the Mongol period. After the Mongol conquest of Turfan, a great number of Uighurs were recruited into the Mongol army; Uighur scholars served as Mongol brains, and Uighur culture became the initial basis of Mongol power. If there was any connection in the spread of printing between Asia and the West, the Uighurs who used both block printing and movable type had good opportunities to play an important role in this introduction.

The Mongol conquests incorporated Persia into an empire of which the capital was in China. Thus the Chinese cultural impact on Persia was manifest during the middle of the 13th and the early part of the 14th century. It was here that printing in China was first reported in literary works and was first used in western Asia. As is generally known, paper money was printed in Tabriz in 1294, exactly following the Chinese system, and even the Chinese word *chhao*[1] for money was adopted, and subsequently incorporated, into the Persian vocabulary.[a] Although this monetary system did not last long in Persia, the wood carvers who had been employed for the enterprise may have been engaged in printing other material not known to us.

The earliest description of the methods of Chinese printing in any literature was given by a Persian scholar-official, Rashīd al-Dīn, prime minister under the Mongol ruler Ghazan Khan, who took ten years, from 1301 to 1311, to complete a history of the world, that included a description of the reproduction and distribution of Chinese books. Rashid said that when any book was desired, a copy was made by a skilful calligrapher on tablets and carefully corrected by proof-readers whose names were inscribed on the back of the tablets. The letters were then cut out

---

[a] See Laufer (1), pp. 559–60.

[1] 鈔

by expert engravers, and all pages of the books consecutively numbered. When completed, the tablets were placed in sealed bags to be kept by reliable persons, and if anyone wanted a copy of the book, he paid the charges fixed by the government. The tablets were then taken out of the bags and imposed on leaves of paper to obtain the printed sheets as desired. In this way, alterations could not be made and documents could be faithfully transmitted.[a] A few years later the same description of Chinese printing was incorporated into a work by an Arab author.[b] Thus the Chinese method of printing, including the various steps of transcribing, proof reading, cutting blocks, printing, and distribution, were for the first time carefully recorded.[c]

Despite the fact that the Islamic religion did not favour printing, some fifty pieces of printed matter, believed to have been made between +900 and +1350, were found in Egypt toward the end of the 19th century.[d] These are all fragments of Islamic prayers, charms, and texts from the Koran in early Arabic script (Fig. 1202). Except for one in red, they are printed on paper in black ink, though not by pressure but by rubbing with a brush in a way similar to the Chinese method.[e] Judging from the materials used, the religious nature of the documents, and the printing techniques used, experts believe that these printed specimens are connected with printing in China and Central Asia rather than an independent development.[f] The time of the transmission to Egypt is uncertain, but scholars incline to a comparatively late date, after the time printing in China had begun to travel across Turkestan to the Arab world during the Mongol conquest. It could have been introduced through Persia or by travellers or traders on other routes, since Chinese intercourse with North Africa was very close in the early part of the +14th century.[g]

After the submission of the Uighurs in 1206, the Jurchens and Koreans in 1231, and the Persians in 1243, the Mongol army moved farther north to overrun Russia in 1240 and to invade Poland again in 1259 and Hungary in 1283. They thus reached the border of Germany where block printing appeared not long after the climax of the Mongol conquests. Along with the military expansion, commercial, diplomatic, and cultural relations developed between Europe and Mongol China during the 13th and early 14th centuries; overland highways connecting China,

[a] Cf. tr. by E. G. Browne (1), pp. 102–3.
[b] *Garden of the Intelligent* by Abu Sulaymán Da-ud of Banákatí, written in 1317, tr. Browne (1), pp. 100–2.
[c] The methods of Chinese printing had never been recorded by any Chinese writer until modern times. This description by Rashīd al-Dīn, although brief, is probably the earliest record of the techniques of Chinese printing in any language including Chinese.
[d] Over 100,000 items of documents in ten different languages, dating from the −4th to the +14th century, on papyrus, parchment, and paper, excavated in the ruins of an ancient city in Egypt, are kept in the Erzherzog Rainer Collection of the Austrian National Library in Vienna, and additional prints are in Heidelberg, Berlin, and Cairo; see Carter (1), pp. 176–8, 181 n. 1.
[e] A Hebrew block print from the late +14th century has recently been found and studied at the Taylor-Schechter Genizah Collection of the Cambridge University Library. It is believed that the Jews of Egypt might also have adopted the method of block printing used by the Egyptians at the time. I am grateful to Dr L. C. Goodrich for calling my attention to the report in *The Jewish Week*, 8 Oct. 1982, p. 26.
[f] Cf. discussion in Carter (1), pp. 179–80.
[g] Cf. the travels of Ibn Batuta (d. 1338) in Yule (2), IV, pp. 1–166; also Duyvendak (8).

Fig. 1202. Printed fragment of Koran in Arabic, *c.* early + 10th century, found at El-Faijûm, Egypt. Erzherzog Rainer Collection, Austrian National Library, Vienna.

Persia, and Russia were built to help the flow of increased traffic in the way of couriers, caravans, craftsmen, and envoys. In 1245 an embassy was sent to the Mongol court by the Pope, who received in reply a letter with a seal carved by a Russian in Chinese characters and impressed in red (Fig. 1203).[a] Soon after, in 1248 and 1253, two other embassies were sent by the King of France and, as we have seen, one of the envoys, William Ruysbroeck, was the first European to report on the use of paper money in China. The same practice, described by Marco Polo in the record of his travels, was somewhat later, but after Polo left China in 1294, John of Monte Corvino, a Roman Catholic missionary, was sent there by the Pope, and stayed for over thirty years until his death in 1328. He and other missionaries worked in Peking, Fukien, and Yangchow, building churches, learning the language, translating the Bible, and preparing religious pictures as aids to preaching Christianity.[b] Since the printing of Buddhist pictures was very common

[a] Carter (1), pp. 159–60. The seal of state, measuring 15 cm. square, bears inscriptions reading: 'Hu Kuo An Min Chih Pao'[1] (Emblem for Protection of the State and Pacification of the People).

[b] Roman Catholic tomb tablets dated to 1332 from Chhüanchow, Fukien and to 1342 and 1344 from Yangchow, Chiangsu were recently found, indicating a sizable European community in China under the Mongol rule, with Catholic converts estimated at 30,000 to 100,000 at the time; see Rouleau (1), pp. 346 ff.; Hsia Nai (7), pp. 532 ff.

[1] 護國安民之寶

Fig. 1203. Impression of the Seal of State in Chinese characters on a letter of +1289 from the Mongol ruler in Persia to the King of France, 15 cm. square. From Carter (1).

in China before and during this time, the use of this simple and convenient method for reproducing the translated Bible and religious pictures would have been natural. As these materials were required in large numbers of copies for circulation both among Chinese Christians and also among non-converts, it would be surprising if they were not printed. If they were, then the sudden appearance of religious prints and block books in Europe in the early 14th century can be reasonably explained.[a]

Before the use of typography in Europe in the middle of the 15th century, various kinds of printed matter were already there, as early perhaps as a century or more before Gutenberg. There were playing cards, printed textiles, prints of religious images, and block books, all of which involved the use of wood blocks for duplication. Among these, playing cards were one of the earliest examples of block printing to appear in Europe, doubtless because of their early and widespread use in the East. For card games were played in China before the +9th century, at the time when books were evolving from paper rolls to paged form,[b] and they spread over much of the Asian continent before the Crusades. Probably they were brought to Europe by the Mongol armies, traders, and travellers, some time in the early 14th century (Fig. 1204),[c] references to their earliest appearance being 1377 in Germany and Spain, 1379 in Italy and Belgium, and 1381 in France.[d] Because card games were played in all sectors of society, their popularity demanded the reproduction of playing cards in great quantity, though the craze for gambling

[a] The close resemblance in techniques and appearances between European and Chinese block printing is further discussed below, pp. 313 ff.

[b] See discussion of playing cards in China on pp. 131 ff. above.

[c] Blum (2), p. 43; De Vinne (1), p. 108.

[d] An Italian writer, Valère Zani, of the 17th century said that Venice was the first European city in which Chinese cards were known; cf. Carter (1), p. 192 n. 24.

Fig. 1204. Chinese playing cards, *c.* +1400, found near Turfan. 9.5 × 3,5 cm.
Museum für Völkerkunde, Berlin.

resulted in their prohibition on economic and moral grounds by government and religious authorities.

The earliest playing cards were manufactured in various forms and in different ways: by painting, by printing outlines to be filled in with colour by hand or with stencils, or by printing from wood blocks or copper engravings. The more expensive ones were printed from a master engraved in intaglio. A recent study reveals that Gutenberg had a major role in the early development of copper engraving to make masters for producing playing cards, because it is suggested that when financial disaster forced the closing of his Mainz workshop, the figures originally intended for decoration of the 42-line Bible were used to print cards instead.[a] Although critics have questioned whether cards exerted any tangible influence on the art of printing,[b] the association between printing books and playing cards does suggest a close relationship between the two.

Printing on textiles has generally been considered one of the forerunners of

---

[a] After comparing some of the designs on playing card with those in the 42-line Bible, Lehmann-Haupt (2), p. 3, says that the designs of the masters were developed by artists connected with Gutenberg's workshop, but that the mechanical means for the multiplication were made by Gutenberg himself.
[b] See Laufer's review of Carter in *JAOS* 47 (1927), p. 76.

printing on paper. Since the method is identical, the transfer from one material to the other is simple, since textile printers and the early block printers in Europe were closely connected. Professional wood carvers could of course, be employed for printing on any material.[a] The technique of carving blocks for printing on textiles was in fact exactly the same as that for paper. The same kind of wood was chosen, the transfer of design from paper to block, the manner of cutting in relief, and of placing cloth on the block and pressing it with a burnisher or pad stuffed with horsehair were all the same. If a piece of paper is substituted for a piece of fabric, the result is printed paper.[b]

The earliest specimens of printed textiles extant in Europe are those from France and Germany dating back to the +6th or +7th century, even earlier than those from Tunhuang and Turfan.[c] However, a recent discovery of silk fabrics at Ma-Wang-Tui, Chhangsha, indicating printing on textiles of a set of continuous patterns, goes back as early as the −2nd century.[d] Whether European textile printing was influenced by the Chinese is not clear, but some patterns of Chinese origin, borrowed by Persian weavers, are said to have been transmitted to Western Europe, and certainly many Chinese decorative motifs had been successfully copied by European makers of figured fabrics before 1500.[e]

Religious pictures and block books provide the closest examples of printing before Gutenberg. Similar in nature, and differing only in format, when single sheets of image prints were collected together, they naturally evolved into book form. The image prints were first produced in southern Germany and Venice and gradually spread over most of central Europe between 1400 and 1450.[f] Their subject-matter is exclusively religious, including pictures of certain sacred personalities or representations of biblical stories with legends in Latin engraved at the foot of the sheet or in cartouches proceeding from the mouths of the principal figures.[g]

Most of the several hundred image prints still in existence are undated, but they are believed to have been produced during the latter part of the 14th and early part of the 15th centuries. Although a few have some artistic merit, most of the pictures are crude in style and workmanship. They were printed in outline and filled in with colour by hand or by stencil, but all the same they may possibly have some connection with Chinese printing,[h] since the use of block prints for Buddhist pictures had long been practised in China. Many such single-sheet prints with Buddhist figures and legends were discovered in Tunhuang, and printing of tens of thousands of such pictures on silk and paper are recorded in literary sources.[i] Then,

---

[a] De Vinne (1), pp. 107–8, cites an expert opinion that engravers made blocks to print cards, images, and other printed matter such as wallpaper; cf. also Carter (1), p. 197.

[b] Cf. Blum (2), p. 50.   [c] Cf. Carter (1), pp. 194–5.   [d] See *KKTH*, 1979 (no. 5), p. 474.

[e] Cf. Lach (5), II:1, pp. 96–7; II:3, p. 405.   [f] Cf. De Vinne (1), p. 75.

[g] Cf. description of individual specimens in De Vinne (1), pp. 69–87.

[h] *Ibid*. pp. 75–6, says the Chinese origin of image prints was suggested, but there were no early specimens to offer as evidence; cf. also Carter (1), p. 206.

[i] Cf. discussion on pp. 158 ff. above.

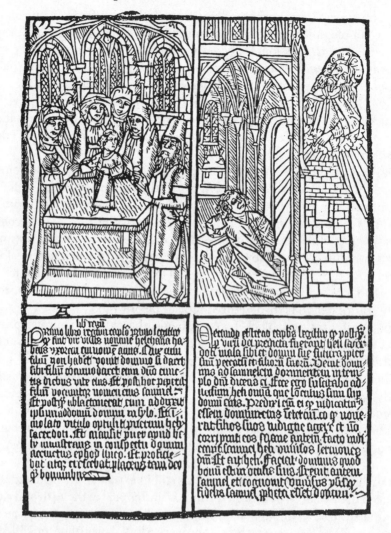

Fig. 1205. The *Book of Kings*, a block-printed book of Europe with illustrations above and text below, printed on one side of the leaf with brownish ink, 18 × 21 cm., similar to the format of religious sheets and illustrated books printed in China.

in the 14th century when European missionaries were sent to China, they made use, it is said, of religious pictures for distribution to the ignorant. It is possible, therefore, that these old practices for religious propagation in China were borrowed for similar purposes in another land.

At first, each picture was printed on a single sheet, or sometimes pictures were paired on one sheet, but later, some were pasted back to back or printed on both sides of the paper and gathered into books. The block prints that have been preserved include some containing pictures with text (Fig. 1205) and some having pictures alone; only very few have text alone. They were prepared not by priests or in monasteries, but by independent printers who sometimes also produced playing cards, image prints, and even textiles. The demand for block printed books was probably very great, as the few such books still in existence are known to have been published in numerous editions,[a] and their production continued after typography came into fashion. This was probably because block books were familiar to users throughout Europe, they cost less to produce, and block carvers were there to continue their traditional way of business until their generation passed away.

The close resemblance between the early block books of Europe and those of China is probably the most convincing evidence that European printers followed Chinese models. Not only were the methods of cutting, printing, and binding similar, but also the materials and the manner in which they were used. It is stated that the wood used in European xylography was cut parallel with the grain in flat blocks. Moreover, the material to be printed was transferred from paper to the woodblock on which it was fastened with rice paste, two pages were engraved on one block, water-based ink was used, the impressions were taken by means of friction on one side of thin paper, and the double pages were put together two by two with the blank sides folded inside.[b] All these procedures were not only exactly the same as the Chinese methods, but were contrary to European practice. The European method was usually to cut wood across the grain, to employ oil-based inks, to print on both sides of the paper, and to use pressure rather than rubbing.[c]

Robert Curzon, Baron de la Zouche (1810–73),has said that the European and Chinese block books are so precisely alike, in almost every respect, that 'we must suppose that the process of printing them must have been copied from ancient Chinese specimens, brought from that country by some early travellers, whose names have not been handed down to our times'.[d] Since all the technical processes are of Chinese rather than European tradition, it seems that the European block printers must not only have seen Chinese samples, but perhaps had been taught by missionaries or others who had learned these un-European methods from Chinese printers during their residence in China.

(4) CHINESE BACKGROUND OF EUROPEAN PRINTING

While Chinese paper was mentioned by European travellers as early as the 13th century, the art of printing in China was not clearly recorded in European literature until some three centuries later. Only after the accomplishments of Gutenberg and other printers became known in the middle of the 16th century, did European writers begin to record the invention and look into the origins of

[a] De Vinne (1), p. 194.
[b] Cf. *Ibid.* pp. 119–20, 203; Isaiah Thomas (1), pp. 75–6; also see pp. 222 ff. above.
[c] Cf. De Vinne (1), pp. 83–4, 203.
[d] Cf. Curzon (2), p. 23; also cited in Yule (1), p. 139.

printing.[a] The fact that printing was used in China several centuries before it was in Europe was acknowledged by historians and other writers, who also offered the opinion that European printing was influenced by the Chinese, and although none of their theories have been substantiated further, neither have they been disproved. On the other hand, these earlier opinions have provided some incentive for later scholars to try to trace possible Chinese connections with European printing, and while no concrete proof has yet been presented, circumstantial evidence is strong. Today, even one who believes that printing in Europe in the 15th century was 'an altogether independent outgrowth of its own times and conditions', has acknowledged that 'Europeans in their varied contacts with the Orient learned something of printing and perhaps even saw documents and books printed on paper'.[b] Almost all defenders of an independent origin of European printing emphasise technical differences between Chinese block printing and typography, but not the cultural considerations offered by many earlier and later writers.

The question of who was the inventor of printing was raised as early as the beginning of the 16th century by Garcia de Recende (c. 1470–1536), a Portuguese poet who, incidentally, refers in a poem to the question of whether printing was first discovered in Germany or China.[c] But it was not until the middle of the century that Europeans began to write books on printing and to notice that printing had been used much earlier in China. The first to make a clear mention of Chinese printing was the Italian historian Paulus Jovius (1483–1552), who noted that printing was invented in China and introduced to Europe through Russia. In his *Historia sui temporis*, published in Venice in 1546, he wrote:

There are there (Canton) printers who print according to our own method, books containing histories and rites on a very long folio which is folded inwards into square pages. Pope Leo has very graciously let me see a volume of this kind, given him as a present with an elephant by the king of Portugal. So that from this we can easily believe that examples of this kind, before the Portuguese had reached India, came to us through the Scythians and Muscovites as an incomparable aid to letters.[d]

Jovius had originally studied medicine but he was close to the political and religious powers in Italy, was sent as ambassador to Moscow not long after Russia was freed from Mongol domination, and wrote a history and several other books about Russia.[e] He may also have had some knowledge of China, as we are told that

---

[a] The first work on the invention of typography, *De Typographiae Inventione*, was written by Matthias Richter and published in Copenhagen in 1566; and no fewer than four similar titles were published in the 16th century; see list in McMurtrie (3), p. 26.

[b] Cf. McMurtrie (1), p. 123.

[c] See Mendes dos Remedios (ed.), *Garcia de Recende: Miscellanea...* (Caimbra, 1917), p. 63, and in which stanza 179 implies the question of priority of this discovery; Lach (7), II:2, pp. 118, 127.

[d] Tr. quoted in Carter (1), p. 159, with original text and comments pp. 164–5, n. 4; also in Lach (5), II:2, p. 227, based on the edition of Ferrero and Visconti (Rome, 1956–64).

[e] Jovius' opinion was apparently used by many European writers of the 16th and 17th centuries without acknowledgement. His name was perhaps first mentioned by Richard Smith in his unpublished manuscript of 1670, titled 'Of the First Invention of the Art of Printing', which states that printing 'was used by the people of Sina or China in the utmost parts of the East some ages before it was known in Europe', and 'that art came unto us by the Scythians and Muscovites before Portugals came into India'; see Smith (1), p. 10.

several Chinese books and maps together with their translations were sent to him by the Portuguese historian João de Barros (1498–1570).[a] With his professional and academic credentials, 'his statements concerning Russia therefore carry considerable weight', as Carter has remarked,[b] and even though he gave no references for his claims about Chinese printing, as a historian he must have had some evidence to support his statements. Of course that evidence might have seemed too obvious to need mention in a general history, for contacts between East and West had been so frequent during the period of the Mongol conquests, not long before his lifetime.

A little later, Gaspar da Cruz and Martin de Rada, the two early visitors to China whose comments on paper were mentioned earlier, also made some remarks on Chinese printing. Cruz said that 'it is over nine hundred years since the Chinese have used printing, and that they not only make printed books but also different figures'.[c] In saying this, he was the first European visitor to China who indicated a period for the earliest use of printing not only for books but also for pictures or illustrations. Theories that printing originated in the Sui or early Thang in the +6th or +7th century were generally spoken of in the latter part of the Ming dynasty,[d] the time when Cruz was in China, while numerous books printed with illustrations or separate sheets of pictures would also have been available to him.

Rada mentioned Chinese printing in his reports and also brought back to Spain a number of Chinese books.[e] He talked with a Chinese official who 'was greatly surprised to learn that we likewise had a script and that we used the art of printing for our books, as they do, because they used it many centuries before we did'.[f] He also acquired many 'printed books of all the sciences, both astrology and astronomy, as [well as] physiognomy, chiromancy, arithmetic, and their laws, medicine, fencing, and of every kind of their games, and of their gods'.[g] Among the books brought from China were eight local gazetteers,[h] in which, he noted, such precious metals as gold and silver were recorded.

From this time on, similar statements were made by many other writers, including Juan Gonzalez de Mendoza, whose most comprehensive and authoritative work on China published in 1585 devoted two full chapters to Chinese books and printing. In one of these, 'The substance and manner of these books that Friar Herrada and his companions brought from China', de Mendoza describes in categories all kinds of books he acquired there. The list includes

---

[a] Cf. Boxer (1), p. lxxxvi.   [b] Cf. Carter (1), p. 165.   [c] Cf. Boxer (1), p. 148.

[d] The earliest suggestion that Chinese printing originated in the Sui dynasty (+581–618) was made by a Ming scholar, Lu Shen[1] (1477–1544), whose theory that wood blocks were carved in +593 had great influence on later opinions; see discussion on pp. 148 ff. above.

[e] Two reports: 'Narrative of his mission to Fukien (June–October, 1575)' and 'Relation of the things of China which is properly called Taybin,' written late 1575 or early 1576, are translated in Boxer (1).

[f] Boxer (1), p. 255.

[g] Ibid. p. 295.

[h] Ibid. p. 261, mentions that 'seven of these books came into our hands', but eight different titles are described in pp. 293–4.

[1] 陸深

history, geography and gazetteers, chronology, navigation, ceremonies and rites, laws and punishments, herbals and medicine, geology, astronomy, biographies of famous persons, games, music, mathematics, architecture, astrology, chiromancy, physiognomy, calligraphy, divination, and military works.[a] All these books must have been read with the help of native Chinese in the Philippines, where the missionaries stayed. As Mendoza wrote, 'they (friars) had bought a good number, out of which are taken the most things that wee haue put in the small historie'.[b] Interestingly, a few Chinese books of the 16th century survive in libraries of Spain and Portugal.[c]

In another chapter, 'Of the antiquitie and manner of printing bookes, vsed in this kingdom, long before the vse in our Europe', he discusses the admirable invention of printing begun in Europe in 1458 by Johann Gutenberg of Germany, whence the same invention was brought into Italy. He added:

But the Chinos doo affirme, that the first beginning was in their countrie, and the inuentiour was a man whome they reuerence for a saint: whereby it is euident that manie years after that they had the vse thereof, it was brought into Almaine by the way of Ruscia and Moscouia, from whence, as it is certaine, they may come by lande, and that some merchants that came from Arabia Felix, might bring some books, from whence this John Cutembergo, whom the histories dooth make authour, had his first foundation.[d]

It is interesting to note that, besides his claim that Gutenberg was influenced by Chinese printing which came by way of Russia, he also mentioned another route, through trade from Arabia by sea. He concluded:

The which beeing of a trueth, as they haue authoritie for the same, it dooth plainlie appeare that this inuention came from them vnto vs: an for the better credite hereof, at this day there are found amongst them many bookes printed 500 years before the inuention began in Almaine: of the which I haue one, and I haue seen others, as well in Spaine and in Italie as in the Indies.[e]

Mendoza's generalisations concerning things Chinese had great influence on some later writers, and throughout the 16th century, such authors as the eminent French historian Louis le Roy (1510–77), Francesco Sansovino (1521–86), a renowned poet and translator; and Michel de Montaigne (1533–92), a brilliant essayist, all repeated the same story that printing originated in China several hundred years before it reached Europe, and inspired Gutenberg's invention.[f]

Besides these opinions, which seem to have all derived from Jovius's account, there is a different view which points to direct and personal contacts with Chinese printing. This alternative theory relates to an Italian engraver, Pamfilo Castaldi (1398–1490), who in 1868 was commemorated by a statue in Lombardy honouring him for having introduced typography to Europe. He is said to have been born at

---

[a] De Mendoza (1), ed. by Staunton, pp. 134–7.   [b] *Ibid.* pp. 133–4.
[c] Cf. Pelliot (66), pp. 45–50; Fang Hao (4), pp. 161–79.
[d] Cf. De Mendoza (1), ed. by Staunton, p. 132.   [e] *Ibid.*
[f] See citations in Lach (5), II: 2, pp. 214, 296, 310–11.

Feltre, a town northwest of Venice, and to have used wooden movable type after having seen Chinese books brought from China by Marco Polo. In 1426 he printed at Venice several broadsides which are said to be preserved among the archives at Feltre.[a] The tradition goes on to say that Gutenberg, whose wife was of the Venetian Contarini family, had seen printing blocks brought from China to Venice, and by development of this inspiration arrived at the invention of printing.[b] This story was given by Robert Curzon, in two accounts to the Philobiblon Society of London in 1854 8, citing a newspaper article by a Dr Jacopo Facen of Feltre in 1843.[c] The same story is included in several editions of Marco Polo, and Henry Yule, the eminent translator of Polo's works, was disinclined to the view that this tradition was correct, though he believed that many a traveller and overland trader may have brought home Chinese wood blocks.[d]

While many authors suggest the Chinese origin of printing and its influence on European typography, there are some who hold a different opinion, not disputing the cultural theories, but basing their contention primarily upon technical differences between Chinese and European methods. An early expression of this view was made by Guido Panciroli (1523–99), an Italian scholar and author, who believed that Gutenberg's movable type differed in technique from Chinese printing. He said that 'typography is old in China, but as found out in Mentz, it is a modern thing'.[e] He did not specify what the differences were between the two, but implied an improvement of modern technology over the old method. As explained by André Blum, a respected author on the origins of paper and printing, 'The essential element in the invention of printing in the West is not that it was derived from wood block printing. . ., but that it consisted rather in the creation of movable characters made from a fusible metal.'[f] He said that three things are needed for typography: a matrix or mould in which the letter is engraved in intaglio, an alloy cast in the matrix, and a reproduction of the character in relief on the punch. Actually, a similar method of casting metal types from punches was used in the Far East at least half a century before Gutenberg,[g] and there are theories that typography could have derived from there.[h] As G. F. Hudson says: 'Since Korean typography underwent so remarkable a development just before the appearance of the process in Europe, and there were possible lines of news transmission between the Far East and Germany, the burden of proof really lies on those who assert the complete independence of the European invention.'[i]

[a] This account is given in Curzon (1), pp. 6 ff.; cited in Yule (1), 1, pp. 138–40.
[b] Curzon (2), p. 23.
[c] Il Gondoliere, no. 103, of 27 December 1843; cited in Yule (1), 1, p. 139.
[d] Yule (1), 1, pp. 139–40.
[e] See English tr. of his Nova reperta, titled The History of Many Things Lost (London, 1715), pp. 342–3.
[f] Cf. Blum (2), pp. 20–1.   [g] See discussion on pp. 327 ff. below.
[h] Dr Fang Chao-ying of Columbia University has suggested in an unpublished paper, 'On Printing in Korea', the possibility that European typography might have come with the knowledge of Korean movable type through the contacts between European residents and Korean students at the Mongol capital of Peking in the 14th century, just as in the case of the meeting between Adam Schall and the Korean crown prince in the 17th century.
[i] Cf. Hudson (1), p. 168.

Another question of controversy is whether typography was an independent invention or merely a combination of existing technology. As Theodore De Vinne remarked, some scholars believe that 'typography was not an original invention, that it was nothing more than a new application of old theories and methods of impression'. According to this view, engraving can be traced back to the Egyptian seals, printing with ink to Roman hand stamps, and the combination of movable letters to the suggestions by Cicero and St Jerome. Gutenberg, therefore, was not the first to print on paper, for printed matter, in the form of playing cards, prints of pictures and printed books, was a merchantable commodity before he was born.[a]

If typography was not an original invention, then the question arose whether existing techniques were derived from the East or the West. A British collector and antiquarian, John Bagford (1659–1716), wrote in 'An Essay on the Invention of Printing':

The general notion of most Authors is, that we had the hint [of printing] from the Chinese; but I am not in the least inclined to be of that opinion, for at that time of day we had no knowledge of them. I think we might more probably take it from the Ancient Romans, their Medals, Seals, and the Marks or Names at the bottom of their sacrificial Pots.[b]

Although this author and some others attributed the existing techniques, including the use of seals, ink, and other materials and facilities, to the root of Western culture and not to the Chinese, it is the reverse that is true, as discussed in detail in the Introduction of this study.[c] All the basic elements prerequisite to printing were available both in the West and in China, but the combination of them led to the early appearance of printing in Chinese culture and not in the West.

After discussing various factors leading to the invention of printing in Europe, Douglas McMurtrie, a modern authority on the history of printing, argues that the Europeans may have learned the idea of printing, though not the processes, from the Orient, but 'an idea is not an invention'.[d] This statement is certainly debatable. Since an invention always involves both novelty and practice, processes carried out without a novel idea cannot qualify as an invention. The materials and facilities for European typography, including the ink, metal, and the press, may be somewhat different from those used in the Orient, but they constitute only an improvement of an already existing idea and procedures to suit different circumstances. If the basic principle of printing is to obtain multiple copies of a positive impression with ink on paper from a mirror image, this very idea suggests an invention.

Based on this principle, block printing is the ancestor of all printing processes, no matter whether wood or metal; block or movable; or plane, intaglio, or in relief. If the technical differences of typography from block printing justify its consideration

---

[a] Cf. De Vinne (1), pp. 50, 67–8.

[b] This article was originally published in the *Philosophical Transactions of the Royal Society*, vol. 25, 1706–7, and reprinted by the Committee on Invention of Printing, Chicago Club of Printing House Craftsmen, Chicago, Illinois, 1940; Douglas McMurtrie says in the Introductory Note that critics considered him not a scholar and 'quite incompetent' to write on this topic.

[c] See above, pp. 3 ff.; also C. R. Miller (1).   [d] Cf. McMurtrie (1), p. 123.

as a separate invention, then all other new methods of printing, such as lithography, offset and photogravure, would have to be regarded as totally unrelated to their predecessors.

In summing up, the origins of printing in Europe seem to have involved three key questions. First, was typography an altogether independent invention, or was it influenced by the principle and practice of block printing? Since block prints and books existed in Europe before and contemporary with the beginning of typography, most opinions agree that European printers were exposed to at least the principle if not the practice of block printing. Secondly, if this was the case, was block printing in Europe introduced from China? For this question, almost all the views which have been expressed cast little doubt about the close relationship between the two, and their near similarity has warranted the belief that European knowledge of engraving on wood must have been taken from China. Thirdly, did the first maker of European typography have direct or indirect access to Chinese printing or metal type from the Far East? While traditions which suggest a particular name or names are doubtful, it is the general belief that samples of printed books, wood blocks, or metal types might have been brought to Europe from the Far East by unknown travellers via land or sea trading routes. All this circumstantial evidence suggests strongly the presence of a Chinese connection in the origins of European printing.

(i) MIGRATION OF PAPER AND PRINTING EASTWARDS AND SOUTHWARDS

AMONG the many neighbours of ancient China, some formed close ties with Chinese civilisation while others did not. To the north and west, the Mongols, Turks, Manchus, and Tibetans, although their histories were interwoven with that of China through wars and conquests, did not assimilate Chinese culture until they took up residence on Chinese territory. To the east and southwest, on the other hand, the Koreans, Japanese, and Vietnamese were clearly identified with the Chinese cultural outlook from very early times. They borrowed the Chinese writing system, followed Confucian thought, modelled their political and social institutions after those of China, and adopted Chinese forms of art and material life. While Japan maintained an independent political relationship with China, both Korea and Vietnam were under Chinese rule or acknowledged the suzerainty of China for prolonged periods. In one way or another, these three nations, and perhaps also Liu-Chhiu, became parts of the domain of Chinese culture, which is the basic element of East Asian civilisation.

(1) EARLY USE OF PAPER AND PRINTING IN KOREA

Korea not only was the earliest nation to borrow many things Chinese for her own, but also formed a cultural bridge between China and Japan before they made direct contact in the +7th century. How early paper and papermaking methods were introduced to Korea is uncertain, but geographical proximity suggests that these dates must have been very early. Since the northern part of Korea, including Lolang,[1] was under Chinese control from −108 and throughout the entire Han period, the importing of paper and paper books to Korea must have been no later than the +3rd century, when paper began to be popular and spread beyond the Chinese border in both the northwest and southeast.[a] From the latter half of the +4th century, Chinese Buddhist missions were sent to Korea, and in the +6th century Korean monks and students were in the Thang capital, Chhang-an, while more Chinese monks, scholars, artisans, and painters went to Korea. Since the crafts of making brushes, ink, and paper were learned by all foreign students in China, and papermaking is said to have been introduced to Japan in +610 by the Korean monk Damjing[2] (in Japanese: Doncho, +579–631),[b] the manufacture of paper in Korea must have been no later than this date, and perhaps began as early as the +6th century.

Korean papermakers used raw materials, tools, and techniques similar to those used by the Chinese. The materials included hemp, rattan, mulberry, bamboo, rice straw, seaweed, and especially paper mulberry (tak[3] in Korean), which has been one of the major materials for papermaking in East Asia. The preparation of the pulp by pounding fibres of paper mulberry bark, boiling, sun-bleaching, and adding mucilaginous liquid was the same as is described in Chinese records.[c] Moreover, the mould was made either of bamboo or a Korean grass (*Miscanthus sp.*), with Chinese methods of construction for the frame, cover, and the two deckle sticks. After examining several hundred Korean papers from the +16th century onward, Dard Hunter said that the 'laid lines' run the narrow way and the 'chain lines', often narrowly spaced and irregular, run the length of the mould. Every sheet of Korean paper carries this marked characteristic.[d]

Some specimens of the earliest Korean paper survive. A piece of glossy white paper made of hemp fibres is reported to have been discovered recently in a site of the Koguryō era (+37–668) in North Korea,[e] and from the Thang dynasty, Korean paper, known as *Chi-Lin chih*[4] (Paper from the Silla Kingdom), was an item of tribute to China, and its fine quality received high praise from Chinese artists and literati. It was described as thick, strong, whitish, and glossy, and was especially good for calligraphy and painting.[f] Korean papers were also used for mounting scrolls and rubbings. A coarser and more durable kind called *teng phi chih*[5]

[a] See discussion on spread of paper on pp. 296 ff. above.
[b] See *Nihongi*, vol. 3, p. 450; tr. Aston (1), vol. II, p. 140.
[c] Cf. *Thien Kung Khai Wu* (*KHCP*), p. 29; tr. Sun & Sun (1), pp. 230–1; also Jeon Song-woon (1), pp. 267–8.
[d] Cf. Hunter (9), pp. 94–6.
[e] Cf. *Choson Munhwasa* (Pyungyang, 1966–   ), vol. 1, p. 50; Kim Hyo-Gun (1), p. 17.
[f] The Ming artist Tung Chhi-Chhang[6] (+1555–1636) is said to have been fond of using Korean paper for his ink-splash style of painting; see *Fei Fu Yü Lüeh*[7] (*TSHCC*), pp. 8–9.

[1] 樂浪　　　[2] 曇徵　　　[3] 楮　　　[4] 鷄林紙　　　[5] 等皮紙
[6] 董其昌　　[7] 飛鳧語略

(leather-like paper) was used for making raincoats and curtains, and for mounting book bindings.[a] A number of such sheets pressed together and oiled were used as floor mats, and a single sheet was used in place of glass for windows.[b] Another kind was large and durable enough to make a tent of several sheets joined together,[c] and a thick and absorbent sort was used by Manchus to make shrouds.[d] As the Ming author Sung Ying-Hsing said: 'It is not known what the "white hammered paper" (*pai chhui chih*[1]) of Korea is made of,'[e] but more recent research makes it seem likely to have been made of paper mulberry bark by repeatedly pounding the long fibres into a fine pulp; this was a unique feature of Korean paper.

The Korean government attended to papermaking with great concern. A special Office of Papermaking (Chojiso[2]) was established in the capital early in the +15th century, and staffed with nearly 200 papermakers, mould makers, carpenters, and other labourers under the charge of three supervisors.[f] Hundreds of local papermakers were also commissioned in various provinces, and when large-scale printing projects occurred the government coordinated the paper supply from all over the country. This happened, for instance, in 1434, when 300,000 sheets were required for printing the general history of China, *Tzu Chih Thung Chien*,[3] and paper mulberry bark was collected from the provinces in 1437 when the *Tripitaka* was printed.[g] With the development of printing and the increasing needs of foreign trade, paper supplies became inadequate and various measures had to be taken by the government. The import of Japanese materials and technology was increased, productive papermakers were rewarded, and a search for new raw materials was encouraged. But this was not a unique instance, for government sponsorship of papermaking paralleled their encouragement of the development of printing and publishing throughout many centuries of Korean history.

Korea also made good ink for writing and printing. Inkmakers were employed in government offices from very early times, and as early as the Thang dynasty, Korean ink was among the annual tributes to China. It was produced by mixing the lampblack of old pine with a special glue obtained from the antlers of the tailed deer (*Cervus davidianus*) and is described as being as black as varnish.[h] For printing with metal type, a high-quality oil was added to make it less heavy and greasy than European ink,[i] and both Korean paper and ink were cherished by Chinese poets, as well as by people of other countries.

Among many landmarks in the history of printing, the Koreans have at least three distinctions: possession of the world's earliest known printed specimen;

---

[a] Cf. Chuang Shen (1), p. 94.   [b] Cf. Hunter (9), pp. 96–7.
[c] Cf. *Liu-Chhiu Kuo Chih Lüeh* (*TSHCC*), p. 166.
[d] Cf. Sohn Pow-Key (1), p. 102. Because the Manchus made heavy levies on this paper, the Koreans had to use thinner paper for printing.
[e] See *Thien Kung Khai Wu*, p. 219; tr. Sun & Sun (1), p. 231.
[f] See Jeon Song-Won (1), p. 267.
[g] See Korean *Veritable Records*, cited in Chang Hsiu-Min (5), p. 130.
[h] Cf. *Wei Lüeh Chi Pen*, ch. 12, p. 1.   [i] Cf. McGovern (1), p. 15.

[1] 白硾紙   [2] 造紙所   [3] 資治通鑑

survival of a complete set of wood blocks which is perhaps the largest and oldest of its kind in the world; and, finally, being the first to use metal type, so antedating Europe by some two hundred years. The earliest extant printing was discovered in Korea in 1966 in a stone stupa at Pulguk-sa[1] in Kyongju, the capital of the Silla Kingdom (+668–935),[a] and provides material evidence that printing existed around +700. This specimen is composed of separate pieces of thick mulberry paper joined together in a continuous scroll about 20 feet long and $2\frac{1}{4}$ inches wide, and mounted on a wooden roller lacquered at each end. The printing was done from a series of twelve woodblocks, each about 20 or 21 inches long and 2 inches wide, with eight characters in each vertical line. The text is a Buddhist sutra in Chinese, *Wu Kou Ching Kuang Ta Tho Lo Ni Ching*[2] (Fig. 1110), translated from the Sanskrit *Raśmivimalavaviśuddhaprabhādhārani* by the monk Mi-Tho-Hsien[3] of Tokhara between +680 and +704, while he was living in the Thang capital Chhang-an. This period corresponds closely with the reign of Empress Wu, who ruled China from +684 to +704, and during whose period on the throne about a dozen new forms of characters were created in Chinese.[b] At least four of these, including *cheng*[4] for proof, *chhu*[5] for beginning, *shou*[6] for to confer, and *ti*[7] for earth, occur in this printed text, the last appearing four times. What is more, the calligraphic style and its variations are very similar to those of the Thang manuscripts from Tunhuang,[c] and it is generally believed that, after completion of the translation in +704, this specimen must have been printed in Thang China and brought to Korea for ceremonial use no later than +751, when the stupa was built.[d]

Korean printing was promoted, as in Vietnam and Japan, first by the spread of Buddhism and, later, by the adoption of the civil service examination system modelled after the Chinese pattern. As early as in the +10th century, several printed sets of the *Tripitaka* were obtained from Sung China and the Liao or Kitan Kingdom,[e] and with these examples as a basis, the first *Tripitaka Koreana*,[f] in some

[a] See Goodrich (31), (32); Ledyard (2); Yi Hong-Jik (1), (2); also discussion on pp. 149 ff. above.

[b] Cf. *Tzu Chih Thung Chien* (reprint, 1956), ch. 204, p. 14; such new forms of characters are found in some 47 rolls of Tunhuang manuscripts of the late Thang period, see Giles (13), p. xvi; sample characters are given in Nghien Toan and Louis Ricard (1), pp. 114–15.

[c] See examples given by Yi Hongjik (1), p. 56; (2), pp. 183 ff.

[d] See Goodrich (32), pp. 376 ff. Some Korean scholars try to prove that this text was printed in Korea, on the ground that the *tak* or paper mulberry was used and the new forms of characters and certain calligraphic variations also appear in a few manuscripts now kept in Japan; see Yi Hong-Jik (1), (2). These arguments seem unconvincing, inasmuch as paper mulberry had been used in China since the +2nd century, and there is no evidence that these peculiar forms of characters were also used in Korea. Furthermore, there is no other record indicating that printing was done in Korea until some 300 years later; see also discussion on pp. 149 ff. above.

[e] Cf. *Sung Shih* (*ESSS/TW*), ch. 487, p. 5a, which says a copy of the *Tripitaka* was requested by Korea in +989 and granted by China two years later; and that no fewer than six or seven copies were obtained from the Kitan Kingdom; see Carter (1), pp. 89, 100; Chang Hsiu-Min (5), p. 105.

[f] The theory that the first *Korean Tripitaka* was cut in the middle of the +9th century is invalid, since an ambiguous statement based on a dream in the source is generally considered a forgery; see Chang (5), pp. 104–5.

[1] 佛國寺   [2] 無垢淨光大陁羅尼經   [3] 彌陀仙   [4] 鑒(證)
[5] 颺(初)   [6] 穮(授)   [7] 埊(地)

願不著色界有願不著聲香味觸法
界無願不著聲香味觸法界有願舍
利子諸菩薩摩訶薩修行般若波羅
蜜多與如是法相應故當言與般若
波羅蜜多相應
復次舍利子諸菩薩摩訶薩修行般
若波羅蜜多不著眼識界有不著眼
識界非有不著耳鼻舌身意識界有
不著耳鼻舌身意識界非有不著眼
識界常不著眼識界無常不著耳鼻
舌身意識界常不著耳鼻舌身意識
界無常不著眼識界樂不著眼識界
苦不著耳鼻舌身意識界樂不著耳
鼻舌身意識界苦不著眼識界我不
著眼識界無我不著耳鼻舌身意識
界我不著耳鼻舌身意識界無我不
著眼識界寂靜不著眼識界不寂靜
不著耳鼻舌身意識界寂靜不著耳
鼻舌身意識界不寂靜不著眼識界
空不著眼識界不空不著耳鼻舌身
意識界空不著耳鼻舌身意識界不
空不著眼識界無相不著眼識界有
相不著耳鼻舌身意識界無相不著

大般若經第五卷 第二十五張 天

耳鼻舌身意識界有相不著眼識界
無願不著眼識界有願不著耳鼻舌
身意識界無願不著耳鼻舌身意識
界有願舍利子諸菩薩摩訶薩修行
般若波羅蜜多與如是法相應故當
言與般若波羅蜜多相應

大般若波羅蜜多經卷第五

丁酉歲高麗國大藏都監奉
勅雕造

大般若經第五卷 第二十六張 天

Fig. 1206. Text of the Eighty Thousand *Tripitaka*, printed in Korea in the +13th century. From *Tripitaka Koreana*, reprint by Dong-Kook University, Seoul, 1957.

5924 chapters, was printed between 1011 and 1082, in fulfilment of a vow for expulsion of the Kitan invaders. In addition, a supplement, consisting of about 4000 chapters of writings by Korean, Kitan, and Sung authors, was compiled and printed by the princely monk Gitan[1] before he died in 1101.[a] This edition was later destroyed when the Mongols invaded the country in 1232, and a new edition in 6791 chapters was printed from 1237 to 1251.[b] This is the famous 'Eighty Thousand *Tripitaka*' (Fig. 1206), so called because it consisted of 81,258 blocks of magnolia

[a] After completion of the supplement, copies were presented to China, Kitan, and Japan; see Paik Nak-Choon (1), pp. 69–70. Some scholars believe that the supplement was never completed; see Chang Hsiu-Min (*5*), pp. 106–7.

[b] Cf. Sohn Pow-Key (1), p. 97; Jeon Song-Woon (1), pp. 107 ff.

[1] 義天

Fig. 1207. Eighty thousand blocks for printing the *Tripitaka Koreana* in the +13th century are preserved intact in the Haein Temple in South Korea.

wood, carved on both sides which are still kept almost intact today at Haein-sa,[1] high on Mt Kaya in southern Korea (Fig. 1207).

The printing of secular works came somewhat later, and on a smaller scale. First, in +1042, three Chinese historical works appeared under imperial auspices, and in +1045 two Confucian classics were printed by the Imperial Library,[a] but the rising tide of scholarship caused the Koreans to turn to Sung China for their supply of books, where some blocks were also engraved and brought to Korea.[b] However, a reluctance developed among some Sung scholar-officials to exporting Chinese

---

[a] See Chang Hsiu-Min (5), p. 110.
[b] Gitan brought printing blocks from China, in addition to some 4000 *chuan* of printed books and manuscripts.

[1] 海印寺

books to Korea for reasons of national security,[a] but this only encouraged further development of printing by the Koreans so that they could become self-sufficient in supplying the books they needed, especially the Confucian classics, Neo-Confucian writings, and medical works.[b] In the +12th century, Koryŏ began extensive printing after the establishment in 1101 of a printing office in the National Academy, which took over the wood blocks from the imperial library, with the result that a member of the Chinese mission to Korea, Hsü Ching[1] (+1091–1153), could report a government collection at Koryŏ which numbered several tens of thousands.[c]

Under the Mongol domination from around 1270, further political and cultural ties were established between Korea and China, and also with Central Asia. In 1290 the Yüan court sent a group of craftsmen to repair the Haein-sa wood blocks, and one copy of the *Korean Tripitaka* was presented to the Yüan court in 1308, and another to Mongolia in +1314.[d] Furthermore, to honour the Yüan emperor, in 1312 a Korean king who had married a Mongol princess ordered fifty copies of the *Tripitaka* to be printed for distribution to various temples.[e] Again, a large collection of over 10,000 Chinese books was brought to Korea in 1314, this in addition to some 4000 volumes donated by the Yüan court.[f] All these activities were related to Buddhism in one way or another, for extensive printing of secular works was not begun until the overthrow of the Koryŏ dynasty at the end of the 14th century. Then the establishment of the Yi dynasty (+1392–1910) brought political stability, social reforms, and cultural vitality to Korea; it promoted Confucianism over Buddhism, adopted the civil service examination system, established the national university, and created an alphabetic script, known as Han-gul, as its national form of writing. It was under this new regime also that the demand for more books promoted the wide application of metal type for printing.

Although the first extant book printed from metal type was made at the beginning of the 15th century, a contemporary record indicates that a copy of the ancient and modern ritual code, *Kogum Sangjong Yemun*,[4] was printed about +1234 on Kanghwa Island, off the west coast of Korea from 'cast characters' (in Korean: *chu cha*[5]).[g] However, at least two wooden movable-type editions, a work on the laws

[a] See Poon Ming-sun (2), pp. 55–63; a Chhüan-chou printer, Hsü Chien,[2] in Hangchow, engraved some 2900 printing blocks for the Koreans, for 3000 taels of silver; he was exiled as the result of a petition by Su Shih;[3] see *Su Tung-Pho Chi* (*KHCP*), ch. 5, pp. 38–40.

[b] Local officials submitted to the court ninety-nine woodblocks of medical books in +1058 and seventy-three in +1059; see Chang Hsiu-Min (5), p. 111.

[c] *Hsüan-Ho Feng Shih Kao-Li Thu Ching* (*TSHCC*), pp. 32, 99.

[d] Cf. Paik Nak-Choon (1), p. 72.

[e] Cf. Chang Hsiu-Min (5), pp. 109–10.

[f] Cf. Carter (1), p. 223.

[g] Mentioned in the collected writings of Yi Kyo-Bo[6] (+1168–1241) as cited in Kim Won-Yong (1), pp. 5–6. A recent claim that a movable type edition of the *Komin Chinbo Taejun*[7] dates from *c.* +1160 has been proved in error through a mistaken attribution of its author, Huang Chien[8] of the +14th century; see Goodrich (38), p. 476.

[1] 徐兢	[2] 徐戩	[3] 蘇軾	[4] 古今詳定禮文
[5] 鑄字	[6] 李奎報	[7] 古文眞寶大全	[8] 黃堅

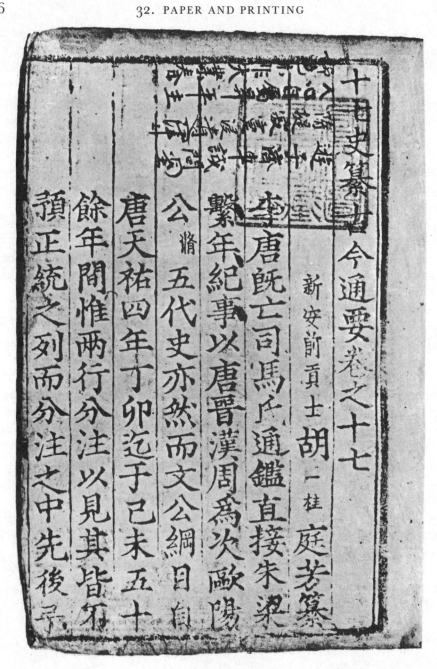

Fig. 1208. Earliest extant specimen printed with bronze movable type in Korea, +1403. From Sohn Pow-Key (2).

and statutes of the Ming dynasty and a biography of the founding fathers of the Yi dynasty, were printed in 1395 and 1397 respectively.[a] Then, in 1403, a new Bureau of Type Casting (Chuja-so[1]) was added to the Office of Publications (Sŏjŏkwŏn[3]). From this time until the middle of the 19th century, no fewer than thirty fonts are known to have been produced, ranging from 60,000 to as many as 300,000 characters each, making a total of two or three million.[b] Except for one made of lead in 1436 and two of iron in +1668 and 1721-4, the metal fonts were bronze; there were also six or seven fonts of wooden type.[c]

All the early fonts were given names from the sexagesimal cycle, for example *kemi-ja*[4] for type cast in 1403 (Fig. 1208); later ones were named after calligraphers, the titles of books to be printed, or places where the fonts were stored. The most elegant calligraphy among all these fonts was that of one made in 1434 and modelled after the style of the famous Chinese calligrapher Madame Wei[5] of the +4th century (Fig. 1209). All the type was of Chinese characters except for a few of the Korean alphabetic scripts (Fig. 1210). Most of the early fonts have now been lost through fire or war, or melted down for recasting,[d] the great loss occurring during the Japanese invasion of 1592-8, when many technicians and their fonts were taken back to Japan which was starting its own movable-type printing.[e] The use of wood and iron instead of bronze in Korea was necessitated by the resulting shortage of copper and the interruption of trade after the war.[f]

The Office of Publications played a major role in printing and book manufacture, especially with movable type. We know, too, from the number of craftsmen employed, that there was a division of labour in this office during the 15th century. The employees included over 100 foundrymen, typecasters, type cutters, wood carvers, typesetters, printers, papermakers, proofreaders, and inspectors, and a very strict system of punishments and rewards was maintained for quality control. Careful work was rewarded with bonuses or official titles; negligence was punished by thirty blows per error.[g] For this reason, Korean

---

[a] A copy of the 1397 printing survives; see Kim Won-Yong (1), pp. 7–8. Chang Hsiu-Min (5), pp. 87–8, reports a copy of the *Yü Shih Tshe*,[2] or Collected Papers Written at the Imperial Palace Examinations, now at the National Peking Library, was printed with Korean paper and movable type similar to those of Korea. He believes it could have been printed with Korean paper in Yüan China. But Fang Chao-Ying has suggested in an unpublished paper, 'On Printing in Korea,' that it was very likely printed in Korea between 1333 and 1368, because Korean students who participated in the examinations at the Yüan court and received the *chin-shih* degree, usually printed their papers as models for study and imitation by other candidates. Some fifty Korean students took that examination from 1315 to 1354 and at least eight of them gained the *chin-shih* degree. If this book was actually printed in Korea, it represents another example of movable type printing known to be extant between 1234 and 1397.

[b] The tables of movable-type fonts from 1403 to 1858 list twenty-six in Kim Won-Yong (1), pp. 12–13; twenty in Chang Hsiu-Min (5), pp. 120–2; and twenty-eight in McGovern (1), pp. 19–20.

[c] See facsimile specimens reproduced in McGovern (1) and Sohn Pow-Key (2).

[d] Over 600,000 such types dating from 1777 are still preserved in the National Museum of Korea; examples are also kept elsewhere in the world.

[e] Cf. Nagasawa Kikuya (1), pp. 135–8; also discussion of Japanese movable-type printing on pp. 341 ff. below.

[f] Cf. Sohn Pow-Key (1), p. 102.

[g] See Korean records cited in Chang Hsiu-Min (5), pp. 124, 126.

---

[1] 鑄字所   [2] 御試策   [3] 書籍院   [4] 癸末字   [5] 衛夫人

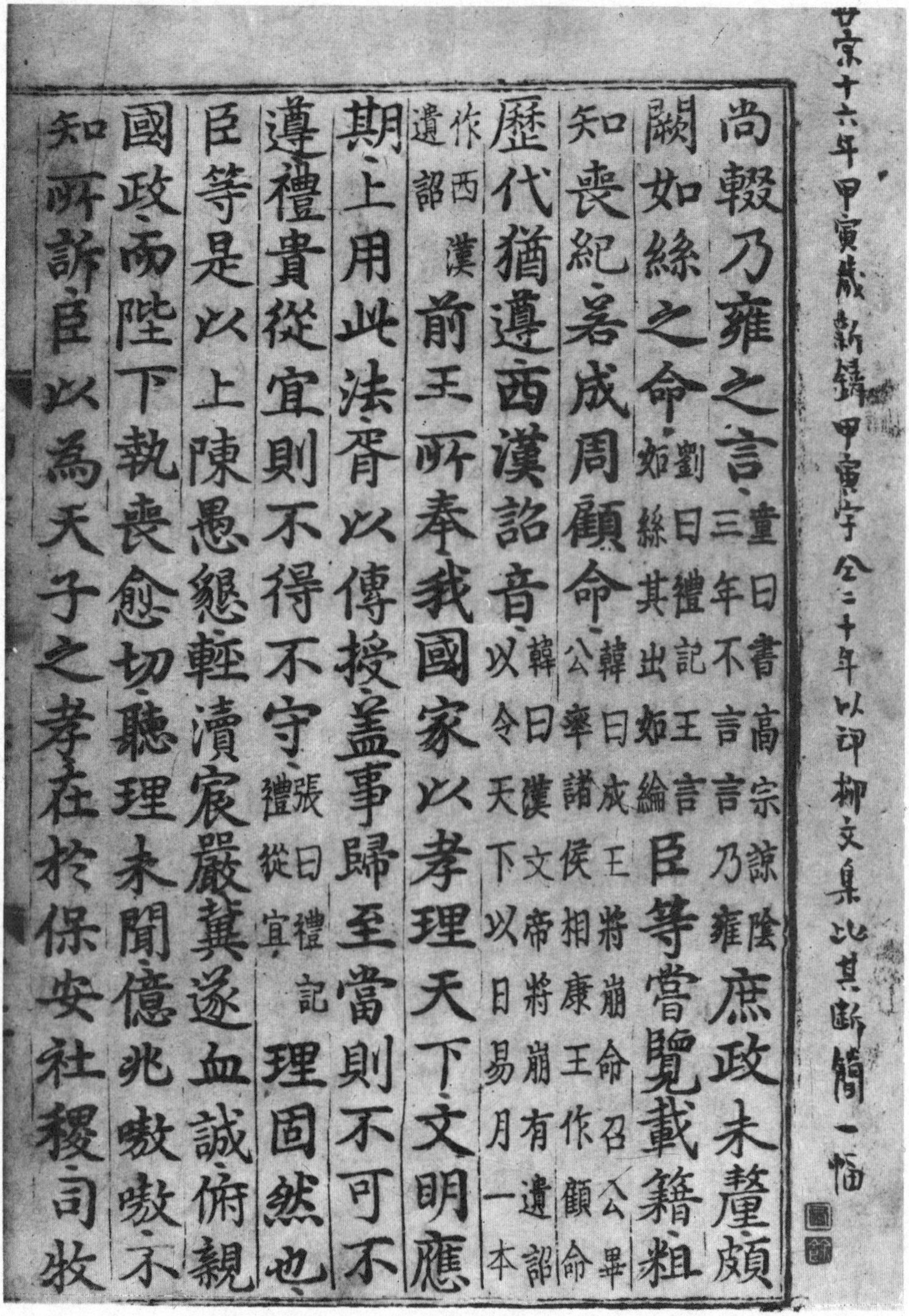

Fig. 1209. A page of the *Yuminjip*, or Collected Works of Liu Tsung-Yüan (+773–819), printed with bronze movable type of +1434 in the calligraphic style of Madame Wei. From British Library.

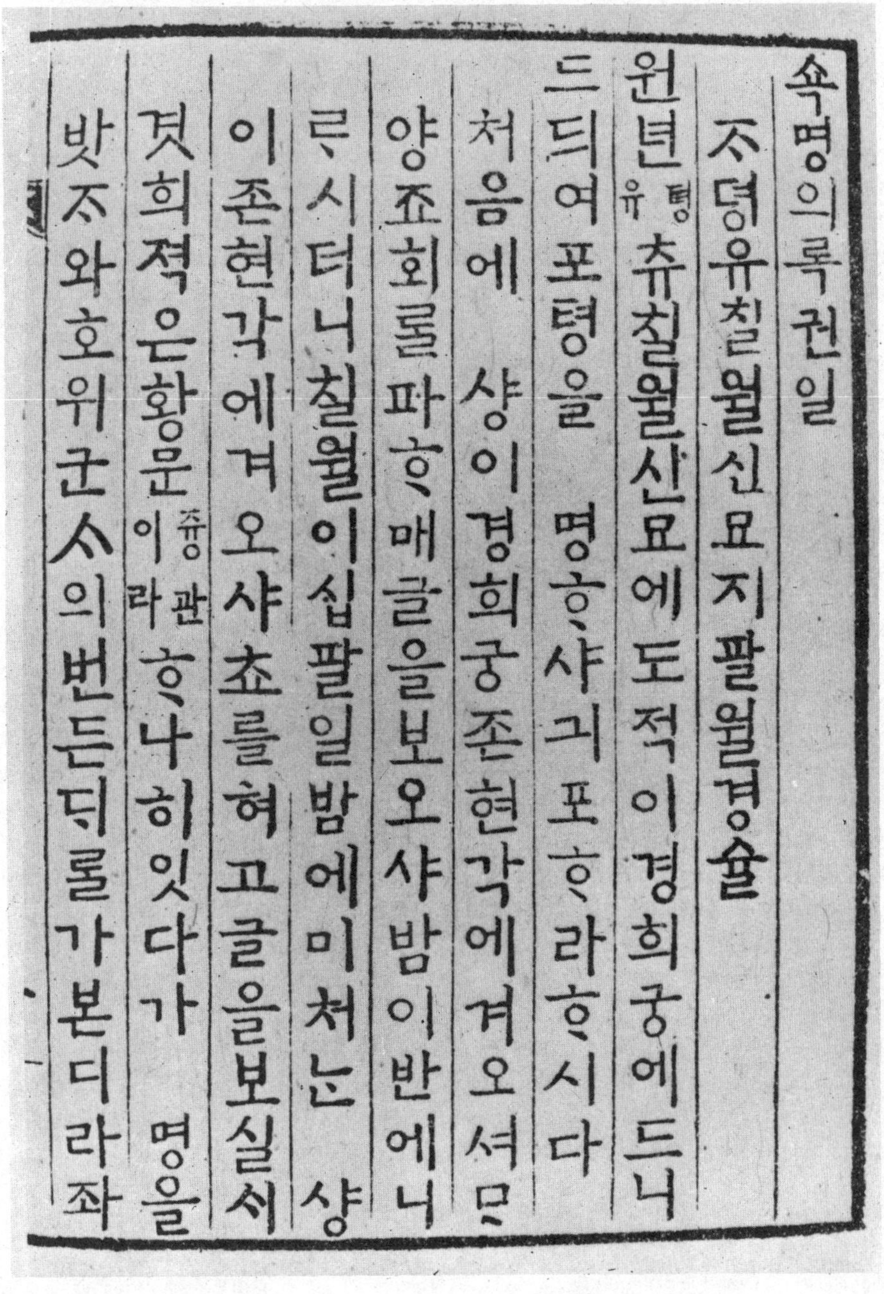

속명의록권일

조뎡유칠월신묘지팔월경슐

원년뎡유츄칠월신묘에도적이경희궁에드니

드듸여포텽을 명ᄒᆞ샤긔포ᄒᆞ라ᄒᆞ시다

처음에 샹이경희궁존현각에겨오셔ᄆᆞ

양죠회ᄅᆞᆯ파ᄒᆞ매글을보오샤밤이반에니

ᄅᆞ시더니칠월이십팔일밤에미처ᄂᆞᆫ 샹

이존현각에겨오샤쵸를혀고글을보실ᄉᆡ

겻희젹은황문즁관이라ᄒᆞ나히잇다가 명을

밧ᄌᆞ와호위군ᄉᆞ의번드ᄂᆞᆫᄃᆡᄅᆞᆯ가본디라좌

Fig. 1210. A Korean work printed with bronze movable type in Han-gul script, +1777.

editions have been known for careful collation and physical qualities compared with those printed in Ming China.

It appears that the use of movable-type printing in Korea was influenced by three major factors, all more or less related to Chinese practice. One was the idea of movable type. This was unquestionably inspired in Korea by the method described in Chinese records, for in the preface to a movable-type edition of *Po Shih Wen Chi*,[1] printed in Korea in +1485, the Korean scholar Kim Jongjik[2] said explicitly that 'the movable type method was begun by Shen Kua and brought to perfection by Yang Wei-Chung'.[a] Although he was mistaken in identifying Shen Kua as the inventor, his acknowledgement of its Chinese origin is clear. How it arrived in Korea is not certain, but it may have been brought back by the princely monk Gitan, who travelled to China and resided in Hangchow in the latter part of the +11th century, at the time and in the very place of Pi Sheng's invention.[b] He could, therefore, have been informed by his contemporaries in China, or through reading Shen Kua's description, which certainly influenced the application of movable type by Korean printers. If so, the use of movable type in Korea must have begun earlier than the generally accepted date of 1234.

Second, the technology of type-casting was apparently adapted from that of casting coins. As described by the Korean scholar Sŏng Hyŏn[4] (+1439–1504), the character was first cut in beech wood which was then pressed into soft clay to make a mould. Molten bronze was poured into the mould to form the type, which was then polished to its final shape.[c] In 1102, the Chinese method of casting coins, known as 'drum-casting' (*ku chu*[5]), was introduced to Korea,[d] and later it was said that the 'clean and even' inscriptions cast by the Chinese method were the indispensable prerequisite for making a clear type.[e]

Third, the demand by the educated elite for more books during the 12th century could only be solved by the use of movable-type printing. With the establishment of the Bureau of Type Casting, printing developed with such vigour that 'no book on any subject was not available in print' and 'the Office of Publications was engaged every day in printing rare books which were brought from China'.[f] Some scholars have suggested that because of the shortage of suitable wood for block carving, copper alloys and other metals were used for casting type.[g] However, this does not seem to have been the major reason, since tens of thousands of woodblocks were carved for printing the *Tripitaka* on several occasions. The high quality of the

[a] See *Yongjae Ch'onghwa*[3] (Collected Essays of Sŏng Hyŏn), tr. Sohn Pow-Key (1), p. 99.
[b] Cf. Hu Tao-Ching (4), pp. 61–3.
[c] Cf. Sohn Pow-Key (1), p. 99.
[d] Sohn (1), p. 99, considers that the *ku chu* method coincides with what was described by Sŏng Hyŏn and not with what the commentator on the *Han Shu* referred to as 'smelting by the use of a fan'.
[e] Cf. Sohn (1), p. 100.
[f] See citations from Korean sources in Chang Hsiu-Min (5), p. 125.
[g] For example, Sohn Pow-Key (1), p. 98.

[1] 白氏文集   [2] 金宗直   [3] 傭齋叢話   [4] 成俔   [5] 鼓鑄

copper produced in Korea may have been the primary factor for the choice of movable type cast in bronze.[a]

(2) BEGINNINGS OF PAPER AND PRINTING IN JAPAN AND LIU-CHHIU

The date when paper was first introduced to Japan is unknown, for though Japanese historical works contain references to early records and documents, it is uncertain whether they were written on paper.[b] However, about the second half of the +4th century a Korean scholar, Wani,[2] who served as tutor to the crown prince, presented a number of Chinese books to the Japanese court;[c] these were obviously of paper. Then, during the +6th century, the Korean kings of Paekche sent books to Japan on several occasions,[d] and in +610 the Korean monk Doncho[3] (in Korean: Damjing) (Fig. 1211) came to Japan: he, it is said, was 'able to make paint, paper, and ink, and moreover to make mills'.[e] Probably the making of mills began at this time. Traditionally this has been taken as the beginning of paper-making in Japan, but it may not necessarily be so, for the account stresses Doncho's varied accomplishments and does not specifically refer to the introduction of papermaking as it does to making mills. Paper could have been produced on a small scale in Japan before +610, and it has been suggested that Chinese and Korean immigrants may have introduced the art of papermaking in the +5th century.[f]

The earliest paper manuscript extant in Japan is probably a commentary on the Lotus *sutra*, *Hoke-kyō Gisho*,[4] said to have been written or annotated by Prince Shōtoku[5] during the period +609 to 616, the paper probably coming from China.[g] The oldest existing specimens of Japanese-made paper, preserved in the Shōsōin Imperial Repository in Nara, are fragments of household registers from three provinces, and are dated +701. The paper was made in the provinces of paper mulberry, and its quality is said to be inferior to that of contemporary Chinese paper.[h] Later, in the middle of the +7th century, the Japanese government adopted many of the features of the Chinese administrative system, of which household registers were one example. It all required numerous documents, so there was a large-scale demand for paper by both central and local governments;

[a] See a report on Korea, *Chhao-Hsien Fu* (*YCTS*), p. 15a, by Tung Yüeh[1], a Chinese envoy to Korea early in 1488, who said that Korea produced its own copper of superior quality. For manufacture of bronze type, copper was also imported from Japan.
[b] See Jugaku Bunshō (*1*), pp. 2–5.
[c] See *Kojiki*, p. 69, tr. Philippi (1), p. 285; *Nihongi*, vol. 2, p. 213, tr. Aston (1), vol. 1, pp. 262–3.
[d] See Jugaku Bunshō (*1*), p. 5.
[e] See *Nihongi*, vol. 3, p. 450, tr. Aston (1), vol. 11, p. 140.
[f] Cf. Jugaku Bunshō (*1*), pp. 19–20.   [g] Cf. Jugaku Bunshō (*1*), pp. 22–4.
[h] See *Shōsōin no kami*, or *Various Papers preserved in Shōsōin* (Tokyo, 1970), pp. 19–21.

[1] 董越   [2] 王仁   [3] 曇徵   [4] 法華經義疏   [5] 聖德

Fig. 1211. Three patron saints of papermaking in Japan. The scroll painting depicts Tshai Lun, the inventor of paper, at the centre; Doncho (Damjing), the introducer of paper to Japan, on the left; and Seibei Mochizuki, one of the early papermakers of Japan, at right. Dard Hunter Paper Museum.

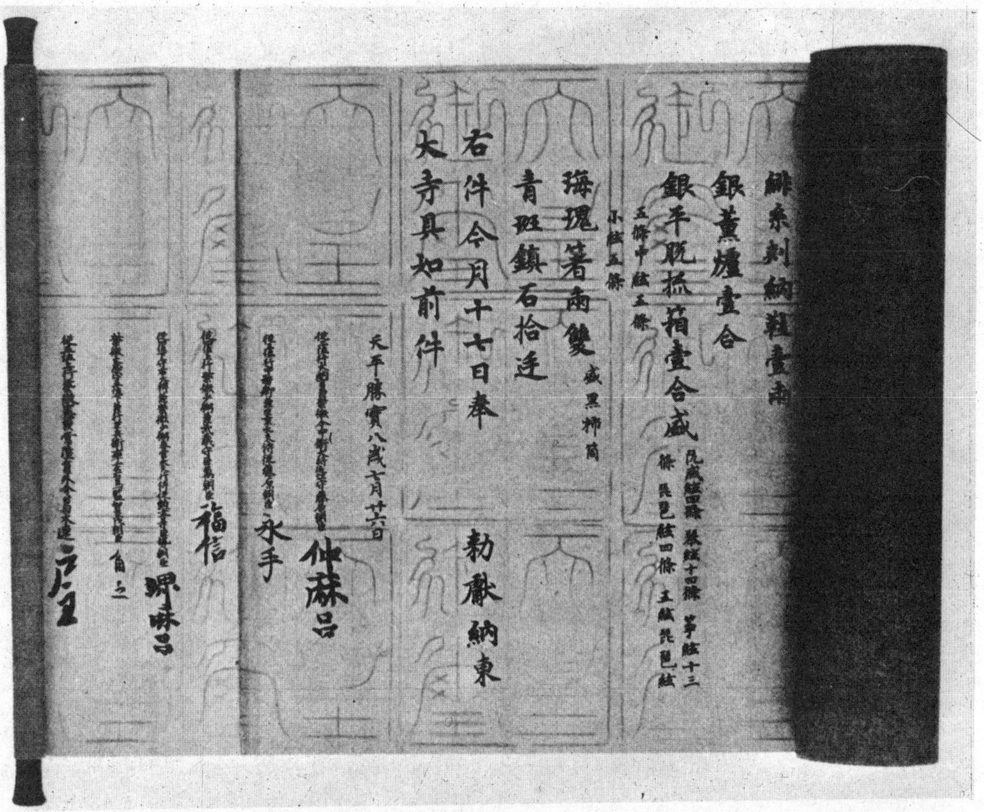

Fig. 1212. List of articles donated to the Todaiji Temple written on paper mulberry paper, dated +756, and now kept at the Shōsō-in, Nara, Japan.

what is more, the court used hundreds of thousands of sheets of paper at a time copying Buddhist *sutras*, thus creating another major demand.[a]

The administrative code of +701 provided for a government office to make paper, and after the capital had been moved to Kyoto, the Kamayain,[1] a paper mill, was set up between +806 and 810 to supply the needs of the court.[b] Today the Shōsōin still preserves many examples of old paper (Fig. 1212), and early government documents there often contain references to paper; indeed, from +727 to 780 some 233 different varieties of paper were referred to.[c] At this time the government used many different papermaking materials. Most early Japanese papers were made of hemp, two types of paper mulberry: *kozo*[2] (*Broussonetia papyrifera*, Vent.) and *kajinoki*[3] (*Broussonetia kazinoki*, Sieb.), and *gampi*[4] (*Wikstroemia*

[a] Cf. Jugaku Bunshō (*1*), p. 40.
[b] Cf. Jugaku Bunshō (*1*), pp. 39–42, 51–2, 207, 319–20.
[c] Cf. Jugaku Bunshō (*1*), pp. 105–10.

[1] 紙屋院   [2] 楮   [3] 梶の木   [4] 雁皮

sikokiana, Franch. et Sav.); the finest papers were made primarily of hemp. From
the Heian period (+794–1185) hemp fell into disuse, and paper mulberry and
gampi were the major materials employed. Much later, *mitsumata*[1] (*Edgeworthia
papyrifera*, Sieb. et Zucc.) was introduced, its first use being recorded in 1598,[a] and
paper mulberry, *gampi*, and *mitsumata* have continued to be the major raw fibres for
handmade paper in Japan to the present time.

The early papers in Japan were produced by the same method as that used
in China, called *tamezuki*[2] (accumulation papermaking), which is still in use for
some papers. Another method, *nagashizuki*[3] (discharge papermaking), that was
developed in the +8th or +9th century, has been used to produce most Japanese
paper in later times;[b] its distinctiveness lies both in the technique of the vatman and
in the addition of a vegetable mucilage to the fibres in the vat. The mucilage
performs a number of functions, such as causing the fibres to be evenly distributed
in the solution, with the result that the paper is stronger, firmer, and glossier. With
the *tamezuki* method the vatman dips the mould into the vat, lifts out the pulp
solution, and then allows the water to drain, but with the *nagashizuki* method the
vatman shakes the mould with the pulp both forward and back and from right to
left, which serves to align the fibres regularly, and rather than allowing the excess
water to drain naturally, he shakes the mould so as to remove the water.[c] One of the
early books on papermaking, *Kamisuki chōhōki*,[4] published in 1798, gives the step-
by-step procedures with illustrations (Fig. 1213).[d]

During the Heian period the most famous papers were those manufactured in the
Kamayain in Kyoto, but towards the end of the period these papers were sometimes
made of recycled materials, and they declined in quality. Danshi,[5] a high-quality
paper made of paper mulberry (*kozo*), originally manufactured in Tohoku and
later at other places also, replaced the Kamayain paper at court. During the Heian
period paper production spread throughout the country as indicated by the fact
that at the beginning of the +9th century levies of paper were exacted from forty-
two provinces.[e] From the Kamakura period (+1192–1333) a variety of papers
came into prominence, some limited to particular places and others produced more
widely, and paper became a popular as well as an aristocratic commodity. During
the 15th century paper guilds and paper markets arose, and later the paper trade
increased with economic growth and the elimination of tariffs. The scale of this
trade may be gauged from the fact that in the 19th century the association of paper
merchants in Osaka consisted of about 70 wholesalers, 155 or 156 brokers, and
about 500 retailers; indeed, paper was one of the most important trade
commodities in both Osaka and Tokyo.[f] Unfortunately, despite its high quality,
aesthetic appeal, and popularity, Japanese handmade paper has gone into a long-

[a] Cf. Jugaku Bunshō (*1*), pp. 25, 29, 81, 104, 322; Hughes (1), pp. 76–83; Hunter (9), pp. 56–8.
[b] Cf. Hughes (1), pp. 84–5; Jugaku Bunshō (*1*), pp. 78–81.
[c] Cf. *Tesuki Washi Taikan*, vol. 1, pp. 31–5.   [d] Cf. tr. C. E. Hamilton (1).
[e] Cf. Jugaku Bunshō (*1*), pp. 205–22.   [f] Cf. Jugaku Bunshō (*1*), pp. 302–11.

[1] 三椏   [2] 溜漉   [3] 流漉   [4] 紙漉重寳記   [5] 檀紙

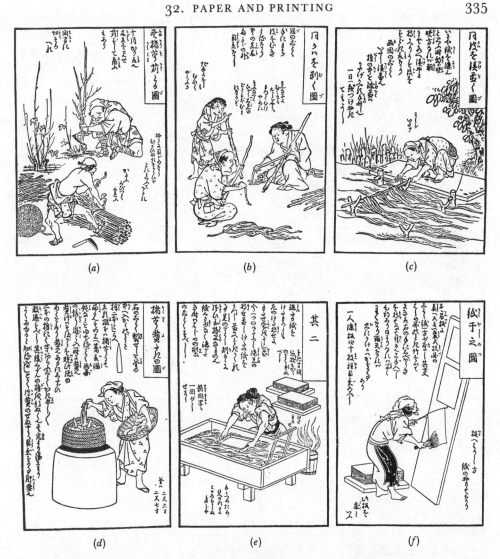

Fig. 1213. Steps in making paper mulberry paper in Japan in the +18th century. (*a*) Cutting paper mulberry trees. (*b*) Stripping the bark. (*c*) Washing the bark after peeling. (*d*) Boiling the fibres into pulp. (*e*) Dipping the mould into the vat. (*f*) Drying the sheets on wooden board. From *Kamisuki Chōhōki*, +1798.

term decline as a result of competition with cheaper machine-made paper, first introduced in the 1870s.[a]

In Japan, besides its use as a material for writing, paper has been put to many such uses as for umbrellas and waterproof coverings, handkerchiefs and toilet paper, for windows and walls, and for clothing.[b] The last came in two types: *kamiko*[1]

[a] Cf. *Tesuki Washi Taikan* (5 vols., Tokyo, 1973–4), I, p. 20.
[b] Cf. Hughes (1), pp. 48–69; Hunter (9), pp. 217–21; Seki Yoshikuni (3), pp. 77–103; *Tesuki Washi Taikan*, IV, pp. 5–21.

[1] 紙衣

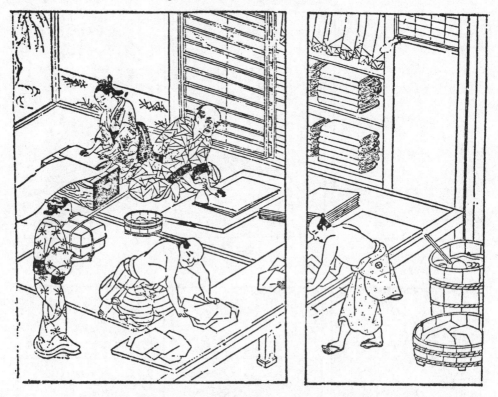

Fig. 1214. Making paper clothing (*kamiko*) in Japan, +1754. From Seki Yoshikuni (*3*).

(Figs. 1087, 1214) made directly of specially treated sheets of paper, and *shifu*,[1] made from thread spun out of paper strips. The former, which originated among Buddhist monks for ritualistic reasons, was most likely an idea imported from China, and its use was recorded first in the +11th century. The weaving of *shifu* is recorded only from 1712.[a]

Chinese books were introduced to Japan at an early date, as discussed above. Following the Taika reform of +645, the institution of the Taiho code in +701, and the spread of Buddhism at the same time, Chinese influence at the court was paramount, and as a result, fifteen official missions travelled to the Thang from +630 to 834, and many monks and students went to China for study, often staying there for many years. With such close cultural and religious contacts between the two countries, it is not surprising that it was during this period that printing appeared in Japan.[b] The earliest extant Japanese printing is certainly the famous

[a] Ōmichi Kōyū (*1*), pp. 6–8; Seki Yoshikuni (*3*), pp. 94–7; *Tesuki Washi Taikan*, IV, pp. 14–16; Seki (*1*), vol. 2, pp. 17, 51; Jugaku Bunshō (*1*), p. 317, notes that *kamiko* may date to the +8th century.
[b] Cf. Kimiya Yasuhiko (*2*), pp. 74–84, 196–214; the monks took back to Japan many books, mostly religious though some were on secular subjects. According to a catalogue of +865, two dictionaries taken back by monk Shūei were printed editions; see above, pp. 152 ff.

[1] 紙布

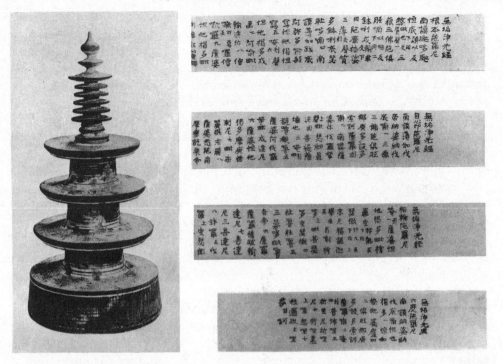

Fig. 1215. Four different versions of the *dharani* charms (*Kompon*, *Jishin-in*, *Sorin*, and *Rikudo*), printed in Japan from *c.* +764 to +770, were inserted in one million small wooden pagodas.

'one million *dharani*' (Fig. 1215), which consists of four different versions of Sanskrit charms transliterated into Chinese characters, and was probably printed between +764 and 770. Each charm, printed on yellowish hemp paper varying in size from 12 to 22 inches long and a little over 2 inches wide, was inserted into a small wooden pagoda. The pagodas were equally divided among ten leading Buddhist temples of the time, seven of which were in Nara.[a] Each of the four versions consists of a text of from 71 to a little more than 200 characters, with fifteen to forty lines to each text. Making copies of *dharani* was a popular means of gaining religious merit; in this case the crisis of a revolt was the occasion for the good work, carried out by the Empress Shotoku. Contemporary records do not refer to printing, but it has been determined from examination that the *dharani* were in fact printed, either from wooden blocks or possibly from stone, porcelain, or copper plates.[b] The characters are uneven and crudely formed when compared with earlier printing found in Korea or the *Diamond Sutra* of 868 from Tunhuang. Despite the early date of this

[a] Cf. Carter (1), 46–53; Hickman (1), pp. 87–93; Nagasawa Kikuya (11), II, pp. 2–3, suggests that the charms and pagodas may actually have been completed only for the Horyuji.

[b] Cf. Kimiya Yasuhiko (2), pp. 4–11 and frontispiece; Nagasawa Kikuya (11), II, pp. 2–3. General opinion leans to wood blocks, while one leading scholar, Kawase Kazuma, favours copper plates; see further discussion on pp. 150 ff. above.

example, there seems to be no question that the technique for printing came from China.[a]

The first known printing of complete books in Japan did not occur until some two centuries after this incident. One stimulus at that time was the importation to Japan of the Chinese imperial Khai-Pao edition of the *Tripitaka*, which was presented by the Sung Emperor Thai-Tsung to the Japanese monk Chonen[1] in +983.[b] Another was the Buddhist custom of making a large number of copies of a sutra as a pious work, often to commemorate a deceased person. As these copies were not meant to be read, there was no need for care in their preparation, and many copies were turned out efficiently by printing. Contemporary records state that one thousand copies of the *Lotus Sutra* were printed in 1009 and again in 1014; these are the earliest examples of such works, which are known as *surikyo*[2] (folded or printed *sutras*).[c] The earliest specimen of this type still in existence is a copy of the *Lotus Sutra* bearing a handwritten date of 1080; it must have been printed in or before that year.[d] Such books are characterised by light ink, and are sometimes almost illegible.

The *sutras* read by monks and others were originally reproduced by hand-copying in the monasteries, and this work continued to be important even after the development of printing. Following the examples of the Sung *Tripitaka* and the *surikyo*, however, printing began to be used to reproduce *sutras* for reading, and the earliest extant example is a Chinese text, *Chheng Wei Shih Lun*[3] (The Doctrine of Mere Consciousness), printed in 1088 by the Kofukuji[4] in Nara (Fig. 1216).[e] From the 11th century through to the end of the Kamakura period (+1192–1333), the printing of sutras was concentrated in the great Buddhist temples of Nara and Kyoto. Though these were almost all reissues of Chinese books in Japan, the calligraphic style followed that of the handwritten copies of *sutras* rather than the square and formalised printing style developed in China during the Sung.[f]

During the Kamakura period Zen Buddhism and Neo-Confucianism were introduced from China to Japan where they became very influential. One result was that from the +13th to the 16th centuries, the major efforts in Japanese printing were carried out in parallel groups of Zen temples in Kyoto and Kamakura, known as the Gozanji.[5] The books published by these temples are known as *Gozanban*[6] and represented several new developments in Japanese printing. First, the calligraphic style of these works made a break with the past; in

---

[a] See Kimiya Yasuhiko (2), pp. 17–29.   [b] Cf. *ibid.* pp. 302–4, 307.
[c] Cf. Kawase Kazuma (1), pp. 6–7; (3), pp. 10–23; Kimiya Yasuhiko (1), pp. 305–6, and (2), pp. 34–7, both give the same convenient table of *surikyo* editions described in contemporary records.
[d] See Kawase Kazuma (1), No. 2, which illustrates this work; Chibbett (1), p. 50, refers to a similar book with a handwritten date of +1053, but Nagasawa Kikuya (11), II, pp. 23–4, points out that since this date is on the reverse side of the paper from the printing, it is not acceptable as a date for the printing of the book.
[e] Kawase Kazuma (1), No. 4.
[f] Cf. Kawase Kazuma (3), pp. 24–7; (4), p. 47; (1), p. 6; Chibbett (1), pp. 39–57.

[1] 奝然   [2] 摺經   [3] 成唯識論   [4] 興福寺   [5] 五山寺
[6] 五山版

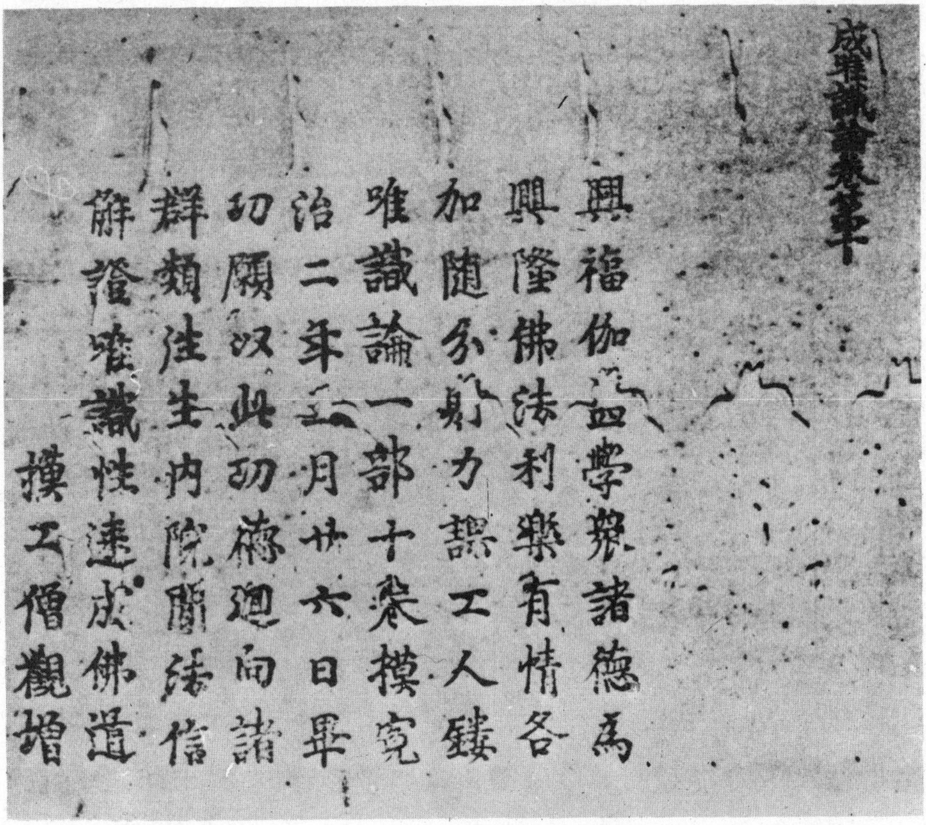
成唯識論卷第十
興福伽藍學衆諸德爲
興隆佛法利樂有情各
加隨分財力課工人鏤
唯識論一部十卷模寛
治二年五月廿六日畢
功願以此功德迴向諸
群類往生内院聞法信
解證證識性速成佛道
摸工僧觀增

Fig. 1216. Earliest extant book of Japanese block printing. Chinese text of the *Chhêng Wei Shih Lun*, printed in Nara in +1088.

place of the earlier styles of handwriting, these copied closely the forms of Sung editions, including their square characters.[a] Secondly, secular works were published for the first time in Japan, and significantly these were all reprints of Chinese books. First among them was the poetry of Han-Shan[1] printed in 1325, while a milestone in the study of Confucian classics was reached with the publication of the *Analects of Confucius* (Fig. 1217) in 1364.[b] Altogether the Gozan temples are known to have published seventy-nine secular Chinese works in addition to almost two hundred editions of religious writings. Over half of the secular works were the collected literary writings of Chinese authors, and the *Gozanban* also included the earliest medical works printed in Japan.[c] Thirdly, it was during the fourteenth century that the Japanese script (*kana*) was used for the

[a] See Kawase Kazuma (*3*), pp. 24–7, pp. 11–12; the basic study of *Gozamban* is Kawase Kazuma (*4*).

[b] *Ibid.* (*3*), pp. 72–3, cites Nagasawa's demonstration of the mistakes in Shimada Kan's earlier attribution of another edition of the *Analects*, an attribution followed in Chang Hsiu-Min (*5*), p. 135.

[c] Cf. Kawase Kazuma (*3*), pp. 70–83; (*4*), pp. 190–211; Kimiya Yasuhiko (*2*), pp. 354–5.

[1] 寒山

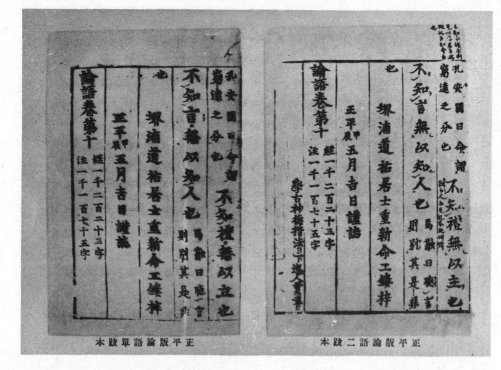

Fig. 1217. The Gozanban edition of *Confucian Analects*, printed for the first time in Japan in +1364. The above is a reprint of the original edition with single commentary on the left and double commentary on the right.

first time in printed books; the first example came out in 1321, followed from 1387 to 1589 by its use in a number of calendars.[a]

In the second half of the 14th century, during the unsettled conditions at the end of the Yüan, many Chinese block carvers migrated to Japan and worked on the *Gozanban*. One group of eight carvers arrived in 1367 and may have included Chhen Meng-Tshai[1] and Chhen Po-Shou,[2] both of whom came from Nan-thai, a suburb of Fuchow.[b] The first book printed by them came out in 1367, and we know that more than thirty Chinese printers were active in Japan for approximately the next thirty years.[c] The names appearing most often in colophons and book margins were Chhen Meng-Jung[3] from Chiang-nan and Yü Liang-Fu[4] from Phu-thien; indeed, the name Yü Liang-Fu appears in seventeen books, indicating that he was by far the most productive of the carvers.[d] This work by Chinese carvers was

[a] Cf. Kawase Kazuma (*3*), pp. 91–2; (*4*), pp. 279–80.
[b] Cf. *ibid.* (*4*), pp. 142–3; Kimiya Yasuhiko (*1*), pp. 485–92.
[c] Cf. Kawase Kazuma (*4*), pp. 147–9; Kimiya Yasuhiko (*1*), pp. 486–8. The number of Yüan printers and the earliness of their activities may have sometimes been exaggerated, for in some reprints the Japanese simply copied the original printers' marks from the Chinese editions.
[d] Kawase Kazuma (*4*), pp. 143, 151–5.

[1] 陳孟才　　[2] 陳伯壽　　[3] 陳孟榮　　[4] 俞良甫

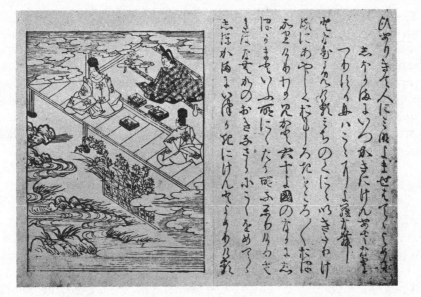

Fig. 1218. *Ise Monogatari*, a Japanese literary classic written in *kana* accompanied with fine illustration and printed on tinted paper in +1606. Far Eastern Library, University of Chicago.

significant because it accompanied a rise in the quality and quantity of printing in Japan. It was also associated with the adoption of the Chinese format in printing, and served to introduce many Chinese works to Japan.

Until the end of the 16th century Japanese printing was entirely dominated by the presses in the Buddhist temples, and the spread of printing outside Buddhist circles only began during the brief flourishing of movable-type printing. From 1592 to 1595 the Japanese warlord Toyotomi Hideyoshi unsuccessfully attempted to conquer Korea, and among the booty he brought back was equipment for movable-type printing; it was used until about 1650, being popular among the court, individuals, and the temples.[a] The most noteworthy examples produced by it were the *Sagabon*,[1] fine editions of famous classical works of Japanese literature:[b] for example, the *Ise Monogatari*[2] (Fig. 1218). This was the first time such works had been printed in Japan, and use was made of an important technical innovation, linked type, which represented more than one *kana* symbol.[c] The movable type brought back from Korea was bronze, but bronze type was used only rarely in Japan, wooden type being the more usual.[d] Printing of many secular works in Chinese continued and though an impressive number of editions were published in

[a] The standard work on movable-type printing is Kawase Kazuma (3); for a summary see the English appendix, pp. 1–17; see also Chibbett (1), pp. 67–78.
[b] See Kawase Kazuma (3), a study of the *Sagabon*.
[c] Cf. *ibid.* (3), pp. 644–9; English appendix pp. 12–14.
[d] Cf. *ibid.* (3), pp. 631–4.

[1] 嵯峨本　　[2] 伊勢物語

a short period with movable type, from 1650 block printing once again became ascendent.[a]

During the same period that movable type was introduced from Korea, the Jesuits brought a printing press from Europe to Japan. The press reached Japan in 1590, accompanied by two Japanese brothers who had been trained in type-casting and printing in Portugal, but because Christianity was already proscribed at this time, the press was moved about among various locations in western Japan, and in 1614 it was sent to Macao.[b] Thirty complete editions in Japanese, romanised Japanese, and European languages published by this Japanese Mission Press are extant; they include religious tracts, dictionaries and aids to language study, as well as works of literature.[c] Yet because of the increasingly strict interdiction on Christianity, the influence of these missionary efforts on Japanese printing was limited.[d]

Beginning in the 17th century wood block prints developed into one of the great arts of Japan. In its early stages this development was related to the importation of Ming books with wood-block illustrations, as is especially evident in the work of Hishikawa Moronobu[1] (c. +1618–94), sometimes considered the founder of ukiyoe,[2] or 'Pictures of the Floating Life'. He first made black and white prints, with colours often applied by hand, a practice continued by his successors. Moronobu not only studied Chinese prints but also reproduced Chinese art books, and an album of erotic colour prints, Feng Liu Chüeh Chhang Thu,[3] published in China in 1606, was copied and published in Japan in the late 17th century by him or his followers and under the same title. Reproduction of other Chinese colour prints followed later.[e] Among many masters of the late 18th century, the work of such artists as Suzuki Harunobu[4] (+1725–70) and Ando Hiroshige[5] (+1797–1858) (see Fig. 1219) was especially noted.[f] Although ukiyoe soon surpassed the artistic level reached in China by wood block prints and became famous for realistic portrayal of contemporary subjects and Japanese life, some elements of Chinese stylistic influence are clear in its formative stages.[g] Even the technique of perspective was perhaps not derived from Dutch paintings, as has been frequently been asserted, but learned indirectly through acquaintance with Chinese block prints influenced by Western works.[h]

[a] Compared with earlier accomplishments, the results in this period were spectacular. In only about sixty years 430 editions were published, and also it was at this period that the Tripitaka was printed in Japan for the first time; this was done with movable type; see Kawase Kazuma (3), p. 327; appendix, pp. 1–53.
[b] Laures (1), pp. 1–101; Lach (5), vol. 2, book 3, pp. 497–501.
[c] See the list in Chibbett (1), pp. 64–5.
[d] Cf. Kawase Kazuma (1), pp. 17–18.
[e] Cf. Shizuya Fujikake (2), in Kokka, No. 486, p. 143.
[f] See works of Harunobu and his age by Waterhouse (1).
[g] Cf. Shih Hsio-Yen (1), p. 99, who says Chinese influence on certain elements such as iconographic types, stylistic conventions, and internal architectural settings can be discerned; see also Shizuya Fujikake (1, 2); Chang Hsiu-Min (5), pp. 145–7.
[h] Cf. Shizuya Fujikake (2), in Kokka, No. 485, pp. 115–17.

[1] 菱川師宣　　[2] 浮世繪　　[3] 風流絕暢圖　　[4] 鈴木春信　　[5] 安藤廣重

Fig. 1219. *Ukiyoe* print, *Tōkaodō Gojūsantsuji*, by Ando Hiroshige, +1832.
Art Department, University of Chicago.

During the Edo period (+1603–1867), as the Japanese economy prospered, more and more books were printed to fulfil the demands of the increasingly sophisticated townspeople. Particularly popular were numerous illustrated stories and novels, for though book illustrations had been used to some degree in earlier periods, it was at this time that they began to reflect the contemporary culture (Fig. 1220). Publishing became increasingly the domain of commercial establishments in the large cities; Tokyo displaced Kyoto as the major publishing centre, even though the latter continued to be important, and Osaka also became active in printing at this time.

Despite these new trends, Chinese classical books and Buddhist works continued to be important elements in publishing during the early part of this period. The government supported Neo-Confucianism and issued many Chinese books for use in schools,[a] while with support of the Shogunate, several temples engaged in large-scale printings of the Buddhist *Tripitaka*, and at least two ambitious projects were completed in the 17th century. One is a complete set of the *Daizōkyō*[1] in 6323 *kan*, printed with wooden movable type by the Tendai Monk Tenkai[2] of the Kan'ei Temple from 1637 to 1648. It was probably the first *Tripitaka* ever printed with movable type.[b] Some thirty years later, another set of the *Tripitaka* was printed with some 60,000 cherry wood blocks from 1669 to 1681 by the monk Tetsugen,[3] a disciple of the Chinese Zen priest Yin-Yüan[4] (in Japanese Ingen) who founded the

[a] Cf. Kawase Kazuma (*1*), pp. 17–18.
[b] Although numerous editions of the Buddhist *Tripitaka* had been printed in China and Korea, none of them were printed with movable type before this set, known as the Tenkai edition; cf. Kawase (*3*), I, pp. 327–8.

[1] 大藏經   [2] 天海   [3] 鐵眼   [4] 隱元

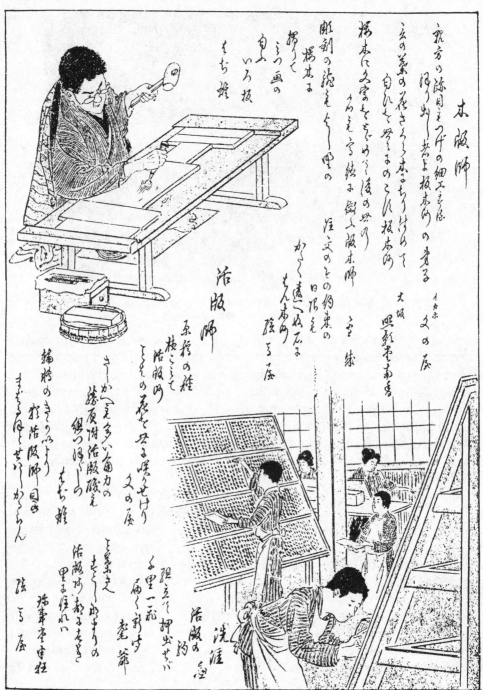

Fig. 1220. Woodblock cutters and movable type-setters operated contemporaneously in Japan in the +19th century, as illustrated in the popular pictorial *Fūzoku Gahō*, +1897.

Mampukuji[1] in Mount Ōbaku,[2] Kyoto, where the entire set of blocks survives intact today.[a]

Close to both China and Japan are the Liu-Chhiu (Ryukyu) Islands, the people of which had their own line of kings, even though they became a tributary of China, from the +14th century on.[b] In 1392 the king sent a son to China to study, and at about the same time the first Ming emperor sent thirty-six Fukienese families of boatmen and artisans to Liu-Chhiu to service the tribute missions. These Chinese settled in a special village called Thang-ying[3] or Chinese Camp, which also became the site of a Confucian and other Chinese temples.[c] Descendants of these Chinese families played important roles in Liu-Chhiu government, in education, and culture, and beginning in 1481 they, rather than native Liu-Chhiu people, provided the students regularly sent to China.[d] From the 13th century the Liu-Chhiu people had begun to use Japanese *kana* to write their language, but after contact with China they also used Chinese characters.[e] One of the fruits of their close relationship with China was that probably by the 14th or 15th century the Liu-Chhiu began to use paper. Before this time, diplomatic documents sent to the Yüan court were written on wooden tablets.

During the 18th century Chinese envoys wrote about Liu-Chhiu paper; and one described various kinds of paper made there from paper mulberry bark. The two best-known Liu Chhiu papers were called *hu shou chih*[5] (paper of longevity) and *wei phing chih*[6] (screen mounting paper), the latter a decorated paper used for windows and walls.[g] From 1723 to 1798 these and one other kind, *tzu hsia chih*[7] (purple cloud paper), were presented as tribute to China on several occasions.[h] It seems then that paper production must have begun in Liu-Chhiu by the 17th century and possibly earlier.

The date when printing began in Liu-Chhiu is unknown, but was probably around 1500. An 18th-century Chinese official teacher of Liu-Chhiu students, Phan Hsiang,[8] recorded valuable notes on printing in Liu-Chhiu (Fig. 1221), and according to these it was during the Cheng-Te period (+1506–21) that the king arranged for the printing of the *Four Books*, the *Five Classics*, as well as works on philology, Neo-Confucianism, and literature; the printing blocks were stored in the palace.[i] As aids in reading, some of these books had printed punctuation marks and

---

[a] According to a brochure given to Joseph Needham when he visited the collection in 1971, the set is known as the Obaku edition, housed in a new building in the Mampuku Temple.

[b] The most important collection of documents on Chinese-Liu-Chhiu relations is the *Li Tai Pao An*,[4] which contains 263 chapters of official papers from 507 years of the tributary period from 1372 to 1879. A copied ms. is now kept in the Taiwan University Library; see selected edition by Wu Fu-Yüan (1).

[c] See *Liu-Chhiu Ju Hsüeh Chien Wen Lu* (repr. 1973), pp. 116–22, 182–3.

[d] See *ibid.* pp. 117–22.   [e] See *ibid.* pp. 76–7.

[f] Thao Tsung-I (fl. +1360), cited in *Liu-Chhiu Ju Hsüeh Chien Wen Lu*, p. 76.

[g] See *Liu-Chhiu Kuo Chih Lüeh* (*TSHCC*), ch. 14, pp. 2b–3a; Chuang Shen (1), pp. 100–2.

[h] Cf. Chhen Ta-Tuan (1), pp. 34–8; Chuang Shen (1), pp. 100–2.

[i] See *Liu-Chhiu Ju Hsüeh Chien Wen Lu*, pp. 77–83; Chang Hsiu-Min (5), pp. 161–6.

---

[1] 萬福寺   [2] 黃檗山   [3] 唐營   [4] 歷代寶案   [5] 護壽紙

[6] 圍屏紙   [7] 紫霞紙   [8] 潘相

Fig. 1221. Teaching Chinese classics to Liu-Chhiu students by Phan Hsiang at the National Academy in Peking, +1760–4. From *Liu-Chhiu Ju Hsüeh Chien Wen Lu, c.* +1764.

kana to show pronunciation beside the characters,[a] but unlike Japanese editions, Liu-Chhiu publications used Chinese reign periods in dating, and were undoubtedly based directly on Chinese editions.[b] Some even contained the original Chinese publication information. For instance, an illustrated edition of the *Four Books* had a picture of the original publisher, Yü Ming-Thai,[1] and a picture of Confucius bore the surname of the original carver, Liu.[2c]

Phan recorded the names of three works about the royal family by native Liu-Chhiu authors, and fourteen or fifteen works by Chinese residents, for the most part literary collections.[d] Five of the latter were by Chheng Shun-Tse,[3] including one book printed in 1708. It seems that the blocks for the latter may have been carved in Fukien where, in that very same year Chheng, as a Liu-Chhiu envoy, had

---

[a] See *Liu-Chhiu Ju Hsüeh Chien Wen Lu*, pp. 78, 81–2, 85.
[b] See *ibid.* p. 78.
[c] See *Liu-Chhiu Ju Hsüeh Chien Wen Lu*, pp. 79–81.
[d] See *Liu-Chhiu Ju Hsüeh Chien Wen Lu*, pp. 83–5.

[1] 余明臺   [2] 劉   [3] 程順則

commissioned the blocks for the *Liu Yü Yen I*[1], the Six Maxims in Colloquial Chinese, for use as a language text in Liu-Chhiu.[a]

The calendar used in Liu-Chhiu was issued annually by the Chinese government, but because of delay in transportation, temporary calendars were printed in Liu-Chhiu.[b] Yet, despite the development of Liu-Chhiu handicrafts, Chinese envoys in 1808 still found it necessary to take with them to Liu-Chhiu block carvers, as well as barbers and tailors.[c] This suggests that Liu-Chhiu printing must have remained quite limited, and that as late as the early 19th century was still dependent on Chinese craftsmen.

(3) DEVELOPMENT OF PAPERMAKING AND PRINTING IN VIETNAM

Like Korea and Japan, Vietnam has been within the cultural sphere of East Asian civilisation, although it is located on the southeastern tip of the Asian continent. The northern part of modern Vietnam, Tongking, known in Chinese history as Chiao-Chou,[2] Chiao-Chih,[3] or Nan-Yüeh,[4] was ruled directly by China as a tributary state from the late −2nd to the early +10th century. Its territory later extended to include Champa or Chan-Chheng[5] in the southern part of the peninsula. For over two thousand years, Vietnam followed Chinese patterns of life and thought, practised Chinese bureaucratic and family systems, adopted Confucian learning as well as Buddhism from China, and used Chinese writing for serious literature. The Vietnamese created their own system of writing, called *nom*[6] in the +14th century, by combining parts of Chinese characters, and adopted a 'national writing' by transcribing their language into Latin letters, from 1600 (Fig. 1222). But the primary system of writing, used in government, scholarship, and religion, consisted of Chinese characters, and this continued up to the early 20th century.

Because of its geographical proximity and its political ties with China, Vietnam must have been introduced to paper and printing very early, and several references in early Chinese literature imply that paper may have been made in Vietnam in the +3rd century. Some 30,000 rolls of a kind of 'honey fragrance paper' (*mi hsiang chih*[7]) are reported to have been brought to China in +284 from 'Ta Chhin'[8], and scholars believe this must have been made of garco bark in Vietnam and shipped to

---

[a] See *Liu-Chhiu Ju Hsüeh Chien Wen Lu*, pp. 83–4; Ogaeri Yoshio (*1*), pp. 141–2, 204–5; appendix, pp. 9–11; *Liu-Chhiu Kuo Chih Lüeh*, ch. 6, p. 8*b*, says the Liu-Chhiu king had had the *Sacred Edict* printed and appointed Chheng Shun-Tse to expound it.

[b] Cf. Chang Hsiu-Min (*5*), p. 165; *Liu-Chhiu Kuo Chih Lüeh*, ch. 4, gives a notice included in the Liu-Chhiu calendar that the Chinese calendar was to be taken as the authoritative version when it arrived.

[c] See *Hsu Liu-Chhiu Kuo Chih Lüeh*, ch. 5; cited in Chang Hsiu-Min (*5*), p. 165.

[1] 六諭衍義   [2] 交州   [3] 交趾   [4] 南越   [5] 占城
[6] 喃   [7] 蜜香紙   [8] 大秦

Quốc Ngữ Quốc Ngữ Quốc Ngữ Quốc Ngữ Quốc Ngữ Quốc Ngữ Quốc Ngữ Quốc Ngữ
ôc Ngữ Quốc Ngữ Quốc Ngữ Quốc Ngữ Quốc Ngữ Quốc Ngữ Quốc Ngữ Quốc Ngữ Quố
Ngữ Quốc Ngữ Quốc Ngữ Quốc Ngữ Quốc Ngữ Quốc Ngữ Quốc Ngữ Quốc Ngữ Quốc
1945 Ngữ Quốc Ngữ Quốc Ngữ Quốc Ngữ Quốc Ngữ Quốc Ngữ Quốc Ngữ Quốc Ngữ Qu
ais franç ngữ Quốc Ngữ Quốc Ngữ Quốc Ngữ Quốc Ng
français Quốc Ngữ Quốc Ngữ Quốc Na
ança s f ngữ Quốc Ngữ Quốc N
rança Quốc Ngữ Quốc
1905 uốc Ngữ
çai Ngữ Quốc
s f Quốc
rac Ngữ
ça Quốc
aruôc
1861
1651
939

Fig. 1222. Evolution of writing system in Vietnam, showing stages in which the Chinese, *nom*, and Latin forms were adopted or replaced. From DeFrancis (1).

China by Alexandrian merchants.[a] During the period from +265 to 290 a tribute of more than 10,000 rolls of an 'intricate filament paper' (*tshe li chih*[1]) made of fern or seaweed was recorded as sent by Nan-Yüeh to China,[b] while another source from the +3rd century says that paper was made by the people of Chiang-Nan[2] by pounding the bark of the paper mulberry tree, which 'was called *ku*,[3] in Ching, Yang, Chiao, and Kuang[4]';[c] Chiao here corresponds to modern Vietnam.[d] From these early sources we may assume that paper could have been produced in the northern part of Vietnam by the +3rd century.

This assumption seems to contradict a later reference that says paper was not made in Vietnam even as late as the first part of the 13th century. The *Chu Fan Chih*,[5] a record of foreign nations and products written by Chao Ju-Kua[6] in 1225, says: 'In Chiao-Chih they do not know how to manufacture paper and writing brushes, so those from our provinces are in demand.'[e] Since this book was based on oral information from Chinese and foreign oversea traders, the above statement may have referred to the central and southern part of Vietnam along the sea coast, while the earlier source perhaps applied to the northern part bordering China. This theory seems to be supported by the fact that, at about this time, the method of papermaking also crossed Chinese borders to both the northeast and the northwest overland, not by the sea route.[f] Even in modern times the Chinese method of papermaking is said to have been closely followed by the papermakers in north Vietnam. According to Dard Hunter, who visited the Tongking area in 1904, all the techniques used by papermakers in the Tongking area early in this century were more closely related to those of China than to those of any other country of Asia.[g]

Chinese sources of later date record that paper and paper products continued to be exported to China from Vietnam, for their tributary relations were maintained after Vietnam's independence in the +10th century. It is said that paper fans were presented to the Chinese emperor by a Vietnamese envoy in 1370, and that an annual tribute of 10,000 paper fans was sent to China from the six provinces of

---

[a] See *Nan Fang Tshao Mu Chuang* (*HWTS*), ch. 2, p. 6a; authorship and dating of this work are controversial but since this historical incident appears also in several other sources it may not be dismissed. For the Vietnamese origin of this paper, see Hirth (1), pp. 274–5; Hirth & Rockhill (1), pp. 205–6, n. 2; Tsien (2), pp. 140–1; also discussion on pp. 43 ff. above.

[b] See *Shih I Chi* (*HWTS*) ch. 9, p. 7a; Tsien (4), p. 517; also discussion on pp. 44, 62 ff. above.

[c] See *Mao Shih Tshao Mu Niao Shou Chhung Yü Su* (*TSHCC*), pp. 29–30; Phan Chi-Hsing (10), pp. 147–8; also discussion on pp. 56 ff. above.

[d] Since Chiao-Chou or Nan-Yüeh at that time included part of China as well as northern Vietnam, some scholars have suggested that this paper may have come from China; see Chi Hsien-Lin (3), p. 133. However, the first two sources both record that these exotic papers were sent in large quantity to the Chinese court as tribute, so they must have been native products from a foreign country.

[e] Hirth & Rockhill (1) (tr.), p. 45.

[f] See discussion in Huang Sheng-Chang (1), pp. 126–7.

[g] Cf. Hunter (9), pp. 110, 141; (12), pp. 42–54.

[1] 側理紙   [2] 江南   [3] 穀   [4] 荊, 揚, 交, 廣
[5] 諸蕃志   [6] 趙汝适

northern Vietnam for more than a decade after 1470.[a] One record says that under the Chhing dynasty some 200 sheets of yellow paper with a golden dragon design, along with inkstones, ink sticks, and brushes, were sent to China in 1730 in return for gifts of books, silk, and jade vessels from the Chinese emperor.[b]

Papermaking in Vietnam was unquestionably a direct transmission from China; its materials, tools, and methods are found to be almost the same. Besides the bark of the daphne tree (*yüeh kuei*,[1] *Daphne involucrata*, Wall.), which is a native product, the raw materials included bamboo, rice straw, and seaweed, and as far as techniques were concerned, not only the woven mould, but also the maceration of paper stock, the construction of vats, and the actual procedures were similar to those used in China. Even the same mucilaginous material used by Chinese women as pomade for the hair was also used as a sizing substance by the Vietnamese papermakers.[c]

While paper was introduced to Vietnam very early, the Vietnamese must also have been exposed to Chinese books at about the same time. Yet the earliest record of access is of the +11th century, when they acquired various kinds of Chinese books—excepting, of course, those whose export was prohibited—and paid for them with their native products and spices. Within eight years after its independence, Vietnam received a gift of three copies of the printed *Tripitaka* and one set of the Taoist canon as a gift from the Sung court. Printing probably began in Vietnam by the 13th century, for the earliest known reference to it concerns population registers printed during the period from 1251 to 1258. In the Tranh dynasty (+1225–1400) a copy of the *Tripitaka* was received from the Yüan court in 1295, to replace one destroyed during the Mongol invasion, and its printing is said to have been arranged, though, apparently, it was not carried out. Four years later, however, a Buddhist liturgy and manuals of writing for official documents are known to have been printed.[d]

Under the Le dynasty (+1418–1789), when Chinese institutions were followed closely, the Confucian Classics were printed in Vietnam for the first time. An edition of the Four Books (*Ssu Shu Ta Chhüan*[4]) was published in 1427 and blocks for the Five Classics (*Wu Ching*[5]) were carved in 1467. Printing flourished especially during the second half of the 15th century, when printing blocks became so numerous that a special house was built at the Confucian temple to store them.[e] In later periods many more government editions of the Confucian Classics, histories, poetry collections, and dictionaries were printed, primarily for the civil service examinations.

The Vietnamese government attempted on certain occasions to control both

[a] See *An-Nan Chih Yüan*,[2] ch. 2, cited in Phan Chi-Hsing (9), p. 148.
[b] See *Yüeh-Nan Chi Lüeh*,[3] ch. 1, cited in Phan (9), p. 148.
[c] Cf. Hunter (9), pp. 110, 141; (12), pp. 42–54.
[d] See Chang Hsiu-Min (5), pp. 152–3.
[e] *Ibid.* pp. 153–4.

[1] 月桂   [2] 安南志原   [3] 越南輯略   [4] 四書大全   [5] 五經

printing and distribution of books. In 1734 scholars were prohibited from buying
Chinese editions of Confucian classics and were restricted to the use of Vietnamese
editions, and at various times there were regulations concerning the distribution of
government publications. In 1796 official editions of the Five Classics and the Four
Books printed at Hanoi were ordered to be distributed throughout the country.[a]
On the basis of a calendrical work obtained in China in 1809, the Vietnamese
calendar was formally inaugurated and the government began issuing an annual
calendar, following exactly the format and content of the Chinese.[b] The official
editions included those printed by the National Academy, the Institute of Worthy
Scholars, the Palace, and the Institute of History, which were similar to the central
publishing agencies in China. Book publishing was concentrated in the capitals,
Hanoi and later Hui, and in Nam-dinh.

Private printings in Vietnam included books similar to the official publications;
Confucian classics, histories, and readers primarily for candidates for the civil
service examination. In addition, literary collections, genealogies, fiction, and
medical works were published, while such Chinese novels as the *Romance of the Three
Kingdoms* were especially popular. Besides Chinese literature, there were many
original books by Vietnamese authors, including some women. The private
publishers and printers came mostly from a single county, Gia-loc in Hai-duong
province, and the block carvers in particular tended to come from two villages
there.[c]

All the earlier Vietnamese editions are of three types: those entirely in Chinese,
those in Vietnamese characters or *nom* (Fig. 1223), and those having Chinese text
with *nom* annotation as an aid to pronunciation. Some idea of these publications
may be derived from the catalogue of books in the library of the former École
Française d'Extrême-Orient, which contains 2258 works in Chinese by Vietnamese, 561 works in *nom*, and 351 Vietnamese editions of Chinese books.[d] Most of the
Vietnamese editions of Chinese works are Buddhist and Taoist writings, with
smaller numbers of Confucian classics, literary works, histories, medical books, and
miscellaneous writings. Although no complete edition of the *Tripitaka* has ever
been printed in Vietnam, many Buddhist works were published and more than 400
printed *sutras* dating from 1652 to 1924 are still preserved in Hanoi, among them
more than twenty written by Vietnamese.[e]

Although the majority of books were printed from wooden blocks, some were
printed with movable type, an early example dating from 1712. Two large sets are
known to have been printed with wooden movable type, which were acquired in
China. A collection of administrative codes was printed in ninety-eight volumes
with this type in 1855, and a collection of imperial poetry and prose was printed in

[a] See *Ibid.* p. 153.
[b] Woodside (1), pp. 123–4, 186–8.
[c] Chang Hsiu-Min (5), pp. 153–4.
[d] See Matsumoto (1), pp. 117–204; Yamamoto (1), pp. 73–130. The catalogue, however, does not usually
distinguish between printed editions and manuscripts.
[e] See the catalogue of the former École Française d'Extrême-Orient at Hanoi, cited in Chang (5), p. 155.

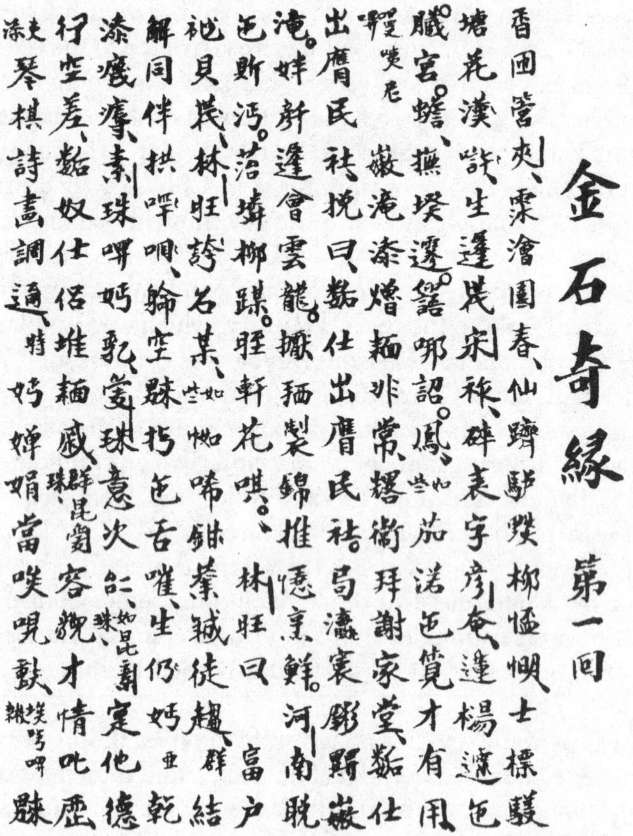

Fig. 1223. Vietnamese *nom* writing mixed with Chinese characters for the dramatic text of
'The Marvellous Union of Gold and Jade'.

sixty-eight volumes in 1877.[a] Bronze movable type may also have been used.
Vietnam has also had a flourishing wood block colour print industry, especially for
New Year pictures with subjects and methods of production similar to those in
China (Fig. 1224).[b]

(4) INTRODUCTION OF PAPER AND PRINTING TO SOUTH AND SOUTHEAST ASIA

The region of the Asian continent and archipelago beyond Vietnam contains a
heterogeneous mixture of racial and cultural elements at all stages of their
development, and has been culturally dominated or at least strongly influenced by
India through the waves of Hinduism, Buddhism, and Mohammedanism before
the coming of European Christians. Moreover, communication between India and

[a] Cf. Chang (5), p. 157.
[b] See a report on modern Vietnamese New Year pictures in Shang I (1), pp. 59–61.

關
公

Fig. 1224. New Year picture from Vietnam printed in colour, with similar theme and method to those used in China. The above shows the picture of the Chinese popular hero Kuan Yü of the +3rd century. From M. Durand (1).

southeast Asia resulted in the mass migration of Indian population and ideas eastward into Burma, Malaysia, Siam (Thailand), Indonesia, and Indochina (modern Vietnam, Cambodia, and Laos) where the Indian and Chinese cultures met and mingled. Generally speaking, the people in the area beyond Vietnam were outside the sphere of East Asian civilisation, although some Chinese influence through trade and migration was felt in some of these countries during various periods.

Despite the cultural divergences, however, one common factor for all lands and peoples in this area seems to have been the lack of a written tradition such as characterised the Chinese culture. Since the sacred texts of India were transmitted primarily through oral tradition and memorisation, written texts were not normally used by the learned men. For the various nations of southeast Asia, few native written records from early times are known, and most of their histories depend upon oral tradition or the records in Chinese and occasionally in Arabic and Persian sources. For this reason, the need of paper and printing for transmission of ideas was negligible in this region, and though paper might have been introduced to this area at an early date, printing was not known until after the coming of Europeans in the 16th century.

In India, before the advent of paper, materials used for writing included tree bark, leaves, wooden boards, leather, cloth, bones, clay, stone, and metal, especially copper—indeed a copper plate of the +9th century bearing Sanskrit inscriptions on both sides was found in 1780 in Eastern India (Fig. 1225).[a] But in Bengal and southern India the commonest were palm leaves while birch bark was used in Kashmir and the northern parts of the country. The palm leaf (Fig. 1226) was cut in a standard shape and written on with an iron stylus, the incisions being filled with a dye; holes were pierced through the leaves, and cords were inserted to hold the pile of leaves together.[b] Although paper production began in India by the 14th century or earlier, palm leaves continued in use even as late as the 19th century,[c] while they were also used in Ceylon (Sri Lanka), Burma, and Siam. In Indonesia both palm leaves and birch bark were used; in the Philippines bamboo, leaves, and bark were all employed, and according to Chinese sources, in Champa (Chan-Chheng[1]) and Cambodia (Chen-La[2]) parchment made of deer and sheep skins was blackened by smoking and written on with bamboo stylus and white powder.[d] Also, in many parts of southeast Asia where true paper was not produced, a kind of quasi-paper called *tapa* was manufactured. In Indonesia, the Philippines, Malaysia, and many Pacific islands, it was made by pounding the inner bark of the paper mulberry, first into small pieces, which were later combined into large sheets

[a] The copper plate recording the gift of the village of Mesikā by King Devapāla to Vihekarātamiśra was found at Mungir, in Bengal Province; see tr. of the text by F. Kielhorn (1).
[b] Cf. Ghori & Rahman (1), pp. 133 ff.; Hunter (9), pp. 11–12.
[c] Cf. Trier (1), p. 137; Priolkar (1), pp. 38–41.
[d] See records of Yüan and Ming authors cited in Chi Hsien-Lin (1), pp. 134–5.

[1] 占城   [2] 眞臘

Fig. 1225. First copper-plate with Sanskrit inscription found at Mungir concerning land grant by King Devapāla, +9th century. The top, surmounted by an ornament, is perhaps a seal. British Museum.

Fig. 1226. Illuminated manuscript from Orissa on palm leaves, *Gitagovinda* by Jayadeva.
From India Office Library.

similar to paper.[a] It was used primarily for clothing and occasionally for writing by people in some of these areas.[b]

Paper appears to have been known and used in India in the second part of the +7th century, as is attested in the writings of a Chinese Buddhist pilgrim, I-Ching,[1] who travelled to India from +671 to 695. In the messages he sent home he said: 'The priests and laymen in India make *Caityas*, or images with earth, or impress the Buddha's image on silk or paper, and worship it with offerings wherever they go.'[c] He also referred to the use of discarded paper for toilet paper and to reinforcing umbrellas or hats with paper, and included the Sanskrit word *kākali* for paper in his one-thousand-character lexicon.[d] Apparently paper was not yet manufactured in this area, for when he was in Sumatra, he requested that paper and ink be sent from China for copying *sutras*.[e] Because previous records, including the detailed account by that earlier pilgrim to India, Hsüan-Tsang,[2] who was back

---

[a] Cf. Hunter (9), pp. 29 ff.
[b] Voorn (1), pp. 31–8, says *tapa* was used as a writing material in Java from the 17th century or earlier.
[c] See *Taisho Daizokyo*, Vol. 54, p. 226, tr. Takakusu; cited in Gode (1), p. 5.
[d] See *Taisho Daizokyo*, Vol. 54, pp. 215, 218; cited in Chi Hsien-Lin (1), pp. 122–7; Huang Sheng-Chang (1), p. 122.
[e] See Huang (1), p. 123.

[1] 義淨   [2] 玄奘

in China by +645, make no mention of paper, it is believed that it must have entered India between +645 and 671.

It seems that paper and papermaking were introduced to India over more than one route and at different times. One way it came was probably from China through Tibet and Nepal to the Bengal region, because we know that papermaking was introduced to Tibet in about +650, when the Tibetan king asked the Thang court to send him silkworms for breeding, and craftsmen for making wine, mills, paper, and ink.[a] Since Nepal was then under Tibetan suzerainty and Nepal-Indian relations were very close, it is likely that paper also entered India at this time.[b] Another route to India was perhaps through Kashmir, for the Muslims, who established power in west India in the +8th century and in north India in the +12th, very likely first imported paper into the country and then fostered its manufacture. Early paper manuscripts from India date from the 11th to the 14th century,[c] while skilled artisans, including papermakers and bookbinders, were brought by a future sultan for Samarkand to Kashmir at the beginning of the 15th century.[d] During the same period, Ma Huan,[1] a Chinese in the mission of Cheng Ho[2] who visited Bengal in 1406, mentioned the manufacture there of paper from bark, which was glossy like deer skin.[e] Clearly, paper was being manufactured in both Kashmir and Bengal no later than about 1400 and possibly as early as the 11th century.

Printing in south and southeast Asia, with the exception of Vietnam, was mostly introduced by Europeans from the 16th century onwards and was used primarily by missionaries, colonial governments, and European residents. The first printing by European techniques in Asia began in the middle of the 16th century,[f] by a press said to have been bound for Ethiopia, but brought to and used in Goa by Jesuit missionaries for printing religious tracts and other literature.[g]

The Portuguese published works in many other Indian cities besides Goa in the 16th and 17th centuries. The first work in a local language was a translation into Malayalam of Xavier's catechism, probably printed in Cochin in 1577, and printing in Tamil began in Panikael in +1587. However, because the native languages were replaced by Portuguese in Goa in the late 17th century, printing ceased there until 1821. Danish missionaries began printing at Tranquebar on the

[a] See *Chiu Thang Shu* (*ESSS*), ch. 196; Huang Sheng-Chang (*1*), p. 116.

[b] A new trade route from China to Tibet and Tibet to Nepal opened for the first time in +650, as a result of the marriage of the Tibetan king to a Nepalese princess in 639 and to a Chinese princess in 641. Cf. Huang (*1*), pp. 114 ff.

[c] Cf. Gode (1), p. 7, which lists a number of old Indian manuscripts dating from +1089 to +1323.

[d] Cf. Ghori and Rahman (1), pp. 135–6, which says that Kashmiri paper earned a reputation for excellence and was presented to other rulers by the sultan.

[e] See *Ying Yai Sheng Lan* (Shanghai, 1935), p. 61; Chi Hsien-Lin (*1*), p. 128; Gode (1), p. 8.

[f] The first printing of loose sheets was the *Coclusões* of Antonio de Quadros in 1556; the first printed book was the *Doctrina Christiana* of Francis Xavier published in 1557; cf. Rhodes (1), pp. 11 ff.

[g] The emperor of Ethiopia had requested a press from the Portuguese king in 1526, but it never reached Ethiopia because of strained relations between the emperor and the missionaries.

[1] 馬歡   [2] 鄭和

east coast of India about 1713, but the British did not begin regular printing until 1778 in Bengal and Hooghly, though an earlier attempt had been made at Bombay in 1674 or 1675 and something may have been printed at that time.[a]

Printing was introduced to the Philippines in the 16th century, but in the 10th century, before the Spanish conquest, trade relations had already been established between China and the islands. Envoys and tribute from Luzon and Mindoro were sent to the Ming court in the 14th and 15th centuries. However, it was in the 16th century that numerous books were brought by Dominican friars to Manila,[b] where a large Chinese community helped not only in their translation but also in introducing wood-block and movable-type printing. Indeed, Chinese printers monopolised the printing industry in the Philippines for over fifteen years before native craftsmen participated in the trade. The earliest extant printing there includes two editions of the *Doctrina Christiana* by the Dominican friar Juan Cobo, one in Spanish and Tagalog, the other in Chinese, entitled *Wu Chi Thien Chu Cheng Chiao Chen Chuan Shih Lu*[1] (Veritable Record of the Authentic Tradition of the True Faith in the Infinite God) (Fig. 1227). Both were printed in 1593 from wood blocks,[c] and both the technique and the Chinese style of the illustrations indicate that these blocks must have been carved by Chinese. Typography was first used in 1604, when two books were printed with locally-made metal types by a Chinese printer named Juan de Vera, whose achievement in cutting punches and striking matrices has been called a 'semi-invention' of typography.[d] His brother, Petro de Vera, and another Chinese named Keng Yong, also printed several books during the following few years. Between 1593 and 1640, there were eight Chinese printers whose names are known,[e] and among fifteen titles printed in Manila between 1593 and 1604, at least five are in Chinese.[f]

The 17th century witnessed the rise of the Dutch at the expense of the Portuguese as the major European power in Asia. The Dutch gradually extended their power over Indonesia, establishing Batavia in 1619, and they took Malacca and Ceylon from the Portuguese in 1641 and 1658. The first Dutch printing in Asia was at Batavia, probably in 1659, but there is more definite evidence of printing under government auspices from 1668. The former may have been an almanac or chronicle, the latter was certainly a peace treaty between the Dutch and the prince of Macassar.[g] Malay vocabularies appeared in the early 18th century, and about 1750 a Malay Bible in Arabic letters was published by a short-lived seminary press. Over the years a variety of presses were set up, government, private, and religious, and the first newspaper appeared in 1744, but was suppressed within two years.[h] In

---

[a] Cf. Rhodes (1), pp. 11–23; Priolkar (2), pp. 1–27, 36–51; for early Indian prints, see Diehl (1).
[b] Cf. Boxer (1), p. 295; also discussion on pp. 316 ff. above.
[c] Cf. Bernard-Maître (19), p. 312; Van der Loon (2), pp. 2–8; Fang Hao (3), pp. 32–3. The only known copy of the book, containing sixty-two leaves, is kept in the Biblioteca Nacional in Madrid.
[d] Cf. Van der Loon (2), p. 25–6.   [e] Cf. Boxer (3), p. 459.   [f] Cf. Van der Loon (2), p. 43.
[g] Cf. McMurtrie (5).   [h] De Graaf (1), pp. 11–28; McMurtrie (5).

[1] 無極天主正教眞傳實錄

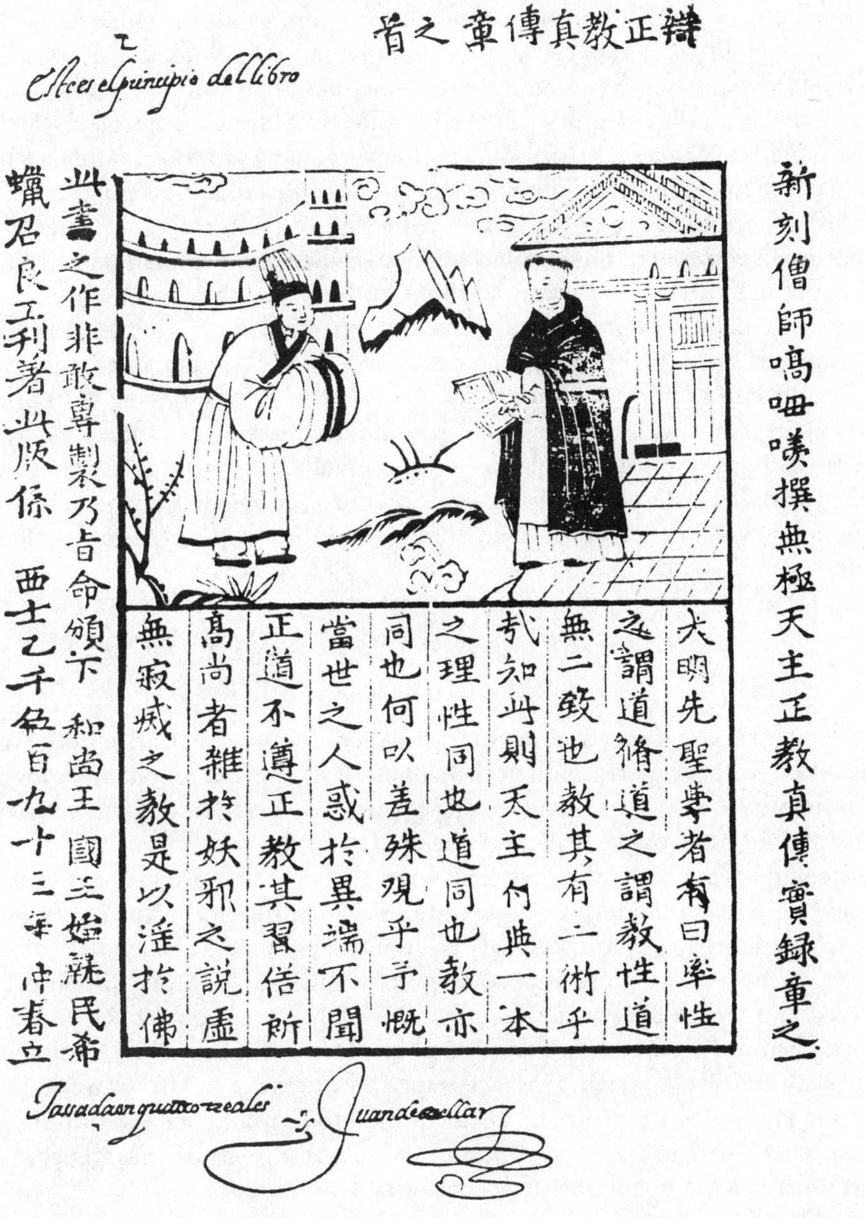

辯正教真傳章之首

Este es el principio del libro

新刻僧師噶呣嘆撰無極天主正教真傳實錄章之一

犬明先聖學者有曰率性之謂道脩道之謂教性道無二致也教其有二術乎哉知此則天主件與一本之理性同也道同也教亦同也何以差殊觀乎予慨當世之人惑於異端不聞正道不導正教其習俗所高尚者雜於妖邪之說虛無寂威之教是以溺於佛

此書之作非敢尊製乃旨命頒下 和尚王國王媱譏民希 蠟召良工刊著此版係 西士乙千五百九十三年仲春立

Tasada en quatro reales *Juan de ...*

Fig. 1227. Earliest extant printing in the Philippines, a Chinese edition of Juan Cobo's *Doctrina Christiana*, +1593. The woodcut represents the Dominican friar showing a book to a Chinese scholar. From Van der Loon (1).

general, the authorities of the various European colonies in Asia kept close control over the press, in order to prevent criticism of the government and to guard against rousing antagonism among the local peoples, due to missionary printing activities. In Ceylon, Dutch missionaries began printing in 1737, beginning with a Sinhalese Prayer Book, and between 1737 and 1767 thirty-four titles are known to have been printed there. Most were religious works and almost all in Sinhalese or Tamil.[a]

Although the British began trade and settlement on a limited scale in Asia in the early 17th century, they did not become the dominant colonial power of the area until the late 18th century. But after they established control of the Straits Settlements in Malaya, they began printing at Penang in 1806, at Malacca in 1815, and at Singapore in 1822. The earliest printing in Penang was commercial and also served government needs, while in the other two cities printing was started by the Baptist missionaries of the London Missionary Society, who had first started their printing activities in the Danish colony of Serampore near Calcutta in 1801.[b] American Baptist counterparts of the British missionaries initiated printing in Burma and Siam, printing in Burmese beginning at Rangoon in 1816, in Siamese at Bangkok in 1836, with some earlier works for the Baptist having been printed at Singapore.[c] In these, as in earlier cases, the main products of missionary presses were religious tracts and Bibles in local languages, but they also published such other works as dictionaries, grammars, and introductions to European knowledge. These missionary presses represented an important step in the early diffusion of printing in this area.

(j) CONTRIBUTION OF PAPER AND PRINTING
TO WORLD CIVILISATION

THE ADVENT of paper and printing reflected a stage of maturity in the progress of civilisation; every step in their development has been a milestone in the history of humanity. Paper may have been discovered by accident, but when it evolved it became the most convenient and cheapest material for writing, and showed its supremacy over every other material wheresoever it had been used. Eventually it replaced writing materials that were more cumbersome or expensive, and penetrated into the fabric of society as an indispensable article of daily life. It was certainly one of the most important prerequisites for printing, which originally served as a mechanical extension of handwriting. But as soon as the printed word multiplied, it had an impact on all aspects of the political, social, economic, and cultural life of mankind. This was especially evident in the transformation of European society from the medieval to the modern age, for the introduction of typography to Europe in the middle of the 15th century has generally been recognised as the turning point in this great transition.

(1) THE ROLE OF PAPER IN CHINESE AND WESTERN CULTURES

Little was written about paper in the West until after its extensive application to printing in the late 15th and 16th centuries.[d] Before that time, however, paper is

[a] Rhodes (1), pp. 67–72; McMurtrie (4); Priolkar (1), pp. 105–29.
[b] Cf. Byrd (1), pp. 2–17; Rhodes (1), p. 29.
[c] Rhodes (1), pp. 79–95; Gammell (1), pp. 187–96; Byrd (1), p. 15.
[d] About 500–1000 printing presses were operating in France in the 16th century and paper mills had to supply some 1500–3000 reams a day, or 450,000 to 900,000 reams a year; see Febvre & Martin (1), p. 40, n. 50.

known to have been used primarily for writing, wrapping, and certain other non-literary purposes, in both Europe as well as the Arab world.[a] Since the latter part of the 18th century, paper became increasingly popular and patented for use in the building of houses, ships, carriages, chairs, tables, and bookcases.[b] By the end of the 19th century, it was being converted into material for almost every conceivable item of personal wear and belongings, for some kitchen utensils, and as a household furnishing; indeed, by then it had become so widely used that a song entitled 'The Age of Paper' (Fig. 1228) was popularly performed in London music halls.[c]

It is now clear that many similar applications of paper and paper products occurred in China at least a thousand years earlier than in Europe. Besides its use for writing, books, and documents from the +1st century onwards, paper was employed extensively for many other purposes. As we have seen, it was made in different colours and designs for stationery and cut or folded into patterns for amusement or for decoration; it was used as a daily necessity for wrapping, for sanitary and medical purposes, and for making such everyday articles as cups, fans, umbrellas, flags, lanterns, kites, and toys. Yet all these uses were developed in China before the close of the +6th century.

Paper is also known to have been used in making hats, turbans, coats, trousers, belts, shoes, bed sheets, mosquito nets, curtains, screens, tiles, together with other household furnishings and appliances, and even for coffins and armour. The symbolic use of paper images in ceremonies and festivals, which consumed a large part of its production, satisfied both the living and the dead without the expense of having to use real objects for sacrifice or burial. The adoption of paper as a medium of exchange in the early 9th century, and its earlier use for business records and transactions brought about a great revolution in economics. All these applications for literary and nonliterary purposes were accomplished before the end of the 9th century, when paper first began to be known to Europeans.[d]

Among numerous uses of paper, a most important contribution to Chinese culture is perhaps its unique role as a medium for Chinese art. Unlike Western art which stresses such forms as sculpture or architecture, the highest form of Chinese art is calligraphy and painting; it has developed primarily through the application of brush and ink on paper. Paper provided the best surface for the free expression by the artists of China, and this has also been true of the main stream of fine art throughout the nations of the East Asian civilisation.

Chinese calligraphy as an art began probably as early as the +2nd century

---

[a] Paper was reported to have been used for wrapping vegetables, spices, and hardware in Cairo in 1035, and for wrapping groceries in food markets in Baghdad in 1140; see Hunter (9), pp. 471–2; it was used in Europe by such tradesmen as box-makers, playing-card makers, and billposters; see Febvre & Martin (1), pp. 39–40.

[b] The first use of paper in Europe for furniture is dated 1772; an English patent for use of paper in building houses, etc., is dated 1788; see Hunter (9), pp. 503, 512.

[c] By 1868 paper was used for such articles as aprons, collars, cuffs, handkerchiefs, hats, petticoats, raincoats, shirt fronts, slippers, vests, boxes, cups, napkins, plates, towels, bowls, carpets, curtains, table tops, roof coverings, and window blinds; see Hunter (9), p. 568.

[d] For the various uses of paper and paper products, see detailed discussion on pp. 84 ff. above.

Fig. 1228. Cover design of a song, 'The Age of Paper', as sung in +1860 by Howard Paul, attired in a suit of paper. Dard Hunter Paper Museum.

during the Later Han.[a] It was not developed into a special form of art until the +3rd or 4th century, after paper had been greatly improved and was extensively used for writing.[b] The basic calligraphic styles—the cursive or running, and standard or regular—which are still prevalent today, were all evolved during this period. The superiority of paper for such art is obvious, for neither bamboo nor wood nor stone has such a smooth and receptive surface. Without paper, the various calligraphic styles could not have developed so perfectly throughout the ages.

Early paintings were made on walls but in the Thang dynasty artists began to paint on paper. By the Sung, calligraphy and painting developed side by side into one art, exemplified by the typical brush works of such influential artists as Su Shih[1] (+1036–1101) and Mi Fu[2] (+1051–1171). The so-called School of Literati in painting, which used brush strokes freely in movement and rapidly in execution, flourished from this time, and it was primarily the use of the whitish, smooth, soft, and absorbent surface of the paper that resulted in such subtle and free expressiveness (Fig. 1229). Although silk has some of these qualities, expense and its other limitations prevented it becoming as popular a medium for art, while some of the tonal effects available on paper such as washing, splashing, or gradations of ink, could not be easily achieved on a silk surface. Most Western artists habitually paint on canvas with oil paint, but the Chinese, as well as all the peoples of East Asia, have found paper the ideal medium for their artistic expression.

Paper is not only preferred for fine arts, but has also been used in applied and decorative arts in the East as well as in the West. The most popular such item is probably wallpaper, which found its way from China to Europe as early as the 16th century and to America in the early part of the 18th.[c] The introduction of this most welcome product of China, which eventually replaced wall hangings of expensive silk, leather, and tapestry in European homes, has enriched the living conditions of ordinary people as well as those of wealthy and royal households.[d] Other popular uses of paper in decorating Chinese houses include folding screens, hanging scrolls, household posters, and New Year pictures, which have made Chinese living quarters much more attractive and enjoyable.[e]

The wonders of paper have been attested by a multitude of literary references which commend its origin, nature, and appearance, as well as utility. The earliest such praise of paper, by the scholar Fu Hsien[3] (+239–93), is expressed in rhymed-

[a] It is usually considered to have been in the time of Tshai Yung[4] (+133–192) that the aesthetic discussion of Chinese writing became prevalent in China.

[b] The earliest examples of Chinese calligraphy on paper surviving today include such specimens as those of Lu Chi[5] (+261–303) and Wang Hsi-Chih[6] (+321–79?) of the Chin period.

[c] Specimens of old wallpapers are said to be still preserved in some of the old colonial mansions of America; see Laufer (48), p. 21.

[d] See discussion on pp. 116 ff. above.

[e] Mounted scrolls of painting and calligraphy were laid flat on tables for examination in early Thang, but later were hung on walls; see van Gulik (9), pp. 142 ff.

[1] 蘇軾　　　[2] 米芾　　　[3] 傅咸　　　[4] 蔡邕　　　[5] 陸璣
[6] 王羲之

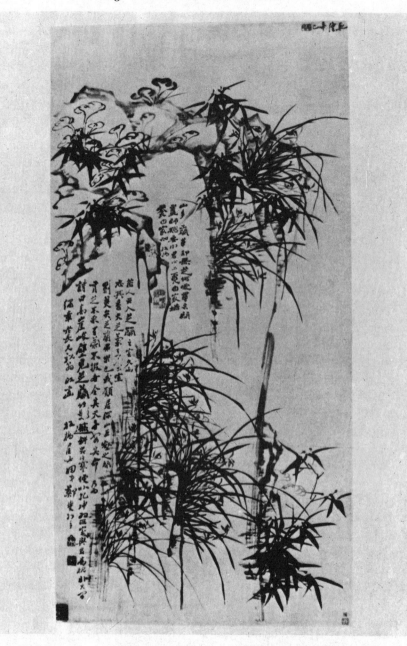

Fig. 1229. Orchids and bamboo with free style of calligraphy by the noted artist Chêng Hsieh (+1693–1765).

prose entitled *Chih fu*[1] (On Paper), the text of which is transcribed in Fig. 1230 and may be translated as follows:[a]

> For order the world requires
> Both crudity and elegance,
> With politeness to balance deficiency and abundance;
> So the tool and the substance alter
> To meet in the changing times.
>
> As the *Book of Changes* used incised symbols
> To substitute for the knotted cords,
> So paper was invented to replace bamboo slips.
> For convenience even plain and thrifty,
> To adapt to changes as time advances.
>
> Lovely and precious is this material,
> Luxury but at a small price;
> Matter immaculate and pure in its nature
> Embodied in beauty with elegance incarnate,
> Truly it pleases men of letters.
>
> It makes new substance out of rags,
> Open it stretches,
> Closed it rolls up,
> Contracting, expanding,
> Secreting, expounding.
>
> To kinship and friendship scattered afar,
> When you are lonely and no one is by,
> You take brush to write on paper
> And the fish and the wild goose[b]
> Carry your affection
> Ten thousand miles ...
> And your thoughts on a corner.[c]

After a lapse of more than a millennium, the first poem on papermaking in a European language appeared in a book about trades by Hans Sachs (+1494 −1576), whose verses mainly describe the technical procedures involved in converting rags to the finished product.[d] It was published in 1568, accompanying a block engraving by Jost Amman, and this picture is the earliest illustration depicting a papermaker at work with his essential tools (Fig. 1199). A little later, in 1588, an account of papermaking in verse form appeared in English. This work, by Thomas Churchyard (*c.* +1520–1604), expresses in its 353 lines an interest in the

[a] This rhymed-prose has been rendered into English by T. H. Tsien and Ming-Sun Poon and paraphrased in verse form by Howard W. Winger; part of the verse in a different translation is cited in Tsien (2), p. 138.
[b] Legends relate to the sending of silk letters by means of fish body and wild goose claw during the Chhin and Han periods.
[c] See *Chhüan Chin Wen* (*CSHK*), ch. 51, p. 5*a*.
[d] See translation of the poem in Dard Hunter (5), p. 14.

[1] 紙賦

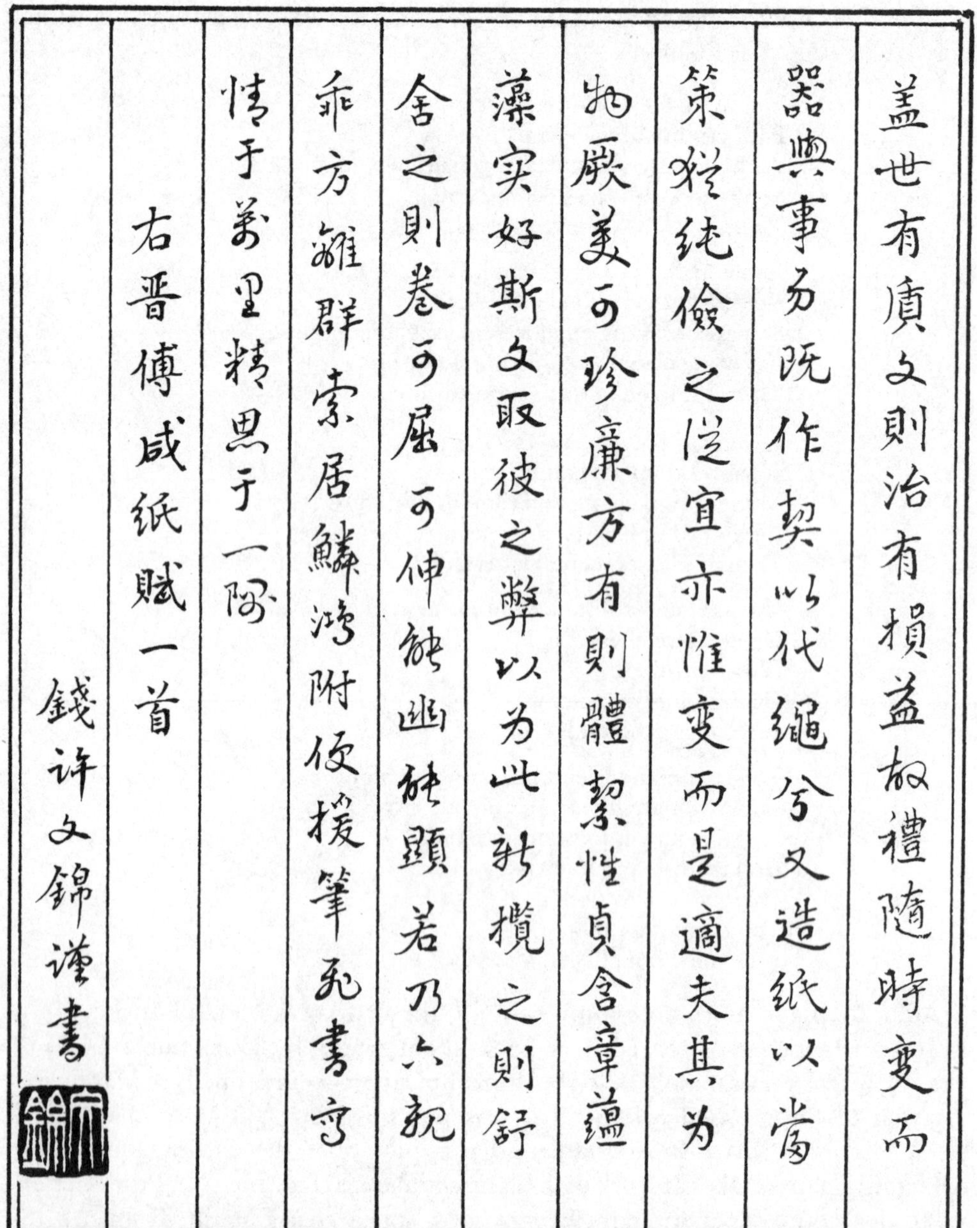

盖世有質文則治有損益故禮隨時変而

器興事勿既作契以代繩今又造紙以當

策猶純儉之從宜亦惟変而是適夫其为

物厥美可珍廉方有則體絜性貞含章蘊

藻実好斯文取彼之弊以为此新攬之則舒

舍之則卷可屈可伸鈍幽姿顯若乃六親

乖方雜群索居鱗鴻附使援筆𠮉書可

情于萬里精思于一隅

右晋傳咸紙賦一首

錢许文錦謹書

Fig. 1230. Text of a rhymed-prose on 'Paper' by Fu Hsien of the +3rd century, written in the running-style calligraphy for the East Asian History of Science Library, Cambridge, by Tsien Hsü Wen-Chin.

nature of paper similar to that of the Chinese verses reproduced above, but relates different details regarding technical matters. A few salient lines on the usefulness of paper are worth quoting:

> I prayse the man that first did Paper make,
> The only thing that sets all virtues forth:
> It shooes new bookes, and keeps old workes awake,
> Much more of price than all the world is worth:
> Though parchment duer a greater time and space,
> Yet can it not put paper out of place:
> For paper, still, for man to man doth go,
> When parchment comes in few men's hands you knowe.[a]

(2) IMPACT OF PRINTING ON WESTERN CIVILISATION

Printing facilitated the economical mass production and distribution of books and had profound effects upon European thought and society in the late 15th and early 16th centuries. It stimulated the spirit of the Renaissance and the Reformation, which in its turn promoted further development of papermaking and printing until there was a flourishing publishing industry. It also helped to establish national languages and indigenous literature, and even to encourage nationalism itself; it popularised education, spread literacy, and increased the chances of social mobility. In short, almost everything in the progress of modern civilisation can be linked in one way or another to the introduction and development of printing in the Western world.[b]

Mass production of texts increased their chance of survival or preservation and reduced the probability of their loss through neglect or the destruction of single collections. But it did more than this. Wider distribution of texts and the enlargement of the reading public, meant that the clergy's monopoly of learning was challenged by laymen, including lawyers, merchants, tradesmen, and artisans, who became important consumers of books. At the same time, the primacy hitherto taken by religious works was gradually superseded by texts of humanist authors.[c] With this increased readership and a broader spectrum of subjects, scholars became more aware of inconsistencies and contradictions in hallowed texts, weakening their faith in the validity of old views, and setting the stage for the advancement of new learning.[d]

[a] The poem, entitled 'A Description and Playne Discourse of Paper . . . 1588,' is cited in Hunter (5), p. 16; (9), p. 120.
[b] For the impact of printing on Western society and thought, see studies by Eisenstein (1) and (2), Febvre & Martin (1), Hirsch (1), and McLuhan (1), and the catalogue of an exhibition, entitled *Printing and the Mind of Man*, held in London in 1963.
[c] Between 1450 and 1500, about twenty million books were printed in Europe, representing some 10,000–15,000 different texts, or 30,000–35,000 editions. Of these incunabula, about seventy per cent are in Latin and the rest in Italian, German, French, and Flemish; forty-five per cent are religious works, thirty per cent are literature, and ten per cent each concern law and scientific subjects; see Steele (1), Lenhart (1), and Febvre & Martin (1), pp. 248–9, n. 344.
[d] Cf. Eisenstein (2), pp. 72 ff.

The standardisation of texts resulting from printing stood in contrast to the inevitable corruption that was bound to be present in all hand-copied texts. The printing press does not guarantee freedom from textual errors, but the requirement for multiple proof-reading before sending to the press and the distribution of errata to correct mistakes after printing, paved the way for the improvement of future editions. The editorial functions of the early printers also brought about a degree of systematisation of book format not to be found in the age of scribes, and this gradually created a habit of systematic thinking by readers, as well as promoting the organisation of knowledge in many diverse fields.[a]

The introduction of printing was very closely related to the religious reformation in Europe, and Martin Luther referred to it as 'God's highest and extremest act of grace, whereby the business of the Gospel is driven forward'.[b] Conditions for the reformation actually came into existence before he launched his protest in 1517, when a number of Bibles were already being printed in the vernacular.[c] This encouraged a belief that Gospel truths could be learned and understood by ordinary men, and made possible national variations in worship in contrast to the international but standard forms of the Roman Church. The original motivation of the Protestant movement was to seek correction of abuses, particularly the Church's sale of indulgences, the scale of which had been enlarged by printing, since it had been used to produce large numbers of indulgences in what became a profit-making enterprise.[d] However, the press enabled Protestant views to be circulated widely in the form of pamphlets, tracts, and manifestoes. Indeed, without the intervention of printing, Protestantism might have remained a local issue and not become a major movement which forever helped to end the priestly monopoly of learning, contributed to the overthrow of ignorance and superstition, and assisted Western Europe to escape from the Dark Ages.[e]

Vernacular literature existed, of course, before the advent of printing, but printing had a profound influence on its development. The spoken languages of Western Europe developed into written languages before the 16th century, and gradually evolved into their modern forms by the 17th, by which time some written languages of the Middle Ages had disappeared, and Latin, once a *lingua franca*, was becoming used less frequently, and was later to become a dying language.[f] The emerging national monarchies and chancelleries favoured this trend towards unified national languages, while authors tried to determine the best style through which to convey what they wanted to say; for their part publishers naturally encouraged the growing use of the vernacular which brought an expanding

---

[a] Cited in Black (1), p. 432.   [b] Cf. *ibid.* pp. 80 ff.

[c] A Bible in High German was first printed in 1466; this ran to nineteen editions before Luther's time; see Febvre & Martin (1), p. 289.

[d] See description of 'A Letter of Indulgence' printed by Caxton in 1476 in *Printing and the Mind of Man* (1963), p. 17.

[e] Cf. Eisenstein (2), pp. 306–7.

[f] For the development of national languages and printing, see Steinberg (1), pp. 120 ff.; Febvre & Martin (1), pp. 319 ff.; Eisenstein (2), pp. 117–18.

market. As books became easier to publish in national languages, printing stabilised the vocabulary, grammar, structure, spelling, and punctuation of each, and, furthermore, promoted its use. Once fiction was printed and widely circulated, the common language became firmly established; this, in its turn, facilitated the eventual growth of specific national literatures and cultures, which in turn led to the realisation of a distinct national consciousness and to nationalism.[a]

The popularisation of education and the spread of literacy were also closely related to the expansion of printing. As books became cheaper and easier to obtain, more people were able to gain access to the printed texts which eventually affected their outlook on the world and their position in it. And, naturally enough, easier access to printed material promoted the rise of literacy, which stimulated a still greater demand for more books. Moreover, early printed manuals and advertisements probably made it clear to many with an artisan background that profits and prestige could be acquired by printing such materials themselves, and this, of course, may have encouraged the spread of literacy among the artisan class.[b] It is evident, too, that some of the manuals were primers for teaching oneself to read and write, thus extending the market for books still further. There is also some biographical evidence which suggests that printing may have opened up opportunities to men of humble origin to advance their social position.[c]

(3) EFFECTS OF PRINTING ON CHINESE BOOK PRODUCTION

In China as in the West, printing made possible more and cheaper books and other material with a wider range of subjects for a larger reading public; naturally enough, this all had a certain amount of influence on the modes of Chinese scholarship and society. When large-scale printing began to emerge in the +10th century, the output was enormous; for instance nearly half a million copies of Buddhist books and pictures are known to have been printed in the eastern part of China in one small area alone over a period of less than half a century.[d] Again, during the Sung dynasty, six or more different editions of the Buddhist *Tripitaka*, which required tens of thousands of blocks for each edition, were printed and distributed throughout the country and abroad.[e] At about the same time, the Taoist canon was also printed.

[a] For stabilisation of language and rise of nationalism, see Chaytor (1), pp. 22 ff.
[b] Cf. Eisenstein (2), p. 242.
[c] Among the priests who spearheaded the Strasbourg reformation, one was the son of a shoemaker and another the son of a blacksmith. Although of humble background, they were steeped in the new learning through their access to the printed page; see Eisenstein (2), p. 372.
[d] These included three versions of the invocation *sutra* at 84,000 copies each, and printings of about a dozen other *sutras*, and charms and pictures, including 140,000 copies of one Buddhist image, amounting to a total of 400,000 copies printed in the Wu-Yüeh Kingdom from +937 to 975; see discussion on pp. 157 ff. above.
[e] These editions, in from 5000 to 7000 *chüan* each containing about 10–15 leaves (double pages) per *chüan*, would require 60,000 to 80,000 blocks for each edition, or about one-half as many if carved on both sides. See printing of Buddhist collections on pp. 159 ff. above.

As everywhere else in the world, religion had proved to be a motivating force for the use of printing. But once the techniques became more sophisticated, the dominance of religious literature was gradually overtaken by secular subjects and the percentage of religious publications declined in China as it was later to do in Europe. Thus as early as the + 10th century Feng Tao borrowed the art of printing from the Buddhists to reproduce standardised Confucian texts, instead of carving them on stone, and since then, the printing of Confucian classics, histories, and other works intensified. Thus when a large printing project was started by the Sung government in +988, only some 4000 wood blocks were kept in the National Academy, but by 1005, when the emperor inquired about the project, Hsing Ping[1] (+930–1010), then Director of the National Academy, reported that the wood blocks at the Academy numbered 100,000, representing all kinds of classics, histories, and their commentaries.[a] This is indeed most impressive; it means that printing by this central government agency alone increased as many as twenty-five times within a period of less than twenty years.

Printing blocks could be used again and again, and sometimes lasted for several hundred years until they either deteriorated or were destroyed. There is very little information about how many copies were printed from the same block, because they were re-used so often, though it appears that the number varied a great deal, from a few in the case of a scholarly work of limited circulation to tens of thousands of a popular text in great demand. However, one modern writer claims that thirty copies were usually printed for the first impression,[b] and each new block could be used to print up to 15,000 copies, or 25,000 after retouching.

In one bronze movable type edition of a Sung encyclopaedia, *Thai Phing Yü Lan*,[2] in 1000 *chüan*, reprinted in 1574, a run of 100 copies is mentioned in the colophon.[c] If this figure for a particular reference work applies to other large sets, no fewer than 100 copies for each new title or each new block may be reckoned to have been produced at that time.[d] This may also be true of other times, since the mode of printing did not change very much until the middle of the 19th century.

While no clear record of copies printed for a block edition is available, we know more precisely the cost of production and how much was saved by the application of printing. There are detailed listings of various costs for certain printed editions, including material, labour, and the charges for renting blocks (Fig. 1231). For example, the complete works of a Sung scholar, Wang Yü-Chheng[3] (+954–1001),

[a] Cf. *Yü Hai* (*CCSC* ed., 1883), ch. 43, p. 18a.
[b] See Lu Chhien (*1*), p. 632.
[c] Cf. description in Chhien Tshun-Hsün (2), p. 15, item 7; a movable type edition was not normally reprinted by the same method but more often was reprinted with woodblocks.
[d] In the case of another encyclopaedia, *Ku Chin Thu Shu Chi Chheng*,[4] which includes over 100 million words in 5020 volumes, sixty-six copies were printed with bronze movable type for the first impression in 1725–6, and 100 copies were reprinted with lithography in + 1890. Another large set, the *Wu Ying Tien Chü Chen Pan Tshung Shu* in 134 titles, was printed in 300 copies in *c.* 1734. Also 400 copies of a collection of poetry were printed with earthenware type by Chai Chin-Sheng in 1847–8; see pp. 203, 209, 216 above.

[1] 邢昺   [2] 太平御覽   [3] 王禹偁   [4] 古今圖書集成

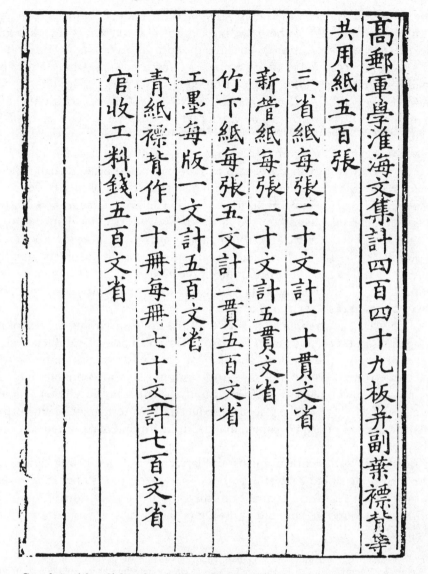

Fig. 1231. Cost of materials and labour for printing the *Huai Hai Chi* by Chhin Kuan of the Sung dynasty, printed *c.* +1173. The account on the back of the book says it used 449 blocks and 500 sheets of paper and was sold for 500 cash. Original copy preserved at the Naikaku Bunko, Tokyo.

Hsiao Chhu Chi,[1] thirty *chüan* in eight *tshe*, printed at Huang-chou in 1147, contains 163,848 characters on 432 blocks. The cost of production included 260 pieces of cash for 448 sheets of printing paper, eleven sheets of blue cover paper, and eight large sheets of paper; 500 cash for ink and for renting blocks and brushes; and 430 cash for binding; with a total cost of 1136 cash for labour, rent, and other expenses,

[1] 小畜集

except for paper. The set was sold for 5000 pieces of cash per copy.[a] These figures give an average unit cost of about one piece of cash for the rent of each block with tools, and another piece of cash for the labour of printing and binding one leaf. The retail price in this case was about 600 pieces of cash per *tshe*, while that of other works during a similar period ranged from 300 to 400 pieces of cash.[b]

Another case is that of a Yüan gazetteer of Nanking, *Chin-Ling Hsin Chih*,[1] fifteen *chüan* in thirteen *tshe*, the printing of which cost a total of 7,179,899 taels of silver in the Chung-Thung currency. There is no mention of how many copies were printed, but the cost seems rather high.[c] In Ming times, printing costs were much lower. Carving was only twenty pieces of cash per 100 characters around 1640, when the thirteen classics and the seventeen standard histories were published by Mao Chin[7] (+1599–1659), who printed in his famous shop Chi Ku Ko[8] some 600 titles with over 200,000 carved blocks.[d] In the latter part of the Chhing dynasty (around 1875), the cost of preparing transcriptions and printing in Hunan was about fifty to sixty pieces of cash per 100 characters; this increased to eighty to ninety later in the 19th century and to 130 at the beginning of the 20th. The pay of women cutters in Hunan, Chiangsi, and Kuangtung is said to have been extremely low, only twenty to thirty pieces of cash per 100 characters.[e]

Printing was of course much cheaper than the time-consuming process of hand copying. For instance, the text (*pai wen*)[9] of a collection of Confucian classics contains nearly a million words, and if a copyist could write as many as 10,000 characters a day, it would still take 100 working days to complete one copy.[f] Certainly, carving on blocks was slower, but in the end it produced more copies far more cheaply. How much the price was reduced by printing may be illustrated by comparing the cost of a hand-copied manuscript with that of a printed edition at a comparable time.

In the early +9th century during the latter part of the Thang dynasty, the charge by a professional copyist was about 1000 pieces of cash per *chüan* of about 5000 to 10,000 characters. This is confirmed by the copying cost of some of the Buddhist *sutras* found in Tunhuang, the charges on which are sometimes given in

---

[a] This information is given in the colophon of the original Sung edition which was reprinted in facsimile in the Ming dynasty; see description in Yeh Te-Hui (2), pp. 144–5. The cost mentioned here seems to indicate that the 448 sheets of paper was the amount used for each copy: the 448 sheets were for 432 blocks, including two blank sheets for fly-leaves for each of the eight *tshe*, and the other papers were for covers and binding; the other costs apparently were for the rent and labour for the entire impression of the book.

[b] The price of the *Ta I Tshui Yen*,[2] twenty *tshe*, printed in 1176, was 8000 pieces of cash (*wen*[3]) per set; and that of the *Han Chün*,[4] two *tshe*, printed in 1184, were 600 pieces of cash per set; see Yeh Te-Hui (2), p. 143.

[c] The total cost of 143 *ting*[5] (silver ingots) and 29,899 *liang*[6] (taels) is mentioned in the preface of the gazetteer; each *ting* equalled fifty taels and each tael was worth 20,000 pieces of cash in the Yüan dynasty. If 100 copies of this book were printed, one set would have cost seventy-two taels or 140,000 pieces of cash. This roused suspicion of corruption by the Yüan officialdom; see Yeh Te-Hui (2), pp. 178–9.

[d] Yeh Te-Hui (2), pp. 185–6.   [e] *Ibid.* p. 186.

[f] See discussion in Poon (2), p. 67.

[1] 金陵新志   [2] 大易粹言   [3] 文   [4] 漢雋   [5] 錠
[6] 兩   [7] 毛晉   [8] 汲古閣   [9] 白文

the colophon and are of the same order.[a] The price for a manuscript copy of the rhymed dictionary hand-copied by the famous woman calligrapher Wu Tshai-Luan[1] (fl. +827–35) also came to this average figure.[b] At about the same time, the Japanese monk Ennin (+793–864) bought in China in 838 a copy of Buddhist *sutra* in four *chüan*, costing 540 pieces of cash.[c] Because of its cheapness in comparison with the copied manuscripts, it is believed that this book must be a printed edition.[d] If so, the average cost was about 100 pieces of cash per *chüan*, indicating that the cost ratio between a printed edition and a copied manuscript was one to ten.

This cost ratio continued with little change in later times. In 1042, for example, the printing of calendars by the Sung government is recorded as costing 30,000 pieces of cash, while copying during the previous years was said to have cost ten times this amount.[e] A Ming author, Hu Ying-Lin[3] (+1551–1602), said that if no printed edition were available on the market, the hand-copied manuscript of a book would cost ten times as much as the printed work; moreover, once a printed edition appeared, the transcribed copy could no longer be sold and would be discarded.[f] All these cases indicate that printing had reduced the cost of a book by as much as ninety per cent before the end of the 16th century although, of course, the price fluctuated from time to time.[g]

The development of printing naturally encouraged greater emphasis on textual criticism so that more reliable texts could be produced than ever before. Because of the permanence and wider dissemination of the text, scholars were more aware of the need for its reliability and correctness through careful collation and proof-reading before it was finally engraved on to blocks. Collation of texts was a preoccupation of many scholars (Fig. 1232), who served in the official agencies or worked independently, and at least four proof-readings were usually required before printing; the latter occurred following transcription, correction, engraving, and the first impression.[h] Because of this careful preparation, a well-collated and printed edition was valued above a copied manuscript, which was likely to contain unintentional errors. Textual accuracy was, therefore, an additional important reason besides its lower cost, for readers to choose a printed edition.[i]

A typical example of such serious effort in preparation is the printing of the Nine Classics and Three Commentaries in *c.* +1230 or 1300. Some twenty-three different editions were used for collation, and a special manual, the *Chiu Ching San Chuan Yen Ko Li*,[5] was prepared to provide guidelines; it gave specifications on such

[a] One Buddhist *sutra* in forty *chüan* cost 30,000 pieces of cash; another in seven *chüan* cost 10,000; see Thai Ching-Lung (*1*), p. 9.
[b] The *Thang Yün*[4] in five *chüan* was sold for 5000 pieces of cash; see Yeh Te-Hui (*2*), pp. 285 ff.
[c] See Reischauer (*4*), p. 48.　　　　　　[d] See Weng Thung-Wen (*2*), pp. 38–9.
[e] See *Hsü Tzu Chih Thung Chien Chhang Pien* (Thaipei, 1961), ch. 102, p. 18.
[f] See *Shao Shih Shan Fang Pi Tshung* (Peking, 1958), ch. 4, p. 57.
[g] See Weng Thung-Wen (*2*), pp. 35 ff.
[h] For the procedures of block printing, see pp. 200 ff. above.
[i] Cf. Poon Ming-Sun (*2*), pp. 72 ff.

[1] 吳彩鸞　　　[2] 圓仁　　　[3] 胡應麟　　　[4] 唐韻　　　[5] 九經三傳沿革例

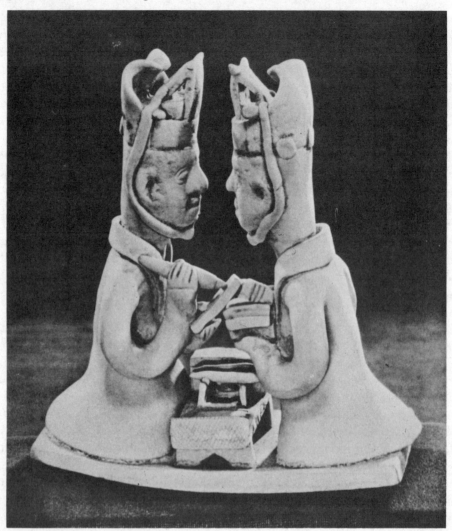

Fig. 1232. A pottery piece of the Western Chin period (+265–316), discovered in Chhangsha in the 1960's, depicts the collation of books by two scribes sitting face to face, perhaps in proofreading. From *Historical Relics Unearthed in New China*, Peking, 1972.

matters as the selection of editions for textual comparison, the use of a standard style of calligraphy, emendations of the original annotations and commentaries, the pronunciation of obsolete characters, demarcation of sentences and punctuation, collation of passages in correct order and, finally, notes on textual variations. This edition of classics, printed in Hsiang-thai, Honan, has been highly praised as the milestone in the transmission of the ancient works and a model of fine printing and careful textual collation.[a]

[a] Cf. description of the specifications by T. H. Tsien in Hervouet (3), p. 53.

Printing did not result in much change in the physical appearance of the book, except perhaps for some special features of the printed page. The evolution from the roll form to the flat binding, in the late +9th or early 10th century, was primarily because of the inconvenience of opening the paper rolls for reading; it was not necessarily a result of printing.[a] The special features that were added to the printed page, and which do not appear on manuscripts except for those copied from a printed edition, include such signs in the body of the block as the centre line, the 'fish tail', and the 'elephant trunk' at the centre of the block to mark the fold. The cutter's name and the number of characters on the block are also sometimes indicated.

An important feature of a printed book is the addition of the printer's colophon (*phai chi*[1]) *or trade mark* in the book, which can normally be found on the back of the title page, or at the end of the table of contents. It is a boxed square either in an oblong form or in designs of a gourd, a tripod, or a round stamp providing such information as the date, place, printer's name and occasionally a note on the process of printing and an advertisement of the printer (Fig. 1233).[b] However, the most visible change in the appearance of a printed book was the calligraphic transformation sometime in the middle of the 16th century of the text from a regular written to a printed style. This printed text, called the Sung style (*Sung thi tzu*[2]), is characterised by a formalised and stereotyped construction, with more straight lines than curves. It is easier for block cutters to carve and has been followed by printers ever since, though with slight variations from time to time. The metal type used in modern printing is developed from this style.

The expansion of printing activities naturally resulted in the establishment of printing centres throughout the country. Wherever skilled block cutters and sponsors were available, more printing facilities were located. Under the Sung, as we have seen, printing centres included Hangchow (in Chekiang) and Khaifeng (in Honan), capitals of the Sung; Chien-an and Chien-yang (in Fukien), where trade editions had been known for centuries; and Mei-shan (in Szechuan), a cultural centre until the Ming.[c] Of some 1,500 Sung editions of which the location of printing can be identified, more than 90 per cent are known to have been printed in provinces where such centres were located.[d] Indeed, it has become clear that factors such as political status, economic strength, cultural tradition, and the availability of materials, were responsible for the prosperity of the printing industry.

[a] Some of the earliest printings of Buddhist works are in the roll form; see discussion of book-binding on pp. 227 ff. above.

[b] See form and translations of such colophons in Poon Ming-Sun (1), pp. 39 ff.

[c] Among fifty-six printing shops under the Sung, fifty-five are known to have been located in Hangchow, Chien-an, and Chien-yang; see Poon Ming-Sun (2), p. 167.

[d] Of the 1478 titles, 614 (42 %) were printed in Liang-che (modern Chekiang), 231 (16 %) in Fukien, 199 (13 %) in Chiang-nan Tung and Hsi (modern Chiangsi and Chiangsu), 171 (12 %), in Ching-chi (modern Honan), and 124 (8 %) in Chhengtu (modern Szechuan); see Poon (2), p. 11, table 1.

[1] 牌記   [2] 宋體字

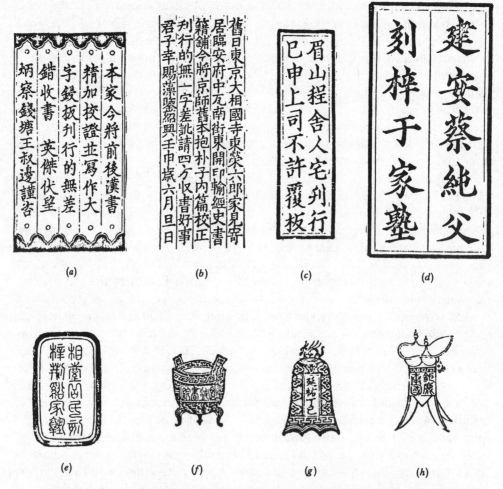

Fig. 1233. Colophons and trade marks of early Chinese printing. (*a*) The printer Wang Shu-Pien at Chien-an, Fukien was formerly from Chhien-Thang (i.e. Lin-An), *c.* +1150–70. (*b*) The printer Jung Liu-Lang at Lin-An (i.e. Hangchow) advertising its old address at the former Sung capital Khai-Feng for prestige, +1152. (*c*) The Chhêng family of Mei-Shan, Szechuan registered the book with the government for prohibiting reprint, *c.* +1190–4. (*d*) The Tshai Family Academy of Chien-An, Fukien, *c.* +1208–24. (*e*) The Yüeh family of Hsiang-Thai, Honan, printed in Ching-chhi (i.e. I-hsin, Chiangsu), *c.* +13th century. Printers of the Yüan dynasty, +14th century, developed such ornamental designs as (*f*) a tripod, (*g*) a bell, and (*h*) a goblet. From Poon Ming-Sun (1).

It should not be a surprise to realise that commercial printers were concerned more with the market demand than with textual accuracy or quality of printing. Such popular materials as handbooks of everyday knowledge, manuals of letter writing, popular novels and drama, as well as basic textbooks and reference works, especially attracted their attention, and they published more books of this kind for the common people than did the official agencies or private families, who hesitated to print such commonplace material. Nevertheless the commercial printers should have the credit for popular education and the spread of literacy, because they made books available at a low price to a large readership that could not otherwise afford to buy or easily gain access to them.[a]

(4) THE INFLUENCE OF PRINTING ON CHINESE SCHOLARSHIP AND SOCIETY

In many respects, Chinese culture has had an extensive literary or bookish tradition, characterised among world civilisations by its productivity, continuity, and universality. It is unique in the volume of its output, the length of the period covered, and its uninterrupted and widespread intellectual transmission. From very ancient times, an enormous amount of literature and documents was produced and transmitted, and the production of historical records and annals has continued almost without interruption to the present day. As to magnitude, a single work rather frequently ran to millions of words. Written words were revered and from antiquity books were assiduously read and studied not only by the Chinese but also by other peoples of East Asia throughout a prolonged period of history, and this bookish tradition contributed to the production of more written and printed pages in Chinese than were produced in the West until about the end of the 17th century.[b] It also contributed to the early invention in China of paper and printing, which became important vehicles for sustaining the Chinese cultural heritage throughout the centuries.

The power of printing in mass production, distribution, and preservation stimulated the revival of classical learning and changed the mode of scholarship and authorship under the Sung dynasty. While the Thang promoted Buddhism and Taoism, and produced poetry that is prized as the gem of the period, the Sung became one of the great ages in Chinese history for its scholarly achievements in critical studies of classical works, art, archaeology, material culture, and science. The restoration of Confucian learning was reflected in such scholarly activities as new interpretations and the large-scale printing of Confucian classics, studies in classical philology, textual criticism, as well as compilation of voluminous general and local histories, bibliographies, and catalogues. The revived interest in Confucianism was an obvious victory of Chinese traditional thought and political philosophy. The work of Chu Hsi[1] (+1130–1200) and other Sung scholars who founded Neo-Confucianism, became the guiding principle of Chinese society until the end of the 19th century, when it was challenged by Western thought and institutions.

---

[a] See commercialisation of Sung printing in Poon (2), pp. 167 ff.

[b] More than 50,000 separate titles, or over half a million volumes (*chüan*), produced before the late Ming period, are known to have been registered in Chinese bibliographies and other sources before printing was widely used in Europe in the late 15th century; see estimate by Yang Chia-Lo (2), p. 27. Both Swingle (13), p. 121, and Latourette (1), p. 770, suggest that by 1700 or even 1800, more pages, written and printed, existed in Chinese than in all other languages put together.

[1] 朱熹

As we have seen, when printing first appeared in China, it was motivated by the need for great quantities of Buddhist texts. Not until two or three hundred years after its invention did the Confucian classics and other scholarly literature appear in print. Two great projects for printing the Confucian classics, which were begun in the 10th century, signalled the revival of Confucian learning. Both the Nine Classics sponsored by the prime minister Feng Tao[1] (+882–954) of the Later Thang, and the printing started privately in +953 by Wu Chao-I[3] (d. +967) of the Later Shu state, were significant in several ways.[a] The initiative of Feng Tao made the National Academy the official agency for printing classics, histories, textbooks, and other volumes to be distributed at the national level. It also made the government one of the most influential agencies in printing. The Wu Chao-I project became a typical case of Confucian scholars preaching the moral values of private enterprise in printing. From this time on, almost all printing engaged in by the government, by private families, and by commercial agencies was dominated by the Confucian scholars.

The revival of Confucian learning gave impetus to the flourishing schools and academies which supplied candidates for the civil service examinations which were themselves based on Confucian writings and ideas. As a result three or more institutions of higher education were set up in the Sung capitals, Khaifeng and Hangchow: the National University (Thai Hsüeh[4]), which had an average enrolment of about 2000 students who had passed the entrance examinations; the National Academy (Kuo Tzu Chien),[2] which enrolled about 200 children of officials; and the School of Four Gates (Ssu Men Hsüeh[5]), which had an average of 500 students from ordinary families.[b] On the local level there were, during the Sung period, over 1000 official and private schools and academies distributed through almost all the prefectures and counties of the country.[c] The National Academy played the key role in the central government's printing, and many of the local schools and academies also engaged in printing textbooks, dictionaries, histories, philosophical writings, and medical works.[d] In fact, under the Sung, no fewer than 250 titles are known to have been printed by the Kuo Tzu Chien, and over 300 such works were printed by local academies.[e]

Printing no doubt also had some positive relationship with the Chinese civil

[a] See discussion of these two projects on pp. 156 ff. above.

[b] The National University student quotas ranged from 1000 in 1071 to 3800 in 1103; the Kuo Tzu Chien had 70 students in 975, 300 in 1044, and 200 in 1078; and the Ssu Men Hsüeh had 450 students in 1058 and 600 in 1062; see Chaffee (1), p. 45, table 6.

[c] A total of 1099 schools were established under the Sung, including 189 at perfectural level, 464 at county level, and 446 private schools; see Chaffee (1), p. 167, table 11.

[d] The prefectural school of Chiang-ning fu (modern Nanking) in c. 1130 had a collection of over 20,000 woodblocks for sixty-eight titles; cf. Chaffee (1), p. 94.

[e] The Kuo Tzu Chien printed 256 titles, including 108 on classics, 61 on history, 83 on philosophy, and 4 literary works; the local academies and other agencies published 324, including 83 on classics, 81 on history, 87 on philosophy, and 73 literary works; private academies contributed 17 titles; and commercial printers 128; see Poon Ming-Sun (2), pp. 123, 134, 154, 170.

[1] 馮道   [2] 國子監   [3] 毋昭裔   [4] 太學   [5] 四門學

service examination which recruited educated personnel for service in the government. The beginning of this system can be traced back to the Han or earlier, but it was not fully utilised until Thang and Sung times.[a] It was, indeed, under the Sung that the system was further perfected, the number of participants greatly increased, and it was at this time that the number in the government of those holding the highest degree, the doctoral graduates (chin-shih[1]), was more than double that of the previous dynasty.[b] The growth of the general scholarly population during the Sung was even more impressive. The number of chin-shih quadrupled from early Sung to late Sung, reaching a total of over 40,000 for the entire period.[c] Yet to become a candidate for the doctoral examination was no easy task; one had to have passed the qualifying examination on the prefectural level and to have received the degree of 'presented man' (chü-jen[2]). Even so, the number of chü-jen is estimated at 200,000 for the 12th century, and 400,000 for the 13th.[d] Including students at the various institutions of higher learning and local schools and academies, the total intellectual population of the Sung must, therefore, have been very large.

Examinations at all levels emphasised literary, historical, and scholarly knowledge, based on Confucian doctrines. The books used in preparation for these examinations included Confucian classics, histories, reference works such as dictionaries and encyclopaedias, as well as examination aids such as model essays, and pocket editions which could be taken into the examination halls. The demand for such materials for reading and study seems to have been one of the major reasons for the large-scale printing of textbooks and other required material during the Sung dynasty. The initiation of the two large printing projects for the Confucian classics, mentioned earlier, was certainly stimulated by the convenience and economy of printing.

The definite relationship between printing and the civil service examination is further attested by the fact that the more successful doctoral candidates there were in a specific region, the larger was the proportion of imprints produced in that area. For example, the top five provinces in the eastern, southeastern, western, and central regions, which produced eighty-four per cent of the holders of the doctoral degree during the Sung, printed ninety per cent of the books during the same period.[e] On the other hand, one poor province in the southwest, which produced

---

[a] For a general history of the Chinese civil service examination system, see Etienne Zi (1), and Teng Ssu-Yü (1).

[b] Cf. Kracke (1), p. 55, tables 3, 60; Chaffee (1), p. 63, table 9, shows 9785 officials in the Sung government during the period 997 to 1022, and more than 42,000 in 1196.

[c] There were 1587 chin-shih in the period 960 to 997, and 6177 in 1241 to 1274, with a total of 41,357 for the entire Sung period; cf. Chaffee (1), p. 354, table 31.

[d] Cf. Chaffee (1), p. 59.

[e] The provinces of Liang-che (modern Chekiang), Fu-chien (modern Fukien), Chhengtu (modern Szechuan), Chiang-nan Hsi (modern Chiangsi), and Chiang-nan Tung (modern Chiangsu) produced 24,172 of a total of 28,926 graduates while printing 1168 of 1303 books during the Sung period; for chih-shih by prefectures, see Chaffee (1), table 37; for numbers of imprints by provinces, see Poon Ming-Sun (2), p. 101, table 6; this uses the figures for two sample years of 1145 and 1256, and these are in general agreement with those for the entire Sung period.

[1] 進士   [2] 舉人

the fewest graduates, also printed the fewest titles.[a] This correlation between books and examinations clearly indicates the way printing contributed to the popularisation of education and advancement of scholarship. The reverse is, however, also true; the examination system promoted the expansion and prosperity of the printing trade.

The civil service examinations provided a fair system for recruiting into government service the qualified people from various social strata and from different geographical regions. An intelligent person of humble origin could climb the ladder of competitive examinations and eventually become one of the top administrators in the Chinese bureaucracy, and the general fairness of the system is confirmed by the fact that a clear majority of the successful candidates under the Sung came from families without any background of official connections.[b] The distribution of books to relatively poor and humble people to a certain extent facilitated their upward movement from the lower strata of society by way of the examination system, and was helped, in spite of the increasing disparity between rich and poor, by the fact that printed books cost much less than manuscripts.[c]

While printing in the West has primarily a business for profit, it had strong moral implications in Chinese society. It was considered a positive merit for an individual to preserve and disseminate knowledge, and rulers were politically rewarded for such action by receiving public support. Reverence for ancient literature was one of the basic elements of Confucian teaching, and the story of the burning of the books by the First Emperor of the Chhin dynasty in -212 has been emphasised by Confucian scholars as the most flagrant crime in history. From Han times onwards, almost all rulers who adopted Confucianism as state doctrine, chose to recover and preserve ancient books when they succeeded to the throne, a procedure that was also considered a virtue which brought honour to the family and to the individual. Yeh Te-Hui,[1] author of the celebrated work on books and printing, opened his book by citing a number of cases in Chinese history to illustrate how printing of books was crucial to gaining prosperity, to preserving fortunes during times of disorder, and to commanding respect from others.[d] Furthermore, attention to the reliability and correctness of a text by block-cutters and printers was particularly emphasised; failure to observe this would be punished, by spiritual if not by human powers.[e] Evidently, printing in China was not primarily for profit; moral

---

[a] The province of Kuei-chou (modern Kueichow) had 103 graduates and printed only two titles.

[b] The names of successful candidates in the two examination lists of 1148 and 1256 show about forty per cent of them were descendants of former office holders in the government, but some sixty per cent came from families with no history of civil service employment in the paternal line for three previous generations; see Kracke (1), p. 69.

[c] For printing and social mobility of the Ming and Chhing periods, see Ho Ping-Ti (2), pp. 212 ff.

[d] The stories include one on Wu Chao-I's private printing of the Confucian classics, which resulted in political status and wealth for his family and descendants after the conquest of the Shu state by the Sung dynasty in 965, while other powerful families were executed or otherwise punished; see Yeh Te-Hui (2), pp. 1 ff.

[e] Hung Mai[2] ($+1123-1202$) said that four block-cutters were struck by lightning for having changed the text of a book on medical prescriptions; see his I Chien Chih (TSHCC), p. 89.

[1] 葉德輝   [2] 洪邁

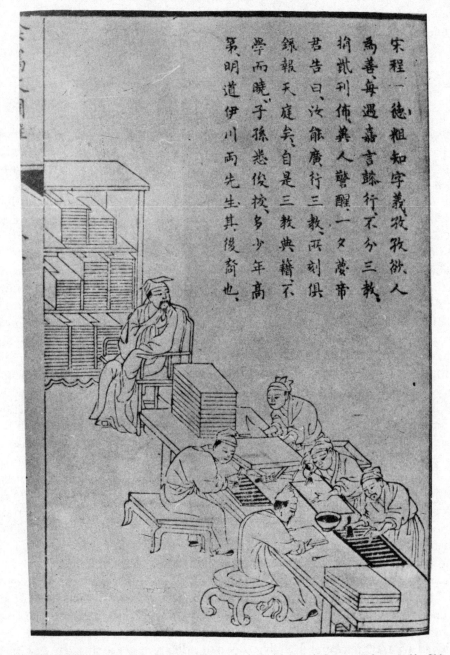

宋程一德粗知字義教牧欲人
為善每遇嘉言蘇行不分三教
掃貲刊佈與人警醒一夕夢帝
君告曰汝能廣行三教所刊俱
錄報天庭矣自是三教典籍不
學而曉子孫恭俊拔多少年高
第明道伊川兩先生其後裔也

Fig. 1234. Moral reward to a sponsor of printing. Picture depicts a private printing enterprise owned by Chheng I-Te of the Sung dynasty, a man with little learning, who was awarded by Heaven with two great scholarly sons, Chheng Hao and Chheng I, because he printed good books from the Confucian, Buddhist, and Taoist canons. From *Ying Chih Wên Thu Chu*.

obligations became an important factor in Chinese society to promote and develop it (Fig. 1234).

(5) SOME CONCLUDING REMARKS

Generally speaking, paper and printing served similar purposes in the East and the West, but had different impact upon their respective societies. Paper, however, seems to have played a more important role in China than in the West. Unlike printing, which was scarcely mentioned in Chinese literature, paper was written about and praised for its qualities and usefulness from very early times. Emperors bestowed it upon scholars and officials as the highest honour and reward, and along with brush, ink, and inkstone, it was one of the scholar's four treasures. Thus paper was no ordinary commodity, but was always associated in Chinese society with prestige and scholarship.

The very early use of paper for writing and for books in China was an epoch-making step in the history of civilisation. Without paper, certainly, no printing could have taken place and, as we have seen, in China paper also had other profoundly important effects; it helped sustain the Chinese cultural tradition, it was instrumental in refining Chinese art, it helped smooth the path of government and transactions in the business world, and played a not insignificant part in improving various household and other daily activities. With the adoption of paper, people everywhere have found their lives much easier, more convenient, better served.

Printing played a similar role in book production in China and in Europe, but the scale and pattern of its influence on the two societies was different. Certainly it made similar contributions in both to the reduction of costs, increase of productivity, and accessibility to a large public, as well as providing the standardisation of texts and a greater chance of their survival. But otherwise printing developed in different directions using different procedures. In the West, the printing press underwent a gradual mechanisation and sophistication, eventually growing into a powerful publishing industry with mass production and distribution; in China, on the other hand, printing was continuously carried on as a handicraft, without significant changes in technology until modern times.

As to the impact of printing on intellectual life and society, there were again some similarities between East and West. In both printing promoted culture, widened the scope of subjects that interested scholars, helped shift the bias from religious to classical learning, it popularised education, spread literacy, and enriched art and literature; though it did so to a different degree in each. But in the West printing also stimulated intellectual unrest and promoted the development of national languages and their use in literature; in China, on the contrary, it facilitated the continuity and universality of the written language and thus became an important vehicle for sustaining the cultural tradition. This is seen especially in the printing of the Confucian classics and similar material for the civil service examinations, and therefore acted as an important element in the relative stability of Chinese culture and society.

There was another difference. While Chinese culture always had an extensive literary tradition, at the close of the Middle Ages Western civilisation possessed only a very limited legacy of books. Yet as Europe emerged from the Dark Ages, the intellectual awakening resulted in a great demand for books. As soon as printing was available, it was utilised for book production to the maximum extent. In this respect, it was unlike printing in China, which has always produced an optimum number of copies for immediate use without excessive accumulation of copies in stock. Furthermore, Chinese printing was generally sponsored by government and private agencies without pecuniary motivation, whereas the European press was usually operated as a trade for profit. The different motives for printing seem also to have been responsible for different effects of the invention on society.

In general, printing in Europe from the 16th century onwards was vigorously expanded, supporting many drastic and radical changes in thought and society. On the other hand, the progress of printing in China and other nations in East Asia was comparatively stable and constant with modest changes occurring within a stable tradition. These diversities reflect the distinctive characters of Eastern and Western cultures, especially their attitudes towards material life. Chinese society has long been dominated by the Confucian doctrine, which concerns itself primarily with proper human relationships and social order by way of moral teachings and ethics rather than with pursuance of material advance and extreme changes in society. The high degree of social and cultural stability over long periods in Chinese history, especially from the 13th to the 19th century, contrasts greatly with the constant turmoil of life and intellectual unrest during the same period in the West. Such different environments in China and the West were certainly bound to affect the role of printing, and in this sense printing was not only shaped by the political and social conditions of the time; it also exerted an equal effect on those conditions.

BIBLIOGRAPHIES

A CHINESE AND JAPANESE BOOKS BEFORE +1800
B CHINESE AND JAPANESE BOOKS AND JOURNAL ARTICLES SINCE +1800
C BOOKS AND JOURNAL ARTICLES IN WESTERN LANGUAGES

In Bibliographies A and B there are two modifications of the Roman alphabetical sequence: transliterated *Chh-* comes after all other entries under *Ch-*, and transliterated *Hs-* comes after all other entries under *H-*. Thus *Chhen* comes after *Chung* and *Hsi* comes after *Huai*. This system applies only to the first words of the titles. Moreover, where *Chh-* and *Hs-* occur in words used in Bibliography C, i.e. in a Western language context, the normal sequence of the Roman alphabet is observed.

When obsolete or unusual romanisations of Chinese words occur in entries in Bibliography C, they are followed, wherever possible, by the romanisations adopted as standard in the present work. If inserted in the title, these are enclosed in square brackets; if they follow it, in round brackets. When Chinese words or phrases occur romanised according to the Wade-Giles system or related systems, they are assimilated to the system here adopted (cf. Vol. I, p. 26) without indication of any change. Additional notes are added in round brackets. The reference numbers do not necessarily begin with (1), nor are they necessarily consecutive, because only those references required for this volume of the series are given.

Korean and Vietnamese books and papers are included in Bibliographies A and B. As explained in Vol. I, pp. 221 ff., reference numbers in italics imply that the work is in one or other of the East Asian languages.

ABBREVIATIONS

The following abbreviations for journals, symposia, *tshung shu* or other collective publications, and a few names of publishers are used in the footnotes and bibliography of this volume. For editions of *tshung shu* consulted, the date is given after the title and an abbreviation is indicated in parentheses following the title in the footnotes and at the end of the entry in the bibliography.

AA	*Artibus Asiae*
AAN	*American Anthropologist*
ACASA	*Archives of the Chinese Art Soc. of America*
ACP	*Annales de Chimie et de Physique*
ACQ	*Asian Culture Quarterly* (Thaipei)
ACTAS	*Acta Asiatica (Bull. of Eastern Culture, Tōhō Gakkai, Tokyo)*
ADVC	*Advances in Chemistry*
AJSLL	*American Journal of Semitic Languages and Literatures*
AM	*Asia Major*
AMP	*American Printer*
APR	*Asian Pacific Record* (Singapore)
ARSI	*Annual Reports of the Smithsonian Institution* (Washington, D. C.)
ARTY	*Ars Typographie*
AS/A	*Chung Yang Yen Chiu Yüan Yüan Khan (Annals of Academia Sinica, Thaipei)*
AS/BIE	*Bulletin of the Institute of Ethnology, Academia Sinica* (Thaipei)
AS/BIHP	*Bulletin of the Institute of History and Philology, Academia Sinica* (Shanghai, Thaipei)
AS/MIE	*Monographs of the Institute of Ethnology, Academia Sinica* (Thaipei)
ASQR	*Asiatic Quarterly Review; Asian Review*
ASS	*Asian Scene; Business Japan*
BALA	*Bulletin of the American Library Association*
BBSK	*Bulletin of the Bibliographical Society of Korea*
BCEAL	*Bulletin (Newsletter) of the Committee on East Asian Libraries, Assoc. for Asian Studies*
BCS	*Bulletin of Chinese Studies (Chung-Kuo Wen Hua Yen Chiu Hui Khan, Chhengtu)*
BCSH	*Pai Chhüan Hsüeh Hai* (1921)
BEFEO	*Bulletin de l'École Française de l'Extrême Orient* (Hanoi)
BGTI	*Beitrage z. Gesh. d. Technik u. Industrie* (continued as *Technik Geschichte*)
BH	*Pai Hai* (1965)
BIB	*Bibliographica*
BIPH	*Bulletin of the International Association of Paper Historians*
BJRL	*Bull. John Rylands Library* (Manchester)
BLAC	*Bulletin of the Library Association of China (Chung-Hua Thu Shu Kuan Hsieh Hui Hui Pao)*
BLSOAS	*Bulletin of the London School of Oriental and African Studies*
BMFA	*Bulletin of the Museum of Fine Arts*
BMFEA	*Bulletin of the Museum of Far Eastern Antiquities* (Stockholm)

BMQ	*British Museum Quarterly*
BNISI	*Bull. National Inst. of Sciences of India*
BSEI	*Bulletin de la Société des Études Indochinoises*
BSVP	*Bull. Soc. le Vieux Papier*
BSZA	*Bull. Soc. Zool. Acclim*
BUA	*Bulletin de l'Université de Aurore* (Shanghai)
BURM	*Burlington Magazine*
BV	*Bharakiya Vidya* (Bombay)
CCB	*Chhung-Chi Bulletin (Chhung-Chi Hsiao Khan, Hongkong)*
CCSC	Che-Chiang Shu Chu
CCSM	*Chang Chü-Sheng Hsien Sheng Chhi Shih Sheng Jih Chi Nien Lun Wen Chi* (A Festschrift in Honour of the 70th Birthday of Mr. Chang Yüan-Chi, Shanghai, 1937)
CCTH	*Cheng Chi Ta Hsüeh Hsüeh Pao (National Cheng Chi University Journal, Thaipei)*
CCUL	*Chinese Culture* (Thaipei)
CCWH	*Che-Chiang Ta Hsüeh Wen Hsüeh Yüan Chi Khan* (Tsun-I, Kweichow)
CFC	*Cahiers Franco-Chinois* (Paris)
CFTS	*Chung Fa Han Hsüeh Yen Chiu So Thu Shu Kuan Kuan Khan (Scripta Sinica: Bulletin Bibliographique, Peking)*
CGJ	*Canadian Geographical Journal*
CHB	*Cambridge History of the Bible*
CHFH	*Chung-Hua Wen Hua Fu Hsing Yüeh Khan (Chinese Cultural Renaissance Monthly, Thaipei)*
CHHP	*Chhing-Hua Ta Hsüeh Hsüeh Pao* (Peking)
CHINE	*La Chine*
CHJ/T	*Chhing-Hua (Tsinghua) Journal of Chinese Studies* (new series, publ. Thaiwan)
CHSC	Chiang-Hsi Shu Chü
CHTH/S	*Chiang-Hsi Ta Hsüeh Hsüeh Pao, She Hui Kho Hsüeh* (Nanchhang)
CHWH	*Chung Han Wen Hua Lun Wen Chi* (Collected Essays on the Chinese and Korean Culture, Thaipei, 1955)
CHY	*Chhen Hsiu Yüan Hsien Sheng I Shu Chhi Shih Erh Chung* (1803)
CIBA/A	*Ciba Review* (Textile Technology)
CINA	*Cina*
CJ	*China Journal of Science and Arts*
CJWH	*Che-Chiang Ta Hsüeh Wen Hsüeh Yüan Chi Khan* (Tsun-I, Kweichow)
CKCP/CT	*Chung-Kuo Chin Tai Chhu Pan Shih Liao* (Materials for the Study of Modern Chinese Publishing, 1862–1918, Shanghai, 1954)

CKCP/HT Chung-Kuo Hsien Tai Chhu Pan Shih Liao (Materials for the Study of Contemporary Chinese Publications, 1919–1949, Peking, 1955)

CKIC Chung-Kuo I Chou (Thaipei)

CKKC Chung-Kuo Kho Hsüeh Chi Shu Fa Ming Ho Kho Hsüeh Chi Shu Jen Wu Lun Chi (Essays on Chinese Discoveries and Inventions of Science and Technology, and on the Men who Made Them. Peking, 1955)

CKLS Chung-Kuo Li Shih Po Wu Kuan Kuan Khan (Peking)

CLC Columbia Library Columns

CLIT Chinese Literature

COJ Contemporary Japan

CPNM Chhen Pai-Nien Hsien Sheng Chih Chiao Wu Shih Chou Nien Chi Pa Shih Ta Chhing Chi Nien Lun Wen Chih (A Festschrift in Honour of the 50th Anniversary of Teaching and 80th Birthday of Mr Chhen Pai-Nien, Thaipei, 1965)

CPTC Chih Pu Tsu Chai Tshung Shu (1921)

CPYK Chhu Pan Yüeh Khan (Thaipei)

CR China Review (Hongkong and Shanghai)

CREC China Reconstructs

CRR Chinese Recorder

CRRR Chinese Repository

CSHK Chhüan Shang Ku San Tai Chhin Han San Kuo Liu Chhao Wen (Complete Collection of Prose Literature from Remote Antiquity through the Chhin and Han Dynasties, the Three Kingdoms, and the Six Dynasties, 1930)

CSOH Chinese Studies of History

CTH Chia Tshao Hsien Tshung Shu (1918)

CTPS Chin Tai Pi Shu (1922)

CUMC Cooper Union Museum Chronicle

CWTM Chhing Chu Chiang Wei-Thang Hsien Sheng Chhi Shih Jung Chhing Lun Wen Chi (A Festschrift in Honour of the 70th Birthday of Dr Chiang Fu-Tshung, Thaipei, 1968)

CWYK Chin Wen Yüeh Khan (Chungking)

CYTSK Chung Yang Thu Shu Kuan Kuan Khan (Bulletin of the National Central Library, Thaipei)

CYWH Chung Yüeh Wen Hua Lun Chi (Collected Essays on Chinese-Vietnamese Culture, Thaipei, 1956)

DAW/MN Denkschriften d. k. Akademie d. Wissenschaften (Math.-Nat. Klasse)

DAW/PH Denkschriften d. k. Akademie d. Wissenschaften (Phil.-Hist. Klasse)

DPN Dolphin

EAST East (Tokyo)

EB Encyclopaedia Brittanica

EHOR Eastern Horizon (Hongkong)

EHSTC Explorations in the History of Science and Technology in China. Shanghai, 1982.

EQ Education Quarterly

ER Eclectic Review

ESCH Essays on the Sources for Chinese History (Canberra)

ESSS/TW Erh Shih Ssu Shih (Thung Wen Shu Chu, 1886)

ETH Ethnos

EYT Erh Yu Thang Tshung Shu (1821)

FCWH Fu-Chien Wen Hua (Foochow)

FCWS Fu-Chien Wen Hsien (Thaipei)

FER Far Eastern Review (London)

FF Forschungen und Fortschritte

FJHC Fu Jen Hsüeh Chih (Peiping)

FMNHP/AS Field Museum of Natural History (Chicago) Publications; Anthropological Series

GB Geibon (Tokyo)

GUF Gutenberg Festschrift

GUJ Gutenberg Jahrbuch

HCH Hsin Chung Hua (Shanghai, Chhungking)

HCTY Hsüeh Chin Thao Yüan (1916)

HGSH Han'guk Sakhoekwahak Nonjip (Collected Essays on Social Sciences of Korea, Seoul, 1968)

HHTP Hua Hsüeh Thung Pao (Peking)

HHYK Hsin Hua Yüeh Pao; Hsin Hua Pan Yüeh Khan (Peking)

HJAS Harvard Journal of Asiatic Studies

HK Hua Kuo (Hongkong)

HLW Hai-Ning Wang Ching-An Hsien Sheng I Shu (Collected Works of Wang Kuo-Wei, 1936)

HNMM Harper's New Monthly Magazine

HPIS Hu-Pei Hsien Cheng I Shu (1923)

HSCK Hsüeh Shu Chi Khan (Thaipei)

HSL Hsüan Lan Thang Tshung Shu (1947)

HSYK Hsüeh Shu Yüeh Khan (Shanghai)

HTD History Today

HTF Hsin Tung Fang Tsa Chih (Shanghai)

HTFH Hsien Tai Fo Hsüeh (Peking)

HTHY Hsien Tai Hsüeh Yüan (Thainan, Thaiwan)

HTSH Hsien Tai Shih Hsüeh (Canton)

HWTS Han Wei Tshung Shu (Shanghai, 1925)

HYHTS Hsi Ying Hsien Tshung Shu (1846)

IAQ Indian Antiquity

IAQR Imperial and Asiatic Quarterly Review; Asian Review

IEC/AE Industrial and Engineering Chemistry; Analytical Edition

IHCC I Hai Chu Chhen

IJHS Indian Journ. History of Science

IMP Imprimature

IQR Irish Quarterly Review

ISHP I Shu Hsüeh Pao (Bulletin of National Taiwan Academy of Art, Thaipei)

ISIS Isis

JA Journal Asiatique

JACU Journal of Asian Culture (University of California, Los Angeles)

JAOS	*Journal of the American Oriental Society*	*LC/QJCA*	*Library of Congress Quarterly Journal of Current Acquisitions; Quarterly Journal of the Library of Congress*
JAS	*Journal of Asian Studies* (Continuation of *Far Eastern Quarterly, FEQ*)		
JFI	*Journal Franklin Institute*	*LIB*	*The Library*
JHI	*Journal of the History of Ideas*	*LIBRI*	*Libri*
JICC/HK	*Journal of the Institute of Chinese Studies, Chinese University of Hong Kong*	*LNHP*	*Lingnan Hsüeh Pao* (Lingnan Journal, Canton)
JMH	*Journal of Modern History*	*LQ*	*Library Quarterly*
JMWH	*Jen Min Wen Hsüeh* (Peking)	*LSCH*	*Li Shih Chiao Hsüeh* (Tientsin)
JOB	*Journal of Occasional Bibliography*	*LSJ*	*Lingnan Science Journal*
JOSHK	*Journal of Oriental Studies* (Hongkong Univ.)	*LSYC*	*Li Shih Yen Chiu* (Peking)
JPOS	*Journal of the Peking Oriental Society*	*MAI/NEM*	*Mémoires de l'Académie des Inscriptions et Belles-Lettres,* Paris (*Notices et Extraits des MSS*)
JQ	*Japan Quarterly*		
JRAS	*Journal of the Royal Asiatic Society*		
JRAS/HK	*Journal of the Hongkong Branch of the Royal Asiatic Society*	*MGSC*	*Memoirs of the Geological Survey of China*
		MID	*Midway*
JRAS/KB	*Journal* (or *Transactions*) *of the Korea Branch of the Royal Asiatic Society*	*MMB*	*Metropolitan Museum Bulletin*
		MMS	*Metropolitan Museum Studies*
JRAS/NCB	*Journal* (or *Transactions*) *of the Royal Asiatic Society* (North China Branch)	*MN*	*Monumenta Nipponica*
		MNANS	*Museum Notes of the American Numismatic Society*
JSHS	*Japanese Studies in the History of Science* (Tokyo)		
		MPYK	*Ming Pao Yüeh Khan* (Hong Kong)
JUB	*Journ. Univ. Bombay*	*MR*	*Modern Review*
JWYK	*Jen Wen Yüeh Khan* (Shanghai)	*MRA*	*Mémoires Relatifs à l'Asie*
		MRDTB	*Memoirs of the Research Dept. of Toyo Bunko* (Tokyo)
KBK	*Kobunka* (Tokyo)		
KHCH	*Kuo Hsüeh Chuan Khan* (Shanghai)	*MS*	*Monumenta Serica*
KHCK	*Kuo Hsüeh Chi Khan* (Peking)	*MSAF*	*Mémoires de la Sociétié* (*Nat.*) *des Antiquaires de France*
KHCP	*Kuo Hsüeh Chi Pen Tshung Shu* (Collected Works of Basic Studies in Sinology, Shanghai, 1935)		
		MSHU	*Mei Shu* (Peking)
		MSPER	*Mitteilungen aus der Sammlung der Papyrus Erzherzog Rainer*
KHHK	*Ku Hsüeh Hui Khan* (Shanghai, 1912)		
KHTP	*Kho Hsüeh Thung Pao* (Peking)	*MSSJ*	*Mémoires de la Société Sinico-Japonaise*
KHTS	*Chung Yang Ta Hsüeh Kuo Hsüeh Thu Shu Kuan Nien Khan; Chiangsu Sheng Li Kuo Hsüeh Thu Shu Kuan Nien Khan* (Nanking)	*MSTS*	*Mei Shu Tshung Shu,* ser. 1–6 (Thaipei)
		MSYC	*Mei Shu Yen Chiu* (Shanghai)
		MTPL	*Min Tsu Phing Lun* (Thaipei)
KI	*Ku I Tshung Shu* (1884)	*NCR*	*New China Review*
KJ	*Korea Journal*	*NCSAS*	*Newsletter of the Midwest Chinese Student and Alumni Services* (Chicago)
KKCK	*Ku Kung Chi Khan* (Thaipei)		
KKHP	*Khao Ku Hsüeh Pao* (Peking)	*NM*	*New Mandarin*
KKT	*Kuan Ku Thang Hui Kho Shu* (1920)	*NPMB*	*National Palace Museum Bulletin*
KKTH	*Khao Ku Thung Hsün; Khao Ku*	*NQCJ*	*Notes and Queries on China and Japan*
KKTS	*Ku Kung Thu Shu Chi Khan* (*Quarterly Journal of Bibliography,* National Palace Library, Thaiwan)	*NRW*	*New Review*
		OEO	*Orient et Occident*
		ORE	*Oriens Extremus* (Hamburg)
KKWW	*Khao Ku Yü Wen Wu* (Sian)		
KKYK	*Ku Kung Po Wu Yüan Yüan Khan* (Peking)	*PA*	*Public Affairs*
KMF	*Korean Mission Field*	*PAM*	*Paper Maker*
KR	*Korean Repository*	*PAP*	*Paper*
KRW	*Korea Review*	*PAPR*	*Das Papier*
KSTK	*Ku Shu Tshung Khan* (1922)	*PBLN*	*Philobiblon* (Nanking)
KT	*Korea Today*	*PC*	*People's China*
KTHP	*Kuo Tshui Hsüeh Pao* (Shanghai)	*PCSH*	*Pei-ching Ta Hsüeh She Hui Kho Hsüeh Chi Khan* (Peiping)
KTHS	*Kuang Tshang Hsüeh Chiung Tshung Shu* (Shanghai, 1916)		
		PDM	*Periodico di Matematiche*
KTPH	*Chung-Kuo Ku Tai Pan Hua Tshung Khan* (Collection of Block-printed Books with Illustrations, Peking, 1961)	*PENA*	*Penrose Annual*
		PG	*Papier-Geschichte*
		PPA	*Przeglad Papierniczy*
		PPI	*Pulp and Paper International*
KTWH	*Kuangtung Wen Hsien Tshung Than* (Hong Kong)	*PPKK*	*Pei-Phing Thu Shu Kuan Kuan Khan* (Peiping)
KTWW	*Kuangtung Wen Wu* (Canton)		

PPMC   *Pulp and Paper Magazine of Canada*
PSHB   *Paeksan Hakbo* (Seoul)
PSL/M   *Philobiblon Society* (London) *Miscellanies*
PTJ   *Printing and Twine Journal*
PTRS   *Philosophical Transactions of the Royal Society*
PULC   *Princeton University Library Chronicle*

QBCB/C   *Quarterly Bulletin of Chinese Bibliography (Chinese Edition), Thu Shu Chi Khan*

RCC   *Revue des Cultures Colonies*
RD   *Renditions; a Chinese-English Translation Magazine* (Hong Kong)
RH   *Revue Historique*
RO   *Rocznik Orientalistyczny*
ROA   *Revue de l'Or. et de l'Alq.*

S   *Sinologica* (Basel)
SCENE   *Scene*
SF   *Shuo Fu* (1927)
SFSTK   *Sung Feng Shih Tshung Khan* (1908–22)
SG   *Shigaku* (Tokyo)
SGZ   *Shigaku Zashi* (Tokyo)
SH   *Shih Huo* (Shanghai, Peiping)
SHCH   *She Hui Kho Hsüeh Chan Hsien*
SHCS   *Shih Hsiang Chan Hsien* (Kunming)
SHHP   *Shan-Hsi Shih Fan Hsüeh Yüan Hsüeh Pao* (Linfeng, Shansi)
SHKH   *She Hui Kho Hsüeh* (Shanghai)
SHTH   *Shan-Hsi Ta Hsüeh Hsüeh Pao; Che Hsüeh She Hui Kho Hsüeh Pu Men (Shanxi University Journal; Philosophy and Social Sciences*, Thai-yüan)
SKCS   *Ssu Khu Chhüan Shu Chen Pen* (1934–)
SKCSTM   *Ssu Khu Chhüan Shu Tsung Mu Thi Yao*
SLHP   *Shih Liao Hsü Pien* (Thaipei, 1968)
SMCK   *Shu Mu Chi Khan* (Thaipei)
SMTP   *Shu Mu Tshung Pien*, ser. 1–5 (Thaipei, 1967–72)
SPPY   *Ssu Pu Pei Yao* (1936)
SPTK   *Ssu Pu Tshung Khan*, ser. 1–3 (1929–36)
SS   *Studia Serica* (Chhengtu)
SSGK   *Shoshigaku* (Tokyo)
SSKTS   *Shou Shan Ko Tshung Shu* (1922)
ST   *Shih Thung* (1937)
STTK   *Shih Ti Tshung Khan* (Shanghai)
SWAW/PH   *Sitzungsberichte d. k. Akad. d. Wissenschaften Wien* (Phil.-Hist. Klasse), Vienna
SWYK   *Shuo Wen Yüeh Khan* (Shanghai, Chungking)

TAPS   *Transactions of the American Philosophical Society*
TAS/J   *Transactions of the Asiatic Society of Japan*
TCKY   *Tsao Chih Kung Yeh* (Peking)
TCULT   *Technology and Culture*
TFTC   *Tung Fang Tsa Chih (Eastern Miscellany,* Shanghai)
TG/K   *Tōhō Gakuhō,* Kyoto
TG/T   *Tōhō Gakuhō,* Tokyo (*Tokyo Journal of Oriental Studies*)

TH   *Thien Hsia Monthly* (Shanghai)
THG   *Tōhōgaku (Eastern Studies),* Tokyo
TKTT   *Tang Kuei Tshao Thang Tshung Shu* (1863–6)
TLTC   *Ta Lu Tsa Chih* (Thaipei)
TMWS   *Tshung Mu Wang Shih I Shu* (Chhangsha, 1886)
TP   *T'oung Pao (Archives concernant l'Histoire, les Langues, la Géographie, l'Ethnographie et les Arts de l'Asie Orientale),* Leiden
TPYL   *Thai Phing Yü Lan* (SPTK)
TSCC   *Thu Shu Chi Chheng* (Thaipei reprint, 1964)
TSHCC   *Tshung Shu Chi Chheng* (1935–7)
TSK   *Thu Shu Kuan* (Peking)
TSKH   *Thu Shu Kuan Hsüeh Chi Khan (Library Science Quarterly,* Peiping)
TSKP   *Thu Shu Kuan Hsüeh Pao (Journal of Library Science,* Thaichung, Thaiwan)
TSKT   *Thu Shu Kuan Kung Tso* (Peking)
TSKTH   *Thu Shu Kuan Hsüeh Thung Hsün* (Peking)
TSPL   *Thu Shu Phing Lun* (Nanking)
TSPY   *Tse Shan Pan Yüeh Khan* (Chhengtu)
TXR   *Textil-Rundschau*
TYCK   *Tzu Yu Chung-Kuo* (Thaipei)
TYG   *Tōyō Gakuhō (Reports of the Oriental Society of Tokyo)*
TYPM   *Chhing Chu Tshai Yüan-Phei Hsien Sheng Liu Shih Wu Sui Lun Wen Chi* (A Festschrift in Honour of the 65th Birthday of Mr Tshai Yüan-Phei, Shanghai, 1933–5)

UNESC   *Unesco Courier*

VBQ   *Visvabharati Quarterly*
VOHD   *Verhandlung der Orientalischen Handschriften in Deutschland*

WE   *West and East* (Thaipei)
WH   *Wen Hsien* (Peking)
WIFH   *Wen I Fu Hsing Yüeh Khan* (Thaipei)
WL   *Wu Lin Chang Ku Tshung Pien* (1883)
WLHP   *Wen Lan Hsüeh Pao* (Hangchow)
WSC   *Wen Shih Che* (Tsingtao)
WSKK   *Washi Kenkyū* (Tokyo)
WWTK   *Wen Wu* (formerly *Wen Wu Tshan Khao Tzu Liao*)
WYTCC   *Wu Ying Tien Chü Chen Pan Chhüan Shu* (Canton, 1899)
WYWK   *Wan Yu Wen Khu* (1930)

YCHP   *Yenching Hsüeh Pao (Yenching University Journal of Chinese Studies)*
YCTS   *Yü Chang Tshung Shu* (1917)
YJSS   *Yenching Journal of Social Studies*
YYT   *Yüeh Ya Thang Tshung Shu* (1853)

ZB   *Zeitschrift f. Bücherfreunde*
ZDMG   *Zeitschrift der Deutschen Morgenländischen Gesellschaft*
ZINB   *Zinbun* (Kyoto)

A. CHINESE AND JAPANESE BOOKS BEFORE +1800

Each entry gives particulars in the following order:

(a) title, alphabetically arranged, with characters;
(b) alternative title, if any;
(c) translation of title;
(d) cross-reference to closely related book, if any;
(e) dynasty;
(f) date as accurate as possible;
(g) name of author or editor, with characters;
(h) title of other book, if the text of the work now exists only incorporated therein; or, in special cases, references to sinological studies of it;
(i) references to translations, if any, given by the name of the translator in Bibliography C;
(j) notice of any index or concordance to the book if such a work exists;
(k) reference to the number of the book in the *Tao Tsang* catalogue of Wieger (6), if applicable;
(l) reference to the number of the book in the *San Tsang* (Tripitaka) catalogues of Nanjio (1) and Takakusu & Watanabe, if applicable;
(m) reference to the special edition or *tshung shu* used in this volume.

Words which assist in the translation of titles are added in round brackets.

Alternative titles or explanatory additions to the titles are added in square brackets.

It will be remembered that in Chinese indexes words beginning *Chh-* are all listed together after *Ch-*, and *Hs-* after *H-*, but that this applies to initial words of titles only.

Where there are any differences between the entries in these bibliographies and those in Vols. 1–4, the information here given is to be taken as more correct.

An interim list of references to the editions used in the series, and to the *tshung-shu* collections in which books are available, has been given in Vol. 4, pt. 3, pp. 913 ff., and is available as a separate brochure.

ABBREVIATIONS

C/Han	Former Han.
E/Wei	Eastern Wei.
H/Han	Later Han.
H/Shu	Later Shu (Wu Tai).
H/Thang	Later Thang (Wu Tai).
H/Chin	Later Chin (Wu Tai).
S/Han	Southern Han (Wu Tai).
S/Phing	Southern Phing (Wu Tai).
J/Chin	Jurchen Chin.
L/Sung	Liu Sung.
N/Chou	Northern Chou.
N/Chhi	Northern Chhi.
N/Sung	Northern Sung (before the removal of the capital to Hangchow).
N/Wei	Northern Wei.
S/Chhi	Southern Chhi.
S/Sung	Southern Sung (after the removal of the capital to Hangchow).
W/Wei	Western Wei.

Ai Jih Chai Tshung Chhao 愛日齋叢鈔.
Miscellaneous Notes from the Ai-Jih Studio.
Sung.
Writer unknown.
SSKTS.

Chan Kuo Tshe 戰國策.
Intrigues of the Warring States.
Chou, c. −5th to −3rd century.
Ed. Liu Hsiang 劉向.
Tr. Crump (1).

Chen Hsi-Shan Wen Chi 眞西山文集.
Collected Works of Chen Te-Hsiu (+1178–1235).
Sung.
Chen Te-Hsiu 眞德秀.
1665 ed.

Cheng Lei Pen Tshao 證類本草.
Reorganised Pharmacopoeia.
Sung, +1108.
Thang Shen-Wei 唐愼微.

Chi Chu Fen Lei Tung-Pho Shih 集註分類東坡詩.
Classified Variorum Edition of Su Shih's Poetry.
Sung.
Su Shih 蘇軾.
Ed. Wang Shih-Pheng 王十朋.
SPTK.

Chi Ho Yüan Pen 幾何原本.
Elements of Geometry (Euclid's) [first six chapters].
Ming, +1607.

Tr. begun by Li Ma-Tou (Matteo Ricci) 利瑪竇 & Hsü Kuang-Chhi 徐光啓.
Completed by Alexander Wylie & Li Shan-Lan (1857).

Chi-Ku-Chai Chung Ting I Chhi Khuan Shih 積古齋鐘鼎彝器款識.
Inscriptions of Bronze Sacrificial Vessels from the Studio of Accumulated Antiques.
Chhing.
Juan Yüan 阮元.

Chi Ku Lu 集古錄.
Collection of Ancient Inscriptions.
Sung, c. +1050.
Ouyang Hsiu 歐陽修.
SF.

Chia-Thai Kuei-Chi Chih 嘉泰會稽志.
Chia-Thai Reign-Period Gazette of Kuei-Chi.
Sung, +1201.
Shih Su 施宿.
1926 ed.

Chiang-Hsi Sheng Ta Chih 江西省大志.
Local History of Chianghsi Province.
Ming, +1556.
Wang Tsung-Mu 王宗沐.

Chiao Chhuang Chiu Lu 蕉窗九錄.
Nine Discussions from the (Desk at the) Banana-Grove Window.
Ming, c. +1575.
Hsiang Yüan-Pien 項元汴.
TSHCC.

Chieh-Tzu-Yüan Hua Chuan 芥子園畫傳.

389

Chieh-Tzu-Yüan Hua Chuan (cont.)
The Mustard-Seed Garden Guide to Painting.
Chhing, +1679.
Li Li-Weng (preface) 李芝翁.
Wang Kai (text and illustrations) 王概.
Tr. Petrucci (1), Sze Mai-Mai (1).
Cf. A. K. Chiu (1).

Chien Chih Phu 牋紙譜.
(=*Shu Chien Phu* 蜀牋譜).
On Paper.
Yüan, +14th century.
Fei Chu 費著.
TSHCC.

Chien Nan Chi 劍南集.
Collected Works of Lu Yu.
Sung.
Lu Yu 陸游.
SPTK.

Chih Chien Phu 紙箋譜.
A Manual on Paper.
Yüan, c. +1300.
Hsien-yü Shu 鮮于樞.
TSHCC.

Chih Mo Pi Yen Chien 紙墨筆硯箋.
On Paper, Ink, Pen and Ink-slab.
Ming, c. +1600.
Thu Lung 屠隆.
MSTS.

Chih-Ya-Thang Tsa Chhao 志雅堂雜鈔.
Notes Taken at the Hall of Refined Temperament.
Sung, late +13th century.
Chou Mi 周密.
YYT.

Chin Hsiang Shuo 巾箱說.
Notes from the Kerchief Box.
Chhing, +18th century.
Chin Chih 金埴.
KHHK.

Chin Hsiu Wan Hua Ku 錦繡萬花谷.
A Splendid Kaleidoscopic Encyclopaedia.
Sung, c. +1188.
Author unknown.
1536; Thaipei reprint, 1969.

Chin Kang Ching 金剛經.
Vajracchedikā Sūtra [Kumarajiva's Condensation of the *Prajñāpāramitā Sūtra*]; Diamond-cutter Sutra.
Chin, +405.
Kumārajīva 鳩摩羅什婆.

Chin Ku Yü Ti Thu 今古輿地圖.
Maps of Geographical Areas in Contemporary and Ancient Times.
Ming, +1643.

Chin-Ling Hsin Chih 金陵新志.
New Gazette of Nanking.
Yüan, c. +1305.
Chang Hsüan 張鉉.
SKCS.

Chin Phing Mei Tzhu Hua 金瓶梅詞話.
Plum Blossoms in a Golden Jar [novel].
Ming, c. +16th century.

Hsiao-Hsiao-Sheng 笑笑生.

Chin Shih Lu 金石錄.
Records of Bronze and Stone Inscriptions.
Sung, +1132.
Chao Ming-Chheng 趙明誠.

Chin Shu 晉書.
History of the Chin Dynasty.
Thang, +644.
Fang Chhiao 房喬 *et al.*
ESSS/TW.

Chin-Su Chien Shuo 金粟箋說.
On the Paper from the Chin-su Monastery.
Chhing, c. +1800.
Chang Yen-Chhang 張燕昌.
TSHCC.

Chin Thai Chi Wen 金臺紀聞.
Accounts in the Hanlin Academy.
Ming, c. +1505-8.
Lu Shen 陸深.
TSHCC.

Ching Chhu Sui Shih Chi 荊楚歲時記.
Annual Folk Customs of the States of Ching and Chhu.
Prob. Liang, c. +550, but perhaps partly Sui, c. +610.
Tsung Lin 宗懍.
SPPY.

Ching Hsing Chi 經行記.
Record of Travels.
Thang, +8th century.
Tu Huan 杜環.
Ed. Wang Kuo-Wei 王國維.
HLW.

Ching Shih Cheng Lei Pen Tshao 經史證類本草.
See *Cheng Lei Pen Tshao.*

Ching Tien Shih Wen 經典釋文.
Textual Criticism of the Classics.
Sui, c. +600.
Lu Te-Ming 陸德明.
SPTK.

Chiu Chang Suan Shu 九章算術.
Nine Chapters on the Mathematical Art.
H/Han, +1st century (containing material from C/Han).
Writer unknown.

Chiu Ching San Chuan Yen Ko Li 九經三傳沿革例.
Specifications for Transmission of the Nine Classics and Three Commentaries.
Sung or Yüan, +1230 or +1300.
Attrib. Yüeh Kho 岳珂, or Yüeh Chün 岳浚.
CPTC.

Chiu Jih Lu 就日錄.
Daily Journal.
Sung.
Mr. Chao 趙氏.
SF.

Chiu Pien Thu Shuo 九邊圖說.
Maps and Description of the Northern Border Regions.
Ming, +1538.
Hsü Lun 許論.

Chiu Pien Thu Shuo (cont.)
 HSL.
Chiu Thang Shu 舊唐書.
 Old History of the Thang Dynasty [+618–906].
 Wu Tai, +945.
 Liu Hsü 劉昫
 ESSS/TW.
Chiu Wu Tai Shih 舊五代史.
 Old History of the Five Dynasties [+907–59].
 Sung, +974.
 Hsüeh Chü-Cheng 薛居正.
 ESSS/TW.
Cho Chung Chih (Lüeh) 酌中志 (略).
 An Enlightened Account of the Life in the
  Imperial Palace of the Ming.
 Ming, c. +1641.
 Liu Jo-Yü 劉若愚.
 TSHCC.
Cho Keng Lu 輟耕錄.
 Talks while the Plough is Resting.
 Yüan, +1366.
 Thao Tsung-I 陶宗儀.
 TSHCC.
Chou I Shuo Lüeh 周易說略.
 Commentary on the *Book of Changes*.
 Chhing, +1719.
 Hsü Chih-Ting 徐志定.
Chou Li 周禮.
 Record of the Rites of (the) Chou (Dynasty)
  [Descriptions of all government posts and
  their duties].
 C/Han, perhaps containing some material from
  the Chou.
 Compilers unknown.
 Tr. E. Biot (1).
 SPTK.
Chou Pei Suan Ching 周髀算經.
 The Arithmetical Classic of the Gnomon and
  the Circular Path (of Heaven).
 Han, ±1st century (probably containing ma-
  terial from Chou).
 Author unknown.
Chu Chhen Tsou I 諸臣奏議.
 Collected Memorials of Northern Sung Officials.
 Sung, +1186.
 Chao Ju-yü 趙汝愚.
Chu Fan Chih 諸蕃志.
 Records of Foreign Peoples (and their Trade).
 Sung, +1225 (or +1242, +1258).
 Chao Ju-Kua 趙汝适.
 Tr. Hirth & Rockhill (1).
 TSHCC.
Chü Lu 橘錄.
 Orange Record [Citrus horticulture].
 Sung, +1178.
 Han Yen-Chih 韓彥直.
 Tr. Hagerty (1).
 BCSH.
Chu Ping Yüan Hou Lun 諸病源候論.
 Discourses on the Origin of Diseases [Systematic
  Pathology].
 Sui, c. +607.

Chhao Yüan-Fang 巢元方.
Chu Shu Chi Nien 竹書紀年.
 The Bamboo Books [Annals].
 Chou, −296 or before.
 HWTS.
Chu-Tzu Chhüan Shu 朱子全書.
 Collected Works of Chu Hsi.
 Sung (ed. Ming), *editio princeps* +1713.
 Chu Hsi 朱熹.
 Ed. Li Kuang-Ti 李光地 (Chhing).
 Partial tr. Bruce (1); Le Gall (1).
 Chiang-Hsi Shu Chü ed.
Chu Wen-Kung Wen Chi 朱文公文集.
 Collected Works of Chu Hsi.
 Sung
 Chu Hsi 朱熹.
 SPTK.
Chuang Huang Chih 裝潢志.
 Methods of Mounting and Treatment of Paper
  Materials.
 Chhing
 Chou Chia-Chou 周嘉冑.
 TSHCC.
Chuang Tzu 莊子.
 [= *Nan Hua Chen Ching*].
 The Book of Master Chuang.
 Chou, c. −290.
 Chuang Chou 莊周.
 Tr. Legge (5); Feng Yu-Lan (5); Lin Yü-Thang
  (1); Wieger (7).
 SPTK.
Chuang Yüan Thu Khao 狀元圖考.
 Illustrated Work of the Highest Graduates [from
  Imperial Examinations in +1436 to +1521].
 Ming, c. +1607.
 Huang Ying-Jui 黃應瑞.
Chün-Chai Tu Shu Chih 郡齋讀書志.
 Memoir on the Authenticities of Ancient Books.
 Sung, c. +1151.
 Chhao Kung-Wu 晁公武.
 SPTK.
Chung Hua Ku Chin Chu 中華古今注.
 Commentary on Things Old and New in China.
 Wu Tai (H/Thang), +923–6.
 Ma Kao 馬縞.
 BCSH.
Chha Ching 茶經.
 Book on Tea.
 Thang.
 Lu Yü 陸羽.
 HCTY.
Chha Phu 茶譜.
 Treatise on Tea.
 Ming, c. +1640.
 Ku Yüan-Chhing 顧元慶.
 SF.
Chhan Yüeh Chi 禪月集.
 Collected Poems of Master Chhan-Yüeh.
 Wu Tai, +923.
 Kuan-Hsiu (monk) 貫休.
Chhang Wu Chih 長物志.
 Records on Cultural Objects.

Chhang Wu Chih (cont.)
Ming.
Wen Chen-Heng 文震亨.
MSTS.

Chhao-Hsien Fu 朝鮮賦.
Impressions of Korea.
Ming, c. +1490.
Tung Yüeh 董樾.
Ed. Wei Yüan-Khuang 魏元曠.
YCTS.

Chheng Shih Mo Yüan 程氏墨苑.
Chheng's Collection of Inkcake Designs.
Ming, +1606.
Chheng Ta-Yüeh 程大約.
1606 ed.

Chheng Wei Shih Lun 成唯識論.
Vijnapti-mātratā-siddhi; Completion of the
Doctrine of Mere Ideation [by Vasu-
bandhu 天親, +5th century, and ten
commentators].
India, late +5th century.
Tr. into Chinese and conflated, Hsüan-Tsang
玄奘, Thang, c. +650.
Tr. de la Vallee Poussin (3).

Chhi Min Yao Shu 齊民要術.
Important Arts for the People's Welfare.
N/Wei (and E/Wei or W/Wei), between +533
and +544.
Chia Ssu-Hsieh 賈思勰.
TSHCC.

Chhieh Yün 切韻.
Dictionary of the Sounds of Characters [rhym-
ing dictionary].
Sui, +601.
Lu Fa-Yen 陸法言.
See *Kuang Yün.*

Chhien Han Shu 前漢書.
History of the Former Han Dynasty [−206 to
−24].
H/Han, c. +100.
Pan Ku 班固.
ESSS/TW.

Chhien Shu 鉛書.
Local History of Chhien-Shan, Chiangsi.
Ming, +1608.
Ed. Tan Wei-Liang 笪維良 & Kho Chung-
Chiung 柯仲烱.
Ming Wan-li ed.

Chhien Shun Sui Shih Chi 乾淳歲時記.
Annual Customs of the Chhien-Tao and Shun-
Hsi period.
Sung, +13th century.
Chou Mi 周密.
SF.

Chhih Pei Ou Than 池北偶談.
Chance Conversations North of Chih(-chow).
Chhing, +1691.
Wang Shih-Chen 王士禎.
1701 ed.

Chhin Ting Jih Hsia Chiu Wen Khao 欽定日下
舊聞考.
See *Jih Hsia Chiu Wen Khao.*

Chhin Ting Ku Chin Thu Shu Chi Chheng 欽定古
今圖書集成.
See *Thu Shu Chi Chheng.*

Chhing I Lu 清異錄.
Records of the Unworldly and the Strange.
Wu Tai, c. +950.
Thao Ku 陶穀.
HYHTS.

Chhing Pi Tshang 清秘藏.
On Connoisseurship.
Ming.
Chang Ying-Wen 張應文.
MSTS.

Chhu Hsüeh Chi 初學記.
Entry into Learning [encyclopaedia].
Thang, +700.
Hsü Chien 徐堅.
Peking, 1962.

Chhu Tzhu 楚辭.
Elegies of Chhu (State).
Chou (with Han additions), c. −300.
Chhü Yüan 屈原 et al.
Partial tr. Waley (23).
SPTK.

*Chhung Hsiu Cheng-Ho Ching Shih Cheng Lei Pei Yung Pen
Tshao* 重修政和經史證類備用本草.
New Revision of the Pharmacopoeia of the
Cheng-Ho Reign-Period; the Classified and
Consolidated Armamentarium (A combi-
nation of the Cheng-Ho ... Cheng Lei ... Pen
Tshao with the Pen Tshao Yen I.).
Yüan, +1249; reprinted many times afterwards,
esp. in the Ming, +1468, with at least seven
Ming editions, the last in +1624 or +1625.
Thang Shen-Wei 唐慎微.
Khou Tsung-Shih 寇宗奭.
Pr. (or ed.) Chang Tshun-Hui 張存惠.

Chhü-Wei Chiu Wen 曲洧舊聞.
Old Stories Heard in Chhü-Wei (Hsin-Cheng,
Honan).
Sung, +12th century.
Chu Pien 朱弁.
TSHCC.

Chhüan Chin Wen 全晉文.
See Yen Kho-Chün (1) under Bibliography B.

Chhüan Thang Wen 全唐文.
Collected Prose Literature of the Thang
Dynasty.
Chhing.
Ed. Tung Kao 董誥 et al.
1818 ed.

Dai-Viêt Su-ký Toàn-thú 大越史記全書.
The Complete Book of the History of Great
Annam.
Vietnam, c. +1479.
Ngô Si-Lien 吳士連.

Engishiki 延喜式.
Collected Texts [on Shinto Ceremonies and
Japanese Customs] of the Engi Period
(+901−2).

Engishiki (*cont.*)
Japan, +927.
Fujiwara Tokihira 藤原時平 & Fujiwara
Tadahira 藤原忠平.
Erh Ya 爾雅.
Literary Expositor [dictionary].
C/Han, c. −200.
Ed. Kuo Phu 郭璞, c. +300.
Yin-Te Index no. (suppl.) 18.
SPTK.

Fa Shu Yao Lu 法書要錄.
Catalogue of Famous Calligraphy.
Thang, c. +847.
Chang Yen-Yüan 張彥遠.
TSHCC.
Fa Yen 法言.
Admonitory Sayings [in admiration, and imi-
tation, of the Lun Yü].
Hsin, +5.
Yang Hsiung 楊雄.
SPTK.
Fa Yüan Chu Lin 法苑珠林.
Forest of Pearls in the Garden of the [Buddhist]
Law.
Thang, +668.
Tao-Shih (monk) 道世.
SPTK.
Fan Han Ho Shih Chang Chung Chu 番漢合時掌
中珠.
Bilingual Glossary of Tangut-Chinese.
Hsi-Hsia, +1190.
Ku-Le 骨勒.
CTH.
Fang Shih Mo Phu 方氏墨譜.
Fang's Collection of Inkcake Designs.
Ming, +1588.
Fang Yü-Lu 方于魯.
1589 ed.
Fei Fu Yü Lüeh 飛鳧語略.
Brief Notes Taken from Quick Moments.
Ming, c. +1600.
Shen Te-Fu 沈德符.
TSHCC.
Feng Shih Wen Chien Chi 封氏聞見記.
Things Seen and Heard by Mr Feng.
Thang.
Feng Yen 封演.
Peking, 1958.
Feng Su Thung I 風俗通義.
Popular Traditions and Customs.
H/Han, +175.
Ying Shao 應劭.
Chung-Fa Index, no. 3.
SPTK.
Fu-Chien Thung Chih 福建通志.
General Topographical History of Fukien
Province.
Chhing, +1737.
Ed. Hao Yü-Lin 郝玉麟.
1737 ed.
Fu Hsüan Yeh Lu 負暄野錄.

Miscellaneous Notes by the Rustic while
Warming Himself under the Sun.
Sung, +13th century.
Chhen Yu 陳樨.
CPTC.

Genji Monogatari 源氏物語.
The Tale of (Prince) Genji.
Japan, +1021.
Murasaki Shikibu 紫式部.

Han Kuan I 漢官儀.
The Civil Service of the Han Dynasty and its
Regulations.
H/Han, +197.
Ying Shao 應劭.
Ed. Chang Tsung-Yüan 章宗源.
TSHCC.
Han Lin Chih 翰林志.
On the Han-Lin Academy.
Thang.
Li Chao 李肇.
BCSH.
Han Shih Wai Chuan 韓詩外傳.
Moral Discourses Illustrating the Han Text of
the *Book of Odes*.
C/Han, c. −135.
Han Ying 韓嬰.
SPTK.
Hou Han Shu 後漢書.
History of the Later Han Dynasty [+25−220].
L/Sung, +450.
Fan Yeh 范曄.
Yin-Te Index, no. 41.
ESSS/TW.
Hua Shih 畫史.
History of Painting.
Sung.
Mi Fu 米芾.
TSHCC.
Hua Yang Kuo Chih 華陽國志.
Record of the Country South of Mount Hua
[Historical Geography of Szechuan].
Chin, +347.
Chhang Chhü 常璩.
SPTK.
Huai Nan Tzu 淮南子.
[= *Huai Nan Hung Lieh Chieh* 淮南鴻
烈解].
The Book of (the Prince of) Huai Nan
[Compendium of Natural Philosophy].
C/Han, c. −120.
Attrib. Liu An (Prince of Huai Nan) 劉安.
Chung-Fa Index, no. 5.
SPTK.
Huang Chhao Li Chhi Thu Shih 皇朝禮器圖式.
Illustrated Description of Sacrificial Articles for
Imperial Rites of the Reigning Dynasty.
Chhing, +1759.
Yün-Lu 允祿 *et al.*
Huang Chhao Shih Shih Lei Yüan 皇朝事實類苑.
Classified Miscellanea of Sung Events.

Huang Chhao Shih Shih Lei Yüan (cont.)
Sung, +1145.
Ed. Chiang Shao-Yü 江少虞.
SFSTK.

Huang Chhing Chih Kung Thu 皇淸職貢圖.
Pictures of Foreigners Paying Tributes to the
Imperial Chhing Court.
Chhing, *c.* +1751.
Fu-Heng 傅恒 *et al.*

Hung Chien Lu 弘簡錄.
A History of the Thang and Sung Dynasties,
including Liao, Chin, and Hsi-Hsia.
Ming, +1557.
Shao Ching-Pang 邵經邦.
SLHP.

Hung Ming Chi 弘明集.
Collected Essays on Buddhism.
(Cf. *Kuang Hung Ming Chi*).
S/Chhi, *c.* +500.
Seng Yu 僧祐.
SPTK.

Hsi Chhi Tshung Yü 西溪叢語.
Western Pool Collected Remarks.
Sung, *c.* +1150.
Yao Khuan 姚寬.
TSHCC.

Hsi Ching Tsa Chi 西京雜記.
Miscellaneous Records of the Western Capital.
Liang or Chhen, mid +6th century.
Attrib. to Liu Hsin 劉歆 (C/Han) and to Ko
Hung 葛洪 (Chin) but probably by Wu
Chün 吳均.
BH.

Hsi-Hu Yu Lan Chih 西湖遊覽志.
Record of Sceneries of the West Lake [in
Hangchow during the Southern Sung Period].
Ming, +1547.
Thien Ju-Chheng 田汝成.
WL.

Hsi Yu Chi 西遊記.
A Journey to the West.
Ming, *c.* +1583.
Wu Chheng-En 吳承恩.
Tr. Waley (17), Yü (1).

Hsiang Lien Shih Tshao 香奩詩草.
Collected Poems from the Trousseau.
Ming.
Sang Chen-pai 桑貞白.
TSHCC.

Hsiang Phu 香譜.
A Treatise on Incense and Perfumes.
Yüan, +1322.
Hsiung Pheng-Lai 熊朋來.

*Hsiang-Thai Shu Shu Khan Cheng Chiu Ching San Chuan
Yen Ko Li* 相臺書塾刊正九經三傳沿
革例.
See *Chiu Ching San Chuan Yen Ko Li.*

Hsiao Chhu Chi 小畜集.
Complete Works of Wang Yü-Chheng.
N/Sung, +12th century.
Wang Yü-Chheng 王禹偁.
SPTK.

Hsiao Erh Fang Chüeh 小兒方訣.
A Collection of Pediatric Prescriptions.
Sung.
Chhien I 錢乙.
Ed. Hsiung Tsung-Li 熊宗立.
Ming.

Hsien-Shun Lin-An Chih 咸淳臨安志.
Hsien-Shun Reign-period Topographic Records
of the Hangchow District.
Sung, +1274.
Chhien Yüeh-Yu 潛說友.
1830 ed.

Hsin-An Hsien Chih 新安縣志.
Gazette of Hsin-An.
Sung, +12th century.
Lo Yüan 羅願.
1888 ed.

Hsin I Hsiang Fa Yao 新儀象法要.
New Design for an Astronomical Clock (lit.
Essentials of a New Device for Making an)
Armillary Sphere and a Celestial Globe
(Revolve) [including a Chain of Years for
Keeping Time and Striking the Hours, the
Motive Power Being a Water-wheel Checked
by an Escapement].
Sung, +1094.
Su Sung 蘇頌.

*Hsin Khan Chhüan Hsiang Chheng Chai Hsiao Ching Chih
Chieh* 新刊全相成齋孝經直解.
Newly Cut and Fully-Illustrated Edition of the
Book of Filial Piety with Commentaries by
Chheng-Chai.
Yüan, *c.* +1308.
Kuan Yün-Shih 貫雲石.
Reprint, 1938.

Hsin Khan Hsiu Chen Fang Ta Chhüan 新刊袖珍
方大全.
Newly Printed Medical Prescriptions.
Ming, +1505.
Chu Su 朱橚.

*Hsin Khan Pu Chu Thung Jen Shu Hsüeh Chen Chiu Thu
Ching* 新刊補註銅人腧穴針灸圖經.
See *Thung Jen Chen Chiu Thu Ching.*

Hsin Pien Fu Jen Liang Fang Pu I Ta Chhüan 新編
婦人良方補遺大全.
A Complete Collection of Good Prescriptions for
Gynaecology, Supplement.
Sung.
Chhen Tzu-Ming 陳自明.
Ed. Hsüeh Chi 薛己 (Ming).

Hsin Thang Shu 新唐書.
New History of the Thang Dynasty
[+618–906].
Sung, +1061.
Ouyang Hsiu & Sung Chhi 歐陽修，宋祁.
ESSS/TW.

Hsin Wu Tai Shih 新五代史.
[=Wu Tai Shih Chi].
New History of the Five Dynasties [+907–59].
Sung, *c.* +1070.
Ouyang Hsiu 歐陽修.
ESSS/TW.

Hsing Li Ta Chhüan (Shu) 性理大全 (書).
Collected Works of (120) Philosophers of the
Hsing-Li (Neo-Confucian) School.
Ming, +1415.
Ed. Hu Kuang 胡廣 *et al.*
Ming ed.

Hsiu The Yü Pien 修慝餘編.
Record of Evil Doings.
Chhing.
Chhen Chin 陳蓋.
TSHCC.

Hsü Tzu Chih Thung Chien Chhang Pien 續資治
通鑑長編.
Collected Data for the Continuation of the
Comprehensive Mirror for Aid in Govern-
ment, covering the Period +960-1127.
Sung, +1183.
Ed. Li Thao 李燾.
Thaipei, 1961.

Hsüan-Ho Feng Shih Kao-Li Thu Ching 宣和奉使高
麗圖經.
Illustrated Record of an Embassy to Korea in
the Hsün-Ho Reign-period.
Sung, +1124 (+1167).
Hsü Ching 徐兢.
TSHCC.

Hsüan-Ho Po Ku Thu Lu 宣和博古圖錄.
Hsüan-Ho Reign-Period Illustrated Record of
Ancient Objects.
Sung, +1111-25.
Wang Fu 王黼 *et al.*
1752 ed.

Hsüan Tu Pao Tsang 玄都寶藏.
Comprehensive Treasures of Taoist Literature
from the Hsüan-Tu Monastery.
J/Chin, c. +1188-91.
Ed. Sun Ming-Tao 孫明道.

Hsüeh Chai Chan Pi 學齋佔畢.
Glancing into Books in a Learned Studio.
Sung, +13th century.
Shih Sheng-Tsu 史繩祖.
BCSH.

Hsüeh Thang Mo Phin 雪堂墨品.
Inks from the Snow Pavilion.
Chhing, +1670.
Chang Jen-Hsi 張仁熙.
TSHCC.

Hsün Chhu Lu 詢芻錄.
Inquiries and Suggestions (concerning Popular
Customs and Usages).
Ming.
Chhen I 陳沂.
TSHCC.

I Chien Chih 夷堅志.
Strange Stories from I-Chien.
Sung, c. +1185.
Hung Mai 洪邁.
TSHCC.

I Ching 易經.
The Classic of Changes [Book of Changes].
Chou with C/Han additions.
Compiler unknown.
Tr. R. Wilhelm (2); Legge (9); de Harlez (1).
Yin-Te Index no. (Suppl.) 10.

I Chüeh Liao Tsa Chi 猗覺寮雜記.
Miscellaneous Records from the I-Chüeh
Cottage.
Sung, c. +1200.
Chu I 朱翌.

I Shu Pien 蛾術編.
The Antheap of Knowledge [Miscellaneous
Essays].
Chhing, c. +1770.
Wang Ming-Sheng 王鳴盛.
1841 ed.

I Wen Lei Chü 藝文類聚.
Literary Records Collected and Classified
[encyclopaedia].
Thang, c. +620.
Ouyang Hsün 歐陽詢.

Ise Monogatari 伊勢物語.
Tale of Ise.
Japan, +10th century.
Author unknown.
Tr. McCullough (1)

Jih Chih Lu 日知錄.
Daily Additions to Knowledge.
Chhing, +1673.
Ku Yen-Wu 顧炎武.
TSHCC.

Jih Hsia Chiu Wen Khao 日下舊聞考.
Archaeological and Historical Descriptions of
the Imperial Precincts in Peking and the
Immediate Dependencies.
Chhing.
Chu I-Tsun 朱彝尊; rev. under imperial
auspices by Yü Min-Chung 于敏中.
1774 ed.

Kai Yü Tshung Khao 陔餘叢考.
Collection of Miscellaneous Studies.
Chhing.
Chao I 趙翼.
1750 ed.

Kamisuki Chōhōki 紙漉重寶記.
A Handy Guide to Papermaking.
Japan, +1798.
Kunisaki Jihei 國東治兵衞.
Tr. C. E. Hamilton (1).

Kao Feng Wen Chi 高峯文集.
Collected Works of Liao Kang.
Sung, c. +12th century.
Liao Kang 廖剛.
SKCS.

Kao Sung Hua Phu 高松畫譜.
Painting Manual of Plants and Birds by Kao
Sung [in several parts].
Ming, c. +1550-4.
Kao Sung 高松.

Keng Chih Thu 耕織圖.
Pictures of Tilling and Weaving.
Sung, presented in MS., +1145, and perhaps
first printed from wood blocks at that time;

Keng Chih Thu (cont.)
　engraved on stone, +1210, and probably
　　then printed from wood blocks in +1237.
　Lou Shou　樓璹; redrawn by Chiao Ping-
　　Chen　焦秉貞 under imperial auspices and
　　printed in +1712.
Ko Chih Ching Yüan　格致鏡原.
　Mirror of Scientific and Technological Origins.
　Chhing, +1735.
　Chhen Yüan-Lung　陳元龍.
　1735 ed.
Ko Ku Yao Lun　格古要論.
　Essential Discussions of Appreciating Antique
　　Objects.
　Ming, +1388.
　Tshao Chao　曹昭.
　TSHCC.
Kogo Shūi　古語拾遺.
　Collected Missing Passages from Classical
　　Languages.
　Japan, +807.
　Inbe Hironari　齋部広成.
Kojiki　古事記.
　Record of Ancient Matters [of Japan].
　Japan, +712.
　Tr. D. L. Philippi (1).
Ku Chin Chu　古今註.
　Commentary on Things Old and New.
　Chin, mid +4th century.
　Tshui Pao　崔豹.
　HWTS.
Ku Chin Shih Wu Khao　古今事物考.
　Origins of Things Ancient and Modern.
　Ming.
　Wang San-Phin　王三聘.
　TSHCC.
Ku Chin Shu Kho　古今書刻.
　Printed Works of Old and Modern Times.
　Ming, c. +1559.
　Chou Hung-Tsu　周宏祖.
　KKT.
Ku Chin Thu Shu Chi Chheng　古今圖書集成.
　See *Thu Shu Chi Chheng.*
Ku Lieh Nü Chuan　古列女傳.
　See *Lieh Nü Chuan.*
Ku Shih Hua Phu　顧氏畫譜.
　Collection of Paintings [by Famous Artists of
　　Successive Dynasties] Compiled by Mr Ku.
　Ming, c. +1603.
　Ku Ping　顧炳.
Ku Yü Thu　古玉圖.
　Illustrated Description of Ancient Jade Objects.
　Yüan, +1341.
　Chu Te-Jun　朱德潤.
　1752 ed.
Kuan Tzu　管子.
　The Writings of Master Kuan.
　Chou and C/Han.
　Attrib. Kuan Chung　管仲.
Kuang-Chhuan Shu Pa　廣川書跋.
　The Kuang-chhuan Bibliographical Notes.
　Sung, c. +1125.

Tung Yu　董逌.
　TSHCC.
Kuang Hung Ming Chi　廣弘明集.
　Further Collections of Essays on Buddhism.
　Cf. *Hung Ming Chi.*
　Thang, c. +660.
　Tao-Hsüan　道宣.
　SPTK.
Kuang-Hsi Thung Chih　廣西通志.
　General Topographical History of Kuangsi
　　Province.
　Chhing, +1800.
　Ed. Hsieh Chhi-Khun　謝啓昆.
　1891 ed.
Kuang Yün　廣韵.
　Enlargement of the *Chhieh Yün* Dictionary of
　　Sounds of Characters.
　Sung, +1011.
　Chhen Pheng-Nien　陳彭年.
　SPTK.
Kuei-Hsin Tsa Shih　癸辛雜識.
　Miscellaneous Information from Kuei-Hsin
　　Street (in Hangchow).
　Sung, late +13th century, perhaps not finished
　　before +1308.
　Chou Mi　周密.
　HCTY.
Kuei Thien Lu　歸田錄.
　On Returning Home.
　Sung, +1067.
　Ouyang Hsiu　歐陽修.
　SF.
Kung Men Pu Fei Chhien Kung Te Lu　公門不費
　　錢功德錄.
　Public Records of Merits without Expenditures.
　Chhing.
　Author unknown.
　TSHCC.
Kuo Shih Pu　國史補.
　See *Thang Kuo Shih Pu.*
Khai Pao Pen Tshao　開寶本草.
　Khai-Pao Reign-Period Pharmacopoeis.
　Sung, c. +970.
　Liu Han　劉翰 & Ma Chih　馬志.
Khao Ku Thu　考古圖.
　Illustrations of Ancient Objects.
　Sung, +1092.
　Lü Ta-Lin　呂大臨.
　1752 ed.
Khao Phan Yü Shih　考槃餘事.
　Further Works by the Recluse.
　Ming, c. +1600.
　Thu Lung　屠隆.
　TSHCC.
Khun Yü Wan Kuo Chhüan Thu　坤輿萬國全圖.
　(= *Yü Ti Shan Hai Chhüan Thu*).
　Map of All Countries in the World.
　Ming, +1584.
　Li Ma-Tou (Matteo Ricci)　利瑪竇.

Lan Thing Khao　蘭亭考.
　Investigation of the Meeting at Orchid Pavilion

Lan Thing Khao (*cont.*)
    [and a preface to it written by Wang Hsi-
    Chih].
    Sung, *c.* +1224.
    Sang Shih-Chhang 桑世昌.
    CPTC.
Lao Hsüeh An Pi Chi 老學庵筆記.
    Notes from the Hall of Learned Old Age.
    Sung, *c.* +1190.
    Lu Yu 陸游.
    HCTY.
Lao Tzu Tao Te Ching 老子道德經.
    See *Tao Te Ching.*
Li Sao Thu 離騷圖.
    Pictures on the Elegy on Encountering Sorrows.
    Chhing, +1645.
    Hsiao Yün-Tshung 蕭雲從.
Li Tai Ming Hua Chi 歷代名畫記.
    Catalogue of Famous Paintings.
    Thang, +847.
    Chang Yen-Yüan 張彥遠.
    TSHCC.
Li Yü Chhüan Chi 李漁全集.
    Collected Works of Li Yü.
    Chhing, +17th century.
    Li Yü 李漁.
    Reprint, Thaipei, 1970.
Liang Chhi Man Chih 梁溪漫志.
    Bridge Pool Essays.
    Sung, +1192.
    Fei Kun 費袞.
    CPTC.
Lieh Nü Chuan 列女傳.
    Lives of Celebrated Women.
    Date uncertain, nucleus probably Han.
    Attrib. Liu Hsiang 劉向.
    KSTK.
Ling-Yen-Ko Kung Chhen Thu Hsiang 凌煙閣功臣
    圖像.
    Portraits of Meritorious Persons in the Hall of
    Ling-Yen.
    Chhing, *c.* +1668.
    Liu Yüan 劉源.
Liu Ching Thu 六經圖.
    Illustrations of Objects Mentioned in the Six
    Classics.
    Sung, *c.* +1155.
    Yang Chia 楊甲.
    1740 ed.
Liu-Chhiu Ju Hsüeh Chien Wen Lu 琉球入學見聞
    錄.
    Records of Liu-Chhiu as Learned from its
    Students in China.
    Chhing, +1764.
    Phan Hsiang 潘相.
    Reprint, Thaipei, 1973.
Liu-Chhiu Kuo Chih Lüeh 琉球國志略.
    Account of the Liu-Chhiu Islands.
    Chhing, +1757.
    Chou Huang 周煌.
    TSHCC.
Liu Thieh 六帖.

The Six Cards [Encyclopaedia].
    Thang, *c.* +800.
    Pai Chü-I 白居易.
    Enlarged in Sung by Khung Chuan 孔傳.
Lo Hsüan Pien Ku Chien Phu 蘿軒變古箋譜.
    Collection of Letter-Papers with Antique and
    New Designs from the Wisteria Pavilion.
    Ming, +1626.
    Wu Fa-Hsiang 吳發祥.
Lu Shih-Lung Wen Chi 陸士龍文集.
    Collected Works of Lu Yün.
    Chin, +3rd century.
    Lu Yün 陸雲.
    SPTK.
Luan Chheng Chi 欒城集.
    Collected Works of Su Chhe.
    Sung.
    Su Chhe 蘇轍.
    KHCP.
Lun Heng 論衡.
    Discourses Weighed in the Balance.
    H/Han, +82 or +83.
    Wang Chhung 王充.
    Chung-Fa Index no. 1.
    SPTK.
Lung Khan Shou Chien 龍龕手鑑.
    Handbook of Khitan-Chinese Glossary.
    Liao, *c.* +997.
    Ed. Hsing-Chün (Khitan Monk) 行均.
    SPTK.

Man-Thang Mo Phin 漫堂墨品.
    Inks from the Boundless Pavilion.
    Chhing, +1684.
    Sung Lo 宋犖.
    TSHCC.
Mao Shih 毛詩.
    Mao's Version of the *Book of Odes.*
    Chou.
    Ed. Mao Heng 毛亨; Cheng Hsüan 鄭玄.
    SPTK.
Mao Shih Tshao Mu Niao Shou Chhung Yü Su 毛詩
    草木鳥獸蟲魚疏.
    On the Various Plants, Birds, Animals, Insects
    and Fishes Mentioned in the *Book of Odes.*
    San Kuo, +3rd century.
    Lu Chi 陸璣.
    TSHCC
Mei Hua Hsi Shen Phu 梅花喜神譜.
    An Album of the Life-like Flowering Plum.
    Sung, +1238.
    Sung Po-Jen 宋伯仁.
    CPTC.
Meng Chhi Pi Than 夢溪筆談.
    Dream Pool Essays.
    Sung, +1086, last supplement dated +1091.
    Shen Kua 沈括.
    TSHCC.
Meng Liang Lu 夢粱錄.
    Dreaming of the Capital while the Rice is
    Cooking.
    Sung, +1275.

Meng Liang Lu (*cont.*)
  Wu Tzu-Mu 吳自牧.
  TSHCC.
Meng Tzu 孟子.
  The Book of Master Meng (Mencius).
  Chou, *c.* −290.
  Meng Kho 孟軻.
  Tr. Legge (3); Lyall (1).
  Yin-Te Index no. (Suppl.) 17.
  SPTK.
Miao Fa Lien Hua Ching 妙法運華經.
  Saddharma-pundarika Sūtra; The Lotus of
    Wonderful Law.
  India, *c.* +200; tr. into Chinese +5th century.
  Ch. tr. Kumarajiva 鳩摩羅什婆.
  Eng. tr. Soothill (3).
Mien Hua Thu 棉花圖.
  Pictures of Cotton Growing and Weaving.
  Chhing, +1765.
  Fang Kuan-Chheng 方觀承.
Min Hsiao Chi 閩小記.
  Notes about Fukien.
  Chhing.
  Chou Liang-Kung 周亮工.
  TSHCC.
Min Tsa Chi 閩雜記.
  See Shih Hung-Pao (1) under Bibliography B.
Ming Chi Pei Lüeh 明季北略.
  Miscellaneous Notes on the Affairs of the
    Northern Capital during the Late Ming
    Dynasty.
  Chi Liu-Chhi 計六奇.
  KHCP.
Ming I Pieh Lu 名醫別錄.
  Informal (or Additional) Records of Famous
    Physicians (on Materia Medica).
  Ascr. Liang, *c.* +510.
  Attrib. Thao Hung-Ching 陶弘景.
  CHY.
Ming Shih 明史.
  History of the Ming Dynasty [+1368 to
    +1643].
  Chhing, begun +1646, completed +1736, first
    pr. +1739.
  Chang Thing-Yü 張廷玉 *et al.*
  ESSS/TW.
Mo Chih 墨志.
  Record of Ink.
  Ming, *c.* +1637.
  Ma San-Heng 麻三衡.
  TSHCC.
Mo Ching 墨經.
  Classic of Ink.
  Sung, *c.* +1100.
  Chao Kuan-Chih 晁貫之.
  MSTS.
Mo Chuang Man Lu 墨莊漫錄.
  Recollections from the Estate of Literary
    Learning.
  Sung, *c.* +1131.
  Chang Pang-Chi 張邦基.
  BH.

Mo Fa Chi Yao 墨法集要.
  Essentials of Inkmaking Methods.
  Ming, +1398.
  Shen Chi-Sun 沈繼孫.
  TSHCC.
Mo Phu Fa Shih 墨譜法式.
  Handbook of Ink Recipes and Inkmakers.
  Sung, +1095.
  Li Hsiao-Mei 李孝美.
  MSTS.
Mo Shih 墨史.
  History of Ink.
  Yüan, *c.* +1330.
  Lu Yu 陸友.
  TSHCC.
Mo Tzu 墨子.
  The Book of Master Mo.
  Chou, −4th century.
  Mo Ti (and disciples) 墨翟.
  Yin-Te Index no. (Suppl.) 21.
  Tr. Mei Yi-Pao (1); Forke (3).
  SPTK.
Mu-An Chi 牧庵集.
  Literary Collections of (Yao) Mu-An.
  Yüan, *c.* +1310.
  Yao Sui 姚燧.
  SPTK.
Mu Thien Tzu Chuan 穆天子傳.
  Account of the Travels of Emperor Mu.
  Chou, before −245.
  Writer unknown.
  SPTK.

Nan Fang Tshao Mu Chuang 南方草木狀.
  An Account of the Plants of the Southern
    Regions.
  Chin, +3rd century or later.
  Attrib. to Hsi Han 嵇含.
  HWTS.
Nan Hai Chi Kuei Nei Fa Chuan 南海寄歸內法傳.
  Record of Buddhist Practices Sent Home from
    the South Seas.
  Thang, *c.* +689.
  I-Ching (monk) 義淨.
Nan Hsün Sheng Tien 南巡盛典.
  Imperial Inspection Tours to Southern
    Provinces from +1751 to +1765.
  Chhing, +1776.
  Kao Chin 高晉 *et al.*
Nan Hua Chen Ching 南華眞經.
  See *Chuang Tzu.*
Ni Wen-Cheng Kung Nien Phu 倪文正公年譜.
  Annalistic Biography of Ni Yüan-Lu (+1593–
    1644).
  Chhing
  Ni Hui-Ting [Yüan-Lu] 倪會鼎 [元璐].
  TSHCC.
Nihongi 日本紀.
  [= *Nihon-shoki.*]
  Chronicle of Japan [from the Earliest Times to
    +696].
  Japan (Nara), +720.

Nihongi (cont.)
Tr. Aston (1).
Nihon-koku Genzai-sho Mokuroku 日本國見在書
目錄.
Bibliography of Extant Books in Japan.
Japan (Heian), *c.* +895.
Fujiwara Sukeyo 藤原佐世.
Nittō-Guhō Junrei Gyōki 入唐求法巡禮行記.
Record of a Pilgrimage to China in Search of
the (Buddhist) Law.
Thang, +838-47.
Ennin 圓仁.
Tr. Reischauer (2).
Nung Cheng Chhüan Shu 農政全書.
Complete Treatise on Agriculture.
Ming, composed +1625-8; printed +1639.
Hsü Kuang-Chhi 徐光啓.
Ed. Chhen Tzu-Lung 陳子龍.
1837 ed.
Nung Shu 農書.
Treatise on Agriculture.
Yüan, +1313.
Wang Chen 王禎.
Peking, 1956.

Pai Shih Liu Thieh 白氏六帖.
See *Liu Thieh*.
Pai Shih Wen Chi 白氏文集.
Literary Collections of Pai Chü-I.
Thang, +824.
Pai Chü-I 白居易.
SPTK.
Pai Yung Thu Phu 百詠圖譜.
Illustrations of 100 Poems.
Ming, +1597.
Pao Phu Tzu 抱朴子.
Book of the Preservation-of- Solidarity Master.
Chin, early +4th century, prob. *c.* +320.
Ko Hung 葛洪.
SPTK.
Pei Hu Lu 北戶錄.
Northern Family Records.
Thang, +875.
Tuan Kung-Lu 段公路.
HPIS.
Pei Thang Shu Chhao 北堂書鈔.
Book Records of the Northern Hall
[encyclopaedia].
Thang, *c.* +630.
Yü Shih-Nan 虞世南.
1888 ed.
Pen Tshao Kang Mu 本草綱目.
The Great Pharmacopoeia.
Ming, +1596.
Li Shih-Chen 李時珍.
Peking, 1975.
Pi Shu Chien Chih 祕書監志.
Records of the Bureau of Publications.
Yüan, *c.* +1350.
Wang Shih-Tien 王士點.
KTHS.
Pi Shu Shan Chuang Shih Thu 避暑山莊詩圖.

Poems and Illustrations on the Summer Resort
[in Jehol].
Chhing, +1712.
Paintings by Shen Yü 沈嵰 with poems by
Emperor Khang-Hsi.
Po Jo Po-Lo-Mi-To Ching 般若波羅蜜多經.
Prajnaparamita Sutra; The Perfection of
Wisdom.
India, *c.* +3rd century; tr. into Chinese +5th
century.
Writer unknown.
Trs. Lamotte (1); Conze (4).
Po Wu Chih 博物志.
Records of the Investigation of Things.
Chin, *c.* +290.
Chang Hua 張華.
HWTS.
Phan-Chou Wen Chi 盤洲文集.
Collected Works of Hung Kua.
Sung, +12th century.
Hung Kua 洪适.
SPTK.
Phing Chih Thieh 評紙帖.
(= *Shih Chih Shuo*).
Commentaries on Paper.
Sung, *c.* +1100.
Mi Fei 米芾.
MSTS.
Phing Ting I-Li Hui Pu Chan Thu 平定伊犂回
部戰圖.
Pictures on Battles and Memorable Events in
the Conquests of Ili and Chinese Turkestan.
Chhing, +1776.

San Fu Chüeh Lu 三輔決錄.
Miscellaneous Notes on the Han Capital.
L/Han, *c.* +200.
Chao Chhi 趙岐.
EYT.
San Fu Ku Shih 三輔故事.
Anecdotes concerning the Han Capital.
Chhing, *c.* +1820.
Ed. Chang Chu 張澍.
EYT.
San Kuo Chih Yen I 三國志演義.
The Three Kingdoms Story [novel].
Yüan.
Lo Kuan-Chung 羅貫中.
Tr. Brewitt-Taylor (1).
San Tshai Thu Hui 三才圖會.
Universal Encyclopaedia.
Ming, +1609.
Wang Chhi 王圻.
1609 ed.
San Tzu Ching 三字經.
Trimetrical Primer.
Sung, *c.* +1270.
Wang Ying-Lin 王應麟.
Shan Lu 剡錄.
Local History of Shan-Chhi [Chheng-Hsien,
Chekiang].
Sung, *c.* +1184.

Shan Lu (cont.)
    Ed. Kao Ssu-Sun　高似孫.
    1870 ed.
Shan Tso Chin Shih Chih　山左金石志.
    Bronze and Stone Inscriptions of Shantung
        Province.
    Chhing.
    Pi Yüan　畢沅 & Juan Yüan　阮元.
    1797 ed.
Shan-Thang Chhün Shu Khao So　山堂羣書考索.
    Critical Compilation from All Books by Shan-
        Thang.
    Sung, +1210.
    Chang Ju-Yü　章如愚.
    1508 ed.
Shang Han Lun　傷寒論.
    Treatise on fevers caused by cold.
    H/Han.
    Chang Chi　張機.
    Ed. Wang Shu-Ho　王叔和 (Chin).
    SPTK.
Shao-Shih-Shan-Fang Pi Tshung　少室山房筆叢.
    Notes from Shao-Shih-Shan-Fang.
    Ming, *c.* +1598.
    Hu Ying-Lin　胡應麟.
    Peking, 1964.
Sheng Chiao Shih Lu　聖教實錄.
    Catechism of Christianity.
    (= *Thien Chu Shih Lu*).
    Ming, +1582.
    Lo Min-Chien (Michele Ruggieri)　羅民堅.
Shih Chi　史記.
    Historical Records.
    C/Han, *c.* −90.
    Ssuma Chhien　司馬遷 and his father Ssuma
        Than　司馬談.
    ESSS/TW.
Shih Chih Shuo　十紙說.
    See *Phing Chih Thieh*.
Shih-Chu-Chai Chien Phu　十竹齋箋譜.
    Ornamental Letter-paper Designs from the Ten
        Bamboo Studio.
    Ming, +1645.
    Hu Cheng-Yen　胡正言.
    Cf. Tschichold (4).
Shih-Chu-Chai Shu Hua Phu　十竹齋書畫譜.
    Manual of Calligraphy and Painting from the
        Ten Bamboo Studio.
    Ming, *c.* 1619–33.
    Hu Cheng-Yen　胡正言.
    Cf. Tschichold (3).
Shih I Chi　拾遺記.
    Memoirs on Neglected Matters.
    Chin, *c.* +370.
    Wang Chia　王嘉.
    HWTS.
Shih-Men Wen Tzu Chhan　石門文字禪.
    Literary Works at Shih-Men Monastery.
    Sung.
    Te-Hung (monk)　德洪.
    TSHCC.
Shih Ming　釋名

Explanation of Names [dictionary].
    H/Han, *c.* +100.
    Liu Hsi　劉熙.
    SPTK.
Shih Shuo Hsin Yü　世說新語.
    New Discourse on the Talk of the Times [notes
        of minor incidents from Han to Chin].
    L/Sung, *c.* +5th century.
    Liu I-Chhing　劉義慶.
    Commentary by Liu Hsün　劉峻 (Liang).
    Tr. Mather (1).
    SPTK.
Shih Thung　史通.
    Generalities on History.
    Thang, +710.
    Liu Chih-Chi　劉知幾.
    SPTK.
Shih Wu Chi Yüan　事物紀原.
    Record of the Origins of Affairs and Things.
    Sung, *c.* +1085.
    Kao Chheng　高承.
    TSHCC.
Shih Yü Hua Phu　詩餘畫譜.
    Manual of Painting with Themes from Lyrical
        Poetry.
    Ming, *c.* +1612.
    Ed. Mr Wang　汪氏.
Shou Shih Thung Khao　授時通考.
    Complete Investigations of the Works and Days
        [Imperially Commissioned; a treatise on
        agriculture, horticulture and all related
        technologies].
    Chhing, +1742.
    Ed. O-Erh-Thai (Ortai)　鄂爾泰, with Chang
        Thing-Yü　張廷玉, Chiang Fu　蔣溥, *et al.*
Shu Chien Phu　蜀箋譜.
    See *Chien Chih Phu*.
Shu Chung Kuang Chi　蜀中廣記.
    Treatise on Szuchuan.
    Ming, *c.* +1600.
    Tshao Hsüeh-Chhüan　曹學佺.
    SKCS.
Shu Shih　書史.
    History of Calligraphy.
    Sung, *c.* +1100.
    Mi Fei　米芾.
    TSHCC.
Shu Yüan Tsa Chi　菽園雜記.
    The Bean Garden Miscellany.
    Ming, +1475.
    Lu Jung　陸容.
    TSHCC.
Shui Hu Chuan　水滸傳.
    The Story of the Lake [novel = All Men are
        Brothers].
    Ming, *c.* +1380.
    Ascr. Shih Nai-An　施耐庵.
    Tr. Buck (1).
Shun Thien Fu Chih　順天府志.
    Local History of Shun-Thien Prefecture, Chili
        Province.
    Chhing.

Shun Thien Fu Chih (*cont.*)
    Ed. Miao Chhüan-Sun 繆荃孫.
    1885 ed.
Shun-Yu Lin-An Chih 淳祐臨安志.
    Shun-Yu Reign-Period Topographical Records
        of the Hangchow District.
    Sung, *c.* +1245.
    Shih E 施諤.
    WL.
Shuo Wen Chieh Tzu 說文解字.
    Analytical Dictionary of Characters.
    H/Han, +121.
    Hsü Shen 許慎.
    SPTK.
Ssu Khu Chhüan Shu 四庫全書.
    Complete Library of the Four Categories (of
        Literature). [Chhing Imperial MS Collec-
        tion.]
    Chhing, +1772–82.
    Reprint of selected titles: *Ssu Khu Chhüan Shu
        Chen Pen* 四庫全書珍本. Series 1–12,
        1934–82.
Ssu-Ming Hsü Chih 四明續志.
    Local History of Ningpo, a supplement.
    Sung, *c.* +13th century.
    Mei Ying-Fa 梅應發.
    1854 ed.
Su Tung-Pho Chi 蘇東坡集.
    Collected Writings of Su Shih.
    Sung, *c.* +11th century.
    Su Shih 蘇軾.
    KHCP.
Sui Shu 隋書.
    History of the Sui Dynasty [+581–617].
    Thang, +636–56.
    Wei Cheng 魏徵 *et al.*
    ESSS/TW.
Sung Chhao Shih Shih 宋朝事實.
    Records of Affairs of the Sung Dynasty.
    Yüan, +13th century.
    Li Yu 李攸.
    TSHCC.
Sung Shih 宋史.
    History of the Sung Dynasty [+960–1270].
    Yüan, *c.* +1345.
    Tho-Tho (Toktaga) 脫脫 & Ouyang
        Hsüan 歐陽玄.
    ESSS/TW.

Ta Chhing Hui Tien Shih Li 大清會典事例.
    See Khun-Kang (*1*) under Bibliography B.
Ta-Kuan Ching Shih Cheng Lei Pen Tshao 大觀經
    史證類本草.
    See *Cheng Lei Pen Tshao.*
Ta Ming Hui Tien 大明會典.
    History of the Administrative Statutes of the
        Ming Dynasty.
    Ming, 1st ed. +1509, 2nd ed. +1587.
    Ed. Shen Shih-Hsing 申時行 *et al.*
    1589 ed.
Ta Ming I Thung Chih 大明一統志.
    Comprehensive Geography of the Chinese

Empire (under the Ming Dynasty).
    Ming, *c.* +1450 (+1461?).
    Ed. Li Hsien 李賢.
Ta Thang Hsi Yü Chi 大唐西域記.
    Records of the Western Countries in the Time of
        the Thang.
    Thang, +646.
    Hsuan-Tsang (monk) 玄奘.
    Ed. Pien Chi 辯機.
    SSKTS.
Ta Thang Liu Tien 大唐六典.
    See *Thang Liu Tien.*
Tao Hsiang Chi 道鄉集.
    Collected Works of [Tsou] Tao-Hsiang.
    N/Sung, *c.* +1100.
    Tsou Hao 鄒浩.
    1831 ed.; reprint, Thaipei, 1970.
Tao Te Ching 道德經.
    Canon of the Virtue of the Tao.
    Chou, before −300.
    Attrib. Li Erh (Lao Tzu) 李耳(老子).
    Tr. Waley (4); Chhu Ta-Kao (2); Lin Yü-
        Thang (1); Wieger (7); and very many others.
Tao Te Ching Kuang Sheng I 道德經廣聖義.
    Study of Emperor Hsüan-Tsung's Commentary
        on *Lao Tzu.*
    Thang, +913.
    Tu Kuang-Thing 杜光庭.
Tao Tsang 道藏.
    Taoist Patrology [containing 1464 Taoist
        works].
    All periods, but first collected and printed in
        the Sung. Also printed in J/Chin (+1186/
        +1191), Yüan, and Ming (+1445, +1598
        and +1607).
    Yin-Te Index no. 25.
    Shanghai, 1924–6.
Tao Tsang Chi Yao 道藏輯要.
    Essentials of the Taoist Patrology.
    All periods, pr. 1906 at Erh-hsien-ssu 二仙寺,
        Chhengtu.
    Writers numerous.
    Ed. Ho Lung-Hsiang 賀龍驤 & Pheng Han-
        Jan 彭瀚然 (Chhing).
    Chhengtu, 1906.
Thai Phing Hui Min Ho Chi Chü Fang 太平惠民
    和劑局方.
    Standard Formularies of the (Government)
        Great Peace People's Welfare Pharmacies
        [based on the *Ho Chi Chü Fang*, etc.].
    Sung, +1151.
    Ed. Chhen Shih-Wen 陳師文, Phei Tsung-
        Yüan 裴宗元, and Chhen Chheng 陳承.
Thai Phing Kuang Chi 太平廣記.
    Copious Records Collected in the Thai-Phing
        Reign-Period [anecdotes, stories, mirabilia
        and memorabilia].
    Sung, +978.
    Ed. Li Fang 李昉.
    Peking, 1959.
Thai-Phing Shan Shui Thu Hua 太平山水圖畫.
    Landscape Pictures of Thai-Phing [in Anhui].

Thai-Phing Shan Shui Thu Hua (*cont.*)
  Chhing, +1648.
  Chang Wan-Hsüan 張萬選; paintings by
    Hsiao Yün-Tshung 蕭雲從.
Thai-Phing Sheng Hui Fang 太平聖惠方.
  Prescriptions Collected by Imperial Benevolence
    during the Thai-Phing Reign-Period.
  Sung, commissioned +982; completed +992.
  Ed. Wang Huai-Yin 王懷隱, Cheng Yen
    鄭彥 *et al.*
Thai Phing Yü Lan 太平御覽.
  Thai-Phing Reign-Period Imperial Encyclo-
    paedia.
  Sung, +983.
  Ed. Li Fang 李昉.
  Yin-Te Index, no. 23.
  SPTK.
Thang Hui Yao 唐會要.
  History of the Administrative Statutes of the
    Thang Dynasty.
  Sung, +961.
  Wang Phu 王溥.
  Peking, 1955.
Thang Kuo Shih Pu 唐國史補.
  Supplements to the History of the Thang
    Dynasty.
  Thang, *c.* +820.
  Li Chao 李肇.
  CTPS.
Thang Liu Tien 唐六典.
  Institutes of the Thang Dynasty.
  Thang, +738 or +739.
  Ed. Li Lin-Fu 李林甫.
  Japanese ed., 1836.
Thang Shih Hua Phu 唐詩畫譜.
  Manual of Painting with Themes from the
    Thang Poetry.
  Ming, *c.* +1600.
  Ed. Huang Feng-Chhih 黃鳳池.
Thang Ta Chao Ling Chi 唐大詔令集.
  Collected Imperial Edicts of the Thang
    Dynasty.
  Sung, +1070.
  Ed. Sung Min-Chhiu 宋敏求.
  Shanghai, 1959.
Thang Yün 唐韵.
  Thang Dynasty Rhyme Sounds.
  Thang, +677.
  Ed. Chhangsun Na-Yen 長孫納言 & Sun
    Mien 孫愐.
Thien Chu Shih Lu 天主實錄.
  See *Sheng Chiao Shih Lu.*
Thien Kung Khai Wu 天工開物.
  The Exploitation of the Works of Nature.
  Ming, +1637.
  Sung Ying-Hsing 宋應星.
  Tr. Sun Jen I-Tu & Sun Hsüeh-Chuan (1).
  KHCP.
Thing-Lin Wen Chi 亭林文集.
  Collected Writings of Ku Yen-Wu.
  Chhing, +17th century.
  Ku Yen-Wu 顧炎武.

  SPTK.
Thu Shu Chi Chheng 圖書集成.
  Imperial Encyclopaedia.
  Chhing, +1726.
  Ed. Chhen Meng-Lei 陳夢雷 *et al.*
  Index by Giles (2).
  Thaipei, 1964.
Thung Chih 通志.
  Historical Collections.
  Sung, *c.* +1150.
  Cheng Chhiao 鄭樵.
  ST.
Thung Jen Chen Chiu Thu Ching 銅人鍼灸圖經.
  Book on the Charts of Tract of the Bronze Man
    for Acupuncture and Moxibustion.
  Sung, +1026.
  Wang Wei-I 王惟一 & Wang Wei-Te
    王惟德.
Thung Meng Hsün 童蒙訓.
  An Admonition to Those Who are Immature
    and Ignorant.
  Sung, first pr. +1215.
  Lü Pen-Chung 呂本中.
  TKTT.
Thung Su Pien 通俗編.
  Thesaurus of Popular Terms, Ideas and
    Customs.
  Chhing, +1751.
  Chai Hao 翟灝.
  1751 ed.; reprint, Thaipei, 1977.
Thung Tien 通典.
  Reservoir of Source Material on Political and
    Social History.
  Thang, *c.* +812.
  Tu Yu 杜佑.
  ST.
Tshe Fu Yüan Kuei 册府元龜.
  Collection of Material on the Lives of Emperors
    and Ministers.
  Sung, +1013.
  Ed. Wang Chhin-Jo 王欽若 & Yang I
    楊億.
  1640 ed.
Tso Chuan 左傳.
  Master Tsochhiu's Enlargement of the *Chhun
    Chhiu* (Spring and Autumn Annals), [dealing
    with the period −722 to −468].
  Chou, *c.* −400 to −250.
  Attrib. Tsochhiu Ming 左邱明.
  SPTK.
Tsun Sheng Pa Chien 遵生八牋.
  Collection of Essays on the Daily Life and
    Interest of a Scholar.
  Ming, +1591.
  Kao Lien 高濂.
  1810 ed.
Tung-Ching Meng Hua Lu 東京夢華錄.
  Dreams of the Glories of the Eastern Capital
    (Khaifeng).
  S/Sung, +1148, first pr. +1187.
  Meng Yüan-Lao 孟元老.
  TSHCC.

Tung-Kuan Han Chi 東觀漢記.
An Official History of the Later Han Dynasty.
L/Han, *c.* +120.
Liu Chen 劉珍.
SPPY.

Tung Nan Chi Wen 東南紀聞.
Miscellaneous Notes about the Southeast, i.e.
Southern Sung.
Yüan, *c.* +1270.
Writer unknown.
SSKTS.

Tung-Pho Chih Lin 東坡志林.
Journal and Miscellany of (Su) Tung-Pho.
Sung, +1097–1101.
Su Shih 蘇軾.
TSHCC.

Tung Thien Chhing Lu (Chi) 洞天清錄(集).
Clarifications of Strange Things [Taoist].
Sung, *c.* +1240.
Chao Hsi-Ku 趙希鵠.
MSTS.

Tzu Chih Thung Chien 資治通鑑.
Comprehensive Mirror (of History) for Aid in
Government [+403–959].
Sung, begun +1065, completed +1084.
Ssuma Kuang 司馬光.
Reprint, 1956.

Tzu-Thao-Hsüan Tsa Chui 紫桃軒雜綴.
Miscellany from the Purple Peach Studio.
Ming.
Li Jih-Hua 李日華.
1768 ed.

Wan-Hsiao-Thang Hua Chuan 晚笑堂畫傳.
Noted Figures in History from the Wan-Hsiao-
Thang.
Chhing, *c.* +1743.
Shangkuan Chou 上官周.
Wan Shou Sheng Tien Thu 萬壽聖典圖.
Illustrations of the Scenes at the Celebration of
Emperor Khang-Hsi's 60th Birthday.
Chhing, +1713–16.
Wang Yüan-Chhi 王原祁.

Wei Lüeh 緯略.
Compendium of Non-Classical Matters.
Sung, +12th century (end).
Kao Ssu-Sun 高似孫.
TSHCC.

Wei Lüeh Chi Pen 魏略輯本.
Memorable Things of the Wei State (San Kuo),
Reconstructed Version.
San Kuo (Wei) or Chin, +3rd or +4th
century.
Yü Huan 魚豢; ed. Chang Pheng-I
張鵬一
1924 ed.

Wei-Nan Wen Chi 渭南文集.
(Collected Works of Lu Yu).
Sung, +1210.
Lu Yu 陸游.
SPTK.

Wei Shu 魏書.

History of the (Northern) Wei Dynasty
[+386–556].
N/Chhi, +554, revised +572.
Wei Shou 魏收.
ESSS/TW.

Wen Fang Ssu Phu 文房四譜.
Collected Studies of the Four Articles for
Writing in a Scholar's Studio.
Sung, +986.
Su I-Chien 蘇易簡.
TSHCC.

Wen Hsien Thung Khao 文獻通考.
Comprehensive Study of (the History of)
Civilisation.
Sung & Yüan, begun perhaps as early as +1270
and finished before +1317, pr. +1322.
Ma Tuan-Lin 馬端臨.
ST.

Wen Hsüan 文選.
General Anthology of Prose and Verse.
Liang, +530.
Ed. Hsiao Thung (prince of the Liang) 蕭統.

Wen Yüan Ying Hua 文苑英華.
The Brightest Flowers in the Garden of
Literature.
Sung, +987; first pr. +1567.
Ed. Li Fang 李昉 *et al.*
Reprint, Chung-Hua, Peking, 1966.

Wu Ching Tsung Yao 武經總要.
Collection of the Most Important Military
Techniques.
Sung (compiled by Imperial Order), +1040
(+1044).
Ed. Tseng Kung-Liang 曾公亮.

Wu Pei Chih 武備志.
Treatise on Armament Technology.
Ming, +1621.
Mao Yüan-I 茅元儀.
1621 ed.

Wu Shuang Phu 無雙譜.
Album of Unique Personalities of Chinese
History.
Chhing, *c.* +1690.
Chin Ku-Liang 金古艮.

Wu Tai Hui Yao 五代會要.
History of the Administrative Statutes of the
Five Dynasties.
Sung, +961.
Wang Phu 王溥.
TSHCC.

Wu Tai Shi Chi 五代史記.
See *Hsin Wu Tai Shih*.

Wu Ying Tien Chü Chen Pan Chheng Shih 武英殿聚
珍版程式.
Printing Manual for Wooden Movable Type.
Chhing, +1776.
Chin Chien 金簡.
Tr. Rudolph (8).
WYTCC.

Yeh Hsi Yüan Chhi 葉戲原起.
Origins of Card-games.

Yeh Hsi Yüan Chhi (cont.)
Chhing, +18th century.
Wang Shih-Han 汪師韓.
TMWS.

Yen Fan Lu 演繁露.
Extension of the String of Pearls (on the Spring and Autumn Annals), [on the Meaning of Many Thang and Sung Expressions].
Sung, +1180.
Chheng Ta-Chhang 程大昌.
SF.

Yen Hsien Chhing Shang Chien 燕閒清賞牋.
The Use of Leisure and Innocent Enjoyments in a Retired Life.
Ming, +1591.
Kao Lien 高濂.
MSTS.

Yen Shih Chia Hsün 顏氏家訓.
Mr. Yen's Advice to his Family.
Sui, *c.* +590.
Yen Chih-Thui 顏之推.
Tr. Teng Ssu-Yü (3).
SPTK.

Yin Shan Cheng Yao 飲膳正要.
Principles of Correct Diet [on deficiency diseases, with the aphorism 'Many diseases can be cured by diet alone'].
Yüan, +1330, reissued by imperial order in +1456.
Hu-Ssu-Hui 忽思慧.
See Lu & Needham (1).
SPTK.

Ying Tsao Fa Shih 營造法式.
Treatise on Architectual Methods.
Sung, +1097; pr. +1103; revised +1141.
Li Chieh 李誡.
Peking, 1925.

Ying Yai Sheng Lan 瀛涯勝覽.
Triumphant Visions of the Ocean Shores [relating to the voyages of Cheng Ho].
Ming, +1451.
Ma Huan 馬歡.
Ed. Feng Chheng-chün 馮承鈞.
Shanghai, 1935.

Yongjae Ch'onghwa 慵齋叢話.
Collected Essays of Song Hyon.
Korea, +15th century.
Song Hyŏn 成俔.
Reprint, Seoul, 1964.

Yu Huan Chi Wen 游宦紀聞.
Things Seen and Heard on My Official Travels.
Sung, *c.* +1233.
Chang Shih-Nan 張世南.
TSHCC.

Yu Yang Tsa Tsu 酉陽雜俎.
Miscellany of the Yu-yang Mountain (Cave) [in S.E. Szechuan].
Thang, +863.
Tuan Chheng-Shih 段成式.
TSHCC.

Yü Chih Wen 御製文.
Collected Imperial Writings of Emperor Khanghsi.
Chhing, +1711.
Ed. Chang Yü-Shu 張玉書 *et al.*

Yü Fu Chih 輿服志.
Monograph on Ceremonies.
H/Han, *c.* +3rd century.
Tung Pa 董巴.
Preserved in *TPYL*.

Yü Hai 玉海.
Ocean of Jade [encyclopaedia].
Sung, +1267.
Wang Ying-Lin 王應麟.
CCSC, 1883.

Yü Phien 玉篇.
Book of Jade [Dictionary].
Liang, +523.
Ku Yeh-Wang 顧野王.
Rev. Chhen Pheng-Nien 陳彭年 (Sung).
SPTK.

Yü Ti Shan Hai Chhüan Thu 輿地山海全圖.
See *Khun Yü Wan Kuo Chhüan Thu*.

Yü Thai Hsin Yung 玉臺新詠.
Anthology of Regulated Verses.
Liang, *c.* +6th century.
Hsü Ling 徐陵.
SPTK.

Yüan-Ho Chün Hsien Thu Chih 元和郡縣圖志.
Yüan-Ho Reign-Period General Geography.
Thang, +814.
Li Chi-Fu 李吉甫.
TSHCC.

Yüan Ming Yüan Ssu Shih Ching Shih Thu 圓明園四十景詩圖.
Poems and Illustrations on 40 Scenes of the Yüan-Ming Garden.
Chhing, +1745.
Sun Ku 孫祜 & Shen Yüan 沈源 by Imperial Order.

Yüan Shu Tsa Chi 宛署雜記.
Records of the Seat of Government at Yüan (-Phing) (Peking).
Ming, +1593.
Shen Pang 沈榜.
Peking, 1961.

Yüeh Chüeh Shu 越絕書.
Lost Records of the State of Yüeh.
H/Han, *c.* +52.
Attrib. Yüan Khang 袁康.
SPTK.

Yüeh Lü Chhüan Shu 樂律全書.
Collected Works on Music and Acoustics.
Ming, *c.* +1606.
Chu Tsai-Yü (prince of the Ming) 朱載堉.

Yün Chhi Yu I 雲溪友議.
Discussions with Friends at Cloudy Pool.
Thang, *c.* +870.
Fan Shu 范攄.
BH.

Yün Hsien San Lu 雲仙散錄.
Scattered Remains of Clouded Immortals.

Yün Hsien San Lu (*cont.*)
Ascr. Thang or Wu Tai, *c.* +901 or +926.
Attrib. Feng Chih 馮贄 but probably by
Wang Chih 王銍.
IHCC.
Yung-Lo Ta Tien 永樂大典.
Grand Encyclopaedia of the Yung-Lo Reign-
Period [only in manuscript].
Amounting to 22,877 chapters in 11,095 vol-
umes, only about 370 being still extant.
Ming, +1407.
Ed. Hsieh Chin 解縉.
See Yüan Thung-Li (*1*).

B. CHINESE AND JAPANESE BOOKS AND JOURNAL ARTICLES SINCE +1800

A Ying (*1*) 阿英.
Chung-Kuo Lien Huan Thu Hua Shih Hua 中國連環圖畫史話.
A History of Picture-Story Comics in China.
Chung-Kuo Ku-Tien I-Shu, Peking, 1957.

A Ying (*2*) 阿英.
Chung-Kuo Nien Hua Fa Chan Shih Lüeh 中國年畫發展史略.
A Brief History of Chinese New Year Pictures.
Chao-Hua Mei-Shu, Peking, 1954.

A Ying (*3*) 阿英.
Min Chien Chhuang Hua 民間窗花.
Folk Paper Cuts.
Mei-Shu Chhu-Pan-She, Peking, 1954.

An Chih-Min (*1*) 安志敏.
Chhangsha Hsin Fa hsien ti Hsi-Han Po Hua 長沙新發現的西漢帛畫.
A Tentative Interpretation of the Western (Former) Han Silk Painting Recently Discovered at Chhangsha (Ma-Wang-Tui, No. 1 Tomb).
KKTH, 1973, no. 1 (no. 124), 43.

An Ch'un-Gun (*1*) 安春根.
Han'guk Sŏjihak 韓國書誌學.
Korean Bibliography.
T'ongmunkwan, Seoul, 1967.

Anon. (*22*).
Ssuchhuan Han Hua Hsiang Chuan Hsüan Chi 四川漢畫像磚選集.
A Selection of Bricks with Stamped Reliefs from Szechuan.
Wen-Wu, Peking, 1957.

Anon. (*43*).
Hsin Chung-Kuo ti Khao Ku Shou Huo 新中國的考古收獲.
Successes of Archaeology in New China.
Wen-Wu, Peking, 1961.

Anon. (*109*).
Chung-Kuo Kao Teng Chih Wu Thu Chien 中國高等植物圖鑑.
Iconographia Cormophytorum Sinicorum (Flora of Chinese Higher Plants).
2 vols.
Kho-Hsüeh, Peking, 1972.

Anon. (*110*).
Chhang Yung Chung Tshao Yao Thu Phu 常用中草藥圖譜.
Illustrated Flora of the Most Commonly Used Drug Plants in Chinese Medicine.
Jen-Min Wei-Sheng, Peking, 1970.

Anon. (*225*).
Che-Chiang Chih Chih Yeh 浙江之紙業.
The Paper Industry of Chekiang.
Hangchow, 1930.

Anon. (*226*).
Chu-Fei Chiu Tshang Ming Chih Mu Lu 竹扉舊藏名紙目錄.
A Catalogue of Famous Papers Originally Collected by Chu-Fei.
Western Union University Museum, Chengtu, 1947.

Anon. (*227*).
Chung-Kuo Ku Tai Pan Hua Tshung Khan 中國古代版畫叢刊.
Collection of Old Illustrated Books of China.
44 vols.
Chung-Hua, Peking, 1961.

Anon. (*228*).
Chung-Kuo Pan Hua Hsüan 中國版畫選.
Selected Specimens of Chinese Woodblock Illustrations (re-engraved in facsimile).
2 vols.
Jung-Pao-Chai, Peking, 1958.

Anon. (*229*).
Chung-Kuo Pan Kho Thu Lu 中國版刻圖錄.
Collection of Facsimile Specimens of Chinese Printing.
Compiled by Peking Library.
8 vols.
Chung-Hua, Peking, 1961.

Anon. (*230*).
Ho-Pei Phing-Shan-Hsien Fa Hsien ti 'Chih-Yüan Thung Hsing Pao Chhao' Thung Pan 河北平山縣發現的至元通行寶鈔銅版.
Bronze Plate for Printing Chih-Yüan Period Paper Currency Discovered in Phing-Shan, Hopei Province.
KKTH, 1973 (no. 1), 42.

Anon. (*231*).
Hsi-An Pan-Pho 西安半坡.
The Neolithic Village at Pan-Pho, Sian.
Institute of Archaeology, Academia Sinica, Peking, 1963.

Anon. (*232*).
Ku Tai Chu Chien ti Tho Shui Chhu Li 古代竹簡的脫水處理.
Dehydration Treatment of Ancient Bamboo Tablets.
KKTH, 1976 (no. 4), 276.

Anon. (*233*).
Liu Chhao Ling Mu Tiao Chha Pao Kao 六朝陵墓調查報告.
Report of Investigations of Six Dynasties Tombs.
National Commission for Preservation of Antiques, Nanking, 1935.

Anon. (*234*)
  Ma-Wang-Tui Han Mu Po Shu 馬王堆漢
  墓帛書.
  Silk Books Discovered in the Han Tomb at Ma-
  Wang-Tui, Chhangsha.
  8 vols.
  Wen-Wu, Peking, 1975.
Anon. (*235*).
  *Shou Kung Yeh Sheng Chhan Ching Yen Hsüan
  Pien—Tsao Chih* 手工業生產經驗選編
  —造紙.
  Selections on Experience in Handicraft—Paper-
  making.
  Peking, 1958.
Anon. (*236*).
  Shui-Hu-Ti Chhin Mu Chu Chien 睡虎地秦墓
  竹簡.
  Bamboo Tablets Discovered in a Chhin Tomb
  at Shui-Hu-Ti, Yün-Meng Hsien, Hupei
  Province.
  Wen-Wu, Peking, 1978.
Anon. (*237*).
  *Tui Ming Chhing Shih Chhi Fang Tu Chih ti Yen-
  Chiu* 對明清時期防蠹紙的研究.
  A Study of Specimens of Ming-Chhing Moth-
  Proof Paper.
  WWTK, 1977 (no. 1), 47.
Anon. (*238*).
  Wang-Tu Han Mu Pi Hua 望都漢墓壁畫.
  Painting of a Han Tomb Discovered at Wang-
  Tu, Hopei Province.
  Historical Museum, Peking, 1955.
Anon. (*239*).
  Wen Wu Khao Ku Kung Tso San Shih Nien 文物
  考古工作三十年.
  Thirty Years of Cultural and Archaeological
  Work, 1949–1979.
  Wen-Wu, Peking, 1979.
Anon. (*240*).
  Chūgoku no Min Shin Jidai no Hanga 中国の明
  清時代の版画.
  Chinese Woodcuts and Etchings of the Ming
  and Chhing Dynasties.
  Yamato Bunkakan, Nara, 1972.
Anon. (*241*).
  Shōsōin no Kami 正倉院の紙.
  Various Papers Preserved in the Shōsōin.
  Nihon Keizai Shinbunsha, Tokyo, 1970.
Anon. (*242*).
  Shōsōin no shoseki 正倉院の書蹟.
  Calligraphy Kept at the Shōsōin.
  Nihon Keizai Shinbunsha, Tokyo, 1964.
Anon. (*243*).
  Tesuki Washi Taikan 手漉和紙大鑑.
  Comprehensive Collection of Handmade
  Japanese Paper.
  Mainichi Shimbun, Tokyo, 1973–74.
  5 vols. with 1,000 mounted samples in 5 boxes.
Anon. (*244*).
  *Tōyō Bunko Chōsembon Bunrui Mokuroku
  (Fu Annanhon Mokuroku)* 東洋文庫朝鮮本
  分類目錄（附安南本目錄）.

A Classified Catalogue of Korean Editions in the
  Collection of Tōyō Bunko (appended with a
  Catalogue of Annamese Editions).
  Tokyo, 1939.
Anon. (*245*).
  Chosŏn munhwasa 朝鮮文化史.
  Cultural History of Korea.
  Comp. by the Institute of History, Academy of
  Social Sciences, Democratic People's Republic
  of Korea.
  Pyongyang, 1966–
Aoyama Arata (*1*) 青山新.
  Shina Kohanga Zuroku 支那古版畫圖錄.
  Collection of Old Chinese Woodcuts.
  Introduction by Kuroda Genji 黑田源次.
  Otsuka Kōgei-Sha, Tokyo, 1932.
Asakura Kamezō (*1*) 朝倉龜三.
  Nihon ko kokusho-shi 日本古刻書史.
  History of Old Japanese Printing.
  Kokusho Kankōkai, Tokyo, 1909.

Chai Chin-Sheng (*1*) 翟金生.
  Ni Pan Shih Yin Chhu Pien 泥版試印初編.
  First Experimental Printing with Earthenware
  Type.
  1844.
Chang Chih-Che (*1*) 張志哲.
  Yin Shua Shu Fa Ming yü Sui Chhao ti Hsin Cheng
  印刷術發明于隋朝的新證.
  New Evidence for the Invention of Printing in
  the Sui Dynasty.
  SHKH, 1979 (no. 3), 154.
Chang Ching-Lu (*1*) 張靜廬.
  Chung-Kuo Chin Tai Chhu Pan Shih Liao 中國近
  代出版史料.
  Materials for a History of Publishing in Modern
  China, 1862–1918.
  2 series.
  Lien-Chhün, Shanghai, 1954.
Chang Ching-Lu (*2*) 張靜廬.
  Chung-Kuo Hsien Tai Chhu Pan Shih Liao 中國
  現代出版史料.
  Materials for a History of Publishing in
  Contemporary China, 1919–49.
  4 series.
  Chung-Hua, Peking, 1955.
Chang Ching-Lu (*3*) 張靜廬.
  Chung-Kuo Chhu Pan Shih Liao Pu Pien 中國出
  版史料補編.
  Further Materials for a History of Publishing in
  China, 1862–1949.
  Chung-Hua, Peking, 1957.
Chang Chung-I (*1*) et al. 張仲一.
  Hui-Chou Ming Tai Chu Chai 徽州明代住宅.
  Ming Dynasty Dwelling-Houses in Hui-chou.
  Architectural Engineering Press, Peking, 1957.
Chang Feng (*1*) 張鳳.
  Han Chin Hsi Chhui Mu Chien Hui Pien 漢晉西
  陲木簡彙編.
  Wooden Tablets of the Han and Chin Dynasties
  Discovered in Chinese Turkestan During
  Aurel Stein's Three Expeditions.

Chang Feng (*1*) (*cont.*)
Shanghai, 1931.

Chang Heng (*1*) 張珩.
Tsen Yang Chien Ting Shu Hua 怎樣鑑定書畫.
How to Appraise the Authenticity of Calligraphy and Painting.
WWTK, 1964 (no. 3); Wen-Wu, Peking, 1966.

Chang Hsin-Chheng (*1*) 張心澂.
Wei Shu Thung Khao 僞書通考.
A Complete Investigation of (Ancient and Medieval) Books of Doubtful Authenticity.
2 vols.
Com. Press, Chhangsha, 1939, repr. 1957.

Chang Hsing-Lang (*1*) 張星烺.
Chung Hsi Chiao Thung Shih Liao Hui Pien 中西交通史料滙編.
Materials for the Study of the Intercourse of China and the West. 6 vols.
Fu-Jen University Press, Peiping, 1928, 1930; reprint, World Book Co., Thaipei, 1962.

Chang Hsiu-Min (*1*) 張秀民.
Chin Yüan Chien Pen Khao 金源監本考.
On the Books Printed by the Government during the Chin Dynasty.
QBCB/C, 1935, **2** (no. 1), 19.

Chang Hsiu-Min (*2*) 張秀民.
Sung Hsiao-Tsung Shih Tai Kho Shu Shu Lüeh 宋孝宗時代刻書述略.
A Brief Sketch of Block Printing during the Reign of Sung Hsiao-Tsung (+1163–89).
TSKH, 1936, **10** (no. 3), 385.

Chang Hsiu-Min (*3*) 張秀民.
Chung-Kuo Yin Shua Shu ti Fa Ming Chi Chhi Tui Ya-Chou Ko Kuo ti Ying Hsiang 中國印刷術的發明及其對亞洲各國的影響.
The Invention of Printing in China and its Influence on Other Asian Countries.
WWTK, 1952 (no. 2), 20.

Chang Hsiu-Min (*4*) 張秀民.
Chhao-Hsien ti Ku Yin Shua 朝鮮的古印刷.
Early Printing in Korea.
LSYC, 1957 (no. 3), 61.

Chang Hsiu-Min (*5*) 張秀民.
Chung-Kuo Yin Shua Shu ti Fa Ming Chi Chhi Ying Hsiang 中國印刷術的發明及其影響.
The Invention of Printing in China and its Influence.
Jen-Min, Peking, 1958.

Chang Hsiu-Min (*6*) 張秀民.
Liao Chin Hsi-Hsia Kho Shu Chien Shih 遼金西夏刻書簡史.
A Short History of Printing in Liao, Chin and Hsi-Hsia.
WWTK, 1959 (no. 3), 11.

Chang Hsiu-Min (*7*) 張秀民.
Chhing Tai Ching-Hsien Chai Shih ti Ni Huo Tzu Yin Pen 清代涇縣翟氏的泥活字印本.
On the Books Printed with Clay Movable Type by the Chai Family of Ching County, An-Hui, in the Chhing Dynasty.
WWTK, 1961 (no. 3), 30.

Chang Hsiu-Min (*8*) 張秀民.
Nan Sung Kho Shu Ti Yü Khao 南宋刻書地域考.
On Printing Centres in the Southern Sung Dynasty.
TSK, 1961 (no. 3), 52.

Chang Hsiu-Min (*9*) 張秀民.
Ming Tai ti Thung Huo Tzu 明代的銅活字.
On Bronze Movable Type of the Ming Dynasty.
TSK, 1961 (no. 4), 55.

Chang Hsiu-Min (*10*) 張秀民.
Chhing Tai ti Thung Huo Tzu 清代的銅活字.
Copper Type in the Chhing Dynasty.
WWTK, 1962 (no. 1), 49.

Chang Hsiu-Min (*11*) 張秀民.
Yüan Ming Liang Tai ti Mu Huo Tzu 元明兩代的木活字.
On Wooden Movable Type of the Yüan and Ming Dynasties.
TSK, 1962 (no. 1), 56.

Chang Hsiu-Min (*12*) 張秀民.
Chhing Tai ti Mu Huo Tzu 清代的木活字.
The Wooden Movable Type of the Chhing Dynasty.
TSK, 1962 (no. 2), 60; (no. 3), 60.

Chang Hsiu-Min (*13*) 張秀民.
Ming Tai Hui Phai Pan Hua Huang Hsing Kho Kung Khao Lüeh 明代徽派版畫黃姓刻工考略.
On the Huang Carvers of Anhui during the Ming Dynasty.
TSK, 1964 (no. 1), 61.

Chang Hsiu-Min (*14*) 張秀民.
Wu-Tai Wu-yüeh Kuo ti Yin Shua 五代吳越國的印刷.
Printing of the Wu-Yüeh Kingdom in the Five Dynasties.
WWTK, 1978 (no. 12), 74.

Chang Hsiu-Min (*15*) 張秀民.
Tshai Lun Chuan 蔡倫傳.
Biography of Tshai Lun.
In *Chung-Kuo Ku Tai Kho Hsüeh Chia*, pp. 18–20.
Science Press, Peking, 1963.

Chang Hsiu-Min (*16*) 張秀民.
Ming Tai Yin Shu Tsui To ti Chien-Ning Shu Fang 明代印書最多的建寧書坊.
The Printers in Chien-Ning (Fukien) Who Printed Most of the Books in the Ming Dynasty.
WWTK, 1979 (no. 6), 76.

Chang Hsiu-Min (*17*) 張秀民.
Tiao Pan Yin Shua Khai Shih Yü Thang Chhu Chen-Kuan Shuo 雕板印刷開始于唐初貞觀說.
The Beginnings of Block Printing in the Chen-Kuan Period (+627–59) of the Early Thang Dynasty.
SHCH, 1979 (no. 3), 345.

Chang Hsiu-Min (*18*) 張秀民.
Ming Tai Pei-Ching ti Kho Shu 明代北京的刻書.
Printing in Peking in the Ming Dynasty.

Chang Hsiu-Min (*18*) (*cont.*)
  WH, 1980 (no. 1), 298.
Chang Hsiu-Min (*19*) 張秀民.
  Ming Tai Nan-Ching ti Yin Shu 明代南京的
  印書.
  Printing in Nanking in the Ming Dynasty.
  WWTK, 1980 (no. 11), 78.
Chang Huai-Li (*1*) 張懷禮.
  Yin Shua Shu ti Fa Ming Ho Yen Chin 印刷術
  的發明和演進.
  The Invention and Development of Printing.
  LSCH, 1955 (no. 7), 43.
Chang I-Hui (*1*) 張貽惠.
  *Fu-Chien Pan Pen tsai Chung-Kuo Wen Hua shang
    chih Ti Wei* 福建版本在中國文化上之
    地位.
  The Position of Fukien Printing in Chinese
    Culture.
  FCWH, 1933, **I** (no. 7), 1.
Chang Kheng-Fu (*1*) 張鏗夫.
  Chung-Kuo Shu Chuang Yüan Liu 中國書
    裝源流.
  The Evolution of Book Formats in China.
  LNHP, 1950, **10** (no. 2), 193.
Chang Kuo-Kan (*1*) 張國淦.
  Li Tai Shih Ching Khao 歷代石經考.
  A History of Stone Classics Engraved during the
    Successive Dynasties.
  Peiping, 1930.
Chang Man-Tho (*1*) 張曼陀.
  Chung-Kuo Chih Chih Yü Yin Shua Yen Ko Khao
    中國製紙與印刷沿革考.
  The Development of Paper-Making and
    Printing in China.
  STTK, 1933, **I**, 1.
Chang Ping-Lun (*1*) 張秉倫.
  *Kuan yü Chai Chin-Sheng ti Ni Huo Tzu Wen Thi ti
    Chhu Pu Yen Chiu* 關于翟金生的泥活字
    問題的初步研究.
  A Preliminary Study of Chai Chin-Sheng's
    Earthenware Movable Type.
  WWTK, 1979 (no. 10), 90.
Chang Ssu-Wen (*1*) 張思溫.
  *Huo Tzu Pan Hsi-Hsia Wen Hua-Yen-Ching Chüan
    Shih I Chih Chüan Shih Wu Chien Chieh* 活字
    版西夏文華嚴經卷十一至卷十五
    簡介.
  A Brief Introduction to a Movable Type Edition
    of a Tangut Translation of the *Avatamsaka
    Sutra*, Chapters 11–15.
  WWTK, 1979 (no. 10), 93.
Chang Te-Chün (*1*) 張德鈞.
  *Kuan yü Tsao Chih tsai Wo Kuo ti Fa Chan Ho Chhi
    Yüan ti Wen Thi* 關於造紙在我國的
    發展和起源的問題.
  On "The Origin and Development of Paper in
    China" (by Yüan Han-Chhing).
  KHTP, 1955 (no. 10), 85.
Chang Ting (*1*) 張仃.
  Thao-Hua-Wu Nien Hua 桃花塢年畫.
  New Year Pictures from the Peach Blossom
    Village of Soochow.

MSHU, 1954 (no. 8), 44.
Chang Tzu-Kao (*2*) 張子高.
  *Chung-Kuo Hua Hsüeh Shih Kao (Ku Tai chih
    Pu)*. 中國化學史稿(古代之部).
  A Draft History of Chinese Chemistry (Ancient
    Section).
  Kho-Hsüeh, Peking, 1964.
Chang Tzu-Kao (*7*) 張子高.
  *Kuan yü Tshai Lun tui Tsao Chih Shu Kung Hsien ti
    Phing Chia* 關於蔡倫對造紙術貢獻的
    評價.
  An Evaluation of Tshai Lun's Contribution to
    Paper-making.
  CHHP, 1960, 7 (no. 2).
Chang Ying-Hsüeh (*1*) 張映雪.
  Yang-Liu-Chhing Mu Kho Nien Hua Hsüan Chi
    楊柳青木刻年畫選集.
  Woodcut New Year Pictures from Yang-Liu-
    Chhing.
  Peking, 1957.
Chang Yung-Hui et al. (*1*) 張永惠.
  Chung-Kuo Tsao Chih Yüan Liao chih Yen Chiu
    中國造紙原料之研究.
  A Study of the Raw Materials for Papermaking
    in China.
  Chungking, 1943.
Chao Hung-Chhien (*1*) 趙鴻謙.
  Sung Yüan Pen Hang Ko Piao 宋元本行格表.
  A Table of the Formats of Some Sung and Yüan
    Editions.
  KHTS, 1928 (no. 1), 1.
Chao Wan-Li (*1*) 趙萬里.
  Liang Sung Chu Shih Chien Pen Tshun I Khao
    兩宋諸史監本存佚考.
  On the Survival of Standard Histories Printed
    by the Government during the Sung Dynasty.
  In *TYPM*, p. 167.
Chao Wan-Li (*2*) 趙萬里.
  Chheng Shih Mo Yüan Tsa Khao 程氏墨苑雜考.
  On Chheng Chün-Fang's Inkcake Designs.
  CFTS, 1946, **2**, 1.
Chao Wan-Li (*3*) 趙萬里.
  *Tshung Chien Tu Wen Hua Shuo Tao Tiao Pan Wen
    Hua* 從簡牘文化說到雕版文化.
  From Bamboo Strips to Block Printing (as the
    Media of Diffusion of Knowledge).
  WWTK, 1951, **2** (no. 2), 21.
Chao Wan-Li (*4*) 趙萬里.
  Chung Ko Yin Pen Shu Chi Fa Chan Chien Shih
    中國印本書籍發展簡史.
  A Short History of the Development of Printed
    Books in China.
  WWTK, 1952 (no. 4), 5.
Chao Wang-Li (*5*) 趙萬里.
  Han Wei Nan-Pei-Chhao Mu Chih Chi Shih
    漢魏南北朝墓志集釋.
  Collected Inscriptions on Grave Tablets from
    the +2nd to +7th Centuries.
  6 vols.
  Kho-Hsüeh, Peking, 1955.
Cheng Chen-To (*6*) (ed.) 鄭振鐸.
  Chung-Kuo Pan Hua Shih Thu Lu 中國版畫史

Cheng Chen-To (6) (ed.) (cont.)
圖錄.
Illustrations to the History of Chinese Woodcuts. Series 1–5.
24 vols.
Shanghai, 1940–47.

Cheng Chen-To (7) 鄭振鐸.
Chung Ko Yin Pen Shu Chi Chan Lan Yin Yen
中國印本書籍展覽引言.
A Foreword to 'An Exhibition of Printed
Chinese Books'.
WWTK, 1952 (no. 4), 1.

Cheng Chih-Chhao & Jung Yüan-Khai (1)
鄭志超, 榮元愷.
Hsi-Han Ma Chih Chih I 兩漢麻紙質疑.
Questions on the Hemp Paper of Western Han.
CHTH/S, 1980 (no. 2), 56.

Cheng Shih Hsü (1) 鄭師許.
Yüan Chhao Ssu Kho Pen Piao 元朝私刻本表.
Table of Private Printing in the Yüan Dynasty.
JWYK, 1935, **6**, (no. 5), 1; (no. 7), 17.
Cf. Nagasawa Kikuya (6).

Chi Hsien-Lin (1) 季羨林.
Chung Yin Wen Hua Kuan Hsi Shih Lun Tshung
中印文化關係史論叢.
Studies on the Interrelationship between
Chinese & Indian Cultures.
Jen Min Chhu Pan She, Peking, 1957.

Chi Hsien-Lin (2) 季羨林.
*Chung-Kuo Chih Ho Tsao Chih Fa Shu Ju Yin-Tu ti
Shih Chien Ho Ti Tien Wen Thi* 中國紙和
造紙法輸入印度的時間和地點問題.
On the Date and Place of the Introduction of
Chinese Paper and Papermaking Methods
to India.
LSYC, 1954 (no. 4), 25.

Chi Hsien-Lin (3) 季羨林.
*Chung-Kuo Chih Ho Tsao Chih Fa Tsui Chhu Shih
Fou Yu Hai Lu Chhuan Tao Yin-Tu Chhü Ti?*
中國紙和造紙法最初是否由海
路傳到印度去的?
Was Paper and Papermaking Method First
Introduced to India via the Sea Route?
In *Chung Yin Wen Hua Kuan Hsi Shih Lun Tshung*,
pp. 130–36.

Chi Shu-Ying (1) 冀叔英.
Hsin Fa Hsien ti Ni Huo Tzu Yin Pen 新發現的
泥活字印本.
Recent Discovery of Books Printed by Clay
Movable Type.
TSKT, 1958 (no. 1), 22.

Chi Shu-Ying (2) 冀叔英.
Than Than Pan Kho Chung ti Kho Kung Wen Thi
談談版刻中的刻工問題.
On the Role of Carvers in Block-printing.
WWTK, 1959 (no. 3), 4.

Chi Shu-Ying (3) 冀叔英.
Pei Sung Kho Yin ti I Fu Mu Kho Hua 北宋
刻印的一幅木刻畫.
On a Picture Printed in the Northern Sung
Dynasty.
WWTK, 1962 (no. 1), 29.

Chia Tsu-Chang & Chia Tsu-Shan (1) 賈祖璋,
賈祖珊.
Chung-Kuo Chih Wu Thu Chien 中國植物圖鑑.
Illustrated Dictionary of Chinese Flora
[Arranged by the Engler System; 2602
entries].
Chung-Hua, Peking, 1936; repr. 1955, 1958.

Chiang Fu-Tshung (1) 蔣復聰.
Chung Han Shu Yüan 中韓書緣.
The Relationships of Books and Publishing
between China and Korea.
In *CHWH*, 1955, vol. 2, p. 275.

Chiang Fu-Tshung (2) 蔣復聰.
Chung Jih Shu Yüan 中日書緣.
The Relationships of Books and Publishing
between China and Japan.
In *CJWH*, 1958, Suppl, vol. 2, p. 337.

Chiang Fu-Tshung (3) 蔣復聰.
*Chung-Kuo Thu Shu Pan Kho ti Chhi Yüan Wen
Thi* 中國圖書版刻的起源問題.
The Question of the Origins of Printing in
China.
In *CPNM*, 1965, p. 301.

Chiang Hsüan-I (1) 蔣玄怡.
Mo Tho Shu 墨拓術.
Methods of Inked Squeezing.
SWYK, 1939, **1**, 785.

Chiang Liang-Fu (1) 姜亮夫.
Tun-Huang—Wei Ta Ti Wen Hua Pao Tshang
敦煌—偉大的文化寶藏.
Tun-Huang—a Great Cultural Treasure.
Ku-Tien Wen-Hsüeh, Shanghai, 1956.

Chiang Yin-Chhiu (1) 蔣吟秋.
Shu Hua Yü Chuang Huang 書畫與裝潢.
On Mounting Calligraphy and Painting.
In Theng Ku, *ed.*, *Chung-Kuo I Shu Lun Tshung*.
Commercial Press, Shanghai, 1938.

Chiang Yüan-Chhing (1) 蔣元卿.
Chung Kuo Shu Chi Chuang Ting Shu ti Fa Chan
中國書籍裝訂術的發展.
On the Development of Chinese Bookbinding.
In *CKCP*/*HT*, v. 4, p. 661.

Chien Po-Tsan (1) 翦伯贊.
Chung-Kuo Shih Kang 中國史綱.
Outline of Chinese History.
2 vols.
Sheng-Ho, Shanghai, 1946; reprint, Peking,
1979.

Chin Chi-Tshang (1) 靳極蒼.
Chung-Kuo Shu Chih Chih Tu yü Ku Shu 中國
書籍制度與古書.
The Format of Chinese Books and Ancient
Books.
HTF, 1940, **1**, (no. 3), 170.

Ching Yü (1) 淨雨.
Chhing Tai Yin Shua Shih Hsiao Chi 清代印刷史
小紀.
A Brief History of Printing During the Chhing
Dynasty.
In *CKCP*/*CT*, v. 2, pp. 339–60.

Chou Chün-Fu (1) 周駿富.
Chung-Kuo Huo Tzu Pan Chhuan Han Khao Pien

Chou Chün-Fu (*1*) (*cont.*)
中國活字版傳韓考辨.
On the Introduction of Movable Type from
China to Korea.
CHFH, 1971, **4** (no. 9), 17.

Chou Fa-Kao (*1*) 周法高.
*Lun Chung-Kuo Tsao Chih Shu Chih Yüan Shih Hou
Chi* 論中國造紙術之原始後記.
Postscript on Lao Kan's 'The Invention of
Paper in China'.
AS/BIHP, 1948, **19**, 499.

Chou I-Liang (*1*) 周一良.
*Chih Yü Yin Shua Shu—Chung-Kuo Tui Shih Chieh
Wen Ming ti Wei Ta Kung Hsien* 紙與印刷術
—中國對世界文明的偉大貢獻.
Paper and Printing—China's Great Contribu-
tions to Civilisation.
HHYK, 1951, **4** (no. 1), 24.

Chou Pao-Chung et al (*1*) 周寶中.
Chhien Tan Fang Tu Chih ti Yen Chiu 鉛丹防蠹
紙的研究.
A Study of Litharge-Treated Paper for
Prevention of Bookworms.
CKLS, 1980 (no. 2), 194.

Chou Shu-Chia (*1*) 周叔迦.
*Pei-Phing Thu Shu Kuan Tshang Hsi-Hsia Wen Fo
Ching Hsiao Chi* 北平圖書館藏西夏文
佛經小記.
On the Hsi-Hsia *Sutras* Preserved in the National
Peiping Library.
FJHC, 1931, **2** (no. 2), 55.

Chou Tsu-Ta (*1*) 周祖達.
*Thaiwan Chhan Hsien Wei Chih Tsao Chih Chiang
Chih Yen Chiu Lun Chi* 臺灣產纖維製造
紙漿之研究論集.
Studies of Pulp from Taiwan's Fibres.
Thaipei, 1966.
With English Abstract.

Chü Chhing-Yüan (*1*) 鞠清遠.
Thang Sung Kuan Ssu Kung Yeh 唐宋官私
工業.
Government and Private Industries of Thang
and Sung Period.
Shanghai, 1934; Thaipei, 1978.

Chu Chhuan-Yü (*1*) 朱傳譽.
Sung Tai Hsin Wen Shih 宋代新聞史.
History of Journalism in the Sung Dynasty.
Thaipei, 1967.

Chu Chia-Lien (*1*) 朱家濂.
*Chhing Tai Thai-Shan Hsü Shih ti Tzhu Huo Tzu
Yin Pen* 清代泰山徐氏的磁活字印本.
On the Books Printed with Clay Movable Type
by the Hsü Family of Thai-Shan, Shantung,
in the Chhing Dyasty.
TSK, 1962 (no. 4), 60.

Chu Shih-Chia (*3*) 朱士嘉.
Chung-Kuo Ti Fang Chih Tsung Lu 中國地方
志綜錄.
A Union Catalogue of Chinese Local Histories.
Commercial Press, Shanghai, 1935; rev. ed.,
Peking, 1958. Reprints: Tokyo, 1968; Thaipei,
1975.

Chuang Wei (*1*) 庄葳.
*Thang Khai-Yuan "Hsin-Ching" Thung Fan Hsi
Thung Pan Pien* 唐開元'心經'銅范系
銅版辨.
On the Use of Bronze Plates for Printing the
Hsin Ching during the Khai-Yüan Reign
Period of the Thang Dynasty.
SHKH (no. 3), 151.

Chuang Yen (*1*) 莊嚴.
*Lei-Feng-Tha Tshang Pao Chhieh Yin Tho-Lo-Ni
Ching Pa* 雷峯塔藏寶篋印陀羅尼經跋.
Postscript on the Dharani *Sutra* Preserved in the
Thunder Peak Pagoda, Hangchow.
TSKH, 1926, **1** (no. 2), 331.

Chung Chhung-Min (*1*) 鍾崇敏.
Ssu-Chhuan Shou Kung Yeh Chih Tiao Chha
四川手工業紙調查.
A Survey of the Hand-made Paper Industry in
Szechuan.
Chungking, 1943.

Chhai Tzu-Ying (*1*) 柴子英.
*Than Shih-Chu-Chai Khan Yin ti Chi Chung Yin
Phu* 談十竹齋刊印的幾種印譜.
On the Manuals Printed by the Ten Bamboo
Studio.
WWTK, 1960 (no. 8/9), 76.

Chhang Pi-Te (*2*) 昌彼得.
Yüan Khan Pen Yen Phin Chih Chien Chi
元刊本贗品知見記.
On Forged Yüan Editions Known or Seen by
the Author.
MTPL, 1959, **10** (no. 16), 8.

Chhang Pi-Te (*3*) 昌彼得.
Thang Tai Thu Shu Hsing Chih ti Yen Pien
唐代圖書形制的演變.
The Evolution of the Physical Format of Books
in the Thang Dynasty.
TSKP, 1964, **6**, 1.

Chhang Pi-Te (*4*) 昌彼得.
*Wo Kuo Pan Pen Hsüeh Shang ti Chi Ko Yu Tai Yen
Chiu ti Kho Thi* 我國版本學上的幾個有
待研究的課題.
Some Problems Concerning Chinese Printing.
SMCK, 1966, **1** (no. 1), 3.

Chhang Pi-Te (*5*) (ed.) 昌彼得.
Ming Tai Pan Hua Hsüan 明代版畫選.
Selections of Ming Woodcuts.
2 vols.
Han-Hua, Thaipei, 1969.

Chhang Pi-Te (*6*) 昌彼得.
Pan Pen Mu Lu Hsüeh Lun Tshung 版本目錄學
論叢.
Collection of Essays on Chinese Printing and
Bibliography.
2 vols.
Hsüeh-Hai, Thaipei, 1977.

Chhang Pi-Te (*7*) 昌彼得.
Ming Fan Kho Shu Khao 明藩刻書考.
A Bibliography of Books Printed by the Ming
Princes.
HSCK, 1955, **3** (no. 3), 142; (no. 4), 139.

Chhang Pi-Te (*8*) 昌彼得.

Chhang Pi-Te (*8*) (*cont.*)
Chung-Yüeh Shu Yüan 中越書緣.
The Relationships of Books and Publishing
between China and Vietnam.
In *CYWH*, 1956, vol. 1, 180.

Chhen Chin-Po [Chan Kam-Po] (*1*) 陳錦波.
Chung-Kuo I Shu Khao Ku Lun Wen So Yin
中國藝術考古論文索引.
Chinese Art and Archaeology; a Classified Index
to Articles Published in Mainland China
Periodicals.
University of Hong Kong, 1966.

Chhen Kuo-Chhing (*1*) 陳國慶.
Ku Chi Pan Pen Chhien Shuo 古籍版本淺說.
Introduction to Old Chinese Editions.
Liao-Ning Jen-Min, Shen-Yang, 1957.

Chhen Meng-Chia (*5*) 陳夢家.
Chi Chung Chu Shu Khao 汲冢竹書考.
A Study of Bamboo Books Discovered in the Wei
Tombs in the +3rd Century.
QBCB/C, 1944, n.s. **5** (no. 2/3), 1.

Chhen Phan (*8*) 陳槃.
I Shui Chi Hsü Wei Phiao Chieh 以水繫絮爲
漂解.
On the Stirring of Refuse Silk in Water.
TLTC, 1951, **3** (no. 8), 21.

Chhen Phan (*9*) 陳槃.
Yu Ku Tai Phiao Hsü Yin Lun Tsao Chih
由古代漂絮因論造紙.
On the Invention of Paper from the Process of
Stirring Refuse Silk.
AS/A, 1954, **1**, 257.

Chhen Pheng-Nien (*1*) 陳彭年.
Kuan Yü Hsüan Chih Wen Thi 關於宣紙問題.
On the Problem of Hsüan Chih.
TCKY, 1957, **2**, 24.

Chhen Pin-Ho 陳彬龢 & Cha Meng-Chi (*1*)
查猛濟.
Chung Kuo Shu Shih 中國書史.
A History of Chinese Books.
Com. Press, Shanghai, 1935.

Chhen Ta-Chhuan (*1*) 陳大川.
Chung-Kuo Tsao Chih I Shu Sheng Shuai Shih
中國造紙藝術盛衰史.
The History of Papermaking in China.
Chung-Wai Press, Thaipei, 1979.

Chhen Ta-Tuan (*1*) 陳大端.
*Yung Chhien Chia Shih Tai Ti Chung Liu Kuan
Hsi* 雍乾嘉時代的中琉關係.
Relations between China and Liu-Chhiu during
the Yung-Cheng, Chhien-Lung and Chia-
Chhing Eras.
Thaipei, 1956.

Chhen Yüan (*4*) 陳垣.
Shih Hui Chü Li 史諱舉例.
On the Taboo Changes of Personal Names in
History; Some Examples.
Chung-Hua, Peking, 1962; repr. 1963.

Chhen Yüan (*5*) 陳垣.
Tun-huang Chieh Yü Lu 敦煌劫餘錄.
Catalogue of Tun-huang Rolls in the National
Library of Peiping.

6 vols.
Institute of History and Philology, Academia
Sinica, Peiping, 1931.

Chheng Su-Lo (*4*) 程溯洛.
*Lun Tun-Huang Thu-Lu-Fan Fa Hsien ti Meng
Yüan Shih Tai Ku Wei Wen Mu Kho Huo Tzu
Ho Tiao Pan Yin Shua Phin Yü Wo Kuo Yin Shua
Shu Hsi Chhuan ti Kuan Hsi* 論敦煌吐魯番
發現的蒙元時代古維文木刻活字和
雕版印刷品與我國印刷術西傳的
關係.
On the Wooden Movable Types in Uighur and
Block-Printing of the Yüan Dynasty dis-
covered at Tun-Huang and Turfan, and
Their Relationship to the Westward Spread of
Printing.
In *CKKC*, pp. 225–35.

Chhiao Yen-Kuan (*1*) 喬衍琯.
Wo Kuo Thao Se Yin Shua Chien Shuo
我國套色印刷簡說
Introduction to Chinese Multi-Colour Printing.
CYTSK, 1971, **4** (no. 1), 17.

Chhiao Yen-Kuan 喬衍琯 & Chang Chin-Lang
(*1*) (ed.) 張錦郎.
Thu Shu Yin Shua Fa Chan Shih Lun Wen Chi
圖書印刷發展史論文集.
Collection of Materials for History of the
Development of Books and Printing.
Wen-Shih-Che, Thaipei, 1975; Supplement,
1977.

Chhien Chi-Po (*1*) 錢基博.
Pan Pen Thung I 版本通義.
Introduction to Book Editions.
Com. Press, Shanghai, 1933.

Chhien Mu (*1*) 錢穆.
Thang Tai Tiao Pan Shu Chih Hsing Chhi
唐代雕版術之興起.
The Rise of Block Printing in the Thang
Dynasty.
TSPY, 1941, **2** (no. 18), 21.

Chhien Tshun-Hsün (*1*) 錢存訓.
Han Tai Shu Tao Khao 漢代書刀考.
A Study of the Book-Knife of the Han Dynasty.
AS/BIHP, 1961, extra vol., no. 4, 997.
Cf. J. Winkelman (*1*) (tr.)

Chhien Tshun-Hsün (*2*) 錢存訓.
Lun Ming Tai Thung Huo Tzu Pan Wen Thi
論明代銅活字版問題.
Bronze Movable Type of Ming China: Problems
of its Origin and Technology.
In *CWTM*, p. 129.

Chhien Tshun-Hsün (*3*) 錢存訓.
*Chung-Kuo tui Tsao Chih chi Yin Shua Shu ti Kung
Hsien* 中國對造紙及印刷術的貢獻.
Chinese Contributions to Papermaking and
Printing.
Tr. Ma Thai-Lai 馬泰來.
MPYK, 1972, **7** (no. 12), 2.

Chhien Tshun-Hsün (*4*) 錢存訓.
Chung-Kuo Ku Tai ti Tsao Chih Yüan Liao
中國古代的造紙原料.
Raw Materials for Old Papermaking in China.

Chhien Tshun-Hsün (4) (cont.)
Tr. Ma Thai-Lai 馬泰來.
JICC/HK, 1974, **7** (no. 1), 27.
Chhien Tshun-Hsün (5) 錢存訓.
Chung-Kuo Ku Tai Shu Shih 中國古代書史.
A History of Writing Materials in Ancient
China.
Chinese University of Hong Kong, 1975.
Cf. Japanese tr. Utsugi Akira et al. (1).
Chhien Tshun-Hsün (6) 錢存訓.
Shu Chi Wen Fang Chuang Shih Yung Chih Khao
Lüeh 書籍文房裝飾用紙考略.
Graphic and Decorative Use of Paper in China.
JICC/HK, 1978, **9** (no. 1), 83.
Chhien Tshun-Hsün (7) 錢存訓.
Chung-Kuo Ku Tai Wen Tzu Chi Lu ti I Chhan
中國古代文字記錄的遺產.
The Legacy of Early Chinese Written Records.
Tr. Chou Ning-Sheng 周寧生.
JICC/HK, 1971, **4** (no. 2), 273.
Chhien Tshun-Hsün (8) 錢存訓.
Chung-Kuo Ku Tai ti Chien Tu Chih Tu
中國古代的簡牘制度.
The System of Bamboo and Wooden Tablets in
Ancient China.
Tr. Chou Ning-Sheng 周寧生.
JICC/HK, 1973, **6** (no. 1), 45.
Chhien Tshun-Hsün (9) 錢存訓.
Yin-Kuo Chien-Chhiao Tshang Pen Chü Lu Thi
Chi 英國劍橋藏本橘錄題記.
On Dating the Edition of the Chü Lu at
Cambridge University.
CHJ/T, 1973, n.s. **10** (no. 1), 111.
Chhien Tshun-Hsün (10) 錢存訓.
Fan I tui Chung-Kuo Hsien Tai Hua ti Ying Hsiang
翻譯對中國現代化的影響.
The Impact of Translation on the Moderniza-
tion of China.
Tr. Tai Wen-Pai 戴文伯.
MPYK, 1974, **9** (no. 8), 2.
Chhien Tshun-Hsün 錢存訓. See also Tsien Tshun-
Hsuin in Bibliography C.
Chhin Chhin-Chih (1) 秦欽峙.
Hua Chhiao tui Yüeh-Nan Ching Chi Wen Hua Fa
Chan ti Kung Hsien 華僑對越南經濟文
化發展的貢獻.
Contributions of Overseas Chinese to the
Economic and Cultural Development of
Vietnam.
LSYC, 1979 (no. 6), 57.
Chhiu Hsi-Kuei (1) 裘錫圭.
Than Than Sui-Hsien Tseng Hou-I Mu ti Wen Tzu
Tzu Liao 談談隨縣曾侯乙墓的文字
資料.
Notes on the Written Documents Found in the
Tomb of Tseng Hou-I at Sui-Hsien, Hupei.
WWTK, 1979 (no. 7), 25.
Chhü Chhi-Chia (1) 瞿啓甲.
Thieh-Chhin-Thung-Chien-Lou Sung Chin Yüan Pen
Shu Ying 鐵琴銅劍樓宋金元本書影.
Facsimile Specimens of Sung, Chin, and Yüan
Printing in the Iron Guitar and Bronze Sword

Pavilion.
9 vols.
Chhangshu, 1922.
Chhu Shih-Pin 初仕賓 & Jen Pu-Yün (1)
任步雲.
Chü-Yen Han Tai I Chih ti Fa Chüeh ho Hsin Chhu
Thu ti Chien Tshe Wen Wu 居延漢代遺址
的發掘和新出土的簡册文物.
The Excavation of the Beacon Fire Site at Chü-
Yen and Inscribed Bamboo Slips of the Han
Dynasty recently Unearthed in Kansu
Province.
WWTK, 1978 (no. 1), 1.
Chhü Wan-Li 屈萬里 & Chhang Pi-Te (1)
昌彼得.
Thu Shu Pan Pen Hsüeh Yao Lüeh 圖書版本學
要略.
Fundamentals of Chinese Bibliography.
Thaipei, 1953.
Chhüan Han-Sheng (4) 全漢昇.
Yüan Tai ti Chih Pi 元代的紙幣.
Paper Money of the Yüan Dynasty.
AS/BIHP, 1948, **15**, 1.

Fang Han-Chheng (1) 方漢城.
Tsao Chih Kai Lun 造紙概論.
Introduction to Papermaking.
Commercial Press, Shanghai, 1924.
Fang Hao (3) 方豪.
Fang Hao Wen Lu 方豪文錄.
Collected Essays of Fang Hao (on Chinese-
Western Cultural Contacts).
Shang-Chih, Peiping, 1948.
Fang Hao (4) 方豪.
Liu Lo Yü Hsi Phu ti Chung-Kuo Wen Hsien
流落於西葡的中國文獻.
Old Chinese Documents Found in Spain and
Portugal.
HSCK, 1952, **1** (no. 2), 149; 1953, **1** (no. 3), 161.
Fang Hao (5) 方豪.
Ming Wan-Li Chien Ma-Ni-La Khan Yin Chih Han
Wen Shu Chi 明萬曆間馬尼拉刊印之
漢文書籍.
Earliest Chinese Books Printed in Philippines.
HTHY, 1967, **4** (no. 8), 1.
Fang Hao (6) 方豪.
Sung Tai Fo Chiao tui Chung-Kuo Yin Shua chi Tsao
Chih chih Kung Hsien 宋代佛教對
中國印刷及造紙之貢獻.
The Contributions of Buddhism on Printing and
Papermaking in the Sung Dynasty.
TLTC, 1970, **41** (no. 4), 15.
Feng Chen-Chhün (1) 馮貞羣.
Yin Fan Shih Thien-I-Ko Shu Mu Nei Pien
鄞范氏天一閣書目內編.
Catalogue of the Thien-I-Ko Library of the Fan
Family of Ningpo.
Commission for Restoration of the Thien-I-Ko,
Ningpo, 1940.
Feng Chheng-Chün (1) 馮承鈞.
Chung-Kuo Nan-Yang Chiao Thung Shih 中國
南洋交通史.

Feng Chheng-Chün (*1*) (*cont.*)
History of the Contacts of China with the South
Sea Regions.
Commercial Press, Shanghai, 1937; repr., Thai-
Phing, Hong Kong, 1963.

Feng Chheng-Chün (*2*) 馮承鈞.
Chu Fan Chih Chiao Chu 諸蕃志校注.
Comments and Notes on *Records of Foreign
Peoples.*
Commercial Press, Shanghai, 1940; repr.,
Peking, 1956.
Cf. Hirth & Rockhill (1) (tr.)

Feng Han-Chi (*2*) 馮漢驥.
*Chi Thang Yin Pen Tho-Lo-Ni Ching Chou ti Fa
Hsien* 記唐印本陀羅尼經咒的發現.
Discovery of Dharani Charms Printed in the
Thang Dynasty.
WWTK, 1957 (no. 5), 48.

Fu Chen-Lun (*1*) 傅振倫.
Tshai Ching-Chung Tsao Chih Khao 蔡敬重
造紙考.
On the Invention of Paper by Tsai Lun.
BLAC, 1933, **8** (no. 4), 1.

Fu Chen-Lun (*2*) 傅振倫.
Chung-Kuo Chih ti Fa Ming 中國紙的發明.
The Invention of Paper in China.
LSCH, 1955 (no. 8), 14.

Fujikake Shizuya (*1*) 藤懸静也.
Shina Hanga to Ukiyoe Hanga 支那版画と
浮世繪版画.
Chinese Woodblock Prints and *Ukiyoe* Prints.
Kokka, 1930, no. 459.

Fujikake Shizuya (*2*) 藤懸静也.
Shina Hanga no Ukiyoe Hanga ni Oyoboseru Eikyo
支那版画の浮世繪版画に及ぼせる
影響.
The Influence of Chinese Woodcuts upon *Ukiyoe*
Prints.
Kokka, 1931, nos. 484–486.

Goto Seikichiro (*1*) 後藤清吉郎.
Nihon no Kami 日本の紙.
Japanese Paper.
2 vols.
Bijutsu Shuppansha, Tokyo, 1958–60.

Goto Seikichiro (*2*) 後藤清吉郎.
Kami no Tabi 紙の旅.
The Travels of Paper (woodcuts of papermaking
process).
Bijutsu Shuppansha, Tokyo, 1964.
Limited ed. of 300.

Harigaya Kanekichi (*1*) *et al* 針ケ谷鐘吉.
Ukiyoe Bunken Mokuroku 浮世繪文獻目録.
Ukiyoe Bibliography (books and catalogues in
languages other than Japanese).
2 vols.
Mitō Shooku, Tokyo, 1972.

Hayashi Taisuke (*1*) 林泰輔.
Chosen no Kappanjutsu 朝鮮の活版術.
Typography of Korea.
SGZ, 1906, *17*, 3.

Higuchi Hiroshi (*1*) 樋口弘.
Chūgoku Hanga Shūsei 中國版画集成.
Collection of Chinese Woodblock Prints.
Mitō Shooku, Tokyo, 1967.
326 plates in 7 series, loose-leaf in 1 box.

Hiraoka Takeo (*1*) 平岡武夫.
Chikusatsu to Shina Kodai no Kiroku
竹冊と支那古代の記録.
Bamboo Tablets as Records of Ancient China.
TG/K, 1943, **13**, 163.

Ho Sheng-Nai (*1*) 賀聖鼐.
Chung-Kuo Yin Shua Shu Yen Ko Shih Lüeh
中國印刷術沿革史略.
A Short History of the Development of Printing
in China.
TFTC, 1928, **25** (no. 18), 59.

Hu Chih-Wei (*1*) (tr.) 胡志偉.
*Chung-Kuo Yin Shua Shu ti Fa Ming chi chhi Hsi
Chhuan* 中國印刷術的發明及其西傳.
Invention of Printing in China and its Spread
Westward.
Comm. Pr., Thaipei, 1968.
Tr. of Carter (1), 1955 ed.

Hu Chih-Wei (*2*) 胡志偉.
Tsao Chih Shu Hsi Chhuan ti Ching Kuo 造紙
術西傳的經過.
A History of the Introduction of Paper to the
West.
TLTC, 1955, **10** (no. 1), 1; **10** (no. 2), 17.

Hu Hou-Hsüan (*6*) 胡厚宣.
*Wu Shih Nien Chia Ku Wen Fa Hsien ti Tsung
Chieh* 五十年甲骨文發現的總結.
A Summary of Fifty Years' Discoveries of Shell
and Bone Inscriptions.
Commercial Press, Shanghai, 1951.

Hu Hou-Hsüan (*7*) 胡厚宣.
Wu Shih Nien Chia Ku Hsüeh Lun Chu Mu
五十年甲骨學論著目.
A Bibiliography of the Study of Shell and Bone
Inscriptions.
Chung Hua, Shanghai, 1952.

Hu Shih (*12*) 胡適.
*Lun Chhu Thang Sheng Thang Huan Mei Yu Tiao
Pan Shu* 論初唐盛唐還沒有雕版書.
On Block Printing Had Not been in Existence in
the First Half of the Thang Dynasty.
TYCK, 1959, **21** (no. 1), 7.

Hu Tao-Ching (*1*) 胡道靜.
Meng Chhi Pi Than Chiao Cheng 夢溪筆談
校證.
Complete Annotated and Collated Edition of
the *Dream Pool Essays* (of Shen Kua, +1086).
2 vols.
Shanghai Pub. Co., Shanghai, 1956.

Hu Tao-Ching (*2*) 胡道靜.
Hsin Chiao Cheng Meng Chhi Pi Than 新校
正夢溪筆談.
New Corrected Edition of the *Dream Pool Essays*
(with additional annotations).
Chung-Hua, Peking, 1958.

Hu Tao-Ching (*4*) 胡道靜.
Huo Tzu Pan Fa Ming Che Pi Sheng Tsu Nien chi

Hu Tao-Ching (*4*) (*cont.*)
 Ti Tien Shih Than 活字版發明者畢昇
 卒年及地點試探.
 On the Time and Place of Pi Sheng's Death.
 WSC, 1957 (no. 7), 61.
Hu Tao-Ching (*5*) 胡道靜.
 *Ku Chin Thu Shu Chi Chheng ti Chhing Khuang, The
 Tien chi chhi Tso Yung* 古今圖書集成的
 情況, 特點及其作用.
 A Study of the Grand Encyclopaedia: Its
 Conditions of Compilation, Specialties and
 Usefulness.
 TSK, 1962 (no. 1), 31.
Hu Yün-Yü (*1*) 胡韞玉.
 Chih Shuo 紙說.
 On Paper.
 In *Phu Hsüeh Chai Tshung Khan*, 1923.
Huang Chieh (*1*) 黃節.
 Pan Chi Khao 版籍考.
 On the Formats of Chinese Books.
 KTHP, 1980 (no. 47), 1; (no. 49), 1.
Huang Sheng-Chang (*2*) 黃盛璋.
 *Kuan Yü Chung-Kuo Chih Ho Tsao Chih Fa Chhuan
 Ju Yin Pa Tzhu Ta Lu tei Shih Chien ho Lu Hsien
 Wen Thi* 關於中國紙和造紙法傳入
 印巴次大陸的時間和路綫問題.
 On the Date and Place of Introduction of
 Chinese Paper and Papermaking Methods to
 the Sub-Continent of India and Pakistan.
 LSYC, 1980 (no. 1), 113.
Huang Tzhu-Po (*1*) 黃慈博.
 *Kuang-Tung Sung Yüan Ming Ching Chi Chhien Pen
 Chi Lüeh* 廣東宋元明經籍槧本紀略.
 On the Confucian Classics Block-Printed in
 Kuangtung from Sung to Ming.
 KTWW, 1941, **1**, 861.
Huang Wen-Pi (*1*) 黃文弼.
 Lo-Pu-Cho-Erh Khao Ku Chi 羅布淖爾
 考古記.
 Archaeology of Lopnor.
 Peking, 1948.
Huang Wen-Pi (*2*) 黃文弼.
 Thu-lu-fan Khao Ku Chi 吐魯番考古記.
 Archeology of Turfan.
 Academia Sinica, Peking, 1954.
Huang Yung-Chhüan (*1*) 黃湧泉.
 Chhen Lao-Lien Pan-Hua Hsüan Chi 陳老
 蓮版畫選集.
 Selections of Woodcuts by Chhen Hung-Shou.
 Peking, 1957.
Hung Kuang 洪光 & Huang Thien-yu (*1*)
 黃天右.
 Chung-Kuo Tsao Chih Fa Chan Shih Lüeh
 中國造紙發展史略.
 A Brief History of the Development of Paper-
 making in China.
 Chhing-Kung-Yeh Chhu-Pan-She, Peking,
 1957.
Hung Yeh (*3*) (ed.) 洪業.
 Tu Shih Yin Te 杜詩引得.
 Concordance to the Poems of Tu Fu.
 Harvard-Yenching Institute, Peiping, 1940;

reprint, Thaipei, 1966.
Hsia Nai (*2*) 夏鼐.
 Khao Ku Hsüeh Lun Wen Chi 考古學論文集.
 Collected Papers on Archaeological Subjects.
 Academia Sinica, Peking, 1961.
Hsia Nai (*7*) 夏鼐.
 *Yang-Chou La-Ting Wen Mu Pei Ho Kuang-Chou
 Wei-Ni-Ssu Yin Pi* 揚州拉丁文基碑和
 廣州威尼斯銀幣.
 Latin Tombstones of Yangchow and the
 Venetian Coins from Kuangchow.
 KKTH, 1979 (no. 6), 532.
Hsiang Ta (*3*) 向達.
 Thang Tai Chhang-An yü Hsi Yü Wen Ming
 唐代長安與西域文明.
 Western Cultures at the Chinese Capital
 (Chhang-an) during the Thang Dynasty.
 YCHP, Monograph series, no. 2, Peiping, 1933.
Hsiang Ta (I-Weng-Sheng) (*6*) 向達(覺蟫生).
 Chi Shih Chu Chai 記十竹齋.
 On the Ten Bamboo Studio.
 QBCB/C, 1935, **2** (no. 1), 39.
Hsiang Ta (*7*) 向達.
 *Chung-Kuo Yin Shua Shu chih Fa Ming chi chhi
 Chhuan Ju Ou-Chou Khao* 中國印刷術
 之發明及其傳入歐洲考.
 The Invention of Printing in China and Its
 Spread Westward.
 PPKK, 1940, 2 (no. 2), 103.
Hsiang Ta (*8*) (tr.) 向達.
 *Hsien Tshun Tsui Ku Yin Pen chi Feng Tao Tiao
 Yin Chhün Ching* 現存最古印本及馮道
 雕印羣經.
 The Printing of the Confucian Classics Under
 Feng Tao, +932 to +953.
 TSKH, 1943, **6** (no. 1), 87.
 Tr. of Carter (*1*), Ch. 9.
Hsiang Ta (*9*) (tr.) 向達.
 *Chung-Kuo Tiao Pan Yin Shua Shu Chih Chhüan
 Sheng Shih Chhi* 中國雕板印刷術之全盛
 時期.
 The High Tide of Chinese Block Printing.
 (+960–1368).
 TSKH, 1942, **5** (nos. 3–4), 367.
 Tr. of Carter (1), ch. 10.
Hsiang Ta (*10*) 向達.
 Lun Yin Chhao Pi 論印鈔幣.
 The Printing of Paper Money.
 TSKH, 1943, **6** (no. 4), 503.
 Tr. of Carter (1), ch. 11.
Hsiang Ta (*11*) (tr.) 向達.
 Thu-Lu-Fan Hui-Ho Jen Yin Shua Shu 吐魯
 番回鶻人印刷術.
 Printing of the Uighurs in Turfan.
 TSKH, 1926, **1** (no. 4), 597.
 Tr. of Carter (1), ch. 14.
Hsiang Ta (*12*) 向達.
 Kao-Li chih Huo Tzu Yin Shua Shu 高麗之
 活字印刷術.
 The Wide Use of Movable Type in Korea.
 TSKH, 1928, **2** (no. 2), 247.
 Tr. of Carter (1), ch. 23.

Hsiang Ta (*13*) 向達.
  Thang Tai Khan Shu Khao 唐代刊書考.
  Printing in the Thang dynasty.
  KHTS, 1928 (no. 1), 1.
Hsiao Han (*1*) 曉菡.
  *Chhang Sha Ma-Wang-Tui Han Mu Po Shu Kai
    Shu* 長沙馬王堆漢墓帛書概述.
  Silk Manuscripts Discovered in Han Tomb No.
    3 at Ma-Wang-Tui, Chhangsha.
  WWTK, 1974 (no. 9), 40.
Hsieh Kuo-Chen (*3*) 謝國楨.
  *Tshung Chhing Wu-Ying-Tien Pan Than tao Yang-
    Chou Shih Chü ti Kho Shu* 從清武英殿
    版談到揚州詩局的刻書.
  On Printing from the Wu-Ying-Tien to the
    Yang-Chou Poetry Bookshop.
  KKYK, 1981 (no. 1), 15.
Hsien Yü-Chhing (*1*) 冼玉清.
  Fo-Shan ti Hsi Chu Huo Tzu Pan 佛山的
    錫鑄活字版.
  The Tin Movable Type of Fo-Shan, Kuang-
    tung.
  In *KTWH*, 1965, p. 73.
Hsiung Cheng-Wen (*1*) 熊正文.
  Chih Tsai Sung Tai ti The Shu Yung Thu
    紙在宋代的特殊用途.
  The Special Usages of Paper in the Sung
    Dynasty.
  SH, 1937, **5** (no. 12), 34.
Hsü Chia-Chen (*1*) 徐家珍.
  Feng Cheng Hsiao Chi 風箏小記.
  A Note on Aeolian Whistles (attached to Kites).
  WWTK, 1959 (no. 2), 27.
Hsü Hsin-Fu (*1*) 徐信符.
  Kuang-Tung Pan Phien Chi Lüeh 廣東版
    片記略.
  Printing Blocks in Kuangtung.
  KTWW, 1941, **1**, 858.
Hsü Kho (*1*) 徐珂.
  Chhing Pai Lei Chhao 清稗類鈔.
  Classified Anecdotes in Chhing Times.
  Commercial Press, Shanghai, 1917.
Hsü Kuo-Lin (*1*) 許國霖.
  Tun-Huang Shih Shih Hsieh Ching Thi Chi
    敦煌石室寫經題記.
  Notes on Tun-Huang Manuscripts.
  Commercial Press, Shanghai, 1937.
Hsü Ming-Chhi (*1*) 許鳴岐.
  Jui-Kuang-Ssu Tha Ku Ching Chih ti Yen Chiu
    端光寺塔古經紙的研究.
  A Study of the Old Paper for Buddhist *Sutras*
    Discovered in the Pagoda of the Jui-Kuang
    Temple in Soochow.
  WWTK, 1979, (no. 11), 34.
Hsü Thung-Hsin (*1*) 許同莘.
  Hua Shih Phu Lüeh (Wu-Hsi Hua Shih Phu Pa)
    華氏譜略(無錫華氏譜跋).
  A Brief Genealogical History of Hua Sui's
    Family in Wu-Hsi, Chiangsu Province, with a
    Postface.
  PPKK, 1934, **8** (no. 4), 73.
Hsü Wei-Nan (*1*) 徐蔚南.

Chung-Kuo Mei Shu Kung I 中國美術工藝.
  Chinese Artistic Handicrafts.
  Chung-Hua, Shanghai, 1940.

I Po (*1*) 易波.
  Jung-Pao-Chai ti Mu Pan Shui Yin Hua
    榮寶齋的木版水印畫.
  The Wood-Block Prints of Jung-Pao-Chai.
  MSHU, 1955 (no. 10), 20.
Ikeda Hideo (*1*) 池田秀男.
  Washi Nempyo 和紙年表.
  A Chronology of Japanese Paper.
  Sancha Shobo, Tokyo, 1974.
Ishida Mikinosuke (*1*) 石田幹之助.
  *Pari Kaichō Kanryū Nenkan Jun Kai Ryōbu Heitei
    Tokushozu ni tsuite* パリ開雕乾隆年間準
    回兩部平定得勝圖に就て.
  A Picture of The Victory over the Chun and
    Hui Tribes Engraved in Paris during the
    Chhien-Lung Period.
  TYG, 1919, **9**, 396.
Ishida Mosaku 石田茂作 & Wada Gunichi
    (*1*) 和田軍一.
  Shōsōin 正倉院.
  The Shōsōin: an Eighth Century Treasure House.
  Mainichi Shimbun, Tokyo, 1954.
  With English summary.

Jao Tsung-I (*4*) 饒宗頤.
  *Tshung Shih Kho Lun Wu Hou chih Tsung Chiao
    Hsin Yang* 從石刻論武后之宗教信仰.
  The Religious Beliefs of Empress Wu as Seen
    from Stone Inscriptions.
  AS/BIHP, 1974, **45** (no. 3), 397.
Juan Yüan (*4*) 阮元.
  Chi-Ku-Chai Chung Ting I Chhi Khuan Shih
    積古齋鐘鼎彝器欵識.
  Inscriptions of Bronze Sacrificial Vessels from
    the Studio of Accumulated Antiques.
  1804.
Jugaku Bunshō (*1*) 壽岳文章.
  Nihon no Kami 日本の紙.
  The Papers of Japan.
  Yoshikawa Kobunkan, Tokyo, 1967.
Jugaku Bunshō (*2*) 壽岳文章.
  Washi no Tabi 和紙の旅.
  The Travels of Japanese Paper.
  Geikusadō, Tokyo, 1973.
Jung Keng (*3*) 容庚.
  Chin Wen Pien 金文編.
  Bronze Forms of Characters [Shang and Chou
    Dynasties].
  Peking, 1925; repr. 1959.
  Supplement [Chhin and Han], 1935.
Jung Keng (*4*) 容庚.
  Shang Chou I Chhi Thung Khao 商周彝器
    通考.
  A General Treatise on the Sacrificial Bronzes of
    the Shang and Chou Dynasties.
  2 vols.
  Harvard-Yenching Institute, Peiping, 1941.

Kai Chhi (*1*) 改琦.
　Hung Lou Meng Thu Yung 紅樓夢圖詠.
　Illustrations for the *Dream of the Red Chamber*.
　1884.
Kao Ming (*1*) 高明.
　Chung-Kuo Pan Pen Hsüeh Fa Fan 中國
　　版本學發凡.
　An Introduction to Chinese Bibliography.
　CCTH, 1966, **14**, 1.
Kawase Kazuma (*1*) (ed.) 川瀨一馬.
　Kyūkan Eifu 舊刊影譜.
　Collection of Facsimile Specimens of Old
　　Printing.
　Nihon Shoshigakukai, Tokyo, 1932.
Kawase Kazuma (*2*) 川瀨一馬.
　Sagabon Zukō 嵯峨本圖考.
　An Illustrated Study of Saga Editions.
　Isseidō, Tokyo, 1932.
Kawase Kazuma (*3*) 川瀨一馬.
　Kokatsujiban no Kenkyū 古活字版之研究.
　A Study of Early Movable Type Editions.
　3 vols.
　Antiquarian Booksellers Association of Japan,
　　Tokyo, 1967.
Kawase Kazuma (*4*) 川瀨一馬.
　Gozamban no Kenkyū 五山版の研究.
　A Bibliographical Study of Gozamban Editions.
　2 vols.
　Antiquarian Booksellers Association of Japan,
　　Tokyo, 1970.
Kawase Kazuma (*5*) 川瀨一馬.
　Nihon Shoshigaku Gaisetsu 日本書誌學概說.
　An Outline of Japanese Bibliography.
　Kodansha, Tokyo, 1950. Rev. ed., 1972.
Khun-Kang (*1*) (ed.) 崑崗.
　Ta Chhing Hui Tien Shih Li 大清會典事例.
　History of the Administrative Statutes of the
　　Chhing Dynasty with Illustrative Cases.
　1899; repr., Thaipei, 1963.
Kim Won-Yong (*1*) 金元龍.
　Han'guk Ko Kwaicha Kaeyo 韓國古活字
　　概要.
　Early Movable Type in Korea.
　Eul-Yu Publishing Co., Seoul, 1954.
　Text in Korean and English.
Kimiya Yasuhiko (*1*) 木宮泰彥.
　Nikka Bunka Kōryu Shi 日華文化交流史.
　A History of Cultural Relations between Japan
　　and China.
　Fuzambō, Tokyo, 1955.
　Abstr *RBS*, 1959, **2**, no. 37.
Kimiya Yasuhiko (*2*) 木宮泰彥.
　Nihon Ko Insatsu Bunka Shi 日本古印刷文
　　化史.
　A Cultural History of Old Japanese Printing.
　Fuzambō, Tokyo, 1932.
Ku Jui-Lan (*1*) 辜瑞蘭.
　Chung-Kuo Shu Yüan Khan Kho Thu Shu Khao
　　中國書院刊刻圖書考.
　Books Printed by Private Schools in China.
　CYTSK, 1976, n.s. **9** (no. 2), 27.
Kume Yasuo (*1*) 久米康生.

Washi No Bunkashi 和紙の文化史.
　A Cultural History of Japanese Paper.
　Mokuji-Sha, Tokyo, 1976.
Kume Yasuo (*2*) 久米康生.
　Shōwa Mingeishi Fu 昭和民芸紙譜.
　Specimens of Locally-made Japanese Paper of
　　the Shōwa Period.
　5 vols. Shibunkaku, Tokyo, 1977.
　Limited ed. of 500.
Kuo Liang (*1*) 郭艮.
　Chih Pi Shih Hua 紙筆史話.
　A Popular History of Paper and Brush.
　Chung-Hua Shu Chü, Hong Kong, 1957.
Kuo Mo-Jo (*13*) 郭沫若.
　Liang Chou Chin Wen Tzhu Ta Hsi 兩周金文
　　辭大系.
　The Western and Eastern Chou Dynasties.
　8 vols.
　Bunkyudo, Tokyo, 1935.
Kuo Mo-Jo (*14*) 郭沫若.
　Shih Ku Wen Yen Chiu 石鼓文研究.
　A Study of Stone Drum Inscriptions.
　Commercial Press, Chhangsha, 1940.
Kuo Mo-Jo (*15*) 郭沫若.
　Kuan yü Wan Chou Po Hua ti Khao Chha
　　關于晚周帛畫的考察.
　A Study of the Silk Painting of the Late Chou
　　Period.
　JMWH, 1953 (no. 11), 113; (no. 12), 108.
Kuo Po-Kung (*1*) 郭伯恭.
　Yung-Lo Ta Tien Khao 永樂大典考.
　A Study of the Yung-lo Encyclopaedia.
　Commercial Press, Shanghai, 1938.
Kuo Wei-Chhü (*1*) 郭味蕖.
　Chung-Kuo Pan Hua Shih Lüeh 中國版畫史略.
　A Brief History of Chinese Woodcuts.
　Chao-Hua Mei-Shu, Peking, 1962.
Kuroda Ryō (*1*) 黑田亮.
　Chōsen Kyūshoku 朝鮮舊書考.
　Studies of Old Korean Books.
　Iwanami Shoten, Tokyo, 1960.

Lai Shao-Chhi (*1*) 賴少其.
　Thao Pan Chien Thieh 套板簡帖.
　Colour Prints for Business Stationery.
　Shanghai, 1964.
Lao Kan (*7*) 勞榦.
　Lun Chung-Kuo Tsao Chih Shu Chih Yüan Shih
　　論中國造紙術之原始.
　The Invention of Paper in China.
　AS/BIHP, 1948, **19**, 489.
Lao Kan (*8*) 勞榦.
　Chü-Yen Han Chien Khao Shih 居延漢簡考釋.
　Decipherment and Critical Studies of the
　　Inscriptions on the Wooden Tablets from
　　Chü-yen.
　4 vols.
　Academia Sinica, Chhungking, 1943-44.
Lao Kan (*9*) 勞榦.
　Tun-huang chi Tun-huang ti Hsin Shih Liao
　　敦煌及敦煌的新史料.
　Tun-huang and New Historical Sources from

Lao Kan (9) (cont.)
Tun-huang.
TLTC, 1950, **1** (no. 3), 6.
Lao Kan (10) 勞榦.
Tshung Mu Chien tao Chih ti Ying Yung 從木
簡到紙的應用.
From Wooden Tablets to Paper.
Tr. Chhiao Yen-Kuan 喬衍琯.
CYTSK, 1967, n.s. **1** (no. 1), 3.
Li Chhiao-Phing 李喬苹. (1)
Chung-Kuo Hua Hsüeh Shih 中國化學史.
A History of Chemistry in China.
Commercial Press, Chhangsha, 1940; 2nd enl.
ed., Thaipei, 1955.
Li Chih-Chung (1) 李致忠.
Shan Pen Chhien Shuo 善本淺說.
An Introduction to Rare Editions.
WWTK, 1978 (no. 12), 69.
Li Chin-Hua (1) 李晉華.
Ming Tai Chhih Chuan Shu Khao 明代敕撰
書考.
Bibliography of Official Publications of the
Ming Dynasty.
Harvard-Yenching Institute, Peiping, 1932;
reprint, Thaipei, 1966.
Li Hsing-Tshai (1) 李興才.
Yin Shua Shu ti Fa Ming yü Hsi Phing Shih Ching
印刷術的發明與熹平石經.
The Hsi-Phing Stone Inscriptions of Confucian
Classics and the Invention of Printing.
CKIC, 1964 (no. 731), 6.
Li Hsüeh-Chhin 李學勤 & Li Ling (1) 李零.
Phing-shan San Chhi yü Chung-Shan-Kuo Shih ti Jo
Kan Wen Thi 平山三器與中山國史的
若干問題.
The Three Bronze Vessels of Phing-shan and
Some Problems Concerning the History of the
State of Chung-shan.
KKHP, 1979 (no. 2), 147.
Li Hua (1) 李樺.
Mu Kho Pan Hua Chi Fa Yen Chiu 木刻板
畫技法研究.
A Study of Woodcut Techniques.
Jen-Min Mei-Shu, Peking, 1954.
Li Kuang-Pi 李光璧 & Chhien Chün-Yeh (1)
(ed.) 錢君曄.
Chung-Kuo Kho Hsüeh Chi Shu Fa Ming ho Kho
Hsüeh Chi Shu Jen Wu Lun Chi 中國科學
技術發明和科學技術人物論集.
Essays on Chinese Discoveries and Inventions in
Science and Technology and on the Men
Who Made Them.
San-Lien Shu-Tien, Peking, 1955.
(Abbr. *CKKC*).
Li Kuang-Thao (1) 李光濤.
Chi Chhao-Hsien Shih Lu Chung chih Chu Tzu
記朝鮮實錄中之鑄字.
On the Casting of Type as Mentioned in the
Korean Veritable Records.
TLTC, 1968, 36 (no. 1), 6.
Li Shu-Hua (4) 李書華.
Thang Tai I Chhien Yu Wu Tiao Pan Yin Shua

唐代以前有無雕版印刷.
On Whether Printing Had Been Invented Before
the Thang Dynasty.
TLTC, 1957, **14** (no. 4), 1.
Li Shu-Hua (5) 李書華.
Yin Shua Fa Ming ti Shih Chhi Wen Thi 印刷
發明的時期問題.
On When Printing Was Invented.
TLTC, 1958, 17 (no. 5), 1; (no. 6), 12.
Li Shu-Hua (6) 李書華.
Tsai Lun Yin Shua Fa Ming ti Shih Chhi Wen
Thi 再論印刷發明的時期問題.
Further Study on when Printing was Invented.
TLTC, 1959, **18** (no. 10), 1.
Li Shu-Hua (7) 李書華.
Wu-Tai Shih Chhi ti Yin Shua 五代時期的
印刷.
Printing in the Five Dynasties.
TLTC, 1960, **21** (no. 3), 1.
Li Shu-Hua (8) 李書華.
Tun-Huang Fa Hsien Yu Nien Tai ti Yin Pen
敦煌發現有年代的印本.
Printed Books with Dates Discovered at Tun-
huang.
TLTC, 1960, **21** (no. 11), 1.
Li Shu Hua (9) 李書華.
Thang Tai Hou Chhi ti Yin Shua 唐代後
期的印刷.
The Printing of Books in the Later Half of the
Thang Dynasty.
CHJ/T, 1961, 2 (no. 2), 18.
Li Shu-Hua (10) 李書華.
Huo Tzu Pan Yin Shua ti Fa Ming 活字版
印刷的發明.
The Invention of Movable Type.
TLTC, 1962, special issue, no. 2, 117.
Li Shu-Hua (11) 李書華.
Chung-Kuo Yin Shua Shu Chhi Yüan 中國印
刷術起源.
The Origin of Chinese Printing.
New Asia College, Hongkong, 1962.
Li Shu-Hua (12) 李書華.
Tsao Chih ti Fa Ming chi chhi Chhuan Po 造紙
的發明及其傳播.
The Invention and Spread of Paper.
TLTC, 1955, **10** (no. 1), 1; (no. 2), 53.
Li Shu-Hua (13) 李書華.
Tsao Chih ti Chhuan Po chi Ku Chih ti Fa Hsien
造紙的傳播及古紙的發現.
The Spread of the Art of Papermaking and the
Discoveries of Old Paper.
HSCK, 1957, **6** (no. 2), 16; National Historical
Museum, Thaipei, 1960. (With English text).
Li Shu-Hua (14) 李書華.
Yin Chang yü Mo Tha ti Chhi Yüan chi chhi tui yü
Tiao Pan Yin Shua Fa Ming ti Ying Hsiang
印章與摹搨的起源及其對於
雕版印刷發明的影響.
The Origins of Seals and Inksqueezing and their
Influence on the Invention of Printing.
AS/BIHP, 1956, 28 (no. 1), 107.
Li Wen-Chhi (1) 李文褀.

Li Wen-Chhi (*1*) (*cont.*)
Chung-Kuo Shu Chi Chuang Ting chih Pien Chhien
中國書籍裝訂之變遷.
A Sketch of the Evolution of Chinese Book-
binding.
TSKH, 1929, **3** (no. 4), 539.
Li Wen-Chhi (*2*) 李文裿.
Pan Pen Ming Chheng Shih Lüeh 板本名稱
釋略.
Some Bibliographical Terms Explained.
TSKH, 1931, **5** (no. 1), 17.
Li Yao-Nan (*1*) 李耀南.
Chung-Kuo Shu Chuang Khao 中國書裝考.
The Evolution of Bookbinding in China.
TSKH, 1930, **4** (no. 2), 207.
Li Yen (*1*) 李埏.
Pei Sung Chhu Pi Chhi Yüan Khao 北宋楮
幣起源考.
The Origin of Paper Money in the Northern
Sung Dynasty.
CCWH, 1944, **4**, 1.
Li Yüan-Chih 李元植.
See Yi Wong-Jik.
Liang Shang-Chhun (*1*) 梁上椿.
Chung-Kuo Ku Ching Ming Wen Tshung Than
中國古鏡銘文叢譚.
Studies of Ancient Chinese Mirror Inscriptions.
TLTC, 1951, **2** (no. 3), 1; (no. 4), 18; (no. 5),
16.
Liang Shang-Chhun (*2*) 梁上椿.
Sui Thang Shih Ching chih Yen Chiu 隋唐式
鏡之研究.
Studies of Mirrors of the Sui and Thang
Dynasties.
TLTC, 1953, **6** (no. 6), 181.
Liang Tzu-Han (*1*) 梁子涵.
Jih-Pen Hsien Tshun Sung Pen Shu Lu 日本現
存宋本書錄.
Bibliography of Sung Editions Preserved in
Japan.
HSCK, 1953, **2** (no. 2), 159; 1954, **2** (no. 3), 152;
2 (no. 4), 114; **3** (no. 1), 179.
Liang Tzu-Han (*2*) 梁子涵.
Ming Tai ti Huo Tzu Yin Shu 明代的活
字印書.
Movable Type Printing in the Ming Dynasty.
TLTC, 1966, **33** (no. 6), 13; (no. 7), 30.
Liang Tzu-Han (*3*) 梁子涵.
Yüan Chhao ti Huo Tzu Pan 元朝的活字版.
On Movable Type of the Yüan Dynasty.
CPYK, 1967, **2** (no. 11), 85.
Liang Tzu-Han (*4*) 梁子涵.
*Chung-Kuo Thu Shu Pan Kho Chhi Yüan Sui Chhao
ti I Thiao Wei Cheng* 中國圖書版刻起源
隋朝的一條偽證.
A Refutation of the Theory that Chinese
Printing Originated in the Sui Dynasty.
TSKP, 1967, **8**, 131.
Liang Tzu-Han (*5*) 梁子涵.
Chien-An Yü Shih Kho Shu Khao 建安余氏
刻書考.
On the Block-printing by the Yü Family of

Chien-an, Fukien.
FCWS, 1968, **1**, 53.
Lin-Chhing (*2*) 麟慶.
Ho Kung Chhi Chü Thu Shuo 河工器具圖說.
Illustrations and Explanations of the Techniques
of Water Conservancy and Civil Engineering.
1836.
Lin Tshun-Ho (*1*) 林存和.
Fu-chien chih Chih 福建之紙.
The Paper Industry of Fukien.
Foochow.
Ling Man-Li (*1*) 凌曼立.
*Thaiwan yü Huan Thai-phing-yang ti Shu Phi Pu
Wen Hua* 台灣與環太平洋的樹皮布
文化.
Bark-cloth in Thaiwan and the Circum-Pacific
Areas.
AS/MIE, 1963 (no. 3), 211.
Ling Shun-Sheng (*7*) 凌純聲.
*Shu Phi Pu Yin Wen Thao yü Tsao Chih Yin Shua
Shu Fa Ming* 樹皮布印文陶與造紙印
刷術發明.
Bark-cloth, Impressed Pottery, and the Inven-
tions of Paper and Printing.
Inst. of Ethnology, Acad. Sin., Nankang,
Thaiwan, 1963.
Ling Shun-Sheng (*21*) 凌純聲.
*Chung-Kuo Ku Tai ti Shu Phi Pu Wen Hua yü Tsao
Chih Shu Fa Ming* 中國古代的樹皮布文
化與造紙術發明.
Bark-cloth Culture and the Invention of Paper-
making in Ancient China.
AS/BIE, 1961 (no. 11), 1.
Ling Shun-Sheng (*22*) 凌純聲.
*Sung Yüan I Hou Tsao Chhu Chhao Fa yü Shu Phi Pu
Chih ti Kuan Hsi* 宋元以後造楮鈔法與
樹皮布紙的關係.
The Relation between Bark-paper and Paper
Money Since the Sung and Yüan Dynasties.
TLTC, 1962, Special issue, no. 2, 259.
Ling Shun-Sheng (*23*) 凌純聲.
*Thang Sung I lai ti Chih Chia Chih I Chih Wei
Khao* 唐宋以來的紙甲紙衣紙幛考.
A Study of Paper Armours, Clothes and Cur-
tains of Thang, Sung and Later Dynasties.
AS/MIE, 1963 (no. 3), 69.
Ling Shung-Sheng (*24*) 凌純聲.
Pei Sung Chhu Nien ti Chin-Su Chien Khao
北宋初年的金粟牋考.
Studies of the Chin-Su Paper of Early North-
ern Sung.
AS/MIE, 1963 (no. 3), 81.
Ling Shun-Sheng (*25*) 凌純聲.
Shu Phi Pu Yin Hua yü Yin Shua Shu Fa Ming
樹皮布印花與印刷術發明.
Decorative Prints on Bark Cloth and the
Invention of Printing.
AS/BIE, 1962, **14**, 193.
Ling Shun-Sheng (*26*) 凌純聲.
*Yin Wen Thao ti Hua Wen chi Wen Tzu yü Yin
Shua Shu Fa Ming* 印文陶的花紋及文字
與印刷術發明.

Ling Shun-Sheng (*26*) (*cont.*)
Designs and Inscriptions on Impressed Pottery
and the Invention of Printing.
AS/BIE, 1963, **15**, 1.

Liu Chia-Pi (*1*) (ed.) 劉家璧.
Chung-Kuo Thu Shu Shih Tzu Liao Chi 中國圖
書史資料集.
Collection of Materials for the History of
Chinese Books.
Lung-Men, Hong Kong, 1974.

Liu Chin-Tsao (*1*) 劉錦藻.
Chhing Chhao Hsü Wen Hsien Thung Khao
清朝續文獻通考.
Supplement to the Collected Institutions of the
Chhing Dynasty.
Compiler's preface, 1921.
Com. Pr., Shanghai, 1936.
(ST).

Liu Chhien (*1*) 劉乾.
Chhien Than Hsieh Kho Pen 淺談寫刻本.
Brief Discussion of Printed Editions with
Regular-Style Calligraphy.
WWTK, 1979, (no. 11), 46.

Liu Hou-Tse (*1*) 劉厚澤.
*Chung-Kuo Shih shang chih Chih Pi yü Thung Huo
Pheng Chang* 中國史上之紙幣與通貨
膨脹.
Paper Money and Inflation in Chinese History.
PCSH, 1943, n.s. **2** (no. 1), 41.

Liu I-Cheng (*1*) 柳詒徵.
Nan Chien Shih Than 南監史談.
On the Standard Histories Printed by the
National Academy at Nanking in the Ming
Dynasty.
KHTS, 1930, **3**, 1.

Liu Jen-Chhing (*1*) 劉仁慶.
Chung-Kuo Ku Tai Tsao Chih Shih Hua 中國古
代造紙史話.
A Brief History of Papermaking in Ancient
China.
Chhing-Kung-Yeh Chhu-Pan-She, Peking,
1978.

Liu Jen-Chhing 劉仁慶 and Hu Yü-Hsi (*1*)
胡玉熹.
Wo Kuo Ku Chih ti Chhu Pu Yen Chiu 我國古紙
的初步研究.
A Preliminary Study of Ancient Chinese Papers.
WWTK, 1976 (no. 5), 74.

Liu Kuo-Chün (*1*) 劉國鈞.
Chung-Kuo Shu ti Ku Shih 中國書的故事.
The Story of the Chinese Book.
Chung-Kuo Chhing-Nien, Peking, 1955. Rev.
by Cheng Ju-Ssu 鄭如斯. 1979.

Liu Kuo-Chün (*2*) 劉國鈞.
Chung-Kuo Shu Shih Chien Pien 中國書史
簡編.
A Brief History of Chinese Books.
Kao-Teng Chiao-Yü, Peking, 1958.

Liu Kuo-Chün (*3*) 劉國鈞.
Chung-Kuo Ku Tai Shu Chi Shih Hua 中國古代
書籍史話.
A History of Books in Ancient China.

Chung-Hua, Peking, 1962.

Liu Lin-Sheng (*1*) (tr.) 劉麟生.
Chung-Kuo Yin Shua Shu Yüan Liu Shih 中國印
刷術源流史.
A History of the Inventions of Chinese Printing.
Com. Pr., Shanghai, 1938.
Abridged tr. of Carter (*1*), 1925 ed.

Liu Ming-Shu (*5*) 劉銘恕.
*Sung Tai Chhu Pan Fa chi tui Liao Chin chih Shu
Chin* 宋代出版法及對遼金之書禁.
Laws Concerning Publication during the Sung
Dynasty and concerning the Prohibition of
Books going to the Khitans and Jurchens.
BCS, 1946, **5** (no. 1), 95.

Liu Ping (*1*) 劉冰.
Chung-Kuo Chuang Ting Chien Shih 中國裝訂
簡史.
A Brief History of Chinese Bookbinding.
Han Hua, Thaipei, 1969.

Liu Thi-Chih (*1*) 劉體智.
Hsiao-Chiao-Ching-Ko Chin Wen Tha Pen
小校經閣金文拓本.
Inked Rubbings of Bronze Inscriptions in the
Liu Collection.
1935.

Liu Ya-Nung (*1*) 劉雅農.
Chuang chhih Chhien Shuo 裝襯淺說.
Introduction to Mounting.
I-Lin-She, Thaipei, 1963.

Liu Yung-Chheng (*3*) 劉永成.
*Chhien-lung Su-Chou Yüan Chhang Wu San Hsien I
Ting Chih Fang Thiao I Chang Chheng Pei*
乾隆蘇州元長吳三縣議定紙坊條
議章程碑.
Notes on the Regulations for Paper Mills Agreed
to by the Three Districts Of Yüan(-Ho),
Chhang (Chou) and Wu during the Chhien-
lung Reign-period [+1736 to +1795]).
LSYC, 1958, (no. 2), 85.

Lo Chen-Yü (*4*) 羅振玉.
Sung-Weng Chin Kao 松翁近稿.
Recent Writings of Lo Chen-Yü.
3 vols. 1925–1928.

Lo Chen-Yü (*5*) 羅振玉.
Yung-Feng Hsiang-Jen Kao 永豐鄉人稿.
Collection of Miscellaneous Writings by Lo
Chen-Yü.
6 vols. 1920.

Lo Chi (*1*) 羅濟.
Chu Lei Tsao Chih Hsüeh 竹類造紙學.
The Art of Making Bamboo Paper.
1935.

Lo Chin-Thang (*1*) 羅錦堂.
Li Tai Thu Shu Pan Pen Chih Yao 歷代圖書
版本志要.
The Evolution of Chinese Books.
National Historical Museum, Thaipei, 1958.

Lo Hsi-Chang (*1*) 羅西章.
*Shan-Hsi Sheng Fu-Feng Chung-Yen-Tshun Fa Hsien
Hsi Han Chiao Tshang Thung Chhi ho Ku
Chih* 陝西省扶風中顏村發現西漢窖
藏銅器和古紙.

Lo Hsi-Chang (*1*) (*cont.*)
  Bronze Objects and Old Paper Specimens
    Found in a Western Han Tomb at Chung-
    Yen Village, Fu Feng County, Shensi
    Province.
  WWTK, 1979, (no. 9), 17.
Lu Chhien (*1*) 盧前.
  Shu Lin Pieh Hua 書林別話.
  Separate Talks on Books.
  In *CKCP/HT*, ser. Ting, pt. 2, pp. 627–36.

Ma Heng (*1*) 馬衡.
  *Chung-Kuo Shu Chi Chih Tu Pien Chhien chih Yen
    Chiu* 中國書籍制度變遷之研究.
  A Study of the Evolution of the Chinese Book.
  TSKH, 1926, **1** (no. 2), 199.
Ma Heng (*2*) 馬衡.
  Shih Ku Wei Chhin Kho Shih Khao 石鼓爲秦
    刻石考.
  An Examination of the Theory that the 'Stone
    Drum' Inscriptions Were Made by the Chhin
    State.
  KHCK, 1923, **1**, 17.
Ma Heng (*3*) 馬衡.
  Han Shih Ching Kai Shu 漢石經概述.
  A Treatise on the Stone Classics of the Han
    Dynasty.
  KKHP, 1955 (no. 5), 1.
Ma Thai-Lai (*1*) 馬泰來.
  Mi Hsiang Chih Pao Hsiang Lü 蜜香紙抱香
    履.
  Honey-Fragrance Paper and Fragrance Sandal.
  TLTC, 1969, **38**, 199.
Ma Thai-Lai (*2*) (tr.) 馬泰來.
  *Chung-Kuo tui Tsao Chih Chi Yin Shua Shu ti Kung
    Hsien* 中國對造紙及印刷術的貢獻.
  Chinese Contributions to Papermaking and
    Printing.
  MPYK, 1972, **7** (no. 12), 2.
  Tr. of Tsien (6).
Ma Thai-Lai (*3*) (tr.) 馬泰來.
  Chung-Kao Ku Tai ti Tsao Chih Yüan Liao
    中國古代的造紙原料.
  Raw Materials for Old Papermaking in China.
  JICC/HK, 1974, **7** (no. 1), 27.
  Tr. of Tsien (8).
Ma Tzu-Yün (*1*) 馬子云.
  Chhuan Tha Chi Fa 傳拓技法.
  Techniques for Making Rubbings.
  WWTK, 1962 (no. 10), 53; (no. 11), 50.
Maema Kyōsaku (*1*) (ed). 前間恭作.
  Kosen Satsufu 古鮮册譜.
  Record of Old Korean Books.
  3 vols.
  Toyo Bunko, Tokyo, 1944–57.
Makino Zembei (*1*) 牧野善兵衞.
  Tokugawa Bakufu Jidai Shoseki Kō 德川幕府
    時代書籍考.
  A Study of Publications of the Tokugawa Period
  Shosekishō Kumiai Jimusho, Tokyo, 1912.
Mao Chhun-Hsiang (*1*) 毛春翔.
  Ku Shu Pan Pen Chhang Than 古書版本常談.

Talks on Old Chinese Books.
Chung-Hua, Peking, 1962.
Maruyama Rinpei (*1*) 丸山林平.
  Teihon Nihon Shoki 定本日本書記.
  Authentic Editions of the Records of Japan.
  4 vols.
  Kodansha, Tokyo, 1966.
Matsumoto Nobuhiro (*1*) 松本信廣.
  *Hanoi Fukkoku Kyokutō Gakuin Shozō Annanbon
    Shomoku* 河內佛國極東學院所藏安南
    本書目.
  A Catalogue of Books in Chinese by Annamese
    Authors at the École française d'Extrême-
    Orient, Hanoi.
  SG, 1934, **13**, 117.
Mizuhara Gyōei (*1*) 水原堯榮.
  Kōyaban no Kenkyū 高野板の研究.
  A Study of Koya Editions.
  Rinkō Shoten, Tokyo, 1932.
Mizutani Futō (*1*) 水谷不倒.
  Kohan Shōsetsu Sōga Shi 古版小說插畫史.
  A History of Illustrations in Old Editions of
    Novels.
  Ōokayama Shoten, Tokyo, 1935.
Mu Hsiao-Thien (*1*) 穆孝天.
  An-Hui Wen Fang Ssu Pao 安徽文房四寶.
  Four Treasures of a Scholar's Study Manu-
    factured in Anhui.
  Shanghai, 1962.
Munemasa Isoo 宗政五十緒 & Wakabayashi
  Shoii (*1*) 若林正治.
  Kinsei Kyoto Shuppan Shiryo 近世京都出版
    資料.
  Materials on Modern Publishing in Kyoto.
  Nihon Kosho Tsūshinsha, Tokyo, 1965.

Nagasawa Kikuya (*3*) 長澤規矩也.
  Wakansho no Insatsu to Sono Rekishi 和漢書
    の印刷とその歷史.
  A History of Japanese and Chinese Printing.
  Yoshikawa Kōbunkan, Tokyo, 1952.
Nagasawa Kikuya (*4*) 長澤規矩也.
  Hampon no Kantei 版本の鑑定.
  A Critical Study of Editions.
  Dai Tōkyū Kinen Bunko, Tokyo, 1960.
Nagasawa Kikuya (*5*) 長澤規矩也.
  Hampon no Kōsatsu 版本の考察.
  A Study of Editions.
  Dai Tōkyū Kinen Bunko, Tokyo, 1960.
Nagasawa Kikuya (*6*) 長澤規矩也.
  Shoshigaku Josetsu 書誌學序說.
  An Introduction to Bibliography.
  Yoshikawa Kobunkan, Tokyo, 1960.
Nagasawa Kikuya (*7*) 長澤規矩也.
  Sokanbon Kokukō Meihyo Shokō 宋刊本刻工
    名表初稿.
  Table of Blockcutters' Names in the Sung
    Printed Editions; a Preliminary Draft.
  SSGK, 1934, **2** (no. 2), 1.
Nagasawa Kikuya (*8*) 長澤規矩也.
  Genkambon Kokukō Meihyo Shokō 元刊本刻工
    名表初稿.

Nagasawa Kikuya (*8*) (*cont.*)
    Table of Blockcutter's Names in the Yüan
      Printed Editions; a Preliminary Draft.
    SSGK, 1934, **2** (no. 4), 35.
Nagasawa Kikuya (*9*) (ed.). 長澤規矩也.
    Mindai Sōzubon Zuroku 明代插圖本圖錄.
    An Illustrated Catalogue of Books of the Ming
      Period.
    Nihon Shosigakukai, Tokyo, 1962.
Nagasawa Kikuya (*10*) 長澤規矩也.
    Shoshigaku Ronkō 書誌學論考.
    Studies in Bibliography.
    Shōundō, Tokyo, 1937.
Nagasawa Kikuya (*11*) 長澤規矩也.
    Zukai Wa-Kan Insatsu Shi 圖解和漢印刷史.
    An Illustrated History of Japanese and Chinese
      Printing.
    2 vols.
    Kyūko Shoin, Tokyo, 1976.
Nagasawa Kikuya (*12*) 長澤規矩也.
    Toshogaku Sankō Zuroku 圖書學參考圖錄,
      1-4.
    An Illustrated Introduction to Japanese and
      Chinese Bibliography.
    5 vols.
    Kyuko Shoin, Tokyo, 1973-6.
Nakada Katsunosuke (*1*) 仲田勝之助.
    Ehon no Kenkyū 繪本の研究.
    A Study of Illustrated Books.
    Bijutsu Shuppansha, Tokyo, 1950.
Nakamura Fusetsu (*1*) 中村不折.
    Shinkyo to Kanshuku no Tanken 新疆と甘肅
      の探険.
    Archaeological Explorations of Sinkiang and
      Kansu.
    Yūbunkaku, Tokyo, 1934.
Nakayama Kyushirō (*1*) 中山久四郎.
    Sekai Insatsu Tsushi 世界印刷通史.
    A Comprehensive World History of Printing.
    2 vols.
    Sanshū-Sha, Tokyo, 1930.
Nawa Toshisada (*3*) 那波利貞.
    Kozurofu shi Hakken Nansō Jidai Hanga コヅロ
      フ氏発見南宋時代版画.
    A Woodcut of the Southern Sung Dynasty
      Discovered by M. Kozlov.
    SGZ, 1929, **5**, 95.
Nishikawa Yasushi (*1*) 西川寧.
    Seian Hirin 西安碑林.
    The Forest of Stone Tablets in Sian.
    Kodansha, Tokyo, 1966.

Oda Kazuma (*1*) 織田一麿.
    Ukiyōe to Sashie Geijutsu 浮世繪と挿繪藝術.
    The Art of *Ukiyōe* and Book Illustration.
    Banrikaku, Tokyo, 1931.
Okuno Hikoroku (*1*) 奧野彦六.
    Edo Jidai no Kohambon 江戶時代の古版本.
    Old Editions of the Edo Period.
    Tōyōdō, Tokyo, 1944.
Ōmichi Kōyū (*1*) 大道弘雄.
    Kamiko 紙衣.
    Paper Clothing.
    Richi Shoten, Osaka, 1955.
Ono Tadashige (*1*) 小野忠重.
    Shina Hanga Sōkō 支那版画叢考.
    Studies in Chinese Woodcuts.
    Sōrin-Sha, Tokyo, 1944.
Otani Kozui (*1*) 大谷光瑞.
    Saiiki Koko Zufu 西域考古圖譜.
    Illustrations of Archaeological Explorations in
      the Northwest Regions.
    2 pts., Kokkasha, Tokyo, 1915.

Phan Chi-Hsing (*1*) 潘吉星.
    Shih Chieh shang Tsui Tsao ti Chih Wu Hsien Wei
      Chih 世界上最早的植物纖維紙.
    The World's Earliest Specimen of Paper Made
      of Plant Fibre.
    WWTK, 1964 (no. 11), 48.
Phan Chi-Hsing (*2*) 潘吉星.
    Tun-huang Shih Shih Hsieh Ching Chih ti Yen Chiu
      敦煌石室寫經紙的研究.
    A Study of the Paper used for Copying *Sutras*
      from the Tun-huang Caves.
    WWTK, 1966 (no. 3), 39.
Phan Chi-Hsing (*3*) 潘吉星.
    Kuan yü Tsao Chih Shu ti Chhi Yüan 關于造紙
      術的起源.
    On the Origins of Papermaking.
    WWTK, 1973 (no. 9), 45.
Phan Chi-Hsing (*4*) 潘吉星.
    Hsin-chiang Chhu Thu Ku Chih Yen Chiu 新疆
      出土古紙研究.
    A Study of Old Paper Specimens Discovered in
      Sinkiang.
    WWTK, 1973 (no. 10), 52.
Phan Chi-Hsing (*5*) 潘吉星.
    Ku Kung Po Wu Yüan Tshang Jo Kan Ku Tai Fa
      Shu Yung Chih chih Yen Chiu 故宮博物院藏
      若干古代法書用紙之研究.
    A Study of Early Papers Used for the Calli-
      graphy Preserved at the Palace Museum.
    WWTK, 1975 (no. 10), 84.
Phan Chi-Hsing (*6*) 潘吉星.
    Tshung Chhu Thu Ku Chih ti Mo I Shih Yen Khan
      Han Tai Tsao Ma Chih Chi Shu 從出土
      古紙的模擬實驗看漢代造麻紙技術.
    Techniques of Making Hemp Paper in the Han
      Dynasty as Observed in Experiments
      Imitating Old Papers Discovered at Ancient
      Sites.
    WWTK, 1977 (no. 1), 51.
Phan Chi-Hsing (*7*) 潘吉星.
    Chung-Kuo Ku Tai Chia Kung Chih Shih Chung
      中國古代加工紙十種.
    On Ten Kinds of Processed Paper in Ancient
      China.
    WWTK, 1979 (no. 2), 38.
Phan Chi-Hsing (*8*) 潘吉星.
    Than Shih Chieh shang Tsui Tsao ti Chih Wu Hsien
      Wei Chih. 談世界上最早的植物纖
      維紙.
    On the World's Earliest Specimen of Paper

Phan Chi-Hsing (8) (cont.)
  Made of Plant Fibres.
  HHTP, 1974 (no. 5), 45.
Phan Chi-Hsing (9)   潘吉星.
  Chung-Kuo Tsao Chih Chi Shu Shih Kao   中國
    造紙技術史稿.
  History of Chinese Papermaking Technology; a
    Draft.
  Wen Wu, Peking, 1979.
Phan Chi-Hsing (10)   潘吉星.
  *Than Han-Tan-Pho Tung-Han mu Chhu Thu ti Ma
    Chih*   談旱灘坡東漢墓出土的麻紙.
  On the Hemp Paper Unearthed from an
    Eastern Han Tomb at Han-Tan-Pho, Kansu
    Province.
  WWTK, 1977 (no. 1), 62.
Phan Chi-Hsing (11)   潘吉星.
  *Hsi Khan Chung-Yen-Tshun Hsi-Han Chiao Tshang
    Chhu Thu ti Ma Chih*   喜看中顏村西漢窖
    藏出土的麻紙.
  Viewing with Relish the Hemp Paper of
    Western Han Unearthed from a Vault at
    Chung-Yen Village, Fufeng, Shensi Province.
  WWTK 1979 (no. 9), 21.
Phan Chi-Hsing (12)   潘吉星.
  Hsi-Han Ma Chih Pu Jung Fou Ting   西漢麻紙
    不容否定.
  Paper Made of Hemp in Western Han Cannot
    Be Denied.
  CHTH/S, 1980 (no. 4), 69.
Phan Chheng-Pi   潘承弼 & Ku Thing-Lung
    (1)   顧廷龍.
  Ming Tai Pan Pen Thu Lu   明代版本圖錄.
  Facsimile Specimens of Ming Editions.
  4 vols.
  Shanghai, 1941.
Phan Mei-Yüeh (1)   潘美月.
  Liang Sung Shu Kho ti The Se   兩宋蜀刻的
    特色.
  Special Features of Szechuan Printing during
    the Northern and Southern Sung Dynasties.
  CYTSK, 1976, **9** (no. 2), 45.
Phan Ming-Shen (1)   潘銘燊.
  Sung Tai Ssu Chia Tshang Shu Khao   宋代私家
    藏書考.
  History of Private Collections in the Sung
    Dynasty.
  HK, 1971, **6**, 201.
Phan Ming-Shen (2)   潘銘燊.
  Sung Kho Shu Khan Chi chih Yen Chiu   宋刻書
    刊記之研究.
  A Study of Printers' Colophons in the Sung
    Dynasty.
  CCB, 1974 (no. 56), 35.
Phan Thien-Chen (1)   潘天禎.
  *Ming Tai Wu-hsi Hui-Thung-Kuan Yin Shu Shih
    Hsi Huo Tzu*   明代無錫會通館印書是
    錫活字.
  Tin Movable Type Used for Printing by the
    Hui-Thung-Kuan of Wu-hsi in the Ming
    Dynasty.
  TSKTH, 1980 (no. 1), 51.

Pheng Hsin-Wei (1)   彭信威.
  Chung-Kuo Huo Pi Shih   中國貨幣史.
  A History of Chinese Currency.
  Jen-Min Chhu-Pan-She, Shanghai, 1958.
Pheng Tse-I (1)   彭澤益.
  *Chung Kuo Chin Tai Shou Kung Yeh Shih Tzu Liao,
    1840–1949*   中國近代手工業史資料.
  Materials for the Study of the History of the
    Chinese Handicraft Industry between 1840
    and 1949.
  4 vols.
  San-Lien, Peking, 1957.
Po-Hsi-Ho [P. Pelliot]   伯希和 & Lu Hsiang
    (1)   陸翔.
  *Pa-Li Thu Shu Kuan Tun-huang Hsieh Pen Shu
    Mu*   巴黎圖書館敦煌寫本書目.
  Catalogue of Tun-huang Manuscripts in the
    Bibliothèque Nationale, Paris.
  PPKK, 1933, **7** (no. 6), 21; 1934, **8** (no. 1), 37.
Poon Ming-Sun   潘銘燊.
  See Phan Ming-Shen.

Seki Yoshikuni (1)   関義城.
  Kokon Tōa Shifu   古今東亞紙譜.
  An Album of Ancient and Modern Paper
    Specimens of East Asia.
  2 vols.
  Tokyo, 1957.
Seki Yoshikuni (2)   関義城.
  Wa-Kan Shi Bunken Ruijū   和漢紙文献類
    聚.
  Classified Documents on Japanese and Chinese
    Papers.
  2 vols.
  Shibunkaku, Kyoto, 1973–6.
Seki Yoshikuni (3)   関義城.
  Kokon Kamisuki Kamiya Zue   古今紙漉紙屋
    圖繪.
  Collection of Pictures on Papermaking and
    Paper Shops in Ancient and Modern Times.
  Mokuji Sha, Tokyo, 1975.
  With English Captions and Preface.
Seki Yoshikuni (4)   関義城.
  Tesukigamishi no Kenkyū   手漉紙史の研究.
  A Study of the History of Handmade Paper.
  Mokuji Sha, Tokyo, 1976.
Seki Yoshikuni (5)   関義城.
  Edo Meiji Kamiya to sono Kōkoku Zushū   江戶明
    治紙屋とその広告図集.
  A Collection of Pictures on Advertisements of
    Paper Shops of the Edo and Meiji Periods.
  2 vols.
  Privately Printed, Tokyo, 1966–7.
Shan-Yin   善因.
  See Yao Tshung-Wu.
Shang-I (1)   尚義.
  Yüeh-Nan Shu Chien   越南書簡.
  Report on Vietnam.
  Jen Min Chhu Pan She, Shanghai, 1957.
Shen Chih-Yü (1)   沈之瑜.
  Chien Chih Than Yüan   剪紙探源.
  The Origins of Paper-cutting.

Shen Chih-Yü (*I*) (*cont.*)
WWTK, 1957 (no. 8), 13.

Shen Chih-Yü (*2*) 沈之瑜.
Pa Lo-hsüan Pien Ku Chien Phu 跋蘿軒變
古箋譜.
A Postcript to the Collection of Letter Papers
with Antique and New Designs from the
Wisteria Pavilion.
WWTK, 1964 (no. 7), 7.

Shen Hsieh-Yüan (*I*) 沈爕元.
Ming Tai Chiangsu Kho Shu Shih Yeh Kai Shu
明代江蘇刻書事業概述.
Printing Industry of Chiangsu Province during
the Ming Dynasty.
HSYK, 1957 (no. 9), 78.

Shibui Kiyoshi (*I*) 澁井清.
Edo no Hanga 江戶の板畫.
Woodcut Prints of the Edo Period.
Tōgensha, Tokyo, 1965.

Shibue Zenzen et al (*I*) 澁江全善.
Keiseki Hōko Shi 經籍訪古志.
Bibliographical Notes on Old Chinese Books.
Reprint from Autograph, 1935.

Shih Chih-Lien (*I*) 石志廉.
Pei-Sung Jen Hsiang Tiao Pan Erh Li 北宋人
像雕版二例.
Two Examples of Woodblock Carving of
Human Figures in Northern Sung.
WWTK, 1981 (no. 3), 70.

Shih Hung-Pao (*I*) 施鴻保.
Min Tsa Chi 閩雜記.
Miscellaneous Records of Fukien.
Min-Yüeh, Thaipei, 1968.

Shih Ku-Feng (*I*) 石谷風.
Than Sung Tai I Chhien ti Tsao Chih Shu 談宋代
以前的造紙術.
The Art of Paper-making Before and During the
Sung Dynasty.
WWTK, 1959 (no. 1), 33.

Shih Mei-Tshen (*I*) 史梅岑.
Chung-Kuo Yin Shua Fa Chan Shih 中國印刷
發展史.
A History of Chinese Printing.
Com. Press, Thaipei, 1966.

Shih Mei-Tshen (*2*) 史梅岑.
Yin Shua Shu ti Yüan Liu Fa Chan 印刷術的
源流發展.
The Origin and Development of Printing.
ISHP, 1966, **1**, 243.

Shih Mei-Tshen (*3*) 史梅岑.
*Chih Chang Fa Ming Hou ti Chhuan Po chi chhi Wen
Hua Kung Yung* 紙張發明後的傳播及其
文化功用.
The Spread of Paper and its Effects on Civili-
zation.
ISHP, 1968, **3**, 1.

Shih Mei-Tshen (*4*) 史梅岑.
*Chung-Kuo Yin Shua ti Yen Chin chi Hsien Tai Kai
Khuang* 中國印刷的演進及現代概況.
The Evolution and Present Condition of Chinese
Printing.
ISHP, 1969, **4**, 52.

Shih Sheng-Han (*I*) 石聲漢.
Chhi Min Yao Shu Chin Shih 齊民要術今釋.
New Commentaries on the *Chhi Min Yao Shu*.
4 vols.
Science Press, Peking, 1957–58.

Shih Tao-Kang (*I*) 師道剛.
*Shui Wen Chih Chih Chheng Nien Tai Wen Thi
Chih I* 水紋紙製成年代問題質疑.
Inquiries into the Dating of Watermarks on
Chinese Paper.
SHTH, 1981 (no. 1), 51.

Shimada Kan (*I*) 島田翰.
Kobun Kyūsho Kō 古文舊書考.
Studies of Old Chinese Books.
Minyū-Sha, Tokyo, 1905.

Shimonaka Yasaburo 下中彌三郎 & Shi-
monaka Kunihiko (*I*) (eds.) 下中邦彦.
Shodō Zenshū 書道全集.
Collected Specimens of Chinese and Japanese
Calligraphy.
25 vols.
Heibon-Sha, Tokyo, 1954–61.

Shōji Sensui (*I*) 庄司淺水.
Shoseki Sōtei no Rekishi to Jissai 書籍裝訂の
歷史と實際.
The Art of Bookbinding; its History and Craft.
Gloria Society, Tokyo, 1929.

Shu-Chhao (*I*) 書巢.
Chi Mo Shu Ssu Chung 記墨書四種.
On Four Books on Ink.
WWTK, 1979 (no. 6), 72.

Shu-Hsüeh (*I*) 舒學.
*Wo Kuo Ku Tai Chu Mu Chien Fa Hsien Chhu Thu
Chhing Khuang (Tzu Liao)* 我國古代竹木
簡發現出土情況 (資料).
A Survey of the Discoveries of Bamboo and
Wooden Tablets of Ancient China; sources.
WWTK, 1978 (no. 1), 44.

Shu-Ying 叔英.
See Chi Shu-Ying.

Sogabe Shizuo (*I*) 曾我部靜雄.
Shihei Hattatsu Shi 紙幣發達史.
The Development of Paper Money.
Insatsu-Chō, Tokyo, 1945.

Sorimachi Shigeo (*I*) (ed.) 反町茂雄.
Washi Kankei Bunken Mokuroku 和紙關係文
獻目錄.
Bibliography of Sources relating to Japanese
Paper (Collected by Frank Hawley).
Kōbunso, Tokyo, 1962.

Su Pai (*I*) 宿白.
Nan-Sung ti Tiao Pan Yin Shua 南宋的雕版
印刷.
Block Printing of the Southern Sung.
WWTK, 1962 (no. 1), 15.

Su Pai (*2*) 宿白.
Chao-Chheng Chin Tsang yü Hung-Fa Tsang
趙城金藏與弘法藏.
The Chao-Chheng Edition of the Chinese
Tripitaka of the Chin Dynasty and the Hung-
Fa Edition of the Chinese Tripitaka.
HTFH, 1964 (no. 2), 13.

Su Pai (3) 宿白.
Thang Wu-Tai Shih Chhi Tiao Pan Yin Shua Shou
Kung Yeh ti Fa Chan 唐五代時期雕版印
刷手工業的發展.
Block Printing and its Development in the
Thang and Five Dynasties.
WWTK, 1981 (no. 5), 65.

Su Ying-Hui (1) 蘇瑩輝.
Tun-huang So Chhu Pei Wei Hsieh Pen Li Jih
敦煌所出北魏寫本曆日.
A Hand-copied Calendar of the Northern Wei
Dynasty from Tun-huang.
TLTC, 1950, 1 (no. 9), 4, 8.

Su Ying-Hui (2) 蘇瑩輝.
Ho-ti Khao 赫蹏考.
What Was Ho-Ti?
TLTC, 1967, 34 (no. 11), 333.

Su Ying-Hui (3) 蘇瑩輝.
Lun Tshai Hou Chih I Chhien chih Chih 論蔡侯
紙以前之紙.
Paper before Tshai Lun.
TSKP, 1967, 8, 81.

Su Ying-Hui (4) 蘇瑩輝.
Tun-huang Hsüeh Kai Yao 敦煌學概要.
An Introduction to Tun-huang Studies.
Thaipei, 1964.

Su Ying-Hui (5) 蘇瑩輝.
Tiao Pan Yin Shu Pu Shih yü Chhu Thang Lun
雕版印書不始於初唐論.
Wood-block Printing Did Not Originate in the
Early Thang Dynasty.
CYTSK, 1970, 3 (no. 2), 28.

Su Ying-Hui (6) 蘇瑩輝.
Lun Chin Shih Kho Hua tui Tiao Pan Yin Shu ti Chhi
Fa 論金石刻畫對雕版印書的啓發.
A Discussion of the Influence of Carving
Pictures on Metals and Stone upon Block-
printing.
TLTC, 1964, 29 (nos. 10/11), 111.

Su Ying-Hui (7) 蘇瑩輝.
Tshung Tsao Chhi Wen Tzu Liu Chhuan ti Kung Chü
Than tao Chung-Kuo Thu Shu ti Hsing Chheng
從早期文字流傳的工具談到中國圖
書的形成.
From the Ancient Tools of Writing to the Forms
of Earlier Chinese Books.
TSKP, 1965, 7, 23.

Su Ying-Hui (8) 蘇瑩輝.
Lun Thung Chhi Ming Wen Wei Shih Kho Hang Ko
chi Chiao Ni Huo Tzu chih Hsien Tao 論銅器
銘文爲石刻行格及膠泥活字之先導.
On Bronze Inscriptions as Forerunner of Stone
Carving and Earthenware movable Type.
KKCK, 1969, 3 (no. 3), 19.

Sun Pao-Ming 孫寶明 & Li Chung-Khai (1)
李仲凱.
Chung-Kuo Tsao Chih Chih Wu Yüan Liao Chih
中國造紙植物原料志.
Plant Materials for Papermaking in China.
Chhing-Kung-Yeh Chhu-Pan-She, Peking,
1959.

Sun Tien-Chhi (1) 孫殿起.

Liu-li-Chhang Hsiao Chih 琉璃廠小志.
Notes on the Liu-li-chhang District of Peking.
Pei-Ching Chhu-Pan-She, Peking, 1962.

Sun Yü-Hsiu (1) 孫毓修.
Chung-Kuo Tiao Pan Yüan Liu Khao 中國雕版
源流考.
The Development of Chinese Printing.
Com. Press; Shanghai, 1916, 1930.

Tabuchi Masao (1) 田淵正雄.
Shindai Mokukatsujiban Insatsu Gijutsu Kō 清代
木活字版印刷技術考.
Techniques of Wooden Movable Type Printing
of the Chhing Dynasty.
Biblia, 1980 (no. 75), 434.

Taga Akigorō (2) 多賀秋五郎.
Sōfu no Kenkyū 宗譜の研究.
An Analytical Study of Chinese Genealogical
Books.
Tōyō Bunko, Tokyo, 1960.

Tai I-Hsüan (1) 戴裔煊.
Chih Pi Yin Shua Khao 紙幣印刷考.
A History of Paper Money Printing.
HTSH, 1933, 1 (no. 4), 206.

Takakusu Junjirō 高楠順次郎 & Watanabe
Kaigyoku (1) (ed.) 渡邊海旭.
Taishō Shinshū Daizōkyō 大正新修大藏經.
Tripitaka Compiled during the Taishō Period.
85 vols.
Tokyo, 1924–35.
Appended with illustrations, 12 vols.; catalogues
of Buddhist literature, 3 vols.; extracts &
indexes, 3 vols.

Takeda Etsudō (1) 竹田悅堂.
Washi Yōroku 和紙要錄.
Notes on Japanese Paper.
Bunkaidō, Tokyo, 1966.

Takeo Eiichi (1) 竹尾榮一.
Tesuki Washi 手漉和紙.
Japanese Handmade Papers.
Takeo Co., Tokyo, 1969.

Takeo Eiichi (2) 竹尾榮一.
Kami 紙.
Handmade Papers of the World.
Takeo Co., Tokyo, 1979.
Text in Japanese and English with specimens in
1 box.

Tanaka Chikayoshi et al. (1) 田中親美.
Nihon no Kogei: Kami 日本の工芸：紙
Japanese Handicrafts: Paper.
Tanōka, Tokyo, 1966.

Tanaka Kei (1) 田中敬.
Detchō Kō 粘葉考.
A Study of Traditional Bookbinding.
Ganshōdō, Tokyo, 1932.

Tanaka Kei (2) 田中敬.
Toshogaku Gairon 圖書學概論.
An Introduction to the History of Books.
Fuzambō, Tokyo, 1924.

Tang Shou-Shan (1) 黨壽山.
Kan-Su Sheng Wu-wei Han-than-pho Tung-Han Mu
Fa Hsien Ku Chih 甘肅省武威旱灘坡東

Tang Shou-Shan (*1*) (*cont.*)
漢墓發現古紙.
Ancient Paper Discovered at Han-than-pho,
Wu-wei, Kan-Su Province.
WWTK, 1977 (no. 1), 59.

Teng Chih-Chheng (*1*) 鄧之誠.
Ku Tung So Chi 骨董瑣記; *Hsü Chi* 續記;
San Chi 三記.
Notes on Antiques and Cultural Objects, ser.
1–3.
Peking, 1933.

Teng Ssu-Yü (*1*) 鄧嗣禹.
Chung-Kuo Khao Shih Chih Tu Shih 中國考試
制度史.
The History of the Chinese Examination
System.
Nanking, 1936; reprint, Thaipei, 1966.

Teng Ssu-Yü (*2*) 鄧嗣禹.
Chung-Kuo Yin Shua Shu Chih Fa Ming chi chhi Hsi
Chhuan (Shu Phing) 中國印刷術之發明
及其西傳(書評).
The Invention of Printing in China and its
Spread Westward; a Review.
TSPL, 1934, **2** (no. 11), 35.

Teng Yen-Lin (*1*) (tr.) 鄧衍林.
Sung Yüan Khan Pen Kho Kung Ming Piao Chhu Khao
宋元刊本刻工名表初稿.
A Table of Blockcutters' Names in Sung and
Yüan Printed Editions; a Draft.
TSKH, 1934, **8** (no. 3), 451.
Cf. Nagasawa Kikuya (*5*, *6*).

Thai Ching-Lung (*1*) 臺靜農.
Than Hsieh Ching Sheng 談寫經生.
On Professional Copyists of the Buddhist *Sutras*.
TLTC, 1950, **1** (no. 9), 9.

Than Tan-Chiung (*3*) 譚旦冏.
Chung-Hua Min Chien Kung I Thu Shuo 中華民
間工藝圖說.
Illustrated Description of Chinese Folk
Handicrafts.
Chung-Hua Tshung-Shu Wei-Yüan-Hui,
Thaipei, 1956.

Thang Lan (*4*) 唐蘭.
Shih Ku Wen Nien Tai Khao 石鼓文
年代考.
On the Dating of the Stone Drum Inscriptions.
KKYK, 1958 (no. 1), 4.

Thang Ling-Ko (*1*) 唐凌閣.
Yung Chih Wen Thi Ti Yen Chiu 用紙
問題的研究.
A Study of the Uses of Paper.
TFTC, 1935, **32** (no. 18), 75.

Thang Ling-Ko (*2*) 唐凌閣.
Chung-Kuo Mo Chih Yen Chiu 中國墨之研究.
A Study of Chinese Ink.
TFTC, 1937, **34** (no. 11), 61.

Thao Hsiang (*1*) 陶湘.
Min Pan Shu Mu 閔板書目.
A Catalogue of Colour Printings by the Min and
Other Families.
Chhangchow, 1933.

Thao Hsiang (*2*) 陶湘.

She-Yüan So Chien Sung Pan Shu Ying 涉園所
見宋版書影.
Facsimile Specimens of Sung Editions in the
She-Yüan Collection, series 1–2.
2 vols.
Chhangchow, 1937.

Thao Hsiang (*3*) (ed.) 陶湘.
She-Yüan Mo Tshui 涉園墨萃.
Books on Ink in the She-Yüan Collection.
Chhangchow, 1927–9.

Thao Hsiang (*4*) 陶湘.
Wu-Ying-Tien Chü Chen Pan Tshung Shu Mu
Lu 武英殿聚珍版叢書目錄.
List of Titles in the Wu-Ying-Tien Collection.
TSKH, 1929, **3** (nos. 1–2), 205.

Thao Hsiang (*5*) 陶湘.
Ku Kung Tien Pen Shu Khu Hsien Tshun Mu
故宮殿本書庫現存目.
Extant Titles in the Collection of Palace
Editions in the Palace Museum.
Peiping, 1933.

Thao Jan (*1*) 陶然.
Chung-Kuo Huo Tzu Pan Khao 中國活字版考.
On Chinese Movable Type.
KHCK, 1926, **1** (no. 1), 45.

Theng Ku (*1*) (ed.) 滕固.
Chung-Kuo I Shu Lun Tshung 中國藝術論叢.
Collected Essays on Chinese Art.
Com. Press, Shanghai, 1938.

Thien Yeh (*1*) 田野.
Shan-Hsi Sheng Pa-chhiao Fa Hsien Hsi-Han ti
Chih 陝西省壩橋發現西漢的紙.
Discovery of C/Han Paper at Pa-Chhiao, Shensi
Province.
WWTK, 1957 (no. 7), 78.

Ting Fu-Pao (*1*) 丁福保.
Shuo Wen Chieh Tzu Ku Lin 說文解字詁林.
Collected Commentaries on the Analytical
Dictionary of Characters.
Medical Bookstore, Shanghai, 1930.

Tokushi Yūshō (*1*) 禿氏祐祥.
Tōyō Insatsu-shi Josetsu 東洋印刷史序說.
Introduction to the History of Oriental Printing.
Heirakuji Shoten, Kyoto, 1951.

Tominaga Makita (*1*) 富永牧太.
Kirishitanban no Kenkyū きりしたん版の
研究.
A Study of Christian Publications.
Tenri Daigaku Shuppan-bu, Tenri, 1973.

Tomori Soshinan (*1*) 外守素心庵.
Tōboku Waboku Zusetsu 唐墨和墨圖說.
Illustrated Description of Chinese and Japanese
Ink-cakes.
Bijutsu Shuppansha, Tokyo, 1953.

Tshai Chi-Hsiang (*1*) 蔡季襄.
Wan Chou Tseng Shu Khao Cheng 晚周繪書考
證.
A Critical Study of the Silk Document of the
Late Chou Dynasty.
I Wen Printing Co., Thaipei, 1944.

Tsien Tsuen-Hsuin 錢存訓.
See Chhien Tshun-Hsün.

Tu Po-Chhiu [J. P. Dubosc] 杜伯秋 & Fu Hsi-Hua 傅惜華 (*1*) (ed.).
Ming Tai Pan Hua Shu Chi Chan Lan Mu Lu 明代版畫書籍展覽目錄.
Exposition d'Ouvrages Illustrés de la Dynastie Ming.
Centre franco-chinois d'études sinologiques, Peking, 1944.
Tung Pho (*1*) 佟坡.
Min Chien Chhuang Hua 民間窗花.
Chinese Folk Lattice Designs.
Jen-Min Mei-Shu, Peking, 1954.
Tung Tso-Pin (*12*) 董作賓.
Chia Ku Wen Tuan Tai Yen Chiu Li 甲骨文斷代研究例.
On the Dating of Shell and Bone Inscriptions.
In *TYPM*, I, pp. 323–418.

Ueda Juzō (*1*) 植田壽藏.
Nansō no Hanga 南宋の版画.
Woodblock Prints of the Southern Sung.
GB, 1916, **7** (no. 2), 119.
Uemura Rokurō (*1*) 上村六郎.
Shina Kodai no Seishi Genryō 支那古代の製紙原料.
On the Materials for Paper Manufacture in Ancient China.
WSKK, 1951, **14**, 15.
Uemura Rokurō (*2*) 上村六郎.
Shina no Kodai Senshoku to Honzōgaku 支那の古代染色と本草学.
The Dyeing Art and Materia Medica in Ancient China.
KBK, 1952, **1**, 9.
Uesato Shunsei (*1*) 上里春生.
Edo Shosekishō Shi 江戶書籍商史.
A History of Edo Booksellers.
Shuppan Taimusu-sha, Tokyo & Osaka, 1930.
Utsugi Akira *et al.* (*1*) (tr.) 宇都木章.
Chūgoku Kodai Shoseki Shi 中國古代書籍史.
A History of Books and Inscriptions in Ancient China.
Hosei University Press, Tokyo, 1980.
Tr. of Tsien (1).

Wada Mankichi (*1*) 和田萬吉.
Kokatsujibon Kenkyū Shiryō 古活字本研究資料.
Materials for the Study of Chinese and Japanese Old Movable Type Editions.
Seikan-sha, Kyoto, 1944.
Wada Tsunashirō (*1*) 和田維四郎.
Sagabon Kō 嵯峨本考.
A Study of Saga Editions.
Shimbi Shoin, Tokyo, 1916.
Wan Ssu-Nien (*1*) 萬斯年.
Thang Tai Wen Hsien Tshung Khao 唐代文獻叢考.
Studies on Some Literary Relics of the Thang Dynasty.
Com. Press, Shanghai, 1957.
Wang Ching-Ju (*1*) 王靜如.

Ho-Hsi Tzu Tsang Ching Tiao Pan Khao 河西字藏經雕版考.
On the Hsi-Hsia Tripitaka.
In *Shishiah* (*Hsi-Hsia*) *Studies*, vol. 1, pt. 1.
Acad. Sin., Peiping, 1932.
Wang Ching-Ju (*2*) 王靜如.
Hsi-Hsia Wen Mu Huo Tzu Pan Fo Ching yü Thung Phai 西夏文木活字版佛經與銅牌.
Buddhist *Sutras* Printed with Wooden Movable Type and Bronze Plate in the Tungut Language.
WWTK, 1972 (no. 11), 8.
Wang Chü-Hua 王菊華 & Li Yü-Hua (*1*) 李玉華.
Tshung Chi Chung Han Chih ti Fen Hsi Chien Ting Shih Lun Wo Kuo Tsao Chih Shu ti Fa Ming 從幾種漢紙的分析鑒定試論我國造紙術的發明.
On the Invention of Papermaking Based on the Analysis and Study of Paper Specimens of the Han Dynasty.
WWTK, 1980 (no. 1), 78.
Wang Chung-Min (*1*) 王重民.
Shuo Chuang Huang 說裝潢.
On Paper Dyeing.
TSKH, 1931, **5**, 39.
Wang Chung-Min (*2*) 王重民.
An Kuo Chuan 安國傳.
Biography of An Kuo.
QBCB/C, 1948, n.s. **9** (no. 1/2), 22.
Wang Fang-Chung (*1*) 王方中.
Sung Tai Min Ying Shou Kung Yeh ti She Hui Ching Chi Hsing Chih 宋代民營手工業的社會經濟性質.
The Social-Economic Nature of Private Handicraft Industries during the Sung Dynasty.
LSYC, 1959 (no. 2), 39.
Wang Feng-Ku (*1*) 王豐穀.
Wo Kuo Li Tai Shu Pan chih Yen Chin 我國歷代書版之演進.
The Development of Chinese Printing During the Successive Dynasties.
CPYK, 1965 (no. 3), 7.
Wang Hung-Yüan (*1*) 王紅元.
San Shih Nien Lai ti Khao Ku Fa Hsien yü Shu Shih Yen Chiu 三十年來的考古發現與書史研究.
Archaeological Discoveries and Studies in Chinese Calligraphy during the Past 30 Years.
WH, 1979 (no. 1), 283.
Wang Kuo-Wei (*1a*) 王國維.
Hai-ning Wang Ching-An Hsien Sheng I Shu 海寧王靜安先生遺書.
Collected Works of Wang Kuo-Wei.
48 vols., 1936.
(Abbr. *HLW*).
Wang Kuo-Wei (*3*) 王國維.
Hsien-Te Khan Pen Pao Chhieh Yin Tho-Lo-Ni Ching Pa 顯德刊本寶篋印陀羅尼經跋.
Postscript to an Edition of a Dharani *Sutra* Printed in the Hsien-Te Period (+954–60).

Wang Kuo-Wei (*3*) (*cont.*)
  In *HLW*, vol. 9.
Wang Kuo-Wei (*4*) 王國維.
  Wu-Tai Kho Pen Pao-Chhieh Yin Tho-Lo-Ni Ching Pa 五代刻本寶篋印陀羅尼經跋.
  Postscript to an Edition of a Dharani *Sūtra* Printed in the Five Dynasties.
  In *HLW*, vol. 12.
Wang Kuo-Wei (*5*) 王國維.
  Wu-Tai Liang Sung Chien Pen Khao 五代兩宋監本考.
  On Books Printed by the Government during the Five Dynasties and the Sung Dynasty.
  In *HLW*, vol. 33.
Wang Kuo-Wei (*6*) 王國維.
  Liang Che Ku Khan Pen Khao 兩浙古刊本考.
  On Early Editions of Books Printed in the Chiangsu-Chekiang Area.
  In *HLW*, vol. 34-5.
Wang Kuo-Wei (*7*) 王國維.
  Chin Khai-Yün Kho Phi-Sha-Men Thien Wang Hsiang Pa 晉開運刻毘沙門天王像跋.
  Postscript on an Icon of Vaisravana Printed in the Khai-Yün Period (+944-6).
  In *HLW*, vol. 12.
Wang Kuo-Wei (*8*) 王國維.
  Yüan Khan Pen Hsi-Hsia Wen Hua-Yen Ching Tshan Chüan Pa 元刊本西夏文華嚴經殘卷跋.
  Postscript on an Incomplete Edition of Buddhāvatamsakam in the Hsi-Hsia Language Printed in the Yüan Dynasty.
  In *HLW*, vol. 9.
Wang Kuo-Wei (*9*) 王國維.
  Shih Shih 釋史.
  An Interpretation of the Character *Shih*.
  In *HLW*, vol. 3.
Wang Kuo-Wei (*10*) 王國維.
  Wei Shih Ching Khao 魏石經考.
  A Study of the Stone Classics Engraved in the +3rd Century.
  In *HLW*, vol. 8.
Wang Kuo-Wei (*11*) 王國維.
  Chien Tu Chien Shu Khao 簡牘檢署考.
  A Study of Bamboo and Wooden Documents and Their System of Sealing.
  In *HLW*, vol. 26.
Wang Ming (*1*) 王明.
  Tshai Lun yü Chung-Kuo Tsao Chih Shu ti Fa Ming 蔡倫與中國造紙術的發明.
  Tshai Lun and the Invention of Paper in China.
  KKHP, 1954 (no. 8), 213.
Wang Ming (*6*) 王明.
  Sui Thang Shih Tai ti Tsao Chih 隋唐時代的造紙.
  Papermaking during the Sui and Thang Dynasties.
  KKHP, 1956 (no. 11), 115.
Wang Po-Min (*1*) 王伯敏.
  Chung-Kuo Pan Hua Shih 中國版畫史.
  A History of Chinese Woodcuts.
  Jen-Min, Shanghai, 1961.

Wang Po-Min (*2*) 王伯敏.
  Hu Cheng-Yen chi chhi Shih-Chu-Chai ti Shui Yin Mu Kho 胡正言及其十竹齋的水印木刻.
  Hu Cheng-Yen and the Multi-colour Printing of his *Shih-Chu-Chai*.
  MSYC, 1957 (no. 3), 77.
Wang Yeh-Chhiu (*1*) 王冶秋.
  Liu-li-chhang Shih Hua 琉璃廠史話.
  A History of the Liu-li-chhang District in Peking.
  San-Lien Shu-Tien, Peking, 1963.
Wang Ying-Wen (*1*) 汪應文.
  Kuo Chih Hsiao Shih 國紙小史.
  A Brief History of Chinese Paper.
  CWYK, 1942, **1** (no. 2), 54.
Wang Yü (*1*) 王抒.
  Ma-wang-tui Han Mu ti Ssu Chih Wu Yin Hua 馬王堆漢墓的絲織物印花.
  Silk Textiles with Printed Patterns Discovered in a Han Tomb at Ma-wang-tui, Chhangsha.
  KKTH, 1979 (no. 5), 474; pl. 12.
Wang Yung (*1*) 王庸.
  Chung-Kuo Ti-Thu Shih Kang 中國地圖史綱.
  A Brief History of Chinese Cartography.
  San-Lien, Peking, 1958.
Watanabe Tadaichi *et al.* (*1*) 渡邊忠一.
  Hui Hua Yen Liao La Pi Mo Chih Chih Tsao Fa 繪畫顏料蠟筆墨汁製造法.
  Methods of Making Pigments, Crayons and Liquid Inks.
  Tr. Tshai Chhi-Min 蔡棄民.
  Com. Press, Shanghai, 1939.
Wei Tzu (*1*) 微子.
  An-Hui Sheng Ching-Hsien Tung-chai-tshun Fa Hsien Ni Huo Tzu Tsung Phu 安徽省涇縣東翟村發現泥活字宗譜.
  A Genealogical Record of the Chai Family Printed with Clay Movable Type Discovered in Ching County, An-Hui.
  WWTK, 1960 (no. 4), 86.
Weng Thung-Wen (*1*) 翁同文.
  Chiu-Ching San-Chuan Kho Tzu Jen Wei Yüeh Chün Khao 九經三傳刻梓人爲岳浚考.
  A Discussion of Yüeh Chün as the Publisher of an Edition of the Nine Confucian Classics with the Three Commentaries to the *Chhun Chhiu*.
  TLTC, 1966, **32** (no. 7), 1.
Weng Thung-Wen (*2*) 翁同文.
  Yin Shua Shu Tui yü Shu Chi Chheng Pen ti Ying Hsiang 印刷術對於書籍成本的影響.
  Decreases in the Costs of Books Following the Invention of Printing.
  CHJ/T, 1967, **6** (no. 1/2), 35.
Weng Thung-Wen (*3*) 翁同文.
  Chin Jen Yin Shu Wu Chao-I Kho Shu Shih Ting Pu 近人引述毋昭裔刻書事訂補.
  On Wu Chao-I's Printing Activity.
  TLTC, 1968, 37 (no. 9), 27.
Wu Chhang-Shou (*1*) (ed.) 吳昌綬.
  Shih Liu Chia Mo Shuo 十六家墨說.
  A Collection of Sixteen Works on Ink.

Wu Chhang-Shou (*1*) (ed.) (*cont.*)
  Jen-Ho, 1922.
Wu Chhi-Chün (*1*) 吳其濬.
  Chih Wu Ming Shih Thu Khao 植物名實圖考.
  Illustrated Study and Description of Botany.
  2 vols.
  Com. Press, Shanghai, 1933.
Wu Fu-Yüan (*1*) 吳幅員.
  Liu-Chhiu Li Tai Pao An Hsüan Lu 琉球歷
  代寶案選錄.
  Selections from the Liu-Chhiu Archives of
    Successive Dynasties.
  Thaipei, 1975.
Wu Ta-Chheng (*1*) 吳大澂.
  Ku Yü Thu Khao 古玉圖考.
  Illustrated Study of Ancient Jades.
  Thung-Wen, Shanghai, 1889.
Wu Tse-Yen (*1*) (tr.) 吳澤炎.
  *Chung-Kuo Yin Shua Shu ti Fa Ming ho Tha ti Hsi
    Chhuan* 中國印刷術的發明和它的
  西傳.
  The Invention of Printing In China and Its
    Spread Westward.
  Peking, 1957.
  Tr. of Carter (*1*), 1925 ed.
Wu Tzu-Lin (*1*) 吳梓林.
  Shui Fa Ming Chih 誰發明紙.
  Who Invented Paper?
  SHCS, 1977 (no. 6), 58.
Wu Yu (*1*) 吳猷.
  Wu Yu-Ju Hua Pao 吳友如畫寶.
  A Collection of Paintings by Wu Yu.
  Shanghai, 1909.

Yahagi Katsumi (*1*) 矢作勝美.
  Minchō Katsuji 明朝活字.
  Movable Types of the Ming Dynasty.
  Heibon-Sha, Tokyo, 1976.
Yamamoto Tatsurō (*1*) 山本達郎.
  *Hanoi Fukkoku Kyokutō Gakuin Shozō Jinanbon
    oyobi Annanban Kanseki Shomoku* 河内仏
  国極東学院所蔵字喃本及び安南
  版漢籍書目.
  A Catalogue of *Nom* Character Editions and
    Annamese Editions of Chinese Books in the
    Ecole française d'Extrême Orient, Hanoi.
  SG, 1938, *16* (no. 4), 73.
Yang Chia-Lo (*1*) (ed.) 楊家駱.
  Ssu Khu Chhüan Shu Hsüeh Tien 四庫全
  書學典.
  Bibliographical Index of the *Ssu Khu Chhüan Shu*
    Encyclopaedia.
  World Book Co., Shanghai, 1946.
Yang Chia-Lo (*2*) 楊家駱.
  *Chung-Kuo Ku Chin Chu Tso Ming Shu Chih Thung
    Chi* 中國古今著作名數之統計.
  An Estimate of Chinese Publications from
    Ancient to Modern Times.
  HCH, 1946, n.s. **4** (no. 7), 22.
Yang Chung-Hsi (*1*) 楊鍾羲.
  Hsüeh-Chhiao Shih Hua Hsü Chi 雪橋詩話
  續集.

Further Discourse about Poetry by [the Recluse
    of] Snow-Bridge.
  1917.
Yang Shou-Ching (*1*) 楊守敬.
  Liu Chen Phu 留眞譜.
  Facsimile of Rare Editions by Woodcut.
  Series 1–2.
  Block-print, 1901–17; repr. in *SMTP* (ser. 5),
    Thaipei, 1972.
Yang Shou-Chhing (*1*) 楊壽清.
  Chung-Kuo Chhu Pan Chieh Chien Shih 中國
  出版界簡史.
  A Brief History of Chinese Publishing.
  Yung-Hsiang, Shanghai, 1946.
Yang Wei-Hsin (*1*) 楊維新.
  Jih-Pen Pan Pen chih Li Shih 日本版本之
  歷史.
  A History of Japanese Printing.
  TSKH, 1929, **3**, 551; 1930, **4**, 23.
Yao Ming-Ta (*1*) 姚名達.
  Chung-Kuo Mu Lu Hsüeh Shih 中國目錄學史.
  A History of Chinese Bibliography.
  Commercial Press, Shanghai, 1938.
Yao Shih-Ao 姚士鰲.
  See Yao Tshung-Wu.
Yao Tshung-Wu (*1*) 姚從吾.
  Chung-Kuo Tsao Chih Shu Shu Ju Ou-Chou Khao
  中國造紙術輸入歐洲考.
  The Introduction of the Chinese Method of
    Papermaking to Europe.
  FJHC, 1928, **1**, 1; *SMCK*, 1966–7, **1** (nos. 2–4).
Yeh Chhang-Chhing (*1*) 葉長青.
  Min Pen Khao 閩本考.
  Books Printed in Fukien.
  TSKH, 1927, **2** (no. 1), 115.
Yeh Kung-Chhao (*1*) 葉恭綽.
  Li Tai Tsang Ching Khao Lüeh 歷代藏經
  考略.
  Chinese Editions of the Tripitaka from Various
    Dynasties.
  In *CCSM*, 1937, pp. 25–42.
Yeh Kung-Chho (*2*) (ed.) 葉恭綽.
  Ssu Chia Tshang Mo Thu Lu 四家藏墨圖錄.
  An Illustrated Catalogue of Four Ink Collec-
    tions.
  [Peking, *c.* 1956].
Yeh Sheng-Thao (*1*) 葉聖陶.
  Jung-pao-chai ti Tshai Se Mu Kho Hua 榮寶
  齋的彩色木刻畫.
  Chinese Colour Wood-block Printing (by Jung-
    pao-chai).
  HKC, 1954 (no. 10), 17.
Yeh Te-Hui (*2*) 葉德輝.
  Shu Lin Chhing Hua 書林清話.
  Plain Talks on Chinese Books.
  Kuan Ku Thang, Changsha, 1920. Reprs.:
    Peking, 1959, 1961; Thaipei, 1960.
Yeh Te-Hui (*3*) 葉德輝.
  Shu Lin Yü Hua 書林餘話.
  Further Talks on Chinese Books.
  Shanghai, 1928. Reprs.: Peking, 1959, 1961;
    Thaipei, 1960.

Yeh Te-Hui (4) 葉德輝.
Tshang Shu Shih Yüeh 藏書十約.
The Bookman's Decalogue.
Ku Tien Wen Hsüeh, Shanghai, 1957.
Cf. tr. by A. Fang [Chih-Thung] (1).

Yen I-Chhüan (1) 燕義權.
Thung Pan ho Thao Se Pan Yin Shua ti Fa Ming yü Fa Chan 銅版和套色版印刷的
發明與發展.
The Invention and Development of Bronze Block and Multi-colour Printing.
In *CKKC*, pp. 205–15.

Yen Kho-Chün (1) (ed.) 嚴可均.
Chhüan Chin Wen 全晉文.
Collected Prose Literature of the Chin Dynasty.
In *CSHK*.

Yen Shih (1) 燕石.
Chi Khuai Yu Kuan Chen Ya Chhuai Fang Jan Chih Fang Shou Kung Kung Jen ti Pei Kho Tzu Liao 幾塊有關鎮壓踹坊染紙坊手工工人
的碑刻資料.
Some Stone Tablets Telling about the Suppression [and Exploitation] of Textile and Paper-mill Workers.
WWTK, 1957 (no. 9), 38.

Yi Hong-Jik (1) 李弘植.
Mokp'an Insoe Rul Chungsim Uro Bon Silla Munhwa 木板印刷를中心으로본新羅文化.
The Silla Culture as Viewed from Block Printing.
HGSH, 1969, **8**, 53.

Yi Hong-Jik (2) 李弘植.
Kyŏngju Pulguksa Sŏkkat'ap Palkyŏn Ŭi Muku-jŏngkwang Taedaranikyong 慶州佛國寺
釋迦塔發現의無垢淨光大陀羅尼經.
The *Dharani* Sutra Recovered from the Sokka Pagoda in Pulguksa Monastery in Kyongju.
PSHB, 1968, **4**, 169.

Yi Song-Ui (1) 李聖儀.
Ko Dong Hwalcha Ch'aek P'yobon Sŏmok 古銅活字册標本書目.
A Catalogue of Old Bronze Type Imprint Specimens.
Seoul, n. d.

Yi Won-Jik (1) 李元植.
Han-Kuo Huo Tzu Pan ti Yen Pien 韓國
活字版的演變.
The Development of Movable Type Printing in Korea.
CCTH, 1964, 9, 521.

Yin Jun-Sheng (1) 尹潤生.
Ming Chhing Liang Tai ti Chi Chin Mo 明清
兩代的集錦墨.
Decorated Ink Sets of the Ming and Chhing Dynasties.
WWTK, 1958 (no. 12), 12.

Yü Chia-Hsi (1) 余嘉錫.
Shu Tshe Chih Tu Pu Khao 書册制度補考.
Further Study on the Evolution of Chinese Books.
In *Yü Chia-Hsi Lun Hsüeh Tsa Chu*, vol. 2, pp. 539–59. Peking, 1963.

Yü Chheng-Hung 喩誠鴻 & Li Yün (1) 李澐.
Chung-Kuo Tsao Chih Yung Chih Wu Hsien Wei Thu Phu 中國造紙用植物纖維圖譜.
An Illustrated Repertory of Plant Fibres for Papermaking in China.
Science Press, Peking, 1955.

Yü Hsia-Shang (1) 于霞裳.
Chin Yüan Shih Chhi Phing-Shui Yin Shua Yeh Chhu Than 金元時期平水印刷業初探.
A Preliminary Study of the Printing Industry at Phing-Shui during the Chin and Yüan Periods.
SHHP, 1958 (no. 1), 27.

Yü Hsing-Wu (1) 于省吾.
Shuang-Chien-I Ku Chhi Wu Thu Lu 雙劍
誃古器物圖錄.
Illustrations of Antique Objects in the Yü Collection.
Peiping, 1940.

Yü Wei-Kang (1) 于爲剛.
'*Yin Shua Shu Fa Ming yü Sui Chhao ti Hsin Cheng' Hsi I* 印刷術發明于隋朝的新証析疑.
Questions concerning the 'New Evidence for the Invention of Printing in the Sui Dynasty' by Chang Chih-Che.
WH, 1981 (no. 4), 231.

Yüan Han-Chhing (1) 袁翰青.
Chung-Kuo Hua Hsüeh Shih Lun Wen Chi 中國
化學史論文集.
Collected Papers on the History of Chemistry in China.
San-Lien, Peking, 1956.

Yüan Han-Chhing (2) 袁翰青.
Tsao Chih Tsai Wo Kuo ti Fa Chan ho Chhi Yüan 造紙在我國的發展和起源.
The Origin and Development of Paper in China.
KHTP, 1954 (no. 12), 62.

Yüan Thung-Li (1) 袁同禮.
Yung-Lo Ta Tien Hsien Tshun Chüan Mu Piao 永樂大典現存卷目表.
Table of Extant Volumes of the *Yung-Lo Ta Tien*.
QBCB/C, 1939, n.s. *I* (no. 3), 246.

C. BOOKS AND JOURNAL ARTICLES IN WESTERN LANGUAGES

A CHING (1). 'Chinese New-Year Pictures.' *CLIT*, 1959, no. 2, 171.

ACKERMAN, P. (1). *Wall Paper; its History, Design and Use.* Stokes, New York, 1923.

AH JUNG (1). 'Art Recreated; the Woodblock Prints of the Jung Pao Chai Studio.' *CLIT*, 1972, no. 2, 102.

AKATSUKA KIYOSHI (1). 'A New Study of the *Shih-Ku-Wen.*' *ACTAS*, 1963, no. 3, 80.

ALDEN, H. M. (1). 'Origin of Printing and Why the Ancients had no Printing Press.' *ER*, 1824, **40**, 366; *IQR*, 1859, **8**, 20; *HNMM*, 1868, **37**, 394, 637.

ALIBAUX, H. (1). 'L'invention du papier.' *GUJ*, 1939, 9.

ANDERSON, D. M. (1). *The Art of Written Forms; the Theory and Practice of Calligraphy.* Holt, Rinehart & Winston, New York, 1969.

ANDERSON, W. (1). *Japanese Wood Engravings; the History, Technique and Characteristics.* Seeley, London; Macmillan, New York, 1895.

ANDREWS, A. E. (1). *Rags; being an Explanation of why They are used in the Making of Paper.* Strathmore Paper Co., Mittineague, Mass., [1927].

ANON. (180). 'Aperçu sur le développement de l'industrie du papier.' *OEO*, 1935, **59**, 32.

ANON. (181). *The Art of Papermaking: A Guide to the Theory and Practice of the Manufacture of Paper; being a Compilation from the Best-known French, German and American Writers.* 2nd ed. London, 1876.

ANON. (182). *Chinese Paper; A Catalogue of the Chu-Fei Collection.* West China Union University Museum, Chhengtu, 1947.

ANON. (183). *Comprehensive Collection of Handmade Japanese Paper.* Mainichi Newspapers, Tokyo, 1973–4. Text in Japanese and English with samples of raw materials and some 1000 paper specimens in 5 boxes.

ANON. (184). *Dard Hunter Paper Museum.* MIT, Cambridge, Mass., 1939; Institute of Paper Chemistry, Appleton, Wisconsin, 1965?

ANON. (185). *The Dictionary of Paper, including Pulp, Paperboard, Paper Properties, and Related Papermaking Terms.* 3d ed. American Paper and Pulp Association, New York, 1965; 4th ed., 1980.

ANON. (186). 'An Exhibition of Painted and Inscribed Fans.' *CLIT*, 1961, no. 7, 134.

ANON. (187). *Fan Paintings by Late Ch'ing Shanghai Masters.* An Exhibit at the Hong Kong Museum of Art, Feb. 15 to March 20, 1977. Urban Council, Hong Kong, 1977.

ANON. (188). *Manuscrits et peintures de Touen-houang; Mission Pelliot 1906–09, Collection de la Bibliothèque et du Musée Guimet.* Editions des Musées nationaux, Paris, 1947.

ANON. (189). *A Millennium of Printing in China, Korea and Japan; an Inaugural Exhibition, Nov.-Dec. 1972.* Royal Library, Stockholm, 1972.

ANON. (190). *Paper and Paper Products in China.* Comp. by U.S. Bureau of Foreign and Domestic Commerce. Government Printing Office, Washington, 1925. (Trade Information Bulletin 309).

ANON. (191). *Papermaking, Art and Craft; an Account Derived from the Exhibition Presented in the Library of Congress, Washington, D.C. and Opened on April 21, 1968.* Library of Congress, Washington, D.C., 1968.

ANON. (192). *Plane Geometry and Fancy Figures; an Exhibition of Paper Folding Held in Cooper Union Museum.* New York, 1959.

ANON. (193). 'Wallpaper; a Picture-book of Examples in the Collection of the Cooper Union Museum.' *CUMC*, 1961, **3** (nos. 1–3), 1.

ANON. (194). 'Who Invented Paper?' *CREC*, 1972, **21** (no. 6), 20. Reprinted in *CSOH*, 1973, **6** (no. 4), 24.

ASHLEY, M. (1). 'On a Method of Making Rubbings.' *AAN*, 1930, n.s., **32**, 578.

ASTON, W. G. (1) (tr.). *Nihongi; Chronicles of Japan from the Earliest Times to A.D. 697.* 2 vols. Kegan Paul, Trench, Trübner, London, 1896; Allen & Unwin, London, 1956.

ASTON, W. G. (2). *A History of Japanese Literature.* Appleton, New York, 1937. Rev. ed. Tuttle, Rutland, Vt., 1972.

ATKYNS, R. (1). *The Origin and Growth of Printing; Collected out of History, and the Records of this Kingdome.* London, 1664.

BACON, ROGER (1). *Opus Majus.* Eng. tr. by Robert B. Burke. Oxford University Press, 1928; University of Pennsylvania Press, Philadelphia, 1938; Russell & Russell, New York, 1962.

BAGFORD, J. (1). 'An Essay on the Invention of Printing.' *PTRS*, 1706–7, 25. Repr., Committee on the Invention of Printing, Chicago, 1940.

BARKER, L. (1). *A Seminar in Handmade Papermaking for the Fine Arts.* Cranbrook Academy of Art, Bloomfield Hills, Mich., 1962.

BARNARD, N. (2) (ed.). *Early Chinese Art and its Possible Influence in the Pacific Basin.* A symposium arranged by the Department of Art History and Archaeology, Columbia University, New York City, Aug. 21–25, 1967. In collaboration with Douglas Fraser. New York, 1972. 3 vols.

BARNARD, N. (3). *The Chhu Silk Manuscript; Translation and Commentary.* National University of Australia, Canberra, 1973.

BARUCH, W. (1). 'The Writing and Language of the Si-hsia.' In *Sino-Tibetan Art*, ed. by Alfred Salmony. Paris, 1933.

BASANOFF, A. (1). *Itinerario della carta dall'Oriente all'Occidente e sua diffusione in Europa.* (Traduzione di Valentina Bianconcini). Edizioni il Palifilo, Milano [1965].

BATTEUX, C. & DE BRÉQUIGNY, L. G. O. F. (1) (eds.). *Mémoires concernant l'histoire, les sciences, les arts, les moeurs, les usages, etc. des Chinois par les missionnaires de Pékin.* 15 vols. Nyon, Paris, 1776–91.

BEATTY, W. B. (1). 'The Handmade Paper of Nepal.' *PAM*, 1962, **31**, 13.

DE BELLECOURT, DUCHESNE (1). 'Note sur l'écorce de l'arbuste à papier du Japon.' *BSZA*, 1865 (2e sér.), **2**, 36.

BENEDETTI-PICHLER, A. A. (1). 'Microchemical Analysis of Pigments Used in the Fossae of the Incisions of Chinese Oracle Bones.' *IEC/AE*, 1937, **9**, 149.

BERGMAN, FOLKE (1). *Archeological Researches in Sinkiang.* Reports of the Sino-Swedish [Scientific] Expedition [to Northwest China], 1939, vol. 7 (pt. 1).

BERGMAN, FOLKE (4). 'Travels and Archeological Field Work in Mongolia and Sinkiang.' In *History of the Expedition in Asia, 1927–1935.* Part 4 (Sino-Swedish Expedition Publications no. 26), 135. Göteborg Elander, Stockholm, 1945.

BERGMAN, FOLKE (5). 'Lou-lan Wood-carvings and Small Finds Discovered by Sven Hedin.' *BMFEA*, 1935, **7**, 71.

BERNARD-MAÎTRE, H. (18). 'Les adaptations chinoises d'ouvrages européens; bibliographie chronologique depuis la venue des Portugais à Canton jusqu'à la mission Française de Pékin, 1514–1688.' *MS*, 1945, **10**, 309.

BERNARD-MAÎTRE, H. (20). 'Les origines chinoises de l'imprimerie aux Philippines.' *MS*, 1942, **7**, 312.

BESTERMAN, T. (1). *Early Printed Books to the End of the Sixteenth Century; a Bibliography of Bibliographies.* Societas Bibliographica, Geneva, 1961.

BEVERIDGE, H. (1). 'The Papermills of Samarkand.' *ASQR*, 1910, 160.

BIELENSTEIN, H. (3). *The Bureaucracy of Han Times.* Cambridge, 1980.

BIGMORE, E. C. & WYMAN, C. W. H. (1). *A Bibliography of Printing.* London, 1880–6; 2nd ed. 1945.

BINYON, L. (5). *Catalogue of Japanese and Chinese Woodcuts Preserved in the Sub-department of Oriental Prints and Drawings in the British Museum.* London, 1916.

BLACK, M. H. (1). 'The Printed Bible.' In *CHB*, vol. 3, p. 408. Cambridge, 1963.

BLANCHET, A. (1). *Essai sur l'histoire du papier et de sa fabrication.* E. Leroux, Paris, 1900.

BLAND, D. (1). *A History of Book Illustration; the Illuminated Manuscript and the Printed Book.* Word Pub. Co., Cleveland, 1958; 2nd rev. ed., University of California Press, Berkeley and Los Angeles, 1969.

BLISS, D. P. (1). *A History of Wood Engraving.* 2nd ed., Spring Books, London, 1964.

BLUM, A. (1). *On the Origin of Paper.* Tr. by H. M. Lydenberg. Bowker, New York, 1934. Tr. of 'Les origines du papier.' *RH*, 1932, **170**, 435.

BLUM, A. (2). *Origins of Printing and Engraving.* Tr. by H. M. Lydenberg. New York, 1940. Tr. of 2nd and 3rd parts of *Les origines du papier, de l'imprimerie et de la gravure*, Paris, 1935.

BOATWALA, M. & MARCIEL, W. (1). 'Handmade Paper in India.' *PENA*, 1964, **57**, 281.

BOCKWITZ, H. H. (1). 'Die internationale Papiergeschichtsforschung und ihr gegenwärtiger Stand.' *FF*, 1945, **25**, 101.

BODDE, D. (12) (tr.) *Annual Customs and Festivals in Peking, as Recorded in the 'Yen Ching Sui Shih Chi' by Tun Li-Chhen.* Henri Vetch, Peiping, 1936; Hongkong, 1965.

BODDE, D. (13). *China's Gift to the West.* American Council on Education, Washington, 1942. (Asiatic Studies in American Education, no. 1).

BODDE, D. (30). *China's Cultural Tradition; What and Whither?* Rinehart & Co., New York, 1957.

BOHN, H. G. (1) 'The Origin and Progress of Printing.' *PSL/M*, 1857–58, **4** (no. 2).

BOJESEN, C. C. & ALLEY, R. (1). 'China's Rural Paper Industry.' *CJ*, 1938, **28**, 233.

BOWYER, W. & NICHOLS, J. (1). *The Origin of Printing in Two Essays, with Occasional Remarks and an Appendix.* 2nd ed. Privately printed, London, 1776.

BOXER, C. R. (1) (ed.). *South China in the Sixteenth Century; being the Narratives of Galeote Pereira, Fr. Gaspar da Cruz, Fr. Martin de Rada (1550–1575).* Hakluyt Society, London, 1953.

BOXER, C. R. (9) 'Manila Galleon, 1565–1815.' *HTD*, 1958, **8**, 538.

BOXER, C. R. (10). 'Chinese Abroad in the Late Ming and Early Manchu Periods, compiled from Contemporary Sources, 1500–1750.' *TH*, 1939, **9** (no. 5), 459.

BOYER, A. (1). *Kharosthi Inscriptions Discovered by Sir Aurel Stein in Chinese Turkestan.* Tr. by E. J. Rapson and A. E. Senart. 2 vols. Clarendon Press, London, 1920–7.

BREASTED, J. H. (2). 'The Physical Processes of Writing in the Early Orient and Their Relation to the Origin of the Alphabet.' *AJSLL*, 1916, **32**, 230.

BRETSCHNEIDER, E. (1) *Botanicon Sinicum; Notes on Chinese Botany from Native and Western Sources.* Repr. from *JRAS/NCB*, n.s., **16, 25, 29** (1882–1895). 3 vols. Trübner, London, 1882–95; Royal Asiatic Society, Tokyo, 1937; Kraus, Nendeln/Liechtenstein, 1967.

BRETSCHNEIDER, E. (10) *History of European Botanical Discoveries in China.* 2 vols. Unveränderter Nachdruck von K. F. Koehlers Antiquarium, Leipzig, 1935.

BRIQUET, C. M. (1). 'Recherches sur les premiers papiers employés en Occident et en Orient du Xe au XIVe siècle.' *MSAF*, 1886, **46**.

BRITTON, R. S. (1). *The Chinese Periodical Press, 1800–1912*. Kelly & Walsh, Shanghai, 1933.

BRITTON, R. S. (2) 'Oracle-bone Color Pigments,' *HJAS*, 1937, **2**, 1.

BRITTON, R. S. (3). 'A Horn Printing Block.' *HJAS*, 1938, **3**, 99.

BROWN, L. N. (1). *Block Printing and Book Illustration in Japan*. Routledge, London, 1924.

BROWNE, E. G. (2). *A Literary History of Persia*. Vol. 3: *The Tartar Dominion (1265–1502)*. Cambridge, 1956.

BROWNING, B. L. (1). *Analysis of Paper*. Marcel Dekker, New York, 1969.

BROWNING, B. L. (2). 'The Nature of Paper.' *LQ*, 1970, **40**, 18.

BUEHLER, C. (1). *The Fifteenth Century Book; the Scribes, the Printers, the Decorators*. University of Pennsylvania Press, Philadelphia, 1960.

BULLEN, H. L. (1). 'History of Printing and Paper in China and Japan.' *FER*, 1915, **12**, 195.

BURGES, FRANCIS (1). *Some Observations on the Use and Origins of the Noble Art and Mystery of Printing*. Norwich, 1701. Reprint in *Harleian Miscellany*, vol. 3, pp. 154–57. London, 1809.

BURKE, R. B. (1) (tr.) *The 'Opus Majus' of Roger Bacon*. 2 vols. University of Pennsylvania Press, Philadelphia, 1938.

BUSHELL, S. W. (5). 'The Stone Drums of the Chou Dynasty.' *JRAS/NCB*, 1874, n.s., **8**, 133.

BUSHELL, S. W. (6) 'Specimens of Ancient Chinese Paper Money.' *JPOS*, 1889, **2**, 308.

BUSHNELL, G. H. (3). *From Papyrus to Print*. Grafton, London, 1947.

BUTLER, P. (1). *Origin of Printing in Europe*. University of Chicago Press, 1940.

BYRD, C. K. (1). *Early Printing in the Straits Settlements, 1806–1858*. Singapore National Library, Singapore, 1970.

CARRÉ, A. (1). *The Travels of the Abbé Carré in India and the Near East, 1672 to 1674*. Tr. from the manuscript journal of his travels in the India Office by Lady Fawcett and ed. by Sir Charles Fawcett with the assistance of Sir Richard Barn. Hakluyt Society, London, 1947.

CARTER, J. W. & MUIR, P. H., (1) (ed.) *Printing and the Mind of Man; a Descriptive Catalogue Illustrating the Impact of Print on the Evolution of Western Civilization during Five Centuries*. Cassell, London, 1967.

CARTER, T. F. (1). *The Invention of Printing in China and its Spread Westward*. Columbia University Press, 1925; revised ed., 1931; 2nd ed. revised by L. Carrington Goodrich. Ronald, New York, 1955, Rev. B. Laufer, *JAOS*, 1927, **47**, 71; A. C. Moule, *JRAS*, 1926, 140; H. H. Frankel, *FEQ*, 1956, **15**, 284.

CARTER, T. F. (2). 'The Westward Movement of the Art of Printing; Turkestan, Persia, and Egypt in the Long Migration from China to Europe.' In A. Waley, ed., *Yearbook of Oriental Art and Culture*, vol. 1, p. 19. Benn, London, 1925.

CARTER, T. F. (3). 'The Chinese Origins of Moveable Types.' *ARTY*, 1925, **2**, 3.

CARTER, T. F. (4). 'The Chinese Background of the European Invention of Printing.' *GUJ*, 1928, 9.

CARVALHO, D. N. (1). *Forty Centuries of Ink; or, a Chronological Narrative concerning Ink and its Backgrounds*. Banks Law Pub. Co., New York, 1904.

CATHERINOT, NICOLAS (1). *An Essay on Writing and the Art and Mystery of Printing*. A translation out of the anthology. London, 1696. Reprint in *Harleian Miscellany*, vol. 1, pp. 526–8. London, 1808.

CERNY, J. (1). *Paper and Book in Ancient Egypt*. H. K. Lewis, London, 1952.

CHAFFEE, J. W. (1). 'Education and Examination in Sung Society (960–1279).' Thesis (PhD.), University of Chicago, 1979.

CHANDRA, LOKESH (1) (ed.). *Buddha in Chinese Woodcuts*. International Academy of Indian Culture, New Delhi, 1973. (Indo-Asian Literatures, vol. 98). Reproduction of *Shih Chia Ju Lai Ying Hua Shih Chi*, probably of the Ming dynasty (1368–1644).

CHANG HSIU-MIN (1). 'A Note on the Date of the Invention of Paper in China.' *PG*, 1959, **9**, 51.

CHANG, LÉON L. Y. (1). *La calligraphie chinoise; un art à quatre dimensions*. Le Club français du livre, Paris, 1971.

CHANG TEH-CHHANG (1). 'Geographical Distribution of Book Printing Trade in the Chhing Dynasty and Its Cultural Significances.' *ACQ*, 1975, **3** (no. 3), 65.

CHAPPELL, W. (1). *A Short History of the Printed Word*. Alfred Knopf, New York, 1970.

CHATTO, W. A. (1). *Facts and Speculations on the Origin and History of Playing Cards*. London, 1948.

CHATTO, W. A. & JACKSON, J. (1) *A Treatise on Wood Engraving, Historical and Practical*. London, 1861; Gale Research Co., Detroit, 1969.

CHAUDHARY, Y. S. (1). *Handmade Paper in India*. Lucknow, India, 1936.

CHAVANNES, É. (12). 'Introduction to "Les documents chinois découverts par Aurel Stein dans les sables du Turkestan Oriental".' *NCR*, 1922, **4**, 341.

CHAVANNES, É. (12a). *Les documents chinois découverts par Aurel Stein dans les sables du Turkestan Oriental*. Oxford, 1913.

CHAVANNES, É. (24). 'Les livres chinois avant l'invention du papier.' *JA*, 1905, **11** (no. 5), 1.

CHAVANNES, É. & PELLIOT, P. (1). *Un traité manichéen*. Paris, 1913.

CHAYTOR, H. J. (1). *From Script to Print; an Introduction to Medieval Vernacular Literature*. Heffer, Cambridge, 1950.

CHEW, N. D. (1). 'Printing in Korea.' *KMF*, 1906, **2** (no. 3), 47.

CHHEN, KENNETH [CH'EN KUAN-SHENG] (8). 'Notes on the Sung and Yüan Tripitaka.' *HJAS*, 1951, **14**, 208.

CHHIEN TSHUN-HSÜN. *See* TSIEN TSUEN-HSUIN.

CHIANG WEI-PU (1). 'Chinese Picture-story Books.' *CLIT*, 1959, **3**, 144.

CHIANG, YEE (1). *Chinese Calligraphy; an Introduction to its Aesthetic and Technique*. Methuen, London, 1938.

CHIBBETT, D. (1). *The History of Japanese Printing and Book Illustration*. Kodansha International Ltd., Tokyo, New York & San Francisco, 1977.

CHIEN, FLORENCE (1). 'The Commercial Press and Modern Chinese Publishing, 1887–1949.' Thesis (MA), University of Chicago, 1970.

CHIEN HSUIN-YUI (1). 'Eine Studie zur Geschichte der chinesischen Druckkunst.' *GUJ*, 1952, 34.

CHIERA, EDWARD (1). *They Wrote on Clay; the Babylonian Tablets Speak Today*. University of Chicago Press, 1938; Phoenix Books, 1956.

CHIU, A. KAI-MING (1). 'The *Chieh Tzu Yüan Hua Chuan* (Mustard Seed Garden Painting Manual); Early Editions in American Collections.' *ACASA*, 1951, 5, 55.

CHOW, A. (1). 'A Survey of Modern Chinese Woodcut.' *JACU*, 1977, 36.

CHUANG LIEN (1). 'The Development of Ancient Book Editions as Revealed by the Exhibition of Rare Books.' *WE*, 1966, 11 (no. 4), 10; 11 (no. 5), 8.

CHUANG SHEN (1). 'Ming and Chhing Exotica; a Reflection of Literary Taste.' *JOSHK*, 1971, 8 (no. 1), 92.

CLAIR, C. (1). *A History of European Printing*. Academic Press, London & New York, 1976.

CLAPPERTON, R. H. (1). *Paper; an Historical Account of its Making by Hand from its Earliest Times down to the Present Day*. Shakespeare Head Press, Oxford, 1934.

CLAVERIE, F. (1). 'L'arbre à papier du Tonkin (Cây gio).' *RCC*, 1904, 175.

CLUNE, G. (1). *The Medieval Gild System*. Browne & Nolan, Dublin, 1943.

COLEMAN, D. C. (1). *The British Paper Industry, 1495–1860*. Clarendon Press, Oxford, 1958.

CONRADY, A. (5) (ed.). *Die chinesischen Handschriften und sonstigen Kleinfunde Sven Hedins in Lou-lan*. Generalstabens Lithografiska Anstalt, Stockholm, 1920.

CORDIER, H. (16). *Ser Marco Polo; Notes and Addenda to Sir Henry Yule's Edition, Containing the Results of Recent Research and Discovery*. New York, 1920.

COURANT, M. (1). *Bibliographie Coréenne*. 3 vols. Paris, 1894–6; reprint, Burt Franklin, New York, 1960. Supplément, 1901.

COURANT, M. (3). *Catalogue des livres Chinois, Coreens, Japonais, etc.* 3 vols. Ernest Leroux, 1900–12.

CRAWFORD, M. DE C. (1). *The Influence of Invention on Civilization*. World Publishing Co., Cleveland and New York, 1942.

CREEL, H. G. (1). *Studies in Early Chinese Culture* (1st series). Waverly, Baltimore, 1938.

CREEL, H. G. (13). 'Bronze Inscriptions of the Western Chou Dynasty as Historical Documents.' *JAOS*, 1936, 56, 335.

CURZON, ROBERT (1). 'A Short Account of Libraries of Italy.' *PSL/M*, 1854, 1, 6.

CURZON, ROBERT (2). 'The History of Printing in China and Europe.' *PSL/M*, 1860, 6, 1.

DAHL, SVEND (1) *The History of the Book*. Scarecrow Press, Metuchen, N.J., 1968.

DAVENPORT, C. (1). *The Book; its History and Development*. Constable, London, 1907; Van Nostrand, New York, 1917.

DAVIS, A. (1). *Ancient Chinese Paper Money as Described in a Chinese Work on Numismatics*. Boston, 1918.

DAVIS, J. F. (1). *The Chinese; a General Description of the Empire of China and its Inhabitants*. 1st ed. 1836, 2 vols. Knight, London, 1844. 3 vols., 1847, 2 vols. French tr. by A. Pichard, Paris, 1837, 2 vols. German tr. by M. Wesenfeld, Magdeburg, 1843, 2 vols. and M. Drugulin, Stuttgart, 1847, 4 vols.

DAWE, E. A. (1). *Paper and its Uses; a Treatise for Printers, Stationers and Others*. C. Lockwood and Son, London, 1919.

DE BARY, W. T. (3) (ed.). *Self and Society in Ming Thought*. Columbia University Press, New York, 1970.

DELAND, J. (1). 'The Evolution of Modern Printing and the Discovery of Movable Metal Types by the Chinese and the Koreans in the Fourteenth Century.' *JFI*, 1931, 212, 209.

DEMIÉVILLE, P. (9). 'Notes additionnelles sur les éditions imprimées du canon bouddhique.' In P. Pelliot, *Les débuts de l'imprimerie en Chine*, pp. 121–38. Paris, 1953.

DE VINNE, T. L. (1). *The Invention of Printing*. F. Hart, New York, 1876.

DIAZ DEL CASTÉLLO, B. (1). *The True History of the Conquest of New Spain*. Ed. and published in Mexico by G. Garcia, tr. into English with intro. and notes by A. P. Mandslay. Hakluyt Society, London, 1980.

DIEHL, K. S. (1). *Early Indian Imprints*. Scarecrow Press, New York & London, 1964.

DIRINGER, D. (1). *The Alphabet, a Key to the History of Mankind*. Hutchinson's Scientific and Technical Publications, London & New York, 1948; 2nd ed., 1949.

DIRINGER, D. (2). *The Hand-produced Book*. Philosophical Library, New York, 1953.

DODGSON, C. (1). *Woodcuts of the Fifteenth Century in the John Rylands Library, Manchester*. Reproduced in facsimile with an introduction and notes. University Press, Manchester, 1915.

DOUGLAS, R. K. (3). 'Chinese Illustrated Books.' *BIB*, 1896, 2, 452.

DOUGLAS, R. K. (4). 'Japanese Book Illustration.' *BIB*, 1897, 3, 1.

DUBS, H. H. (1). 'The Reliability of Chinese History.' *JAS*, 1946, 6, 23.

DUREAU DE LA MALLE, A.J.C.A. (1). 'Mémoire sur le papyrus et la fabrication du papier chez les anciens.' *MAI/NEM*, 1851, 19, 140.

DUTT, A. K. (1). 'Papermaking in India; a Resumé of the Industry from the Earliest Period until the Year 1949.' *PAM*, 1955, **24**, 11.

DUYVENDAK, J. J. L. (8). *China's Discovery of Africa; Lectures Given at the University of London, 1949.* Probsthain, London, 1949. (Lectures given at London University, Jan. 1947; rev. P. Paris, *TP*, 1951, **40**, 366.)

DUYVENDAK, J. J. L. (22). 'Coster's Chinese Ancestors.' *NM*, 1926, 1 (no. 3).

DYE, D. S. (1). *A Grammar of Chinese Lattice.* 2 vols. Harvard-Yenching Institute, Cambridge, Mass., 1937 (Harvard-Yenching Monograph Series, nos. 5, 6).

ECKE, TSENG YU-HO (1). *Chinese Calligraphy.* Philadelphia Museum of Art, 1971.

ECKE, TSENG YU-HO (2). *Chinese Folk Art in American Collections, from the Early 15th Century to the Early 20th Century.* Honolulu, Hawaii, 1977.

EDGREN, S. (1). 'The Printed Dhāraṇī-sūtra of A.D. 956.' *BMFEA*, 1972, no. 44, 141.

EDKINS, J. (18). 'On the Origin of Paper Making in China.' *NQCJ*, 1867, **1** (no. 6), 67.

EDKINS, J. (19). 'Paper—A Chinese Invention.' *CR*, 1900, **24**, 269.

EDKINS, J. (20). *Chinese Currency.* Presbyterian Mission Press, Shanghai, 1901.

EISEN, G. A. (1). *Ancient Oriental Cylinder and Other Seals with a Description of the Collection of Mrs. William H. Moore.* University of Chicago Press, 1940. (University of Chicago Oriental Institute Publications, vol. 47.)

EISENSTEIN, E. L. (1). 'Some Conjectures about the Impact of Printing on Western Society and Thought: a Preliminary Report.' *JMH*, 1968, **40**, 1.

EISENSTEIN, E. L. (2). *The Printing Press as an Agent of Change; Communications and Cultural Transformation in Early Modern Europe.* 2 vols. Cambridge University Press, 1979.

D'ELIA, PASQUALE (2) (ed.). *Fonti Ricciane; Storia dell'Introduzione del Cristianesimo in Cina.* 3 vols. Libreria dello stato, Rome, 1942–49.

ENGEL, S. (1). *An Inquiry into the Origin of Printing in Europe, by a Lover of the Art.* Gibson, London, 1752.

ENTWISLE, E. A. (1). *The Book of Wallpaper; a History and an Appreciation.* Barker, London, 1954.

ENTWISLE, E. A. (2). *A Literary History of Wall Paper.* Batsford, London [1960].

ERKES, E. (13). 'Der Druck des taoistischen Kanon...'. *GUF*, 1925, 326.

ERKES, E. (23). 'Buch und Buchdruck in China.' *GUF*, 1925, 338.

ERKES, E. (24). 'Zur ältesten Geschichte des Siegels in China.' *GUJ*, 1934, 67.

FANG, A. [CHIH-THUNG] (2) (tr.). 'Bookman's Decalogue.' *HJAS*, 1950, **13**, 132. Tr. Yeh Te-Hui, *Tshang Shu Shih Yüeh*.

FANG, A. [CHIH-THUNG] (3). 'On the Authorship of the *Chiu-ching san-chuan yen-ko-li*.' *MS*, 1946, **11**, 65.

FANG CHAO-YING (1). 'Some Notes on Metal Types.' *BBSK*, 1960, **2**, 28.

FANG CHAO-YING (2). *The Asami Library; a Descriptive Catalogue.* Ed. Elizabeth Huff. University of California Press, Berkeley, 1969.

FARROKH, ROKN OD DIN HOMAYUN (1). *History of Books and the Imperial Libraries of Iran.* Tr. Abutaleb Saremi. Ministry of Culture and Art, Tehran, 1968.

FEBVRE, L. & MARTIN, H. (1). *The Coming of the Book; the Impact of Printing, 1450–1800.* NLB, London; Humanities Press, Atlantic Highlands, 1976. Tr. of *L'apparition du livre* by David Gerard. Albin Michel, Paris, 1958.

FINEGAN, M. H. (1). 'Urbanism in Sung China; Selected Topics on the Society and Economy of Chinese Cities in a Premodern Period.' Thesis (Ph.D.), University of Chicago, 1976.

FITZGERALD, C. P. (11). *The Southern Expansion of the Chinese People.* Barrie and Jenkins, London, 1972.

FLUG, K. K. (1). *Istoriia Kitaĭskoĭ Pechatnoĭ Knigi Sunskoĭ Epokhi X–XIII vv.* Izdatel stvo Akademii Nauk SSSR, Moscow-Leningrad, 1959. Partial English tr. by Sidney O. Fosdick (1).

FOSDICK, S. O. (1). 'Chinese Book Publishing during the Sung Dynasty (A.D. 960–1279); a Partial Translation of *Istoriia Kitaĭskoĭ Pechatnoĭ Knigi Sunskoĭ Epokhi* by Konstantine Konstantinovich Flug with Added Notes and an Introduction.' Thesis (MA), University of Chicago, 1968.

FOSS, T. N. (1). 'A Jesuit Encyclopedia for China; a Guide to Jean-Baptiste DuHalde's Description ... de la Chine (1735).' Thesis (Ph.D.), University of Chicago, 1979.

FOSTER, W. (1). *Early Travels in India, 1583–1619.* S. Chand & Co., India, 1968. Reprinted with permission of the Oxford University Press, Bombay.

FRANKE, H. (19) (ed.) *Sung Biographies.* 3 vols. Steiner, Wiesbaden, 1976.

FRANKE, H. (27). 'A Mongol (Yüan) Calendar Fragment from Turfan.' *CINA*, 1964, **8**, 32.

FRANKE, H. (28). *Kulturgeschichtliches über die chinesische Tusche.* Bayerische Akad. d. Wiss., München, 1962.

FRANKE, W. (4). *An Introduction to the Sources of Ming History.* University of Malaya Press, Kuala Lumpur, 1968.

FRANKFORT, H. (4). *Cylinder Seals.* Macmillan, London, 1939.

FREY, J. P. (1). *Craft Unions of Ancient and Modern Times.* Washington, D.C., 1945.

FU SHEN (1). *Traces of the Brush; Studies in Chinese Calligraphy.* Yale University Art Gallery, 1977.

FUCHS, W. (2). *Der Jesuiten-Atlas der Kang-Hsi-Zeit.* Peking, 1943.

FUCHS, W. (8). 'Rare Chhing Editions of the *Keng-Chih-Thu*.' *SS*, 1947, **6**, 149.

FUCHS, W. (9). 'Der Kupferdruck in China vom 10 bis 19 Jahrhundert.' *GUJ*, 1950, 67.

FUCHS, W. (10) *Die Bilderalben für die Südreisen des Kaisers Kienlung im 18. Jahrhundert*. Franz Steiner Verlag, Wiesbaden, 1976.

FUHRMANN, O. W. (1). 'The Invention of Printing.' *DPN*, 1938, **3**, 25.

FUJIEDA AKIRA (1). 'The Tunhuang Manuscripts; a General Description.' *ZINB*, 1966, **9**, 1; 1969, **10**, 17.

FUJIEDA AKIRA (2). 'The Tunhuang Manuscripts.' In *ESCH*, 1975, 120.

GALLAGHER, L. J. (1) (tr.). *China in the 16th Century; the Journal of Matteo Ricci, 1583–1610*. Random House, New York, 1953.

GAMMELL, W. (1). *A History of American Baptist Missions in Asia, Africa, Europe and North America*. Boston, 1849.

GAN TJIANG-TEK (1). 'Some Chinese Popular Block-prints.' In Rijksmuseum voor Volkenkunde, *The Wonder of Man's Ingenuity*, pp. 26–36. Brill, Leiden, 1962.

GARDNER, K. (1). 'The Book in Japan.' In H. D. Vervliet, ed., *Liber Librorum; the Book through 5000 Years*, pp. 129–38. Arcade, Bruxelles, 1973.

GARNETT, R. (1). 'Early Arabian Paper Making.' *LIB*, 1903, 2nd ser., **4**, 1.

DE GAULLE, J. (1). Des végétaux employés au Japon pour la fabrication du papier.' *BSZA*, 1872 (2e ser.), **9**, 287.

GELB, I. J. (1). *A Study of Writing*. University of Chicago Press, 1952; rev. ed., 1963.

GENTRY, H. and GREENHOOD, D. (1). *Chronology of Books and Printing*. Rev. ed. Macmillan, New York, 1936.

GERNET, J. (6). 'La découverte d'un livre vieux de deux mille ans. *CFC*, 1961, **11**, 86.

GERNET, J. & WU CHI-YU (1) (eds.). *Catalogue des manuscrits chinois de Touen-houang (Fonds Pelliot chinois)*. Vol. 1, nos. 2001–2500. Bibliothèque nationale, Paris, 1970.

GHORI, S. A. K. and RAHMAN, A. (1). 'Paper Technology in Medieval India.' *IJHS*, 1966, **1** (no. 2), 133.

GILES, L. (1). 'A Note on the *Yung-Lo Ta Tien*.' *NCR*, 1920, **2**, 137.

GILES, L. (2). *An Alphabetical Index to the Chinese Encyclopaedia (Chhin Ting Ku Chin Thu Shu Chi Chhēng)*. British Museum, London, 1911.

GILES, L. (13). *Descriptive Catalogue of the Chinese Manuscripts from Tunhuang in the British Museum*. British Museum, London, 1957.

GILES, L. (15). 'Chinese Printing in the Tenth Century.' *JRAS*, 1925, 513.

GILES, L. (16). 'Early Chinese Printing.' *BMQ*, 1929, **4**, 86.

GILES, L. (17). 'Dated Chinese Manuscripts in the Stein Collection.' *BLSOAS*, 1933/35, **7**, 809; 1935/37, **8**, 1; 1937/39, **9**, 1; 1940/42, **10**, 317; 1943, **11**, 148.

GILROY, C. G. (1). *History of Silk, Cotton, Linen, Wool, and Other Fibrous Substances; including Observations on Spinning, Dyeing and Weaving; also an Account of the Pastoral Life of the Ancients, their Social State and Attainments in the Domestic Arts; with Appendices on Pliny's Natural History, on the Origin and Manufacture of Linen and Cotton Paper; on Felting, Netting, etc*. Harper, New York, 1845.

GLAISTER, G. (1). *An Encyclopedia of the Book; Terms Used in Papermaking, Printing, Bookbinding and Publishing*. World Pub. Co., Cleveland, 1960.

GLUB, J. (1). *A Short History of the Arab Peoples*. Hodder and Stoughton, London, 1969.

GODE, P. K. (8). 'The Migration of Paper from China to India.' In *Studies in Indian Cultural History*, vol. 3, pp. 1–12. Poona, 1964.

GODE, P. K. (9). 'Studies in the Regional History of the Indian Paper Industry.' *BV*, 1944, **5**, 87.

GOLDSCHMIDT, E. P. (1). *Medieval Texts and their First Appearance in Print*. Oxford, 1943.

GOLDSCHMIDT, E. P. (2). *The Printed Book of the Renaissance; Three Lectures on Type, Illustration, Ornament*. Cambridge, 1950.

GOODRICH, L. C. (2). *The Literary Inquisition of Chhien-Lung*. American Council of Learned Societies, New York, 1935; reprint with addenda & corrigenda, Paragon, New York, 1966.

GOODRICH, L. C. (7). 'The Revolving Bookcase in China.' *HJAS*, 1942–3, **7**, 130.

GOODRICH, L. C. (27). 'A Bronze Block for the Printing of Chinese Paper Currency (ca. 1287).' *MNANS*, 1950, **4**, 127.

GOODRICH, L. C. (28). 'The Development of Printing in China and its Effects on the Renaissance under the Sung Dynasty (960–1279).' *JRAS/HK*, 1963, **3**, 36.

GOODRICH, L. C. (29). 'Earliest Printed Editions of the Tripitaka.' *VBQ*, 1953–54, **19**, 215.

GOODRICH, L. C. (30). 'The Origin of Printing.' *JAOS*, 1962, **82** (no. 4), 557.

GOODRICH, L. C. (31). 'Printing; a New Discovery.' *JRAS/HK*, 1967, **7**, 39. 2 ill.

GOODRICH, L. C. (32). 'Printing: Preliminary Report on a New Discovery.' *TCULT*, 1967, **8** (no. 3), 376.

GOODRICH, L. C. (33). 'Paper; a Note on its Origin.' *ISIS*, 1951, **42**, 145.

GOODRICH, L. C. (34). 'More on the *Yung-Lo Ta-Tien*.' *JRAS/HK*, 1970, **10**, 17.

GOODRICH, L. C. (35). 'Two Notes on Early Printing in China.' *P. K. Gode Commemoration Volume*, 1960, no. 93, 117 (Poona Oriental Series).

GOODRICH, L. C. (36). 'Two New Discoveries of Early Block Prints.' In Hans Widmann, ed., *Der gegenwärtige Stand der Gutenberg-Forschung*, Bd. 1, p. 214. Anton Hiersemann, Stuttgart, 1972.

GOODRICH, L. C. (37). 'Tangut Printing.' *GUJ*, 1976, 64.

GOODRICH, L. C. (38). 'Movable Type Printing; Two Notes.' *JAOS*, 1974, **94**, 476.

GOODRICH, L. C. (39) (ed.). *Illustrated Chinese Primer; Hsin Pien Tui Hsiang Ssu Yen.* University of Hong Kong, 1967; reprint, 1976.

GOODRICH, L. C. & FANG CHAO-YING (1). *Dictionary of Ming Biography, 1368–1644.* 2 vols. Columbia University Press, New York, 1976.

GOSCHKEWITSCH, J. (2). *Die Methode der Tuschbereitung nebst einem Anhange über die Schminke.* Berlin, 1858.

DE GRAAF, H. J. (1). *The Spread of Printing, Eastern Hemisphere: Indonesia.* Vangendt, Amsterdam, 1969.

GRAY, W. S. (1). *The Teaching of Reading and Writing; an International Survey.* Unesco, Paris, 1956.

GREEN, J. B. (1). *Notes of Making Hand-made Paper.* Maidstone School of Arts and Crafts, Maidstone, England, 1945.

GRINSTEAD, E. D. (1). *Title Index to the Descriptive Catalogue of Chinese Manuscripts from Tunhuang in the British Museum.* British Museum, London, 1963.

GRINSTEAD, E. D. (2). *Guide to Chinese Decorative Script.* Drawn by Yoshihiko Zizuka. Studentlitteraturi, Lund, 1970. (Scandinavian Institute of Asian Studies Monograph series no. 4.)

VAN GULIK, R. H. (2). *Mi Fu on Ink-stones.* Henri Vetch, Peiping, 1938.

VAN GULIK, R. H. (3). *'Pi Hsi Thu Khao'; Erotic Colour-Prints of the Ming Period, with an Essay on Chinese Sex Life from the Han to the Chhing Dynasty (− 206 to + 1644).* 3 vols. in case. Privately printed, Tokyo, 1951 (limited edition, 50 copies). Crit. W. L. Hsü, *MN*, 1952, **8**, 455; H. Franke, *ZDMG*, 1955 (NF) **30**, 380.

VAN GULIK, R. H. (9). *Chinese Pictorial Art as Viewed by the Connoisseur.* Instituto Italiano per il Medio ed Estremo Oriente, Rome, 1958. (Serie Orientale Roma 19.)

VAN GULIK, R. H. (10). 'Yin-ting (Silver Nails) and Yin-ting (Silver Ingots).' *ORE*, **2**, 204.

VAN GULIK, R. H. (11). 'A Note on Inkcakes.' *MN*, 1955–56, **11**, 84.

GUPPY, H. (2). *Stepping-stones to the Art of Typography*, with Fourteen Facsimiles. Manchester University Press, 1928.

GUPPY, H. (3). 'The Evolution of the Art of Printing.' *BJRL*, 1940, **24**, 198.

DU HALDE, J. B. (1). *Description Géographique, Historique, Chronologique, Politique et Physique de l'Empire de la Chine et de la Tartarie Chinoise.* 4 vols. Paris, 1735, 1739; The Hague, 1736. Eng. tr. R. Brookes, London, 1736, 1741.

HALL, H. R. (1). *Scarabs.* London, 1929.

HAMILTON, C. E. (1) (tr.) *Kamisuki Chōhōki; a Handy Guide to Papermaking, after the Japanese edition of 1798*, by Kunisaki Jihei; illus. by Seichūan Tōkei (pseud.). Book Arts Club, University of California, Berkeley, 1948. English and Japanese texts on opposite pages.

HANSARD, T. (1). *Typographia; an Historical Sketch of the Origin and Progress of the Art of Printing.* London, 1825.

HARBIN, R. (1). *Origami; the Art of Paper Folding.* Hodder Paperbacks, London, 1971.

HARDERS-STEINHÄUSER, M. (1). 'Microscopic Study of Some Ancient East Asian Tun-huang papers.' Eng. tr. from *PAPR*, 1969, 23 (no. 4), 210; (no. 5), 272.

HARDERS-STEINHÄUSER, M. & JAYME, G. (1). 'Study of the Paper of Eight Different Na-Khi Manuscripts.' Eng. tr. of 'Untersuchung des Papiers acht verschiedener alter Na-Khi Handschriften auf Rohstoff und Herstellungsweise.' *VOHD*, 1963, *suppl. 2*, 50.

HARGRAVE, C. P. (1). *A History of Playing Cards and a Bibliography of Cards and Gaming.* Houghton Mifflin, Boston, 1930.

HARRIS, J. (3). *A Pleasant and Compendious History of the First Inventors and Instituters of the Most Famous Arts, Mysteries, Laws, Customs and Manners in the Whole World.* London, 1686.

HART, H. H. (1). *Marco Polo, Venetian Adventurer.* University of Oklahoma Press, Norman, 1967.

HAYES, J. R. *Invention; its Attributes and Definitions.* Addison-Wesley Press, Cambridge, Mass., 1942.

HEDIN, S. (1). *Reports from the Scientific Expedition to the Northwestern Province of China under the Leadership of Dr. Sven Hedin.* Goteborg Elander, Stockholm, 1943–45. (Sino-Swedish Expedition Publications nos. 23–26.)

HEJZLAR, J. (2). *Chinese Paper Cut-outs.* Photographs by W. and B. Forman. Tr. I. Havlu. Spring House, London, 1960.

HEJZLAR, J. (3). *Early Chinese Graphics.* Octopus Books, London, 1973.

HELD, M. (1). 'China as Illustrated in European Books, 1705–1810.' Thesis (MA), University of Chicago, 1973.

HELLER, J. (1). *Papermaking.* Watson-Guptill Publications, New York, 1978.

HENNING, W. B. (2). 'The Date of the Sogdian Ancient Letters.' *BLSOAS*, 1948, **12**, 601.

HERBERT, T. (1). *Travels in Persia, 1627–1629.* Abridged and ed. by Sir W. Foster, C. I. E., with an introduction and notes. Routledge, London, 1928.

HERRING, R. (1). *Paper and Paper Making, Ancient and Modern.* Longmans, London, 1855.

HERRMANN, A. (13). *An Historical Atlas of China.* New ed. by N. Ginsburg, with prefatory essay by Paul Wheatley. Aldine, Chicago, 1966.

HERVOUET, Y. (2). 'Les manuscrits chinois de l'École Française d'Extrême Orient.' *BEFED*, 1955, **47** (fasc. 2), 435.

HERVOUET, Y. (3) (ed.). *A Sung Bibliography (Bibliographies des Sung).* Chinese University Press, Hong Kong, 1978.

HESSIG, W. (1). *Catalogue of Mongol Books, Manuscripts and Xylographs.* The Royal Library, Copenhagen, 1971.

HICKMAN, B. F. (1). 'A Note on the Hyakumanto Dharani.' *MN*, 1975, **30** (no. 1), 87.

HIND, A. M. (1). *An Introduction to a History of Woodcut; with a Detailed Survey of Work Done in the Fifteenth Century*. 2 vols. Constable, London, 1935.

HIRSCH, R. (1). *Printing, Selling, and Reading, 1450–1550*. Harrassowitz, Wiesbaden, 1967.

HIRTH, F. (1). *China and the Roman Orient*. Kelly & Walsh, Shanghai, 1885; Alfred A. Knopf, 1925.

HIRTH, F. (28). 'Western Appliances in the Chinese Printing Industry.' *JRAS/NCB*, 1886, **20**, 163.

HIRTH, F. (29). 'Die Erfindung des Papiers in China.' *TP*, 1890, **1**, 1.

HIRTH, F. & ROCKHILL, W. W. (1) (tr.). *Chau Ju-Kua; his Work on the Chinese and Arab Trade in the + 12th and + 13th Centuries, Entitled Chu-Fan-Chih*. Imperial Academy of Sciences, St. Petersburg, 1911; Paragon Book Reprint Corp., New York, 1966.

HITTI, P. K. (1). *History of the Arabs from the Earliest Times to the Present*. 10th ed. Macmillan, London; St. Martin's Press, New York, 1970.

HO PING-TI (2). *The Ladder of Success in Imperial China; Aspects of Social Mobility, 1368–1911*. Columbia University Press, 1962.

HOERNLE, A. F. R. (1). 'Who Was the Inventor of Rag-paper?' *JRAS*, 1903, 663.

HOERNLE, A. F. R. (2). 'Note on the Invention of Rag-paper.' *JRAS*, 1904, 548.

HOH SHAI-WONG (1). 'Supplement to McClure; Native Paper Industry in Kwangtung.' *LSJ*, 1938, **17**, 71.

HOLLOWAY, O. E. (1). *Graphic Art of Japan; the Classical School*. London, 1957.

HOLT, W. S. (1). 'The Mission Press in China.' *CRR*, 1879, **10**, 206.

HONDA ISAO (1). *The World of Origami*. Japan Publication Trading Co., Tokyo & New York, 1965.

HORNE, C. (1). 'Paper Making in the Himalayas.' *IAQ*, 1877, **6**, 94.

HOSIE, A. (5). *Szechwan; its Products, Industries and Resources*. Kelly and Walsh, Shanghai, 1922.

HOWORTH, H. H. (1). *History of the Mongols, from the 9th to the 19th Century*. London and New York, 1888.

HSIAO CHING-CHANG (1). 'Chinese Wood-block Printing.' *CLIT*, 1962, **1**, 98.

HU SHIH (12). 'The Gest Oriental Library at Princeton University.' *PULC*, 1954, **15**, 113.

HUDSON, G. F. (1). *Europe and China; a Survey of their Relations from the Earliest Times to 1800*. Arnold, London, 1931; Beacon Press, Boston, 1961.

HUGHES, S. (1). *Washi; the World of Japanese Paper*. Kodansha International, Tokyo, 1978.

HULBERT, H. B. (4). 'Xylographic Art in Korea.' *KRW*, 1901, 97.

HÜLLE, H. (2). *Über den alten chinesischen Typendruck und seine Entwickelung in den Ländern des Fernen Ostens*. Berlin, 1923.

HULME, E. W. (1). *Statistical Bibliography in Relation to the Growth of Modern Civilization*. Grafton, London, 1923.

HUMMEL, A. W. (2) (ed.). *Eminent Chinese of the Chhing Period, 1644–1912*. 2 vols. Government Printing Office, Washington, D.C., 1943–4.

HUMMEL, A. W. (23). 'The Development of the Book in China.' *JAOS*, 1941, **61**, 71.

HUMMEL, A. W. (24). 'Movable Type Printing in China; a Brief Survey.' *LC/QJCA*, 1943, **1**, 18.

HUMMEL, A. W. (25). 'The printed Herbal of 1249 A.D.'' *ISIS*, 1941, **33**, 439.

HUMPHREYS, H. N. (1). *A History of the Art of Printing, from its Invention to its Widespread Development in the Middle of the Sixteenth Century*. Quaritch, London, 1867.

HUNT, C. (1). 'King Yung-Lo Stole a March on Gutenberg.' *SCENE*, 1952, **3**, 20.

HUNTER, D. (1). *Primitive Papermaking*. Mountain House, 1927. (Limited ed. of 200.)

HUNTER, D. (2). *Chinese Ceremonial Paper*. Mountain House, 1937. (Limited ed. of 125.)

HUNTER, D. (3). *Handmade Paper and its Watermarks; a Bibliography*. The Mill, Marlborough-on-Hudson, 1917.

HUNTER, D. (4). 'Laid and Wove.' *ARS* (for 1921), 1922, 587.

HUNTER, D. (5). *The Literature of Papermaking, 1390–1800*. Mountain House, 1925. (Limited ed. of 190.)

HUNTER, D. (6). *The Romance of Watermarks*. Stratford Press, Cincinnati, 1939 (Limited ed. of 210.)

HUNTER, D. (7). *Old Papermaking*. Mountain House, 1923. (Limited ed. of 200.)

HUNTER, D. (8). *Old Papermaking in China and Japan*. Mountain House, 1932. (Limited ed. of 200.)

HUNTER, D. (9). *Papermaking; the History and Technique of an Ancient Craft*. A. A. Knopf, New York, 1943; 2nd ed., 1947; Dover Publications, New York, 1978.

HUNTER, D. (10). *Papermaking by Hand in America*. Mountain House, 1950. (Limited ed. of 200.)

HUNTER, D. (11). *Papermaking by Hand in India*. Pynson Printers, New York, 1939. (Limited ed. of 370.)

HUNTER, D. (12). *Papermaking in Indo-china*. Mountain House, 1947. (Limited ed. of 182.)

HUNTER, D. (13). *Papermaking in Pioneer America*. University of Pennsylvania Press, 1952.

HUNTER, D. (14). *Papermaking in Southern Siam*. Mountain House, 1936. (Limited ed. of 115.)

HUNTER, D. (15). *Papermaking in the Classroom*. Manual Arts Press, Peoria, 1931.

HUNTER, D. (16). 'Papermaking Moulds in Asia.' *GUJ*, 1940, 9.

HUNTER, D. (17). *A Papermaking Pilgrimage to Japan, Korea and China*. Pynson Printers, New York, 1936. (Limited ed. of 370.)

HUNTER, D. (18). *Papermaking through Eighteen Centuries*. Wm. F. Rudge, New York, 1930.

HUNTER, D. (19). 'Papermaking in the South Seas.' *PPMC*, 1927, **25**, 580.

HUNTER, D. (20). 'Sacred Papers of the Orient.' *PTJ*, 1943, **16**, 17.

ISHIDA MOSAKU (1). *Japanese Buddhist Prints*. Tr. Charles S. Terry. Kodansha International, Tokyo, 1974.

IVINS, W. M., JR. (1). 'A Neglected Aspect of Early Print-making.' *MMB*, 1948, **7**, 51.
IVINS, W. M., JR. (2). *Prints and Visual Communication*. Harvard University Press, 1953.
IWASAKI TAKEO (1). 'Printed Culture.' *ASS*, 1963, **8**, 44.

JACKSON, J. B. (1). *An Essay on the Invention of Engraving and Printing in Chiaro Oscuro, as Practised by Albert Durer, Hugo di Carpi, & c. and the Application of It to the Making of Paper Hangings of Taste, Duration and Elegance*. A. Millar, London, 1754.
JAGGI, O. P. (1). *Science and Technology in Medieval India*. Atma Ram & Sons, Delhi, 1977. (History of Science and Technology in India, Vol. 7.)
JAMETEL, M. (1). *L'encre de Chine; son histoire et sa fabrication d'après des documents chinois*. Ernest Leroux, Paris, 1882.
DE JANCIGNY, P. B. (1). 'Le papier en Chine.' *TP*, 1908, **9**, 589.
JEON SONG-WOON (1). *Science and Technology in Korea; Traditional Instruments and Techniques*. MIT Press, Cambridge, Mass., 1974.
JOHNSON, P. (1). *Creating with Paper; Basic Forms and Variations*. Seattle, 1958.
JONES, G. H. (2). 'Printing and Books in Asia.' *KR*, 1898, **5** (no. 2), 55.
JORDANUS, C. (1). *Mirabilia descripta; Wonders of the East*. By Friar Jordanus of the Order of Preachers and Bishop of Columbum in India the Greater, ca. 1330. Tr. from the Latin original and published in Paris in 1839 in the *Recueil de Voyages et de Mémoires*, of the Society of Geography, with the addition of a commentary by Colonel H. Yule. Hakluyt Society, London, 1858.
JUGAKU BUNSHO (1). *Paper-making by Hand in Japan*. Meiji Shobo, Tokyo, 1959.
JUGAKU BUNSHŌ (2). 'Where They Still Make Paper by Hand.' *JQ*, 1957, **4**, 249.
JULIAN, A. L. (1). 'A Printing Millennary.' *HTD*, 1954, **4**, 668.
JULIEN, S. (12). 'Documents sur l'art d'imprimer a l'aide de planches au bois, de planches au pierre et de types mobiles.' *JA*, 1847 (4e ser.), **9**, 508.
JULIEN, S. (13). 'Fabrication du papier du bambou.' *ROA*, 1856, **20**, 74.
JULIEN, S. (14). *Industries anciennes et modernes de l'empire chinois*. Eugene Lacroix, Paris, 1869.
JULIEN, S. & CHAMPION, P. (2). 'Procédés des chinois pour la fabrication de l'encre.' *AVP*, 1833, **53**, 308–14.

KAGITCI, M. A. (1). *Historical Study of the Paper Industry in Turkey*. Gradik Sanatlar Matbaasi, Istanbul, Turkey, 1976.
KARABACEK, J. (1). *Der Papyrusfund von El-Faijûm*. Wien, 1882.
KARABACEK, J. (2). *Das Arabische Papier; eine Historisch-Antiquarische Untersuchung*. Wien, 1887.
KARABACEK, J. (3). 'Neue Quellen zur Papiergeschichte.' *MSPER*, 1888, **4**, 75.
KARLGREN, B. (1). *Grammata Serica; Script and Phonetics on Chinese and Sino-Japanese*. BMFEA, 1940, **12**, 1.
KARLGREN, B. (18). 'Early Chinese Mirror Inscriptions.' *BMFEA*, 1934, **6**, 9.
KAWASE KAZUMA (1). *An Introduction to the History of Pre-Meiji Publishing; a History of Wood-block Printing in Japan*. Tr. and annotated by Yukihisa Suzuki and May T. Suzuki. Yushodo Booksellers Ltd., Tokyo, 1973.
KAWASE KAZUMA (2). *Old Printed Books in Japan*. Japan Foundation, Tokyo, 1979.
KECSKES, LILY CHIA-JEN (1). 'A Study of Chinese Inkmaking; Historical, Technical and Aesthetic.' Thesis (M.A.); University of Chicago, 1981.
KEIGHTLEY, D. N. (1). *Sources of Shang History; the Oracle-Bone Inscriptions of Bronze Age China*. University of California Press, Berkeley, 1978.
KELLING, R. et al. (2). *Zum chinesischen Stempel- und Holztafeldruck, nebst vermischten Beiträgen aus dem Gesamtgebiete der Schrift- und Buchgeschichte*. Harrassowitz, Leipzig, 1940.
KENMORE, A. H. (1). 'Bibliographié coréene.' *KR*, 1897, **6** (no. 6).
KENYON, F. G. (1). *Ancient Books and Modern Discoveries*. Caxton Club, Chicago, 1927.
KHAN, M. S. (1). 'Early History of Bengali Printing.' *LQ*, 1962, **32**, 51.
KHAN, M. S. (2). *The Early Bengali Printed Books*. Gutenberg, 1966.
KIELHORN, F. (1). 'The Mungir Copper-plate Grant of Devapaladeva.' *IAQ*, 1892, **31**, 253.
KIM DOO-JONG (1). 'History of Korean Printing (until the Yi Dynasty).' *KJ*, 1963, **3**, 22.
KIM DOO-JONG (2). 'Movable Types of the Yi Dynasty Seen from Calligraphic Form.' *BBSK*, 1960, **1**, 17.
KIM HYO-GUN (1). 'Printing in Korea and its Impact on Her Culture.' Thesis (M.A.), University of Chicago, 1973.
KIM WON-YONG (1). *Early Movable Type in Korea*. Eul-yu Pub., Seoul, 1954.
KIM WON-YONG (2). 'Supplementary Notes on the Kemi-ja Type.' *BBSK*, 1960, **1**, 35.
KING, S. (1). 'Essai de bibliographie en vue d'une "Historie du livre chinois".' *BUA*, 1938/39 (2e ser.), **38**, 69.
KING, S. (2). 'L'invention du papier chinois, d'après les sources chinois.' *BUA*, 1933 (2e ser.), **25**, 14.
KLAPROTH, J. (1). *Lettre à M. le Baron A. de Humboldt sur l'invention de la boussole*. Paris, 1834.
KLAPROTH, J. (10). 'Sur l'origine du papier-monnaire.' *MRA*, 1824, 375.
KOOIJMAN, S. (1). *Ornamented Bark-cloth in Indonesia*. Leiden, 1963.
KOOPS, M. (1). *Historical Account of the Substances Which Have Been Used to Describe Events, and to Convey Ideas, from the Earliest Date to the Invention of Paper*. Jacques, London, 1800; 2nd ed., 1801.

KRACKE, E. A., JR. (1). *Civil Service in Early Sung China, 960–1067.* Harvard University Press, 1953.

KRACKE, E. A., JR. (2). 'Sung Society; Change within Tradition.' *JAS*, 1955, **14**, 479.

KRACKE, E. A., JR. (3). 'Family vs. Merit in the Examination System.' *HJAS*, 1947, **10**, 103.

KRACKE, E. A., JR. (5). 'Region, Family, and Individual in the Chinese Examination System.' In J. K. Fairbank, ed., *Chinese Thought and Institutions*, pp. 251–68. University of Chicago Press, 1957.

KUNISAKI JIHEI (1). *An Abridged Reproduction of 'Kamisuki Chōchōki'; the Handbook for Papermaking, Originally Published at Osaka in 1798.* IIS Craft, Tokyo, 1963.

LABARRE, E. J. (1) (ed.). *A Dictionary of Paper and Paper-making Terms with Equivalents in French, German, Dutch and Italian.* N. V. Swets & Zeitlinger, Amsterdam, 1937.

LABARRE, E. J. (2) (ed.). *Monumenta chartae papyraceae historiam illustrantia; or Collection of Works and Documents Illustrating the History of Paper.* 1– . The Paper Publications Society, Hilversum (Holland), 1950– .

LACH, D. F. (5). *Asia in the Making of Europe.* 2 vols. in 6 parts. University of Chicago Press, 1965– .

LAKSHMI, R. (1). 'Handmade Paper in India.' *PAM*, 1957, **26**, 31.

LALOU, M. (1). 'The Most Ancient Tibetan Scrolls Found at Tunhuang.' *RO*, 1957, **21**, 150.

LALOU, M. (2). *Inventaire des manuscrits tibetains de Touen-houang conservés à la Bibliothèque Nationale (Fonds Pelliot tibétain).* 3 vols. Bibliothèque Nationale, Paris, 1939–61.

LANG JU-HENG (1). 'The Four Treasures of the Study.' *WE*, 1970, **15** (no. 3), 6.

LAO KAN (1). 'From Wooden Slip to Paper.' *CCUL*, 1967, **8** (no. 4), 80.

LATOUR, A. (1). 'Paper; a Historical Outline.' *CIBA/T*, 1949, **72**, 2630.

LATOURETTE, K. S. (1). *The Chinese; their History and Culture.* 3rd ed. Macmillan, New York, 1946.

LAUFER, B. (1). *Sino-Iranica; Chinese Contributions to the History of Civilisation in Ancient Iran. FMNHP/AS*, 1919, 15 (no. 3). (Pub. no. 201.) Rev. and crit. Chang Hung-Chao, *MGSC*, 1925 (ser. B), no. 5.

LAUFER, B. (4). 'Pre-history of Aviation.' *FMNHP/AS*, 1928, *18* (no. 1). (Pub. no. 253.)

LAUFER, B. (24). 'The Early History of Felt.' *AAN*, n.s., 1930, **32**, 1.

LAUFER, B. (48). *Paper and Printing in Ancient China.* Caxton Club, Chicago, 1931; repr. Burt Franklin, New York, 1973.

LAUFER, B. (49). 'Papier und Druck im alten China.' *IMP*, 1934, **5**, 65.

LAUFER, B. (50). *Descriptive Account of the Collection of Chinese, Tibetan, Mongol, and Japanese Books in the Newberry Library.* Newberry Library, Chicago, 1913.

LAUFER, B. (51). 'History of the Finger-print System.' *ARSI* (for 1912), 1913, 631.

LAURES, J. (1). *The Ancient Japanese Mission Press.* Monumenta Nipponica, Tokyo, 1940.

LECOMTE, LOUIS (1). *Nouveaux mémoires sur l'état présent de la Chine.* Anisson, Paris, 1696. Eng. tr. *Memoirs and Observations ...* London, 1697.

VON LECOQ, A. (1). *Buried Treasures of Chinese Turkestan; an Account of the Activities and Adventures of the 2nd and 3rd German Turfan Expeditions.* Allen & Unwin, London, 1928. Eng. tr. by A. Barwell, *Auf Hellas Spuren in Ost-Turkestan.* Berlin, 1926.

VON LECOQ, A. (2). *Von Land und Leuten in Ost-Turkestan.* Hinrichs, Leipzig, 1928.

VON LECOQ, A. (3). 'Exploration archéologique à Tourfan.' *JA*, 1909 (10e ser), **14**, 321.

VON LECOQ, A. (4). 'Origin, Journey and Results of the First Royal Prussian Expedition to Turfan.' *JRAS*, 1909, 299.

LEDYARD, G. K. (1). 'Two Mongol Documents from the Koryo Sa.' *JAOS*, 1963, **83**, 225.

LEDYARD, G. K. (2). 'The Discovery in the Monastery of the Buddha Land.' *CLC*, 1967, **16** (no. 3), 3.

LEE, S. B. & KIM, W. Y. (1). *A History of the Korean Alphabet and Movable Types.* Ministry of Culture and Information, Seoul, 1970.

LEGGE, J. (8) (tr.). *The Chinese Classics, etc.* Vol. 4, pts. 1 and 2. *The Book of Poetry.* Land Crawford, Hong Kong, 1871; Trübner, London, 1871. Repr. Commercial Press, Shanghai, n.d.; Peiping, 1936; Hongkong University Press, 1960.

LEGMAN, G. (1). 'Bibliography of Paper-folding.' *JOB*, 1952, 3.

LEHMANN-HAUPT, H. (1). *Seventy Books about Book-making; a Guide to the Study and Appreciation of Printing.* Columbia University Press, New York, 1941.

LEHMANN-HAUPT, H. (2). *Gutenberg and the Master of the Playing Cards.* Yale University Press, 1966.

LEIF, I. P. (1). *An International Sourcebook of Paper History.* Shoe String Press, Hamden, Conn., 1978.

LENHART, J. M. (1). *Pre-Reformation Printed Books; a Study in Statistical and Applied Bibliography.* New York, 1935. (Franciscan Studies, no. 14.)

LESLIE, D. D., MACKERRAS, C. & WANG GUNGWU (1) (ed.). *Essays on the Sources of Chinese History.* Australian National University Press, Canberra, 1973; University of South Carolina, Columbia, S.C., 1975.

LEVEY, M. (1). 'Chemical Technology in Mediaeval Arabic Bookmaking.' *TAPS*, 1962, **52** (no. 4), 5.

LEWIS, B. (1). *The Arabs in History.* Harper & Brothers, New York, 1960.

LEWIS, J. (1). *The Anatomy of Printing; the Influence of Art and History on its Design.* Faber and Faber, London, 1970.

LEWIS, N. (1). *L'industrie du papyrus dans l'Egypt Gréco-Romain.* Librarie L. Rodstein, Paris, 1934.

LI CHHIAO-PHING (1). *The Chemical Arts of Old China.* Journal of Chemical Education, Easton, Pa., 1948; repr. AMS Press, New York, 1979.

Li Chun (2). 'A Woodcut Artists' Group.' *CLIT*, 1964, **11**, 110.

Li Shao-Yen (1). 'Writing-paper with Art Designs.' *CLIT*, 1961, no. 6, 135.

Li Shu-Hua (4). 'The Early Development of Seals and Rubbings.' *CHJ/T*, 1958, **1** (no. 3), 61.

Li Shu-Hua (5). *The Spread of the Art of Paper-making and the Discovery of Old Paper*. National Historical Museum, Taipei, 1958. Text in Chinese and English.

Libiszowski, S. (1). 'Papiernia-Muzeum.' *PPA*, 1971, **27**, 53.

Lin Yü-Thang (1). *A History of the Press and Public Opinion in China*. Kelly & Walsh, Shanghai, 1936; University of Chicago Press, 1936; Greenwood Press, New York, 1968.

Lipman, M. (1). *How Men Kept Their Records*. Nelson, New York, 1934.

Liu Kuo-Chün (1). 'Paper-making and Printing.' *PC*, 1954, no. 12, 19.

Liu Kuo-Chün (2). *The Story of the Chinese Book*. Foreign Language Press, Peking, 1958.

Liu Kuo-Chün (3). 'Ancient Chinese Book Production.' *CLIT*, 1962, no. 5, 72.

Liu Tshun-Jen (4). *Chinese Popular Fiction in Two London Libraries*. Lung-men Bookstore, Hongkong, 1967.

Liu Tshun-Jen (5). 'The Compilation and Historical Value of the Tao-tsang.' In *ESCH*, 1975, 104.

Lo Chin-Thang (1). *The Evolution of Chinese Books*. National Historical Museum, Taipei, 1960? Text also in Chinese.

Loeber, E. G. (1). 'History of Paper-making in Australia.' *PG*, 1958, **8**, 74.

Loeber, E. G. (2). 'Prehistoric Origins of Paper.' *BIPH*, 1979, **13** (no. 4), 87.

Loehr, M. (1). *Chinese Landscape Woodcuts from an Imperial Commentary to a Tenth Century Printed Edition of the Buddhist Canon*. Harvard University Press, 1968.

Loewe, M. (1). 'Some Notes on Han-time Documents from Chüyen.' *TP*, 1959, **47**, 294.

Loewe, M. (4). *Records of Han Administration*. Cambridge, 1967.

Loewe, M. (13). 'Some Notes on Han-time Documents from Tun-huang.' *TP*, 1963, **50**, 150.

Loewe, M. (14). 'Manuscripts Found Recently in China; a Preliminary Survey.' *TP*, 1978, **63** (nos. 2–3), 7.

van der Loon, P. (2). 'The Manila Incunabula and Early Hokkein Studies.' *AM*, 1966, n.s., **12** (no. 1), 1.

Löwenthal, R. (1). 'Printing paper; its Supply and Demand in China.' *YJSS*, 1938, **1**, 107.

McClure, F. A. (1). 'Native Paper Industry in Kwangtung.' *LSJ*, 1927, **5**, 255.

McClure, F. A. (2). 'Some Chinese Papers Made on the Ancient "Wove" Type of Screen.' *LSJ*, 1930, **9**, 115.

McCune, G. M. (1). 'The Yi Dynasty Annals of Korea.' *JRAS/KB*, 1939, **29**, 57.

McDowell, R. H. (1). *Stamped and Inscribed Objects from Seleucia on the Tigris*. University of Michigan Press, Ann Arbor, 1935.

McGovern, M. P. (1). *Specimen Pages of Korean Movable Types*. Dawson's Book Shop, Los Angeles, 1966. Crit. T. H. Tsien, *LQ*, 1967, **37**, 40.

McGovern, M. P. (2). 'Early Western Presses in Korea.' *KJ*, 1967, **7**, 21.

McIntosh, G. (1). *The Mission Press in China; being a Silver Jubilee Retrospect of the American Presbyterian Mission*. American Presbyterian Mission Press, Shanghai, 1895.

McIntosh, G. (2). 'Printing in China.' *PENA*, 1923, **25**, 61.

McLuhan, H. M. (1). *The Gutenberg Galaxy; the Making of Typographic Man*. University of Toronto Press, 1962.

McMurtrie, D. C. (1). *The Book; the Story of Printing and Bookmaking*. Covici-Friede, New York, 1937; 3rd ed. rev., Oxford, 1943.

McMurtrie, D. C. (2). *The Golden Book*. Covici, Chicago, 1927.

McMurtrie, D. C. (3) (ed.). *The Invention of Printing; a Bibliography*. Chicago Club of Printing House Craftsmen, Chicago, 1942.

McMurtrie, D. C. (4). *Memorandum on the First Printing in Ceylon, with Bibliography of Ceylonese Imprints of 1737-1767*. Chicago, 1931.

McMurtrie, D. C. (5). *Memorandum on the History of Printing in the Dutch East Indies*. Chicago, 1935.

Ma Keh (1). 'The Weifang New Year Pictures.' *CLIT*, 1960, no. 8, 153.

Ma Thai-Lai [Ma Tai-Loi] (1). 'The Authenticity of the *Nan-Fang Ts'ao-Mu Chuang*.' *TP*, 1978, **64**, 218.

Mahdihassan, S. (49). 'Chinese Words in the Holy Koran; 5, Quirtas, Meaning Paper, and its Synonym, Kagaz.' *JUB*, 1955, **24**, 148.

Major, R. H. (2). *India in the Fifteenth Century; Being a Collection of Narratives of Voyages to India, in the Century Preceding the Portuguese Discovery of the Cape of Good Hope; from Latin, Persian, Russian and Italian Sources*. Hakluyt Society, London, 1857.

Mangamma, J. (1). *Book Printing in India; with Special Reference to the Contribution of European Scholars to Telugu (1746–1857)*. Bangorey Books, Nellore, S. India, 1975.

Marchlewska, J. (1). 'Das Polnische Papiermuseum in Duszniki (Reinerz).' *PG*, 1972, **22**, 27.

Marchlewska, J. (2). 'Niemieckie Muzeum Papieru.' *PPA*, 1973, **29**, 22.

Martin, G. T. (1). *Egyptian Administrative and Private-Name Seals*. Ashmolean Museum, Griffith Institute, Oxford, 1971.

Martinique, E. (1). 'The Binding and Preservation of Chinese Double-leaved Books.' *LQ*, 1973, **43**, 227.

Martinique, E. (2). *Chinese Traditional Bookbinding; a Study of its Evolution and Techniques*. Thesis (MA). University of Chicago, 1972; Chinese Materials Center, San Francisco & Thaipei, 1983. (Asian Library Series, No. 19.)

MASPERO, H. (29). *Les documents de la troisième expédition de Sir Aurel Stein en Asie centrale.* British Museum, London, 1953.

MATTICE, H. A. (1). *English-Chinese-Japanese Lexicon of Bibliographical, Cataloguing and Library Terms.* New York Public Library, 1944.

MEDINA, J. C. (1). *La imprenta en Manila desde sus origenes hasta 1810.* Impreso y Grabads en Casa de Autor, Chile, 1896.

MEI YI-PAO (1) (tr.). *The Ethical and Political Works of Motse.* Probsthain, London, 1929.

DE MENDOZA, J. G. (2). *The History of the Great and Mighty Kingdom of China and Situation Thereof.* Ed. by George T. Staunton with an introduction by R. H. Major. Burt Franklin, New York, 1970. (Reprint of Hakluyt Society Works.) 2 vols.

MENZEL, J. M. (1) (ed.). *The Chinese Civil Service; Career Open to Talent?* Heath, Boston, 1963.

MICHELL, C. A. & HEPWORTH, T. C. (1). *Inks; their Composition and Manufacture.* Griffin, London, 1916.

MIDDLETON, T. C. (1). *Some Notes on the Bibliography of the Philippines.* The Free Library, Philadelphia, 1900. (Bulletin of the Free Library of Philadelphia, no. 4.)

MILLER, C. R. (1). 'An Inquiry into the Technical and Cultural Pre-requisites for the Invention of Printing in China and the West.' Thesis (MA), University of Chicago, 1975.

MILLER, J. A. (1). *Pulp and Paper History; a Selected List of Publications on the History of the Industry in North America.* St Paul, Minn., 1963.

MITCHELL, C. H. (1). *The Illustrated Books of Naga, Maruyama, Shijo and Other Related Schools of Japan; a Bibliography.* Dawson's Bookshop, Los Angeles, 1972.

MONTELL, G. (4). 'To the History of Writing and Printing.' *ETH*, 1942, **7**, 166.

DE MORGA, A. (2). *The Philippine Islands, Moluccas, Siam, Cambodia, Japan, and China at the Close of the Sixteenth Century,* tr. from the Spanish, with Notes and a Preface. Hakluyt Society, London, 1858.

MORRISON, H. M. (1). 'Making Books in China.' *CGJ*, 1949, **39**, 234.

MOULE, A. C. (18). 'Chinese Printing of the Tenth Century.' *JRAS*, 1925, 716.

MOULE, A. C. & PELLIOT, P. (1) (tr. and annot.). *Marco Polo (+1254–1325); the Description of the World.* 2 vols. Routledge, London, 1938. Further notes by P. Pelliot (posthumously pub.). 2 vols. Impr. Nat., Paris, 1960.

MUNDY, P. (1). *The Travels of Peter Mundy in Europe and Asia, 1608–1667.* Ed. R. C. Temple. Hakluyt Society, London, 1907–36.

MUNSELL, J. (1). *Chronology of the Origin and Progress of Paper and Paper Making.* 5th. ed. Albany, privately printed, 1876.

MURRAY, J. (1). *Practical Remarks on Modern Paper, with an Introductory Account of its Former Substitutes, also Observations on Writing Inks, the Restoration of Illegible Manuscripts, and the Preservation of Important Deeds from the Effects of Damp.* T. Cadell, London, 1829.

MURRAY, W. D. & RIDNEY, F. J. (1). *Fun with Paper-folding.* Revell, New York, 1928, 1953.

NA CHIH-LIANG (1). 'Chinese Woodcut Illustration.' *NPMB*, 1970, **5** (no. 5), 6.

NACHBAUR, A. (1). 'Le papier en Chine.' *CHINE*, 1923, **42**, 573.

NARAYANASWAMI, C. K. (1). *The Story of Handmade Paper Industry.* 2nd ed. Khadi and Village Industries Commission, Bombay, India, 1961.

NARITA KIYOFUSA (1). 'Suminagashi.' *PAM*, 1955, **24**, 27.

NARITA KIYOFUSA (2). 'The First Machine-made Paper in Japan.' *PAM*, 1957, **26**, 11.

NARITA KIYOFUSA (3). 'Japanese Paper and Paper Products.' *PG*, 1959, **9**, 84.

NARITA KIYOFUSA (4). 'Making Paper by Hand in Japan.' *PAM*, 1962, **31**, 38.

NARITA KIYOFUSA (5). 'A Brief History of Papermaking by Hand in Japan.' *PAM*, 1965, **34**, 5.

NARITA KIYOFUSA (6). *A Life of Tshai Lun and Japanese Papermaking.* Rev. ed. Dainihon Press, Tokyo, 1966.

NARITA KIYOFUSA (7). 'Paper Museum in Tokyo (Japan).' *PG*, 1954, **4**, 29.

NEEDHAM, JOSEPH (47). 'Science and China's Influence on the West.' Art. in *The Legacy of China*, ed. R. N. Dawson. Oxford, 1964, p. 234.

NEEDHAM, JOSEPH (58). 'The Chinese Contribution to Science and Technology.' Art. in *Reflections on our Age* (Lectures delivered at the Opening session of UNESCO at Sorbonne, Paris, 1964), ed. D. Hardman & S. Spender. Wingate, London, 1948, p. 211.

NEEDHAM, JOSEPH (64). *Clerks and Craftsmen in China and the West* (Collected Lectures and Addresses). Cambridge, 1970.

NEEDHAM, JOSEPH (65). *The Grand Titration; Science and Society in China and the West.* (Collected Addresses.) Allen & Unwin, London, 1969.

NEWBERRY, P. E. (2). *Scarabs; an Introduction to the Study of Egyptian Seals and Signet Rings.* Archibald Constable, London, 1908.

NEWBERRY LIBRARY, CHICAGO. *Dictionary Catalogue of the History of Printing.* 6 vols. G. K. Hall, Boston, 1961.

NGHIEN TOAN & RICARD, L. (1). 'Wou Tsö T'ien.' *BSEI*, n.s., **34** (no. 2), 114.

NIIDA NOBORU (1). 'A Study of Simplified Seal-marks and Finger-seals in Chinese Documents.' *MRDTB*, 1939, **11**, 79.

NORDSTRAND, O. K. (1). 'Chinese Double-leaved Books and Their Restoration.' *LIBRI*, 1967, **17**, 104.
NORDSTRAND, O. K. (2). 'The Introduction of Paper in Ceylon.' *PG*, 1961, **11**, 67.
NORRIS, F. H. (1). *Paper and Paper Making*. Oxford, 1952.

ODY, K. (1). *Paper Folding and Paper Sculpture*. Emerson, New York, 1905.
OLSCHKI, L. (7). *The Myths of Felt*. University of California Press, Berkeley, 1949.
OTTLEY, W. Y. (1). *An Inquiry concerning the Invention of Printing*. London, 1834–40.

PAIK NAK-CHOON (1). 'Tripitaka Koreana.' *JRAS/KB*, 1951, **32**, 62.
PAINE, R. T. (1). 'The Ten Bamboo Studio.' *ACASA*, 1951, **5**, 39.
PAK SHI-KYUNG (1). 'Ancient Metal Types.' *KT*, 1960, **44**, 39.
PANCIROLI, GUIDO (1). *The History of Many Memorable Things Lost*. London, 1715.
PARKER, E. H. (11). 'Paper and Printing in China.' *IAQR*, 1908 (3rd ser.), **25** (nos. 49–50), 349.
PARKER, T. (1). *A Short Account of the First Rise and Progress of Printing*. London, 1763.
PAWAR, H. H. (1). 'Glimpses of Japanese Book Publishing.' *EQ*, 1970, **22** (no. 1), 15.
PEAKE, C. H. (2). 'The Origin and Development of Printing in China in the Light of Recent Research.' *GUJ*, 1935, 9.
PEAKE, C. H. (3). 'Additional Notes and Bibliography on the History of Printing in the Far East.' *GUJ*, 1939, 57.
PELLIOT, P. (28). 'La peinture et la gravure européennes en Chine au temps de Mathieu Ricci.' *TP*, 1921, **20**, 1.
PELLIOT, P. (41). *Les débuts de l'imprimerie en Chine*. (Oeuvres posthumes 4.) Imprimerie nationale, Paris, 1953.
PELLIOT, P. (47). *Notes on Marco Polo*. Ouvrage posthume. 3 vols. Impr. Nat. and Maisonneuve, Paris, 1959–73.
PELLIOT, P. (60). 'Une bibliothèque mediévale retrouvée au Kan-sou.' *BEFEO*, 1908, **8**, 501.
PELLIOT, P. (61). 'Les documents chinois trouvés par la Mission Koslov à Khara-Khoto.' *JA*, 1914, **3**, 503.
PELLIOT, P. (62). *Un traité Manichéen*. Paris, 1913.
PELLIOT, P. (63). 'Les conquêtes de l'Empereur de la Chine.' *TP*, 1921, **20**, 183.
PELLIOT, P. (64). 'Livres reçus.' *TP*, 1931, **28**, 150.
PELLIOT, P. (65). 'Un recueil de pièces imprimées concernant la "Question des rites."' *TP*, 1924, **23**, 347.
PELLIOT, P. (66). 'Notes sur quelques livres ou documents conservés en Espagne.' *TP*, 1929, **26**, 43.
PENALOSA, F. (1). 'The Mexican Book Industry.' Thesis (Ph.D.), University of Chicago, 1956.
PERKINS, P. D. (1). *The Paper Industry and Printing in Japan*. Privately printed, New York, 1940.
PETRUCCI, R. (3) (tr.). *Encyclopédie de la peinture chinoise*. H. Laurence, Paris, 1918. Tr. of *Chieh Tzu Yüan Hua Chüan*.
PEUVRIER, A. (1). 'Les origines de l'imprimerie dans l'Extrême Orient.' *MSSJ*, 1887, **6**.
PFISTER, L. (1). *Notices Biographiques et Bibliographiques sur les Jésuits de l'Ancienne Mission de Chine, 1552–1773*. 2 vols. Imprimerie de la Mission Catholique, Shanghai, 1932–4. (Variétés sinologiques, nos. 59–60.)
PHILIPPI, D. L. (1) (tr.). *Kojiki*. University of Tokyo Press, 1968.
POON MING-SUN (1). 'The Printer's Colophon in Sung China, 960–1279.' *LQ*, 1973, **43**, 39.
POON MING-SUN (2). 'Books and Printing in Sung China, 960–1279.' Thesis (Ph.D.), University of Chicago, 1979.
POOR, R. (1). 'Notes on the Sung Dynasty Archaeological Catalogs.' *ACASA*, 1965, **19**, 33.
PRIMROSE, J. B. (1). 'The First Press in India and its Printers.' *The Library*, 1939, **12**, 241.
PRIOLKAR, A. K. (1). *The Printing Press in India; its Beginnings and Early Development*. Bombay, 1958.
PROSERPIS, L. (1). 'The First Printing Press in India.' *NRW*, 1935, **2** (no. 10), 321.
PRUNNER, G. (1). *Papiergötter aus China*. Hamburgisches Museum für Völkerkunde, Hamburg, 1973.
PULLEYBLANK, E. G. (2). 'The Date of the Staël-Holstein Roll.' *AM*, 1954, n.s., **4** (pt 1), 90.
PUTNAM, G. H. (1). *Books and their Makers during the Middle Ages*. Putnam's Sons, New York, 1896–97.

REICHWEIN, A. (1). *China and Europe; Intellectual and Artistic Contacts in the Eighteenth Century*. Kegan Paul, London, 1925. Tr. from the German edition, Berlin, 1923.
REINAUD, J. T. (1) (tr.). *Relation des voyages faits par les Arabes et les Persans dans l'Inde et la Chine dans le 9e siècle de l'ère Chrétienne*. 2 vols. Paris, 1845.
REISCHAUER, E. O. (4) (tr.). *Ennin's Diary; the Record of a Pilgrimage to China in Search of the Law*. Ronald Press, New York, 1955.
[RENAUDAT, EUSEBIUS](1) (tr.). *Anciennes relations des Indes et de la Chine de deux voyageurs Mohométans ...* Paris, 1718. Eng. tr., *Ancient Accounts of the Travels of Two Mohammedans through India and China in the +9th Century*. In John Pinkerton, ed., *A General Collection of the Best and Most Interesting Voyages and Travels*. London, 1808–14, Vol. 7 (1811).
RENKER, A. (1). *Papier und Druck im Fernen Osten*. Gutenberg-Gesellschaft, Mainz, 1936; Berlin, 1937.
RENKER, A. (2). *Papiermacher und Drucker; ein Gespräch über alte und neue Dinge*. Druck der Mainzer Presse, Mainz, 1934.
RENKER, A. (3). 'Die Forschungsstelle Papiergeschichte im Gutenberg-Museum zu Mainz.' In Horst Kunze, ed., *Buch und Papier*. Otto Harrossowitz, Leipzig, 1938, pp. 80–9.
RETANA, W. E. (1). *Tablas cronologica y alfabetica de imprentas e impresores de Filipinas (1593–1898)*. Libreria General de Victoriana Suarez, Madrid, 1908.
RETANA, W. E. (2). *Origenes de la Imprenta Filipina*. Manila, 1911.

RETANA, W. E. (3). *La imprenta en Filipinas; adiciones y observaciones á la imprenta en Manila de D. J. T. Medina.* Madrid, 1917.

RHODES, D. E. (1). *The Spread of Printing; Eastern Hemisphere, India, Pakistan, Ceylon, Burma and Thailand.* Van Gendt, Amsterdam, 1969.

RICHTER, M. (1). *De Typographiae Inventione.* Copenhagen, 1566.

ROBINSON, S. (1). *A History of Printed Textiles.* Studio Vista, London, 1969.

ROHRBACH, K. (1). 'Papiergeschichte im Deutschen Museum in München.' *PG*, 1974, **24**, 1–6.

ROULEAU, F. A. (3). 'The Yangchow Latin Tombs as a Landmark of Medieval Christianity in China.' *HJAS*, 1954, **17**, 346.

ROW, S. (1). *Geometric Exercises in Paper Folding.* Ed. and rev. by Beman & Smith. Chicago, 1901, 1941.

ROYDS, W. M. (1). 'Introduction to Courant's Bibliographie Coréenne.' *JRAS/KB*, 1936, **25**, 1.

RUDOLPH, R. C. (9) (tr.). 'Illustrated Botanical Works in China and Japan.' In *Bibliography and Natural History; Essays Presented at a Conference Convened in June 1964 by Thomas R. Buckman*, p. 103. University of Kansas Libraries, Lawrence, 1966.

RUDOLPH, R. C. (13). 'A Reversed Chinese Art Term.' *JAOS*, 1946, **66**, 15.

RUDOLPH, R. C. (14). 'Chinese Movable Type Printing in the 18th Century.' In *Silver Jubilee Volume of the Zinbun-Kagaku Kenkyusyo*, p. 317. Kyoto University, 1954.

RUDOLPH, R. C. (15) (tr.). *A Chinese Printing Manual, 1776.* Typophiles, Los Angeles, 1954. Tr. of Chin Chien, *Wu Ying Tien Chü Chen Pan Chheng Shih.*

RÜHL, I. (1). *Die Papierwirtschaft in China, Japan und Manchukuo.* Diss., Univ. Erlangen, Erlangen, 1942.

RUPPEL, A. L. (1). *Haben die Chinesen und Koreaner die Buchdruckerkunst erfunden?* Verlag der Gutenberg-Gesellschaft, Mainz, 1954. (Kleiner Druck der Gutenberg-Gesellschaft, nr. 56.)

SANBORN, K. (1). *Old Time Wall-papers; an Account of the Pictorial Papers on our Forefathers' Walls, with a Study of the Historical Development of Wall-paper Making and Decoration.* Literary Collector Press, New York, 1905.

SANDERMANN, W. (1). 'Old Papermaking Techniques in Southeast Asia and the Himalayan Countries.' *PG*, 1968, **18**, 29.

SANETOW KEISHU (1). 'Japan's Influence on Chinese Printing.' *COJ*, 1942, **11**, 241.

SANG, C. (1). *Primitivas relaciones de Espana con Asia y Occania.* Libreria General, Victionians Sularez, Madrid, 1958.

SANSOM, G. B. (1). *Japan; a Short Cultural History.* Rev. ed. Appleton-Century-Crofts, New York, 1962.

SARTON, G. (1). *Introduction to the History of Science.* Vol. 1, 1927; Vol. 2, 1931 (2 parts); Vol. 3, 1947 (2 parts). William & Wilkins, Baltimore. (Carnegie Institution Pub. no. 376.)

SARTON, G. (15). *A Guide to the History of Science; a First Guide for the Study of the History of Science, with Introductory Essays on Science and Tradition.* Chromica Botanica Co., Waltham, Mass., 1952.

SARTON, G. and HUMMEL, A. W. (1). 'The Printed Herbal of 1249 A.D.' *ISIS*, 1941, **33**, 439.

SASAKI, S. (1) (ed.). *Publishing in Japan; Present and Past.* Japan Book Publishers Association, Tokyo, 1969.

SATOW, E. M. (3). 'On the Early History of Printing in Japan.' *TAS/J*, 1882, **10** (pt 1), 48.

SATOW, E. M. (4). '*Further Notes on Movable Types in Korea and Early Japanese Printed Books.*' *TAS/J*, 1882, **10** (pt 2), 252.

SATOW, E. M. (5). *The Jesuit Mission Press in Japan, 1591–1610.* Privately printed, London, 1888.

SAUVAGET, J. (2) (tr.). *Relation de la Chine et de l'Inde, rédigée en −857.* (Akhbar al-Sin wa'l-Hind.) Belles Lettres, Paris, 1948. (Budé Association, Arab Series.)

SCHAFER, E. H. (13). *The Golden Peaches of Samarkand; a Study of Thang Exotics.* University of California Press, Berkeley and Los Angeles, 1963.

SCHAFER, E. H. (16). *The Vermilion Bird; Thang Images of the South.* University of California Press, Berkeley and Los Angeles, 1967.

SCHAFER, E. H. & WALLACKER, B. E. (1). 'Local Tribute Products of the Thang Dynasty.' *JOSHK*, 1957, **4**, 213.

SCHÄFER, G. (1). 'The Development of Papermaking.' *CIBA/T*, 1949, **6**, 2641.

SCHÄFER, G. (2). 'The Paper Trade before the Invention of the Paper-machine.' *CIBA/T*, 1949, **6**, 2650.

SCHÄFFER, J. C. (1). *Versuche und Muster ohne alle Lumpen oder doch mit einem geringen Zusatze derselben Papier zu machen.* Regensburg, 1765–71. 6 vols.

SCHEFER, C. (2). 'Notices sur les relations des Peuples Mussulmans avec les Chinois depuis l'Extension de l'Islamisme jusqu'à la fin du 15e Siècle.' In *Volume Centénaire de l'Ecole des Langues Orientales Vivantes, 1795–1895.* Leroux, Paris, 1895, pp. 1–43.

SCHINDLER, B. (4). 'Preliminary Account of the Work of Henri Maspero concerning the Chinese Documents on Wood and Paper Discovered by Sir Aurel Stein on his Third Expedition in Central Asia.' *AM*, 1949, n.s., **1**, 216.

SCHMIDT, A. (1). 'Der Chinesische Buchdruck.' *ZB*, 1927, n.s., **19**, 11.

SEDGWICK, E. (1). 'A Chinese Printed Scroll of the Lotus Sutra.' *LC/QJCA*, 1949, **6** (no. 2), 6. With a note by A. W. Hummel.

SEMEDO, A. (1). *The History of that Great and Renowned Monarchy of China.* I. Crook, London, 1655.

SEN, S. N. (1). 'Hand-made Paper of Nepal.' *MR*, 1940, **67**, 459.

SEN, S. N. (2). 'Transmission of Scientific Ideas between India and Foreign Countries in Ancient and Medieval Times.' *BNISI*, 1963, **21**, 8.

SEUBERLICH, W. (1). 'Ein neues russisches Werk zur Geschichte des chinesischen Buchdrucks der Sung-Zeit.' In *Studia Sino-Altaica; Festschrift für Erich Haenisch zum 80, Geburtstag, im Auftrag der Deutschen Morgenländischen Gesellschaft*, pp. 183–6. Franz Stein Verlag, Wiesbaden, 1961.

SHAFER, R. (1). 'Words for "Printing Block" and the Origin of Printing.' *JAOS*, 1960, **80**, 328.

SHAW, SHIOW-JYU LU (1). 'The Imperial Printing of Early Chhing China, 1644–1805.' Thesis (M.A.), University of Chicago, 1974.

SHEN, PHILIP (1). 'Introducing Chinese Paper-folding.' *NCSAS*, 1958, n.s., **2** (no. 1), 7.

SHIH HSIO-YEN (1). 'On the Ming Dynasty Book Illustration.' Thesis (M.A.), University of Chicago, 1958.

SIRR, H. C. (1). *China and the Chinese*. W. S. Orr & Co., London, 1849.

SITWELL, S. (1). *British Architects and Craftsmen; a Survey of Taste, Design and Style during Three Centuries, 1600 to 1830*. Pan Books, London, 1945; rev. ed., 1960.

SMITH, RICHARD (1). 'On the First Invention of the Art of Printing.' Manuscript, *c.* 1670. St. Bride Foundation Library, London; copy at Cambridge University Library.

SOHN POW-KEY (1). 'Early Korean Printing.' *JAOS*, 1959, **79**, 96.

SOHN POW-KEY (2). *Early Korean Typography*. Korean Library Science Research Institute, Seoul, 1971. Text in Korean and English.

SOMMARSTRÖM, B. (1). *Archaeological Researches in the Edsen-Gol Region, Inner Mongolia*. 2 pts. Stockholm, 1956–58.

SOONG, MAYING (1). *The Art of Chinese Paper-folding*. Harcourt, Brace, New York, 1948, 1955.

SPEAR, R. L. (1). 'Research on the 1593 Jesuit Mission Press Edition of Esop's Fables.' *MN*, 1964, **19** (no. 3–4), 222.

STARR, K. (1). 'Inception of the Rubbing Technique; a Review.' In *Symposium in Honor of Dr. Li Chi on His Seventieth Birthday*, Taipei, 1965, 281.

STARR, K. (2). 'Rubbings; an Ancient Chinese Art.' *NCSAS*, 1966, n.s., **9** (no. 3–4), 1.

STARR, K. (3). 'An "Old Rubbing" of the Later Han *Chang Chhien-Pei*.' In D. Roy & T. H. Tsien (eds.), *Ancient China: Studies in Early Civilization*. The Chinese University Press, Hongkong, 1978, pp. 283–314.

STEELE, R. (5). 'What 15th Century Books Are About.' *LIB*, n.s., 1903, **4**, 337; 1904, **5**, 337; 1905, **6**, 137; 1907, **8**, 225.

STEIN, SIR AUREL (1). *Ruins of Desert Cathay*. 2 vols. Macmillan, London, 1912.

STEIN, SIR AUREL (2). *Innermost Asia*. 2 vols. text, 1 vol. plates, 1 box maps. Oxford, 1928.

STEIN, SIR AUREL (4). *Serindia; Detailed Report of Exploration in Central Asia and Western-most China*. 4 vols. Clarendon Press, Oxford, 1921.

STEIN, SIR AUREL (6). 'Notes on Ancient Chinese Documents Discovered among the Han Frontier Wall in the Desert of Tun-huang.' *NCR*, 1921, **3**, 243.

STEIN, SIR AUREL (7). 'A Chinese Expedition across the Pamirs and Hindukush, A.D. 747.' *NCR*, 1922, **4**, 161.

STEIN, SIR AUREL (11). *Ancient Khotan; Detailed Report of an Archaeological Exploration in Chinese Turkestan*. 2 vols. Clarendon Press, Oxford, 1907.

STEIN, SIR AUREL (12). 'Early Papermaking in the Far East Ruins of Desert Cathay.' *PAP*, 1912, **10**.

STEINBERG, S. H. (1). *Five Hundred Years of Printing*. Criterion Books, New York, 1959.

STEVENS, R. T. (1). *The Art of Papermaking in Japan*. Privately printed, New York, 1909.

STEVENSON, A. H. (1). *Observations on Paper as Evidence*. University of Kansas Press, Lawrence, Kansas, 1961. (University of Kansas Annual Public Lectures on Books and Bibliography, 7.)

STOFF, F. (1). *The Valentine and its Origins*. London, 1969.

ŠTOVIČKOVÁ-HEROLDOVÁ, D. (1). 'Chinese Books.' *EHOR*, 1965, **4** (no. 8), 50.

STREHLNEEK, E. A. (1). *Chinese Pictorial Art*. Commercial Press, Shanghai, 1914.

STÜBE, R. (1). 'Die Erfindung des Druckes in China und seine Verbreutung in Ostasien.' *BGTI*, 1918, **8**, 88.

STUDLEY, V. (1). *The Art and Craft of Handmade Paper*. Van Nostrand Reinhold Co., New York, 1977.

SU YING-HUI (1). 'An Inquiry into the Chinese Pen.' *APR*, 1972, **3** (no. 4), 24.

SUGIMOTO MASAYOSHI & SWAIN, D. L. (1). *Science and Culture in Traditional Japan, A.D. 600–1854*. MIT Press, Cambridge, Mass., 1978.

SUN JEN I-TU & SUN HSÜEH-CHUAN (1) (tr.). *'Thien Kung Khai Wu', Chinese Technology in the Seventeenth Century, by Sung Ying-Hsing*. Pennsylvania State University Press, University Park and London, 1966.

SUN MEI-LAN (1). 'Illustration for Children's Books.' *CLIT*, 1959, no. 3, 144.

SUTERMEISTER, E. (1). *The Story of Papermaking*. S. D. Warren, Boston, 1954.

SWINGLE, W. T. (13). 'Chinese Books and Libraries.' *BALA*, 1917, **11**, 121.

SWINGLE, W. T. (14). 'Chinese Books; their Character and Value and their Place in the Western Library.' In *Essays Offered to Herbert Putnam by his Colleagues and Friends on His 30th Anniversary as Librarian of Congress, 5 April, 1929*. Ed. W. W. Bishop and Andrew Keough, p. 429. Yale University Press, New Haven, 1929.

SZE MAI-MAI (1) (tr.). *The Tao of Painting*. Bollingen Foundation, New York, 1956; Modern Library, New York, 1959. Tr. of *Chieh-Tzu-Yüan Hua Chuan*.

TAAM CHEUK-WOON (1). *The Development of Chinese Libraries under the Ch'ing Dynasty, 1644–1911.* Private ed., distributed by the University of Chicago Libraries, 1935.

TAKEO EIICHI (1) (ed.). *Handmade Papers of the World.* Takeo Co., Tokyo, 1979. Text in English and Japanese with handmade paper specimens from 23 countries in one box.

TAUBERT, S. (1). *Bibliopola; Bilder und Texte aus der Welt des Buchhandels* (Pictures and Texts about the Book Trade; Images et Textes sur la Librairie). 2 vols. Ernst Hauswedell, Hamburg, 1966.

TENG KWEI (1). 'Chinese Ink-sticks.' *CJ*, 1936, **24**, 9.

TENG SSU-YÜ (3) (tr.). *Family Instructions for the Yen Clan; 'Yen Shih Chia Hsün' by Yen Chih-Thui (+531–91).* Brill, Leiden, 1968. (Monograph du T'oung Pao, IV.)

TENG SSU-YÜ (4). 'Chinese Influence on the Western Examination System.' *HJAS*, 1943, **7**, 267.

THOMAS, ISAIAH (1). *The History of Printing in America.* 2 vols. Worcester, Mass., 1810; 2nd ed., John Munsell, Albany, New York, 1874.

TILLEY, R. (1). *Playing Cards.* Putnam's Sons, New York, 1967.

TINDALE, T. K. & TINDALE, H. R. (1). *The Handmade Papers of Japan.* 3 vols., Tuttle, Rutland, 1952.

TING WEN-YÜAN (1). 'Von der alten chinesischen Buchdruckerkunst.' *GUJ*, 1929, 9.

TODA KENJI (1). *Descriptive Catalogue of Japanese and Chinese Illustrated Books in the Ryerson Library at the Art Institute of Chicago.* Chicago, 1931.

TOKURIKI TOMIKICHIRO (1). *Wood-block Print Primer.* Japan Publications, Tokyo, 1970.

TOUSSAINT, A. (1). *The Spread of Printing, Eastern Hemisphere; Early Printing in Mauritius, Reunion, Madagascar and the Seychelles.* Vangendt, Amsterdam, 1969.

TRIER, J. (1). *Ancient Paper of Nepal; Results of Ethnotechnological Field Work on its Manufacture, Uses and History—with Technical Analyses of Bast, Paper and Manuscripts.* Jutland Archaeological Society, Copenhagen, 1972.

TROLLOPE, M. N. (1). 'Korean Books and their Authors.' *JRAS/KB*, 1932, **21**, 1.

TROLLOPE, M. N. (2). 'Book Production and Printing in Corea.' *TRAS/KB* 1936, **25**, 101.

TSAI JO-HUNG (1). 'New Year's Pictures; a People's Art.' *PC*, 1950, **1** (no. 4), 12.

TSCHICHOLD, J. (1). *Chinesische Farbendrucke der Gegenwart.* Holbein-Verlag, Basel, 1945.

TSCHICHOLD, J. (2). *Der Erfinder des Papiers, Ts'ai Lun, in einer alten chinesischen Darstellung* (Neujahrsgabe). Zurich, 1955.

TSCHICHOLD, J. (3). *Chinese Colour Prints from the Ten Bamboo Studio.* Tr. by Katherine Watson. Lund Humphries, London, 1972. Original ed., *Die Bildersammlung der Zehnbambushalle.* Rentsch Verlag, Erlenbach, Switzerland, 1970. Selected reproductions of the *Shih-Chu-Chai Shu Hua Phu.*

TSCHICHOLD, J. (4). *Chinesische Gedichtpapier vom Meister der Zehnbambushalle.* Holbein-Verlag, Basel, 1947. Selected reproductions from the *Shih-Chu-Chai Chien Phu.*

TSCHICHOLD, J. (5). *Chinese Color-prints from the Painting Manual of the Mustard Seed Garden.* Tr. E. C. Mason. Allen & Unwin, London, 1951. Selected reproductions from the *Chieh-Chih-Yüan Hua Phu.*

TSCHICHOLD, J. (6). *Early Chinese Color Prints.* Tr. E. C. Mason. Beechhurst Press, New York, 1953.

TSCHICHOLD, J. (7). *Chinese Color Printing of Today.* Tr. E. C. Mason. Beechhurst Press, New York, 1953.

TSCHUDIN, W. F. (1). 'The Oldest Papermaking Processes in the Far East.' *TXR*, 1958, **13**, 679.

TSCHUDIN, W. F. (2). *The Ancient Paper-mills of Basel and their Marks.* Paper Publications Society, Hilversum, Holland, 1958.

TSCHUDIN, W. F. (3). 'The Paper Museum at Basel, Switzerland.' *PAM*, 1957, **26**, 1.

TSIEN TSUEN-HSUIN (2). *Written on Bamboo and Silk; the Beginnings of Chinese Books and Inscriptions.* University of Chicago Press, 1962. See also Chinese and Japanese revised eds. under Chhien Tshun-Hsün in Bibliography B.

TSIEN TSUEN-HSUIN (3). 'On Dating the Edition of *Chü-lu* (Record of Oranges) at Cambridge University.' *CHJ/T*, n.s., 1973, **10**, 106.

TSIEN TSUEN-HSUIN (4). 'Raw Materials for Old Papermaking in China.' *JAOS*, 1973, **93**, 510.

TSIEN TSUEN-HSUIN (5). 'A History of Bibliographic Classification in China.' *LQ*, 1952, **22**, 307.

TSIEN TSUEN-HSUIN (6). 'Silk as Writing Material.' *MID*, 1962, **11**, 92.

TSIEN TSUEN-HSUIN (7). 'Terminology of the Chinese Book and Bibliography.' Institute for Far Eastern Librarianship, University of Chicago, 1969.

TSIEN TSUEN-HSUIN (8). 'A Study of the Book-Knife of the Han Dynasty.' Tr. by John H. Winkelman. *CCUL*, 1971, **12** (no. 1), 87.

TSIEN TSUEN-HSUIN (9). 'China; True Birthplace of Paper, Printing and Movable Type.' *UNESC*, 1972, **12**, 4; *PPI*, 1974, **2**, 50.

TSIEN TSUEN-HSUIN (10). *China; an Annotated Bibliography of Bibliographies.* G. K. Hall, Boston, 1978.

TSIEN TSUEN-HSUIN (11). 'Rare Chinese-Japanese Materials in American Libraries as Reported in 1957.' *BCEAL*, 1966, no. 16, 10.

TSIEN TSUEN-HSUIN (12). 'Western Impact on China through Translation.' *JAS*, 1954, **14**, 305.

TSIEN TSUEN-HSUIN (13). 'Why Paper and Printing were Invented First in China and Used Later in Europe.' In *EHSTC*, PP. 459–70.

TSIEN TSUEN-HSUIN. *See also* CHHIEN TSHUN-HSÜN in Bibliography B.

TUNG TSO-PIN (2). 'Ten Examples of Early Tortoise-Shell Inscriptions.' *HJAS*, 1948, **11**, 119.

USHER, A. P. (1). *A History of Mechanical Inventions*. Rev. ed. Harvard University Press, 1962.

VACCA, G. (7). 'Della piegatura della carta applicata alla Geometria.' *PDM*, 1930 (ser. 4), **10**, 43.
DE LA VALLÉE-POUSSIN, L. (10). *Catalogue of the Tibetan Manuscripts from Tun-huang in the India Office Library*. With an appendix on the Chinese manuscripts by Kazuo Enoki. Oxford University Press, 1962.
VANDERSTAPPEN, H. A. (1) (ed.). *The T. L. Yuan Bibliography of Western Writing on Chinese Art and Archaeology*. Mansell, London, 1975.
VAUDESCAL, LE C. (1). 'Les pierres gravées du Che King Chan et le Yün Kiu Sseu.' *JA*, 1914, 374.
VÉBER, M. (1). *Le papier*. Fédération Nationale des Maîtres Artisans du Livre, Paris, 1969.
VERGIL, POLYDORE (1). *De Rerum Inventoribus*. 1512. English tr. by Thomas Langley with an account of the author and his works by William A. Hammond. Agathynian Club, New York, 1868.
VERVLIET, H. D. (1) (ed.). *Liber Librorum; The Book through Five Thousand Years*. Areade, Bruxelles, 1973.
VISSERING, W. (1). *On Chinese Currency, Coin, and Paper Money*. E. J. Brill, Leiden, 1877.
VIVAREY, H. (1). 'Vieux papiers de Corée.' *BSVP*, 1900, **1**, 76.
VOORN, H. (1). 'Papermaking in the Moslem World.' *PAM*, 1959, **28**, 31.
VOORN, H. (2). 'Batik Paper.' *PAM*, 1966, **35**, 5.
VOORN, H. (3). 'Javanese Deloewang Paper.' *PAM*, 1968, **37**, 32.
VOORN, H. (4). 'A Paper Mill and Museum in Holland.' *PAM*, 1967, **36**, 3.

WALEY, A. (29). 'Note on the Invention of Woodcuts.' *NCR*, 1919, 413.
WALEY, A. (30). *A Catalogue of Paintings Recovered from Tunhuang by Sir Aurel Stein, Preserved in the Sub-department of Oriental Prints and Drawings in the British Museum and in the Museum of Central Asian Antiquities*, Delhi. British Museum and Government of India, London, 1931.
WANG CHI (1). 'New Chinese Woodcuts.' *CREC*, 1959, **5**, supp., unpaged.
WANG CHI (2). 'Three Woodcut Artists.' *CLIT*, 1964, **4**, 99.
WANG CHI-CHEN (2). 'Notes on Chinese Ink.' *MMS*, 1930/31, **3**, 114.
WANG CHING-HSIEN & HO YUNG (1). 'Chinese Book Illustration.' *CLIT*, 1958, no. 7/8, 134.
WANG I-THUNG (2). 'The Origins of Chinese Books; a Review Article.' *PA*, 1964/65, **37**, 436.
WANG YÜ-CHHÜAN (1). *Early Chinese Coinage*. American Numismatic Society. New York, 1951.
WARD, J. (1). *The Sacred Beetle; a Popular Treatise on Egyptian Scarabs in Art and History*. London, 1902.
WATERHOUSE, D. B. (1). *Harunobu and his Age; the Development of Colour Printing in Japan*. British Museum, London, 1964.
WATSON, B. (1) (tr.). *'Records of the Grand Historian of China,' Translated from the 'Shih Chi' of Ssuma Chhien*. 2 vols. Columbia University Press, 1961.
WATSON, J. (1). *The History of the Art of Printing*. Edinburgh, 1713.
WATTS, ISAAC (1). *The Improvement of the Mind; Containing a Variety of Remarks and Rules for the Attainment and Communication of Useful Knowledge in Religion, in the Sciences, and in Common Life*. London, 1785.
WEAVER, A. (1). *Paper, Wasps and Packages; the Romantic Story of Paper and its Influence on the Course of History*. Container Corp. of America, Chicago, 1937.
WEEKS, L. H. (1). *A History of Paper-manufacturing in the United States, 1690–1916*. New York, 1916.
WEIR, T. S. (1). 'Some Notes on the History of Papermaking in the Middle East.' *PG*, 1957, **7**, 43.
WEITENKAMP, F. (1). *The Illustrated Book*. Harvard University Press, 1938.
WENG THUNG-WEN (WONG T'ONG-WEN) (1). 'Le véritable éditeur du Kieau-king san-tchuan.' *TP*, 1964, **51**, 429.
WERNER, E. T. C. (1). *Myths and Legends of China*. Harrap, London, 1922.
WEST, C. J. (1). *Bibliography of Pulp and Paper Manufacture*. Technical Association of the Pulp and Paper Industry, New York, 1947; 2nd ed., 1956.
WEST, C. J. (2). *Classification and Definitions of Paper*. Rev. ed. Lockwood Trade Journal Co., New York, 1928.
WHITE, W. C. (4). *Bone Culture of Ancient China*. University of Toronto Press, 1945.
WHITE, W. C. (3). *Bronze Culture of Ancient China*. University of Toronto Press, 1956.
WIBORG, F. B. (1). *Printing Ink; a History with a Treatise on Western Methods of Manufacture and Use*. Harper, New York, 1926.
WIESNER, J. (1). 'Mikroskopische Untersuchung der Papiere von El-Faijûm.' *MSPER*, 1886, **1**, 45.
WIESNER, J. (2). Die faijûmer und uschmûneiner Papiere. Eine naturwissenschaftliche, mit Rücksicht auf die Erkennung alter und moderner Papiere und auf die Entwicklung der Papierbereitung durchgeführte Untersuchung.' *MSPER*, 1887, **2/3**, 179.
WIESNER, J. (3). *Mikroskopische Untersuchungen alter ostturkestanischer und anderer asiatischer Papiere nebst histologischen Beiträgen zur mikroskopischen Papieruntersuchung*. *DAW/MN*, 1902, **72**.
WIESNER, J. (4). 'Ein neuer Beitrag zur Geschichte des papiers.' *SWAW/PH*, 1904, **148** (pt. 6).
WIESNER, J. (5). 'Über die ältesten bis jetzt aufgefundenen Hadernpapiere.' *SWAW/PH*, 1911, **168** (pt. 5).
WILKINSON, W. H. (2). 'The Chinese Origin of Playing Cards.' *AAN*, 1895, **8**, 61.

WILLIAMS, J. C. (1) (ed.). *Preservation of Paper and Textiles of Historic and Artistic Value.* A symposium sponsored by the Cellulose, Paper, and Textile Division at the 172nd Meeting of the American Chemical Society, San Francisco, California, Aug. 30–31, 1976. American Chemical Society, Washington, D. C., 1977. (Advances in Chemistry Series, 164.)

WILLIAMS, S. W. (1). *The Middle Kingdom; a Survey of the Geography, Government, Education, Social Life, Arts, Religion, etc. of the Chinese Empire and its Inhabitants.* 2 vols. Wiley, New York, 1848; later eds., 1861, 1900, London, 1883.

WILLIAMS, S. W. (2). 'Brief Statement relative to the Formation of Metal Types for the Chinese Language.' *CRRR,* 1834, **2**, 477.

WILLIAMS, S. W. (3). 'Chinese Metallic Types.' *CRRR,* 1835, **3**, 528.

WILLIAMS, S. W. (4). 'Movable Types for Printing Chinese.' *CRR,* 1875 **6**, 26.

WINKELMAN, J. H. (1) (tr.). 'A Study of the Book-knife of the Han Dynasty.' *CCU,* 1971, **12** (no. 1), 87. Tr. of Chhien Tshun-Hsuin (1).

WINKELMAN, J. H. (2). *The Imperial Library in Southern Sung China, 1127–1279; a Study of the Organization and Operation of the Scholarly Agencies of the Central Government.* American Philosophical Society, Philadelphia, 1974. (*TAPS,* n.s., vol. 64, pt 8.)

WINTER, J. (1). 'Preliminary Investigations on Chinese Ink in Far Eastern Paintings.' *ADVC,* 1975, **138**, 207.

WISEMAN, D. J. (1). *Cylinder Seals of Western Asia.* Batchwork Press, London, 1958.

WOLF, E. (1) (ed.). *Doctrina Christiana.* The first book printed in the Philippines, Manila, 1593. A facsimile of the copy. Library of Congress, Washington, 1974.

WONG T'ONG-WEN. See WENG THUNG-WEN.

WONG VI-LIEN (1). 'Libraries and Book-collecting in China from the Epoch of the Five Dynasties to the End of Ch'ing.' *TH,* 1939, **8**, 327.

WOODBURY, G. E. (1). *A History of Wood Engraving.* Harper & Bros., New York, 1883.

WOODSIDE, A. B. (1). *Vietnam and the Chinese Model; a Comparative Study of Nguyen and Chhing Civil Government in the First Half of the 19th Century.* Harvard University Press, 1971.

WRIGHT, A. (9). 'The Study of Chinese Civilization.' *JHI,* 1960, **21**, 233.

WROTH, L. (1) (ed.). *History of Printed Books; being the Third Number of the Dolphin.* Limited Editions Club, New York, 1938.

WU KWANG-TSING (1). 'The Chinese Book; its Evolution and Development.' *TH,* 1936, **3**, 25.

WU KWANG-TSING (2). 'The Development of Printing in China.' *TH,* 1936, **3**, 137.

WU KWANG-TSING (3). 'Cheng Chhiao; a Pioneer in Library Method.' *TH,* 1940, **10**, 129.

WU KWANG-TSING (4). 'Colour Printing in the Ming Dynasty.' *TH,* 1940, **11**, 30.

WU KWANG-TSING (5). 'Scholarship, Book Production, and Libraries in China, 618–1644.' Thesis (Ph.D.), University of Chicago, 1944.

WU KWANG-TSING (6). 'Ming Printing and Printers.' *HJAS,* 1942, **7**, 203.

WU KWANG-TSING (7). 'Chinese Printing under Four Alien Dynasties, 916–1368.' *HJAS,* 1950, **13**, 447.

WU KWANG-TSING (8). 'The Development of Typography in China during the Nineteenth Century.' *LQ,* 1952, **22**, 288.

WU KWANG-TSING (9). 'Illustrations in Sung Printing.' *LC/QJCA,* 1971, **28**, 173.

WULFF, H. E. (1). *Traditional Crafts of Persia; their Development, Technology, and Influence on Eastern and Western Civilizations.* MIT Press, Cambridge, Mass., 1967.

WYLIE, A. (1). *Notes on Chinese Literature; with Introductory Remarks on the Progressive Advancement of the Art and a List of Translations from the Chinese into Various European Languages.* American Presbyterian Mission Press, Shanghai, 1867; reprints, Vetch, Peiping, 1939; Taipei, 1964.

YANG LIEN-SHENG (3). *Money and Credit in China; a Short History.* Harvard University Press, 1952. (Harvard-Yenching Institute Monograph Series, XII.)

YANG LIEN-SHENG (9). *Studies in Chinese Institutional History.* Harvard University Press, 1963. (Harvard-Yenching Institute Studies, XX.)

YANG LIEN-SHENG (14). 'The Form of the Paper Note Hui-tzu of the Southern Sung Dynasty.' *HJAS,* 1953, **16**, 363.

YEH KUNG-CHHO (1). 'Chinese Editions of the Tripitaka.' *PBLN,* 1946, **2**, 26.

YEH SHEN-TAO (1). 'Chinese Colour Wood-block Printing.' *PC,* 1954, **17**, 25.

YETTS, W. P. (19). *The George Eumorfopoulos Collection of the Chinese and Corean Bronzes, Sculpture, Jades, Jewelry, and Miscellaneous Objects.* 3 vols. Ernest Benn, London, 1929.

YETTS, W. P. (20). 'Bird Script on Ancient Chinese Swords.' *JRAS,* 1934, 547.

YI SANG-BECK (1). *The Origin of the Korean Alphabet.* T'ongmunkwan, Seoul, 1957.

YULE, SIR HENRY (1) (ed.). *The Book of Sir Marco Polo the Venetian, concerning the Kingdoms and Marvels of the East.* 3rd ed. rev. by H. Cordier. 2 vols. Murray, London, 1903.

YULE, SIR HENRY (2). *Cathay and the Way Thither; being a Collection of Medieval Notices of China.* Hakluyt Society Pubs. (2nd ser.), 1913–15. (1st ed. 1866.) Revised by H. Cordier. 4 vols.

YUYAMA AKIRA (1). *Indic Manuscripts and Chinese Blockprints (non-Chinese Texts) of the Oriental Collection of the Australian National University Library.* Center of Oriental Studies, Australian National University, Canberra, 1967.

ZEDLER, G. (1). 'Die Erfindung Gutenbergs und der chinesische und frühholländische Büchdruck.' *GUJ*, 1928, 50.

ZI, ÉTIENNE (1). *Practique des Examens Littéraires en Chine.* Mission Catholique, Shanghai, 1894. (Variétés sinologiques, 7.)

ZIGROSSER, C. (1). *Prints and their Creators; a World History; an Anthology of Printed Pictures and Introduction to the Study of Graphic Art in the West and East.* 2nd ed. rev. Crown Publishers, New York, 1974.

ZOLBROD, L. (1). 'Yellow-back Books; the Chapbooks of Late 18th Century Japan.' *EAST*, 1965, **1** (no. 5), 26.

ADDENDA TO BIBLIOGRAPHY C

ANON. (116). *Historical Relics unearthed in New China* (album), Foreign Language Press, Peking, 1972.

BERNARD-MAÎTRE, H. (19). 'Les adaptations chinoises d'ouvrages européens; bibliographie chronologique depuis la fondation de la mission française de Pékin jusqu'à la morte de l'empereur K'ien-long, 1689–1799.' *MS*, 1960, **19**, 349.

BINYON, L. (3). 'A Note of Colour-printing in China and Japan.' *BURM*, 1907, **2**.

BINYON, L. (4). 'Chinese Colour prints of the XVIIth Century.' *BMQ*, 1932, **7**, 36.

BIOT, E. (1) (tr.). *Le Tcheou-Li; ou Rites des Tcheou* [Chou]. 3 vols. Imp. Nat., Paris, 1951. (Photographically reproduced, Wentienko, Peiping, 1930.)

BLOY, C. H. (1). *A History of Printing Ink.* London, 1967.

BOSCH-TEITZ, S. C. (1). 'Chinese Prints: *Chieh Tzu Yüan.*' *MMB*, 1924, **19**, 92.

BREWITT-TAYLOR, C. H. (1) (tr.). *San Kuo, or the Romance of the Three Kingdoms.* Kelly & Walsh, Shanghai, 1926.

BRUCE, J. P. (1) (tr.). *The Philosophy of Human Nature*, translated from the Chinese with notes. Probsthain, London, 1922. (Chs. 42–8, inclusive, of *Chu Tzu Chhüan Shu.*)

BUCK, P. (1) (tr.). *All Men are Brothers (The Shui Hu Chuan).* New York, 1933.

ČAPEK, A. (1). *Chinese Stone-pictures, a Distinctive Form of Chinese Art.* Spring Books, London, 1962.

CHAU, D. H. S. (1). 'Woodblock Printing, an Essential Medium of Cultural Inheritance in Chinese History.' *JRAS/HB*, 1978, **18**, 75.

CHHU TA-KAO (2) (tr.). *Tao Te Ching, a New Translation.* Buddhist Lodge, London, 1937.

CONZE, E. (4) (tr.). *Selected Sayings from the 'Perfection of Wisdom;' Prajñāpāramitā.* Buddhist Society, London, 1955.

CRUMP, J. (1) (tr.). *Chan Kuo Tshe.* Clarenden, Oxford, 1970.

DEFRANCIS, J. (1). *Colonialism and Language Policy in Viet Nam.* Mouton, Hague, 1977.

DITTRICH, E. (1). *Das Westzimmer; Hsi-Hsiang-Chi, chinesische Farbholzschnitt von Min Ch'i-Chi, 1640.* Museum für Ostasiatiche Kunst, Köln, 1977.

DRÈGE, JEAN-PIERRE (1). 'Papiers de Dunhuang; Essai d'analyse morphologique des manuscrits Chinois datés.' *TP*, 1981, **67**, 305.

DUBOSC, J. P. & FOU HSI-HOUA (1). *Exposition d'ouvrages illustrés de la dynastie Ming.* Centre franco-chinois d'études sinologiques, Peking, 1944.

DUFF, E. G. (1). *Early Printed Books.* Kegan Paul, Trench, Trübner, London, 1893.

DURAND, M. (1). *Imagerie populaire Vietnamienne.* École française d'Extrême-Orient, Paris, 1960.

FENG YU-LAN (5) (tr.). *Chuang Tzu; a New Selected Translation with an Exposition of the Philosophy of Kuo Hsiang.* Commercial Press, Shanghai, 1933.

FORKE, A. (3) (tr.). *Me Ti [Mo Ti] des Sozialethikers und seiner Schüler philosophische Werke.* Berlin, 1922. (*MSOS*, Beiband, **23** to **25**.)

GRANT, J. (1). *Books and Documents.* London, 1937.

VON HAGEN, V. W. (1). *The Aztec and Maya Paper-makers.* With an Introduction by Dard Hunter and an Appendix by Paul C. Standley. Augustin, New York, 1944. Enlarged Eng. tr. of *La Fabrication del Papel entre los Aztecas y los Mayas.* Nuevo Mundo, Mexico City, 1935.

HAGERTY, M. J. (1) (tr.). 'Han Yen-Chih's *Chü Lu* (Monograph on the Oranges of Wenchow, Chekiang),' with introduction by P. Pelliot. *TP*, 1923, **22**, 63.

HAMBIS, L. (2). 'Chinese Woodblock-prints.' In *Encyclopaedia of World Art*, vol. 4, pp. 780 ff. London, 1961.

DEHARLEZ, C. (1). *Le Yi-king [I Ching], Texte Primitif Rétabli, Traduit et Commenté.* Hayez, Bruxelles, 1889.

KERR, G. H. (1). *Okinawa: the History of an Island People.* Charles E. Tuttle, Rutland, Va.; Tokyo, Japan, 1958.

LAMOTTE, E. (1) (tr.). '*Mahāprajñāpāramitā Sutra*'; *le Traité (Mādhyamika) de la Grande Vertu de Sagesse, de Nāgārjuna.* 3 vols. Louvain, 1944. (rev. P. Demiéville, *JA*, 1950, **238**, 375.)

LEGGE, J. (3) (tr.). *The Chinese Classics, etc.*: Vol. 2. *The Works of Mencius.* Legge, Hongkong, 1861; Trübner, London, 1861.

LEGGE, J. (5) (tr.). *The Texts of Taoism.* (Contains (*a*) *Tao Te Ching,* (*b*) *Chuang Tzu,* (*c*) *Thai Shang Kan Ying Phien,* (*d*) *Chhing Ching Ching,* (*e*) *Yin Fu Ching,* (*f*) *Jih Yung Ching.*) 2 vols. Oxford, 1891; Photolitho reprint, 1927. (*SBE*, nos. 39 and 40.)

LEGGE, J. (9) (tr.). *The Texts of Confucianism,* Pt. II. *The 'Yi King' ('I Ching').* Oxford, 1882, 1899. (*SBE*, no. 16.)

LEGGE, J. (11) (tr.). *The Chinese Classics, etc.*: Vol. 5, Pts. I and II. *The 'Ch'un Ts'eu' with the 'Tso Chuen' ('Chhuan Chhiu and Tso Chuan').* Lane Crowford, Hongkong, 1872; Trübner, London, 1872.

LIN YU-THANG (1) (tr.). *The Wisdom of Laotze* [and *Chuang Tzu*], translated, edited and with an introduction and notes. Random House, New York, 1948.

LYALL, L. A. (1) (tr.). *Mencius.* Longmans Green, London, 1932.

MATHER, R. B. (1) (tr.). *Shih Shuo Hsin Yü; a New Account of Tales of the World.* University of Minnesota Press, 1976.

MATSUMOTO SOGO (1). 'Introduction to Chinese Prints.' In *Magazine of Art,* 1937, **30** (no. 7), 410.

MCCLELLAND, N. (1). *Historic Wall-papers from their Reception to the Introduction of Machinery.* Lippincott, Philadelphia & London, 1924.

MCCULLOUGH, H. C. (1) (tr.). *Tales of Ise; Lyrical Episodes from Tenth-Century Japan.* Stanford University Press, 1968.

MEDHURST, W. H. (1). *China; Its State and Prospects.* Boston, 1838.

MITCHELL, A. A. (1). *Inks; Their Composition and Manufacture.* London, 1937.

NACHBAUR, A. & WANG NGEN-JOUNG (1). *Les Images Populaires Chinoises.* Peiping, 1926.

PAINE, R. T. (2). 'Wen Shu and Phu Hsien, Chinese Woodblock Prints of the Wan-li Era.' *AA,* 1961, **24**, 87.

PELLIOT, P. (32). 'Des Artisans Chinois à la Capitale Abbaside en +751/+762.' *TP,* 1928, **26**, 110.

POMMERANZ, G. -L. (1). *Chinesische Neujahrsbilder.* Dresden, 1961.

REED, R. (1). *Ancient Skins, Parchments, and Leathers.* London, 1972.

SOOTHILL, W. E. (3) (tr.). '*Saddharma-pundararika Sūtra'; The Lotus of the Wonderful Law.* Oxford, 1930.

TING, S. P. (1). *A Brief Illustrated History of Chinese Military Notes and Bonds.* Thaipei, 1982.

TSCHICHOLD, J. (8). 'Color Registering in Chinese Woodblock Prints.' In *Printing and Graphic Arts,* 1954, **2**, 1.

TUNG TSO-PIN (3). *An Interpretation of the Ancient Chinese Civilization.* Thaipei, 1952.

TWITCHETT, D. (11). *Printing and Publishing in Medieval China.* Frederic C. Beil, New York, 1983.

DE LA VALLÉE POUSSIN (3) (tr.). *La Siddhi de Hiuen Tsang* [Hsüan Chuang]. Paris, 1928.

WALEY, A. (4) (tr.). *The Way and its Power; a Study of the Tao Te Ching and its Place in Chinese Thought.* Allen & Unwin, London, 1934.

WALEY, A. (17) (tr.). *Monkey,* by Wu Chheng-En. Allen & Unwin, London, 1942.

WALEY, A. (23). *The Nine Songs; a Study of Shamaism in Ancient China* [the *Chiu Ko* attributed traditionally to Chhü Yüan]. Allen & Unwin, London, 1955.

WANG, M. (1). 'Chinesische Farbendrucke aus der Bildersammlung der Zehnbambushalle.' *Alte und Neue Kunst,* 1956, **7**, 24.

WIEGER, L. (7) (tr.). *Taoism.* Vol. 2. *Les Pères du Système Taoiste* (tr. selections of *Lao Tzu, Chuang Tzu, Lieh Tzu*) Mission Press, Hsienhsien, 1913.

WILHELM, R. (2) (tr.). *I Ging* [*I Ching*]; *Das Buch der Wandlungen.* 2 vols. (3 books, pagination of 1 & 2 continuous in first volume). Diederichs, Jena, 1924. Eng. tr. C. F. Baynes (2 vols.), Bollingen, Pantheon, New York, 1950

YU, D. (1). 'The Printer Emulates the Painter, the Unique Chinese Water-and-Ink Woodblock Print.' *Renditions,* 1976, **6**, 95.

YÜ KUO-FAN [ANTHONY C. YÜ] (1) (tr.). *The Journey to the West* [*Hsi Yu Chi*]. 4 vols. University of Chicago Press, 1977–82.

GENERAL INDEX

A Ying, (2) 288 (a) (c) (d), (3) 126 (a)
acacia pods, for washing paper, 80
Ackerman, (1) 116 (c), 118 (b)
Africa, 3, 37, 298, 303
Ahmed Sibad Eddin (writer, +1245–1338), 49
Ai Jih Chai Tshung Chhao (Miscellaneous Notes from the Ai-Jih Studio), 152 (a)
Akatsuka Kiyoshi, (1) 28 (b)
Alexandria, 38
Alexandrian merchants, 44, 349
Alibaux, (1) 14, 63 (g)
alphabetic languages and movable-type printing, xxii, 8, 220, 306
  Korean adoption of an alphabetic script, 325
alum, 80, 82, 144, 247
amaranth (*hsien tshai*, Amaranthus tricolor, L.), in ink, 247
America, 3, 99 (c), 116, 135 (d)
  Central, 37
  paper in the New World, 302–3
An, Emperor (r. +1086–1100), 130
An-chhi, government paper factory at, 48, 98
An-chi, Chekiang, 161
An family (as printers), 212–13, 219
An Kuo (+1481–1534), 20 (c), 180, 213, 220;
  Kuei Pho Kuan, 213
An-Nan Chih Yüan, 350
Analects of Confucius (Japanese reprint, +1364), 339
Anderson, (1) 8 (d)
Ando Hiroshige (artist, +1797–1858), 342
Anhui, 13, 45, 47, 87
Anyang, finds from, 25 (b), 137
Aoyama Arata, (1) 254 (a), 264 (c), 286 (d), 288 (b) (c)
apprentices in papermaking, 51
Arab world, 5, 49, 123, 293, 296
  Western paper monopoly, 3
  early paper, 11
  ink, 236
  arrival of Chinese paper, 297
  early paper-making centres, 297–8; in Spain, 299
  as transmitter of printing, 316
armour, of paper, 114–16
Asakura Kamezo, (1) 21
Ashley, Margaret, (1) 8 (b)
Asia
  Western, 3, 5, 236, 303, 306
  South and Southeast, 38; introduction of paper and printing into, 352–60
  Central, 43, 303, 307
  East, spread of paper and printing to, 319–46
  Southeast, lack of a written tradition in, 354
Asia Minor, 6
Athens, 38
Australia, 3, 303
Aztecs, 302

Bacon, Roger, (1) 293
Baeyer, *Museum Sinicum*, 192 (a)
Bagford, John (antiquarian, 1659–1716), 318
Baghdad, as a paper-making centre, 297
bamboo, 4 (d), 47, 52, 53, 56, 90
  pre-paper writing material, 2, 24–5, 354; bamboo tablets, 29–33; replaced by paper, 43
  in screen mould construction, 43, 66–7, 68, 72; in Korea, 320
  in papermaking, 59–61; process described in the *Thien Kung Khai Wu*, 69–71; in Yang Chung Hsi, 71–2; in Korea, 320; in Vietnam, 350
  for bark paper, 72
  types of bamboo paper, 89, 105, 121, 123
Bambyx, Syrian paper-making centre, 297–8
Banákatí, Abu Sulaymán Da-ud of, xxii (c), 307 (b)
Barbaro, Josafar (Venetian emissary, +15th century), 293
bark cloth (*tapa*), 14 and (d), 36, 58
  made from paper mulberry in South China 4, 37, 41
  tha pu or *ku pu*, varieties of, 110
  for clothing, 110–12
  in Southeast Asia, 354–6
bark paper, 4, 14 (d), 37, 40–1, 42, 61–2
  methods of making, 72–3
  See also mulberry, paper mulberry
Batteux & de Bréquigny (1) 4 (e)
Belgium, 309
Benedetti-Pichler, (1) 238 (b)
Bernard-Maître, (18) 175 (c), (19) 175 (c), 358 (c)
Beveridge, (1) 297 (f)
Black, (1) 368 (a) (b)
blockprinting, 1, 2, 3, 304, 307 (e)
  surviving examples of, 19
  beginnings of, 146–59; key terms examined, 148; prevalence of, 190–2
  printing processes, 200–1; Persian and Arab descriptions, 307
  advantages for Chinese over movable type, 220–2
  in Europe, 309–13; European knowledge of Chinese, 313–17; as Chinese origin of European printing? 317–19
  gives way to movable type in Korea, 325
  ascendancy in Japan, 342; Vietnam, 351
  in the Philippines, 358
  See also colour printing, woodblocks, woodcuts
blocks, image, 149
blocks, printing
  bronze, 17–18
  horn, 18, 196 (b)
  See also woodblocks
Blum, (1) 4 (f), 5 (a) (b), 14, 298 (c), 299 (c) (d) (e) (f), 300 (b); (2) 7 (c), 309 (c), 311 (b), 317 and (f)
Bodde, (12) 104 (f), 106 (c) (d) (e)

Germany, 5 (a), 99 (c), 286 (a)
  early paper industry, 300
  appearance of blockprinting, 307; playing cards, 309; printed textiles, 311; typography, 314
Ghori & Rahman, (1) 354 (b), 357 (d)
Gia-loc, in Hai-duong province, Vietnam, 351
Giles, (1) 174 (b); (2) 185 (d), 219 (b); (6) 86 (f); (7) 86 (f); (13) 10 (c), 75 (c), 322 (b); (15) 157 (c), 158 (b); (17) 54 (d), 75 (c), 151 (b) (c)
Gilroy, (1) 36 (e), 293 (b)
ginger, for patterned paper, 94
Gitan (princely monk), 323, 324 (b), 330
glue, 144, 247
  in papermaking, 73-4, 76; Italian, 299
  in inkmaking, 234, 242, 243, 245, 246, 279; Korean, 1
Goa, Jesuit press in, 357
Gode, (1) 356 (c), 357 (c) (e)
gods, pictures of, 106-9
Goodrich, 22, 263 (b); (26) 174 (b); (28) 154 (a); (29) 169 (c), 304 (b); (30) 22, 213 (a) (b); (31) 22, 149 (f), 322 (a); (32) 22, 149 (f), 151 (a), 322 (d); (34) 174 (c); (38) 325 (g); (39) 251 (a)
Goschkewitsch, J., (1) 16
government, and papermaking
  official positions for artisans, 75
  development of paper money, 96-102
  production of court toilet paper, 123
  in Korea, 321, 333; in Japan, 331-3
  See also factories, requisitioning, tribute
government, and printing and publishing
  medical books at cost, 163
  issue of calendars, 347; in Vietnam, 351
  influential role, 378
  See also government agencies, paper money
government agencies for printing and publishing
  of money, 97-8
  Sung, 162, 165; Chin (Jurchen), 170; Yüan, 171; Ming, 175-8; Chhing, 184-8
  in Korea, 325, 327, 330
  in Vietnam, 351
  See also National Academy, Wu Ying Tien
Gozanban, 21, 338-41
Gozanji (Zen temples in Kyoto and Kamakura), 338
de Graaf, (1) 358 (h)
graphic arts, paper for the, 89-91
Gray, W.S., (1) 32 (c)
Greece, Greeks, 5, 9
  ink, 236
Grijalva, Juan de, 302
guilds, 9, 334
van Gulik, (9) 17, 75 (b), 77 (f), 78 (d), 79 (a) (c), 80 (a) (b) (c), 82 (c), 88 (c), 121 (a), 363 (e); (11) 17, 283 (b)
gum, 5, 73
Gutenberg, Johann, xxii, 203
  and the printing of playing cards, 310
  possible knowledge of Chinese printing, 316, 317
gypsum, 73

Haein-sa, woodblocks at, 162 (a), 324, 325
du Halde, (1) 237, 241-2, 247 (e)

on Chinese paper, 295
han (protective cases for books), 233
Han dynasty, 52, 77, 96, 122, 363
  paper fragments, 10 (a)
  writing surfaces, 24
  inscriptions, 26
  beginnings of paper, 38-42; for writing, 85; for wrapping, 122
  paper mould, 66
  burials, 102
  paper clothing, 110, 112
  seals, 137
  scripts, 223
  ink, 247, 248
  paper imports to Korea, 320
  beginnings of the examination system, 379
Han-gul, Korean alphabetic script, 325
Han Hsin, Lord of Huai-ying (d.-196), 36
Han Lin Chih (On the Han-Lin Academy) by Li Chao, 55 (c)
Han-Shan (poet), published in Japan, 339
Han Shih Wai Chuan (Moral Discourses Illustrating the Han Text of the Book of Odes) by Han Ying, 110, 238 (e)
Han-Than-Pho. See Wu-wei
hand-colouring of prints, 19 (f), 159, 280, 288, 311, 342
hand-copying, 9-10, 135-6, 185
  laborious in Chinese, 8
  of the Yung-Lo Ta Tien, 174
  of sutras, 338
  comparative costs, 372-3
Hangchow, 48, 55, 89, 128, 157, 158, 224, 330, 378
  bookshops, 167
  manuscript library, 185 (a)
  printing and publishing centre, 89, 154, 159, 169, 171, 184, 203 (a), 209, 216, 263, 266, 375
  See also Lin-an
Hanoi, 351
Hao-an hsien hua, miscellaneous notes by Chang Erh-Chhi, 203
Happelius, Everhard, 295
Harders-Steinhäuser, (1) 11, 86 (f)
Hayton, Prince of Armenia, 293
Hedin, Sven, 296
Hejzlar, (1) 254 (a), 288 (a); (2) 23
hemp, 3, 11 (a), 14, 37, 38 and (c), 40, 41, 47, 52-3, 86
  for papermaking, 53-4; Korean, 320; Japanese, 333-4
Henning, (2) 43 (e)
Herrada (friar), 315
Hervouet, (3) 167 (b), 374 (a)
hibiscus, in papermaking, 62, 74, 93
Hickman, (1) 337 (a)
Higuchi Hiroshi, (1) 254 (a), 264 (c), 269 (b)
Hirsch, (1) 367 (b)
Hirth, 297 (c); (1) 44 (b), 349 (a); (28) 192 (d), 194 (a), 217 (b); (29) 14
Hirth & Rockhill, (1) 349 (a) (e)
histories, the standard, 163, 171, 177, 181
Ho-kan Chi (government official, fl. +847-59), 152

TABLE OF CHINESE DYNASTIES

夏 HSIA kingdom (legendary?)   c. −2000 to c. −1520
商 SHANG (YIN) kingdom   c. −1520 to c. −1030

周 CHOU dynasty (Feudal Age)
- Early Chou period   c. −1030 to −722
- Chhun Chhiu period 春秋   −722 to −480
- Warring States (Chan Kuo) period 戰國   −480 to −221

First Unification 秦 CHHIN dynasty   −221 to −207

漢 HAN dynasty
- Chhien Han (Earlier or Western)   −202 to +9
- Hsin interregnum   +9 to +23
- Hou Han (Later or Eastern)   +25 to +220

三國 SAN KUO (Three Kingdoms period)   +221 to +265

First Partition
- 蜀 SHU (HAN)   +221 to +264
- 魏 WEI   +220 to +265
- 吳 WU   +222 to +280

Second Unification 晉 CHIN dynasty:
- Western   +265 to +317
- Eastern   +317 to +420

劉宋 (Liu) SUNG dynasty   +420 to +479

Second Partition — Northern and Southern Dynasties (Nan Pei chhao)
- 齊 CHHI dynasty   +479 to +502
- 梁 LIANG dynasty   +502 to +557
- 陳 CHHEN dynasty   +557 to +589
- 魏 Northern (Thopa) WEI dynasty   +386 to +535
- Western (Thopa) WEI dynasty   +535 to +556
- Eastern (Thopa) WEI dynasty   +534 to +550
- 北齊 Northern CHHI dynasty   +550 to +577
- 北周 Northern CHOU (Hsienpi) dynasty   +557 to +581

Third Unification 隋 SUI dynasty   +581 to +618
唐 THANG dynasty   +618 to +906

Third Partition 五代 WU TAI (Five Dynasty period) (Later Liang, Later Thang (Turkic), Later Chin (Turkic), Later Han (Turkic) and Later Chou)   +907 to +960

遼 LIAO (Chhitan Tartar) dynasty   +907 to +1124
West LIAO dynasty (Qarā-Khiṭāi)   +1124 to +1211
西夏 Hsi Hsia ('Tangut Tibetan) state   +986 to +1227

Fourth Unification 宋 Northern SUNG dynasty   +960 to +1126
宋 Southern SUNG dynasty   +1127 to +1279
金 CHIN (Jurchen Tartar) dynasty   +1115 to +1234
元 YUAN (Mongol) dynasty   +1260 to +1368
明 MING dynasty   +1368 to +1644
清 CHHING (Manchu) dynasty   +1644 to +1911
民國 Republic   +1912

N.B. When no modifying term in brackets is given, the dynasty was purely Chinese. Where the overlapping of dynasties and independent states becomes particularly confused, the tables of Wieger (1) will be found useful. For such periods, especially the Second and Third Partitions, the best guide is Eberhard (9). During the Eastern Chin period there were no less than eighteen independent States (Hunnish, Tibetan, Hsienpi, Turkic, etc.) in the north. The term 'Liu chhao' (Six Dynasties) is often used by historians of literature. It refers to the south and covers the period from the beginning of the +3rd to the end of the +6th centuries, including (San Kuo) Wu, Chin, (Liu) Sung, Chhi, Liang and Chhen. For all details of reigns and rulers see Moule & Yetts (1).

ROMANISATION CONVERSION TABLES

BY ROBIN BRILLIANT

PINYIN/MODIFIED WADE–GILES

Pinyin	Modified Wade–Giles	Pinyin	Modified Wade–Giles
a	a	chou	chhou
ai	ai	chu	chhu
an	an	chuai	chhuai
ang	ang	chuan	chhuan
ao	ao	chuang	chhuang
ba	pa	chui	chhui
bai	pai	chun	chhun
ban	pan	chuo	chho
bang	pang	ci	tzhu
bao	pao	cong	tshung
bei	pei	cou	tshou
ben	pên	cu	tshu
beng	pêng	cuan	tshuan
bi	pi	cui	tshui
bian	pien	cun	tshun
biao	piao	cuo	tsho
bie	pieh	da	ta
bin	pin	dai	tai
bing	ping	dan	tan
bo	po	dang	tang
bu	pu	dao	tao
ca	tsha	de	tê
cai	tshai	dei	tei
can	tshan	den	tên
cang	tshang	deng	têng
cao	tshao	di	ti
ce	tshê	dian	tien
cen	tshên	diao	tiao
ceng	tshêng	die	dieh
cha	chha	ding	ting
chai	chhai	diu	tiu
chan	chhan	dong	tung
chang	chhang	dou	tou
chao	chhao	du	tu
che	chhê	duan	tuan
chen	chhên	dui	tui
cheng	chhêng	dun	tun
chi	chhih	duo	to
chong	chhung	e	ê, o

Pinyin	Modified Wade–Giles	Pinyin	Modified Wade–Giles
en	ên	jia	chia
eng	êng	jian	chien
er	êrh	jiang	chiang
fa	fa	jiao	chiao
fan	fan	jie	chieh
fang	fang	jin	chin
fei	fei	jing	ching
fen	fên	jiong	chiung
feng	fêng	jiu	chiu
fo	fo	ju	chü
fou	fou	juan	chüan
fu	fu	jue	chüeh, chio
ga	ka	jun	chün
gai	kai	ka	kha
gan	kan	kai	khai
gang	kang	kan	khan
gao	kao	kang	khang
ge	ko	kao	khao
gei	kei	ke	kho
gen	kên	kei	khei
geng	kêng	ken	khên
gong	kung	keng	khêng
gou	kou	kong	khung
gu	ku	kou	khou
gua	kua	ku	khu
guai	kuai	kua	khua
guan	kuan	kuai	khuai
guang	kuang	kuan	khuan
gui	kuei	kuang	khuang
gun	kun	kui	khuei
guo	kuo	kun	khun
ha	ha	kuo	khuo
hai	hai	la	la
han	han	lai	lai
hang	hang	lan	lan
hao	hao	lang	lang
he	ho	lao	lao
hei	hei	le	lê
hen	hên	lei	lei
heng	hêng	leng	lêng
hong	hung	li	li
hou	hou	lia	lia
hu	hu	lian	lien
hua	hua	liang	liang
huai	huai	liao	liao
huan	huan	lie	lieh
huang	huang	lin	lin
hui	hui	ling	ling
hun	hun	liu	liu
huo	huo	lo	lo
ji	chi	long	lung

Pinyin	Modified Wade–Giles	Pinyin	Modified Wade–Giles
lou	lou	pa	pha
lu	lu	pai	phai
lü	lü	pan	phan
luan	luan	pang	phang
lüe	lüeh	pao	phao
lun	lun	pei	phei
luo	lo	pen	phên
ma	ma	peng	phêng
mai	mai	pi	phi
man	man	pian	phien
mang	mang	piao	phiao
mao	mao	pie	phieh
mei	mei	pin	phin
men	mên	ping	phing
meng	mêng	po	pho
mi	mi	pou	phou
mian	mien	pu	phu
miao	miao	qi	chhi
mie	mieh	qia	chhia
min	min	qian	chhien
ming	ming	qiang	chhiang
miu	miu	qiao	chhiao
mo	mo	qie	chhieh
mou	mou	qin	chhin
mu	mu	qing	chhing
na	na	qiong	chhiung
nai	nai	qiu	chhiu
nan	nan	qu	chhü
nang	nang	quan	chhüan
nao	nao	que	chhüeh, chhio
nei	nei	qun	chhün
nen	nên	ran	jan
neng	nêng	rang	jang
ng	ng	rao	jao
ni	ni	re	jê
nian	nien	ren	jên
niang	niang	reng	jêng
niao	niao	ri	jih
nie	nieh	rong	jung
nin	nin	rou	jou
ning	ning	ru	ju
niu	niu	rua	jua
nong	nung	ruan	juan
nou	nou	rui	jui
nu	nu	run	jun
nü	nü	ruo	jo
nuan	nuan	sa	sa
nüe	nio	sai	sai
nuo	no	san	san
o	o, ê	sang	sang
ou	ou	sao	sao

Pinyin	Modified Wade–Giles	Pinyin	Modified Wade–Giles
se	sê	wan	wan
sen	sên	wang	wang
seng	sêng	wei	wei
sha	sha	wen	wên
shai	shai	weng	ong
shan	shan	wo	wo
shang	shang	wu	wu
shao	shao	xi	hsi
she	shê	xia	hsia
shei	shei	xian	hsien
shen	shen	xiang	hsiang
sheng	shêng, sêng	xiao	hsiao
shi	shih	xie	hsieh
shou	shou	xin	hsin
shu	shu	xing	hsing
shua	shua	xiong	hsiung
shuai	shuai	xiu	hsiu
shuan	shuan	xu	hsü
shuang	shuang	xuan	hsüan
shui	shui	xue	hsüeh, hsio
shun	shun	xun	hsün
shuo	shuo	ya	ya
si	ssu	yan	yen
song	sung	yang	yang
sou	sou	yao	yao
su	su	ye	yeh
suan	suan	yi	i
sui	sui	yin	yin
sun	sun	ying	ying
suo	so	yo	yo
ta	tha	yong	yung
tai	thai	you	yu
tan	than	yu	yü
tang	thang	yuan	yüan
tao	thao	yue	yüeh, yo
te	thê	yun	yün
teng	thêng	za	tsa
ti	thi	zai	tsai
tian	thien	zan	tsan
tiao	thiao	zang	tsang
tie	thieh	zao	tsao
ting	thing	ze	tsê
tong	thung	zei	tsei
tou	thou	zen	tsên
tu	thu	zeng	tsêng
tuan	thuan	zha	cha
tui	thui	zhai	chai
tun	thun	zhan	chan
tuo	tho	zhang	chang
wa	wa	zhao	chao
wai	wai	zhe	chê

Pinyin	Modified Wade–Giles	Pinyin	Modified Wade–Giles
zhei	chei	zhui	chui
zhen	chên	zhun	chun
zheng	chêng	zhuo	cho
zhi	chih	zi	tzu
zhong	chung	zong	tsung
zhou	chou	zou	tsou
zhu	chu	zu	tsu
zhua	chua	zuan	tsuan
zhuai	chuai	zui	tsui
zhuan	chuan	zun	tsun
zhuang	chuang	zuo	tso

MODIFIED WADE–GILES/PINYIN

Modified Wade–Giles	Pinyin	Modified Wade–Giles	Pinyin
a	a	chhio	que
ai	ai	chhiu	qiu
an	an	chhiung	qiong
ang	ang	chho	chuo
ao	ao	chhou	chou
cha	zha	chhu	chu
chai	chai	chhuai	chuai
chan	zhan	chhuan	chuan
chang	zhang	chhuang	chuang
chao	zhao	chhui	chui
chê	zhe	chhun	chun
chei	zhei	chhung	chong
chên	zhen	chhü	qu
chêng	zheng	chhüan	quan
chha	cha	chhüeh	que
chhai	chai	chhün	qun
chhan	chan	chi	ji
chhang	chang	chia	jia
chhao	chao	chiang	jiang
chhê	che	chiao	jiao
chhên	chen	chieh	jie
chhêng	cheng	chien	jian
chhi	qi	chih	zhi
chhia	qia	chin	jin
chhiang	qiang	ching	jing
chhiao	qiao	chio	jue
chhieh	qie	chiu	jiu
chhien	qian	chiung	jiong
chhih	chi	cho	zhuo
chhin	qin	chou	zhou
chhing	qing	chu	zhu

Modified Wade–Giles	Pinyin	Modified Wade–Giles	Pinyin
chua	zhua	huan	huan
chuai	zhuai	huang	huang
chuan	zhuan	hui	hui
chuang	zhuang	hun	hun
chui	zhui	hung	hong
chun	zhun	huo	huo
chung	zhong	i	yi
chü	ju	jan	ran
chüan	juan	jang	rang
chüeh	jue	jao	rao
chün	jun	jê	re
ê	e, o	jên	ren
ên	en	jêng	reng
êng	eng	jih	ri
êrh	er	jo	ruo
fa	fa	jou	rou
fan	fan	ju	ru
fang	fang	jua	rua
fei	fei	juan	ruan
fên	fen	jui	rui
fêng	feng	jun	run
fo	fo	jung	rong
fou	fou	ka	ga
fu	fu	kai	gai
ha	ha	kan	gan
hai	hai	kang	gang
han	han	kao	gao
hang	hang	kei	gei
hao	hao	kên	gen
hên	hen	kêng	geng
hêng	heng	kha	ka
ho	he	khai	kai
hou	hou	khan	kan
hsi	xi	khang	kang
hsia	xia	khao	kao
hsiang	xiang	khei	kei
hsiao	xiao	khên	ken
hsieh	xie	khêng	keng
hsien	xian	kho	ke
hsin	xin	khou	kou
hsing	xing	khu	ku
hsio	xue	khua	kua
hsiu	xiu	khuai	kuai
hsiung	xiong	khuan	kuan
hsü	xu	khuang	kuang
hsüan	xuan	khuei	kui
hsüeh	xue	khun	kun
hsün	xun	khung	kong
hu	hu	khuo	kuo
hua	hua	ko	ge
huai	huai	kou	gou

Modified Wade–Giles	Pinyin	Modified Wade–Giles	Pinyin
ku	gu	mu	mu
kua	gua	na	na
kuai	guai	nai	nai
kuan	guan	nan	nan
kuang	guang	nang	nang
kuei	gui	nao	nao
kun	gun	nei	nei
kung	gong	nên	nen
kuo	guo	nêng	neng
la	la	ni	ni
lai	lai	niang	niang
lan	lan	niao	niao
lang	lang	nieh	nie
lao	lao	nien	nian
lê	le	nin	nin
lei	lei	ning	ning
lêng	leng	niu	nüe
li	li	niu	niu
lia	lia	no	nuo
liang	liang	nou	nou
liao	liao	nu	nu
lieh	lie	nuan	nuan
lien	lian	nung	nong
lin	lin	nü	nü
ling	ling	o	e, o
liu	liu	ong	weng
lo	luo, lo	ou	ou
lou	lou	pa	ba
lu	lu	pai	bai
luan	luan	pan	ban
lun	lun	pang	bang
lung	long	pao	bao
lü	lü	pei	bei
lüeh	lüe	pên	ben
ma	ma	pêng	beng
mai	mai	pha	pa
man	man	phai	pai
mang	mang	phan	pan
mao	mao	phang	pang
mei	mei	phao	pao
mên	men	phei	pei
mêng	meng	phên	pen
mi	mi	phêng	peng
miao	miao	phi	pi
mieh	mie	phiao	piao
mien	mian	phieh	pie
min	min	phien	pian
ming	ming	phin	pin
miu	miu	phing	ping
mo	mo	pho	po
mou	mou	phou	pou

Modified Wade–Giles	Pinyin	Modified Wade–Giles	Pinyin
phu	pu	tên	den
pi	bi	têng	deng
piao	biao	tha	ta
pieh	bie	thai	tai
pien	bian	than	tan
pin	bin	thang	tang
ping	bing	thao	tao
po	bo	thê	te
pu	bu	thêng	teng
sa	sa	thi	ti
sai	sai	thiao	tiao
san	san	thieh	tie
sang	sang	thien	tian
sao	sao	thing	ting
sê	se	tho	tuo
sên	sen	thou	tou
sêng	seng, sheng	thu	tu
sha	sha	thuan	tuan
shai	shai	thui	tui
shan	shan	thun	tun
shang	shang	thung	tong
shao	shao	ti	di
shê	she	tiao	diao
shei	shei	tieh	die
shên	shen	tien	dian
shêng	sheng	ting	ding
shih	shi	tiu	diu
shou	shou	to	duo
shu	shu	tou	dou
shua	shua	tsa	za
shuai	shuai	tsai	zai
shuan	shuan	tsan	zan
shuang	shuang	tsang	zang
shui	shui	tsao	zao
shun	shun	tsê	ze
shuo	shuo	tsei	zei
so	suo	tsên	zen
sou	sou	tsêng	zeng
ssu	si	tsha	ca
su	su	tshai	cai
suan	suan	tshan	can
sui	sui	tshang	cang
sun	sun	tshao	cao
sung	song	tshê	ce
ta	da	tshên	cen
tai	dai	tshêng	ceng
tan	dan	tsho	cuo
tang	dang	tshou	cou
tao	dao	tshu	cu
tê	de	tshuan	cuan
tei	dei	tshui	cui

Modified Wade–Giles	Pinyin	Modified Wade–Giles	Pinyin
tshun	cun	wang	wang
tshung	cong	wei	wei
tso	zuo	wên	wen
tsou	zou	wo	wo
tsu	zu	wu	wu
tsuan	zuan	ya	ya
tsui	zui	yang	yang
tsun	zun	yao	yao
tsung	zong	yeh	ye
tu	du	yen	yan
tuan	duan	yin	yin
tui	dui	ying	ying
tun	dun	yo	yue, yo
tung	dong	yu	you
tzhu	ci	yung	yong
tzu	zi	yü	yu
wa	wa	yüan	yuan
wai	wai	yüeh	yue
wan	wan	yün	yun